Lecture Notes in Computer Science   8008

Commenced Publication in 1973
Founding and Former Series Editors:
Gerhard Goos, Juris Hartmanis, and Jan van Leeuwen

Lecture Notes in Computer Science

Masaaki Kurosu (Ed.)

Human-Computer Interaction

Towards Intelligent and Implicit Interaction

15th International Conference, HCI International 2013
Las Vegas, NV, USA, July 21-26, 2013
Proceedings, Part V

 Springer

Volume Editor

Masaaki Kurosu
The Open University of Japan
2-11 Wakaba, Mihama-ku, Chiba-shi 261-8586, Japan
E-mail: masaakikurosu@spa.nifty.com

ISSN 0302-9743   e-ISSN 1611-3349
ISBN 978-3-642-39341-9   e-ISBN 978-3-642-39342-6
DOI 10.1007/978-3-642-39342-6
Springer Heidelberg Dordrecht London New York

Library of Congress Control Number: 2013941394

CR Subject Classification (1998): H.5, H.3, I.4, I.2.6, I.2.11, I.5

LNCS Sublibrary: SL 3 – Information Systems and Application, incl. Internet/Web
and HCI

Typesetting: Camera-ready by author, data conversion by Scientific Publishing Services, Chennai, India

Printed on acid-free paper

Springer is part of Springer Science+Business Media (www.springer.com)

Foreword

The 15th International Conference on Human–Computer Interaction, HCI International 2013, was held in Las Vegas, Nevada, USA, 21–26 July 2013, incorporating 12 conferences / thematic areas:

Thematic areas:

- Human–Computer Interaction
- Human Interface and the Management of Information

Affiliated conferences:

- 10th International Conference on Engineering Psychology and Cognitive Ergonomics
- 7th International Conference on Universal Access in Human–Computer Interaction
- 5th International Conference on Virtual, Augmented and Mixed Reality
- 5th International Conference on Cross-Cultural Design
- 5th International Conference on Online Communities and Social Computing
- 7th International Conference on Augmented Cognition
- 4th International Conference on Digital Human Modeling and Applications in Health, Safety, Ergonomics and Risk Management
- 2nd International Conference on Design, User Experience and Usability
- 1st International Conference on Distributed, Ambient and Pervasive Interactions
- 1st International Conference on Human Aspects of Information Security, Privacy and Trust

A total of 5210 individuals from academia, research institutes, industry and governmental agencies from 70 countries submitted contributions, and 1666 papers and 303 posters were included in the program. These papers address the latest research and development efforts and highlight the human aspects of design and use of computing systems. The papers accepted for presentation thoroughly cover the entire field of Human–Computer Interaction, addressing major advances in knowledge and effective use of computers in a variety of application areas.

This volume, edited by Masaaki Kurosu, contains papers focusing on the thematic area of Human–Computer Interaction, and addressing the following major topics:

- Adaptive, Personalised and Context-Aware Interaction
- Computational Vision in HCI
- Emotions in HCI
- Biophysiological Aspects of Interaction

The remaining volumes of the HCI International 2013 proceedings are:

- Volume 1, LNCS 8004, Human–Computer Interaction: Human-Centred Design Approaches, Methods, Tools and Environments (Part I), edited by Masaaki Kurosu
- Volume 2, LNCS 8005, Human–Computer Interaction: Applications and Services (Part II), edited by Masaaki Kurosu
- Volume 3, LNCS 8006, Human–Computer Interaction: Users and Contexts of Use (Part III), edited by Masaaki Kurosu
- Volume 4, LNCS 8007, Human–Computer Interaction: Interaction Modalities and Techniques (Part IV), edited by Masaaki Kurosu
- Volume 6, LNCS 8009, Universal Access in Human–Computer Interaction: Design Methods, Tools and Interaction Techniques for eInclusion (Part I), edited by Constantine Stephanidis and Margherita Antona
- Volume 7, LNCS 8010, Universal Access in Human–Computer Interaction: User and Context Diversity (Part II), edited by Constantine Stephanidis and Margherita Antona
- Volume 8, LNCS 8011, Universal Access in Human–Computer Interaction: Applications and Services for Quality of Life (Part III), edited by Constantine Stephanidis and Margherita Antona
- Volume 9, LNCS 8012, Design, User Experience, and Usability: Design Philosophy, Methods and Tools (Part I), edited by Aaron Marcus
- Volume 10, LNCS 8013, Design, User Experience, and Usability: Health, Learning, Playing, Cultural, and Cross-Cultural User Experience (Part II), edited by Aaron Marcus
- Volume 11, LNCS 8014, Design, User Experience, and Usability: User Experience in Novel Technological Environments (Part III), edited by Aaron Marcus
- Volume 12, LNCS 8015, Design, User Experience, and Usability: Web, Mobile and Product Design (Part IV), edited by Aaron Marcus
- Volume 13, LNCS 8016, Human Interface and the Management of Information: Information and Interaction Design (Part I), edited by Sakae Yamamoto
- Volume 14, LNCS 8017, Human Interface and the Management of Information: Information and Interaction for Health, Safety, Mobility and Complex Environments (Part II), edited by Sakae Yamamoto
- Volume 15, LNCS 8018, Human Interface and the Management of Information: Information and Interaction for Learning, Culture, Collaboration and Business (Part III), edited by Sakae Yamamoto
- Volume 16, LNAI 8019, Engineering Psychology and Cognitive Ergonomics: Understanding Human Cognition (Part I), edited by Don Harris
- Volume 17, LNAI 8020, Engineering Psychology and Cognitive Ergonomics: Applications and Services (Part II), edited by Don Harris
- Volume 18, LNCS 8021, Virtual, Augmented and Mixed Reality: Designing and Developing Augmented and Virtual Environments (Part I), edited by Randall Shumaker
- Volume 19, LNCS 8022, Virtual, Augmented and Mixed Reality: Systems and Applications (Part II), edited by Randall Shumaker

- Volume 20, LNCS 8023, Cross-Cultural Design: Methods, Practice and Case Studies (Part I), edited by P.L. Patrick Rau
- Volume 21, LNCS 8024, Cross-Cultural Design: Cultural Differences in Everyday Life (Part II), edited by P.L. Patrick Rau
- Volume 22, LNCS 8025, Digital Human Modeling and Applications in Health, Safety, Ergonomics and Risk Management: Healthcare and Safety of the Environment and Transport (Part I), edited by Vincent G. Duffy
- Volume 23, LNCS 8026, Digital Human Modeling and Applications in Health, Safety, Ergonomics and Risk Management: Human Body Modeling and Ergonomics (Part II), edited by Vincent G. Duffy
- Volume 24, LNAI 8027, Foundations of Augmented Cognition, edited by Dylan D. Schmorrow and Cali M. Fidopiastis
- Volume 25, LNCS 8028, Distributed, Ambient and Pervasive Interactions, edited by Norbert Streitz and Constantine Stephanidis
- Volume 26, LNCS 8029, Online Communities and Social Computing, edited by A. Ant Ozok and Panayiotis Zaphiris
- Volume 27, LNCS 8030, Human Aspects of Information Security, Privacy and Trust, edited by Louis Marinos and Ioannis Askoxylakis
- Volume 28, CCIS 373, HCI International 2013 Posters Proceedings (Part I), edited by Constantine Stephanidis
- Volume 29, CCIS 374, HCI International 2013 Posters Proceedings (Part II), edited by Constantine Stephanidis

I would like to thank the Program Chairs and the members of the Program Boards of all affiliated conferences and thematic areas, listed below, for their contribution to the highest scientific quality and the overall success of the HCI International 2013 conference.

This conference could not have been possible without the continuous support and advice of the Founding Chair and Conference Scientific Advisor, Prof. Gavriel Salvendy, as well as the dedicated work and outstanding efforts of the Communications Chair and Editor of HCI International News, Abbas Moallem.

I would also like to thank for their contribution towards the smooth organization of the HCI International 2013 Conference the members of the Human–Computer Interaction Laboratory of ICS-FORTH, and in particular George Paparoulis, Maria Pitsoulaki, Stavroula Ntoa, Maria Bouhli and George Kapnas.

May 2013

Constantine Stephanidis
General Chair, HCI International 2013

Organization

Human–Computer Interaction

Program Chair: Masaaki Kurosu, Japan

Jose Abdelnour-Nocera, UK
Sebastiano Bagnara, Italy
Simone Barbosa, Brazil
Tomas Berns, Sweden
Nigel Bevan, UK
Simone Borsci, UK
Apala Lahiri Chavan, India
Sherry Chen, Taiwan
Kevin Clark, USA
Torkil Clemmensen, Denmark
Xiaowen Fang, USA
Shin'ichi Fukuzumi, Japan
Vicki Hanson, UK
Ayako Hashizume, Japan
Anzai Hiroyuki, Italy
Sheue-Ling Hwang, Taiwan
Wonil Hwang, South Korea
Minna Isomursu, Finland
Yong Gu Ji, South Korea
Esther Jun, USA
Mitsuhiko Karashima, Japan

Kyungdoh Kim, South Korea
Heidi Krömker, Germany
Chen Ling, USA
Yan Liu, USA
Zhengjie Liu, P.R. China
Loïc Martínez Normand, Spain
Chang S. Nam, USA
Naoko Okuizumi, Japan
Noriko Osaka, Japan
Philippe Palanque, France
Hans Persson, Sweden
Ling Rothrock, USA
Naoki Sakakibara, Japan
Dominique Scapin, France
Guangfeng Song, USA
Sanjay Tripathi, India
Chui Yin Wong, Malaysia
Toshiki Yamaoka, Japan
Kazuhiko Yamazaki, Japan
Ryoji Yoshitake, Japan
Silvia Zimmermann, Switzerland

Human Interface and the Management of Information

Program Chair: Sakae Yamamoto, Japan

Hans-Jorg Bullinger, Germany
Alan Chan, Hong Kong
Gilsoo Cho, South Korea
Jon R. Gunderson, USA
Shin'ichi Fukuzumi, Japan
Michitaka Hirose, Japan
Jhilmil Jain, USA
Yasufumi Kume, Japan

Mark Lehto, USA
Hiroyuki Miki, Japan
Hirohiko Mori, Japan
Fiona Fui-Hoon Nah, USA
Shogo Nishida, Japan
Robert Proctor, USA
Youngho Rhee, South Korea
Katsunori Shimohara, Japan

Michale Smith, USA
Tsutomu Tabe, Japan
Hiroshi Tsuji, Japan

Kim-Phuong Vu, USA
Tomio Watanabe, Japan
Hidekazu Yoshikawa, Japan

Engineering Psychology and Cognitive Ergonomics

Program Chair: Don Harris, UK

Guy Andre Boy, USA
Joakim Dahlman, Sweden
Trevor Dobbins, UK
Mike Feary, USA
Shan Fu, P.R. China
Michaela Heese, Austria
Hung-Sying Jing, Taiwan
Wen-Chin Li, Taiwan
Mark A. Neerincx, The Netherlands
Jan M. Noyes, UK
Taezoon Park, Singapore

Paul Salmon, Australia
Axel Schulte, Germany
Siraj Shaikh, UK
Sarah C. Sharples, UK
Anthony Smoker, UK
Neville A. Stanton, UK
Alex Stedmon, UK
Xianghong Sun, P.R. China
Andrew Thatcher, South Africa
Matthew J.W. Thomas, Australia
Rolf Zon, The Netherlands

Universal Access in Human–Computer Interaction

Program Chairs: Constantine Stephanidis, Greece, and Margherita Antona, Greece

Julio Abascal, Spain
Ray Adams, UK
Gisela Susanne Bahr, USA
Margit Betke, USA
Christian Bühler, Germany
Stefan Carmien, Spain
Jerzy Charytonowicz, Poland
Carlos Duarte, Portugal
Pier Luigi Emiliani, Italy
Qin Gao, P.R. China
Andrina Granić, Croatia
Andreas Holzinger, Austria
Josette Jones, USA
Simeon Keates, UK

Georgios Kouroupetroglou, Greece
Patrick Langdon, UK
Seongil Lee, Korea
Ana Isabel B.B. Paraguay, Brazil
Helen Petrie, UK
Michael Pieper, Germany
Enrico Pontelli, USA
Jaime Sanchez, Chile
Anthony Savidis, Greece
Christian Stary, Austria
Hirotada Ueda, Japan
Gerhard Weber, Germany
Harald Weber, Germany

Virtual, Augmented and Mixed Reality

Program Chair: Randall Shumaker, USA

Waymon Armstrong, USA
Juan Cendan, USA
Rudy Darken, USA
Cali M. Fidopiastis, USA
Charles Hughes, USA
David Kaber, USA
Hirokazu Kato, Japan
Denis Laurendeau, Canada
Fotis Liarokapis, UK

Mark Livingston, USA
Michael Macedonia, USA
Gordon Mair, UK
Jose San Martin, Spain
Jacquelyn Morie, USA
Albert "Skip" Rizzo, USA
Kay Stanney, USA
Christopher Stapleton, USA
Gregory Welch, USA

Cross-Cultural Design

Program Chair: P.L. Patrick Rau, P.R. China

Pilsung Choe, P.R. China
Henry Been-Lirn Duh, Singapore
Vanessa Evers, The Netherlands
Paul Fu, USA
Zhiyong Fu, P.R. China
Fu Guo, P.R. China
Sung H. Han, Korea
Toshikazu Kato, Japan
Dyi-Yih Michael Lin, Taiwan
Rungtai Lin, Taiwan

Sheau-Farn Max Liang, Taiwan
Liang Ma, P.R. China
Alexander Mädche, Germany
Katsuhiko Ogawa, Japan
Tom Plocher, USA
Kerstin Röse, Germany
Supriya Singh, Australia
Hsiu-Ping Yueh, Taiwan
Liang (Leon) Zeng, USA
Chen Zhao, USA

Online Communities and Social Computing

Program Chairs: A. Ant Ozok, USA, and Panayiotis Zaphiris, Cyprus

Areej Al-Wabil, Saudi Arabia
Leonelo Almeida, Brazil
Bjørn Andersen, Norway
Chee Siang Ang, UK
Aneesha Bakharia, Australia
Ania Bobrowicz, UK
Paul Cairns, UK
Farzin Deravi, UK
Andri Ioannou, Cyprus
Slava Kisilevich, Germany

Niki Lambropoulos, Greece
Effie Law, Switzerland
Soo Ling Lim, UK
Fernando Loizides, Cyprus
Gabriele Meiselwitz, USA
Anthony Norcio, USA
Elaine Raybourn, USA
Panote Siriaraya, UK
David Stuart, UK
June Wei, USA

Augmented Cognition

Program Chairs: Dylan D. Schmorrow, USA, and Cali M. Fidopiastis, USA

Robert Arrabito, Canada
Richard Backs, USA
Chris Berka, USA
Joseph Cohn, USA
Martha E. Crosby, USA
Julie Drexler, USA
Ivy Estabrooke, USA
Chris Forsythe, USA
Wai Tat Fu, USA
Rodolphe Gentili, USA
Marc Grootjen, The Netherlands
Jefferson Grubb, USA
Ming Hou, Canada

Santosh Mathan, USA
Rob Matthews, Australia
Dennis McBride, USA
Jeff Morrison, USA
Mark A. Neerincx, The Netherlands
Denise Nicholson, USA
Banu Onaral, USA
Lee Sciarini, USA
Kay Stanney, USA
Roy Stripling, USA
Rob Taylor, UK
Karl van Orden, USA

Digital Human Modeling and Applications in Health, Safety, Ergonomics and Risk Management

Program Chair: Vincent G. Duffy, USA and Russia

Karim Abdel-Malek, USA
Giuseppe Andreoni, Italy
Daniel Carruth, USA
Eliza Yingzi Du, USA
Enda Fallon, Ireland
Afzal Godil, USA
Ravindra Goonetilleke, Hong Kong
Bo Hoege, Germany
Waldemar Karwowski, USA
Zhizhong Li, P.R. China

Kang Li, USA
Tim Marler, USA
Michelle Robertson, USA
Matthias Rötting, Germany
Peter Vink, The Netherlands
Mao-Jiun Wang, Taiwan
Xuguang Wang, France
Jingzhou (James) Yang, USA
Xiugan Yuan, P.R. China
Gülcin Yücel Hoge, Germany

Design, User Experience, and Usability

Program Chair: Aaron Marcus, USA

Sisira Adikari, Australia
Ronald Baecker, Canada
Arne Berger, Germany
Jamie Blustein, Canada

Ana Boa-Ventura, USA
Jan Brejcha, Czech Republic
Lorenzo Cantoni, Switzerland
Maximilian Eibl, Germany

Anthony Faiola, USA
Emilie Gould, USA
Zelda Harrison, USA
Rüdiger Heimgärtner, Germany
Brigitte Herrmann, Germany
Steffen Hess, Germany
Kaleem Khan, Canada

Jennifer McGinn, USA
Francisco Rebelo, Portugal
Michael Renner, Switzerland
Kerem Rızvanoğlu, Turkey
Marcelo Soares, Brazil
Christian Sturm, Germany
Michele Visciola, Italy

Distributed, Ambient and Pervasive Interactions

Program Chairs: Norbert Streitz, Germany, and Constantine Stephanidis, Greece

Emile Aarts, The Netherlands
Adnan Abu-Dayya, Qatar
Juan Carlos Augusto, UK
Boris de Ruyter, The Netherlands
Anind Dey, USA
Dimitris Grammenos, Greece
Nuno M. Guimaraes, Portugal
Shin'ichi Konomi, Japan
Carsten Magerkurth, Switzerland

Christian Müller-Tomfelde, Australia
Fabio Paternó, Italy
Gilles Privat, France
Harald Reiterer, Germany
Carsten Röcker, Germany
Reiner Wichert, Germany
Woontack Woo, South Korea
Xenophon Zabulis, Greece

Human Aspects of Information Security, Privacy and Trust

Program Chairs: Louis Marinos, ENISA EU, and Ioannis Askoxylakis, Greece

Claudio Agostino Ardagna, Italy
Zinaida Benenson, Germany
Daniele Catteddu, Italy
Raoul Chiesa, Italy
Bryan Cline, USA
Sadie Creese, UK
Jorge Cuellar, Germany
Marc Dacier, USA
Dieter Gollmann, Germany
Kirstie Hawkey, Canada
Jaap-Henk Hoepman, The Netherlands
Cagatay Karabat, Turkey
Angelos Keromytis, USA
Ayako Komatsu, Japan

Ronald Leenes, The Netherlands
Javier Lopez, Spain
Steve Marsh, Canada
Gregorio Martinez, Spain
Emilio Mordini, Italy
Yuko Murayama, Japan
Masakatsu Nishigaki, Japan
Aljosa Pasic, Spain
Milan Petković, The Netherlands
Joachim Posegga, Germany
Jean-Jacques Quisquater, Belgium
Damien Sauveron, France
George Spanoudakis, UK
Kerry-Lynn Thomson, South Africa

Julien Touzeau, France
Theo Tryfonas, UK
João Vilela, Portugal

Claire Vishik, UK
Melanie Volkamer, Germany

External Reviewers

Maysoon Abulkhair, Saudi Arabia
Ilia Adami, Greece
Vishal Barot, UK
Stephan Böhm, Germany
Vassilis Charissis, UK
Francisco Cipolla-Ficarra, Spain
Maria De Marsico, Italy
Marc Fabri, UK
David Fonseca, Spain
Linda Harley, USA
Yasushi Ikei, Japan
Wei Ji, USA
Nouf Khashman, Canada
John Killilea, USA
Iosif Klironomos, Greece
Ute Klotz, Switzerland
Maria Korozi, Greece
Kentaro Kotani, Japan

Vassilis Kouroumalis, Greece
Stephanie Lackey, USA
Janelle LaMarche, USA
Asterios Leonidis, Greece
Nickolas Macchiarella, USA
George Margetis, Greece
Matthew Marraffino, USA
Joseph Mercado, USA
Claudia Mont'Alvão, Brazil
Yoichi Motomura, Japan
Karsten Nebe, Germany
Stavroula Ntoa, Greece
Martin Osen, Austria
Stephen Prior, UK
Farid Shirazi, Canada
Jan Stelovsky, USA
Sarah Swierenga, USA

HCI International 2014

The 16th International Conference on Human–Computer Interaction, HCI International 2014, will be held jointly with the affiliated conferences in the summer of 2014. It will cover a broad spectrum of themes related to Human–Computer Interaction, including theoretical issues, methods, tools, processes and case studies in HCI design, as well as novel interaction techniques, interfaces and applications. The proceedings will be published by Springer. More information about the topics, as well as the venue and dates of the conference, will be announced through the HCI International Conference series website: http://www.hci-international.org/

General Chair
Professor Constantine Stephanidis
University of Crete and ICS-FORTH
Heraklion, Crete, Greece
Email: cs@ics.forth.gr

Table of Contents – Part V

Adaptive, Personalised and Context-Aware Interaction

Computational Vision in HCI

Emotions in HCI

Biophysiological Aspects of Interaction

Part I

Adaptive, Personalised and Context-Aware Interaction

Development of a Virtual Keyboard System Using a Bio-signal Interface and Preliminary Usability Test

Kwang-Ok An, Da-Hey Kim, and Jongbae Kim

Korea National Rehabilitation Research Institute(KNRRI), Seoul, Korea
anko04@korea.kr, dlfpdls87@naver.com, jbkim@nrc.go.kr

Abstract. People with severe speech or language problems rely on augmentative and alternative communication (AAC) to supplement existing speech or replace speech that is not functional. However, many people with severely motor disabilities are limited to use AAC, because most of AAC use the mechanical input devices. In this paper, to solve the limitations and offer a practical solution to disabled person, a virtual keyboard system using a bio-signal interface is developed. The developed system consists of bio-signal interface, training and feedback program, connecting module and virtual keyboard. In addition, we evaluate how well do subjects control the system. From results of preliminary usability test, the usefulness of the system is verified.

Keywords: augmentative and alternative communication, bio-signal interface, preliminary usability test, virtual keyboard system.

1 Instruction

Supporting disabled person's communication is an important factor that influences a person's ability to express their needs, participate in decision-making about their care, and maintain meaningful relationship within their community. Augmentative and alternative communication (AAC) can to compensate for communication problems related to disabilities thereby contributing to increased quality of life and independence. In the majority of existing AAC, mechanical input devices such as switch, keyboard and joystick have widely used because those are widespread available, robust and operationally simple [1]. However, many people with severe motor disabilities such as the amyotrophic lateral sclerosis, brainstem stroke, cerebral palsy, and spinal cord injury are limited, uncontrollable, or no hand or arm movement to use the mechanical input devices.

Recently, many advances have been made in interface technologies, by means of bio-signal [2-8]: infrared sensing, electromyography, electrooculogram, computer vision, and brain wave. Especially, brain computer interface (BCI) can be an alternative approach to those who are lack any useful muscle control or even locked-in. It can be served by using various imaging technologies such as electro/magneto-encephalography (E/MEG), positron emission tomography (PET), functional

M. Kurosu (Ed.): Human-Computer Interaction, Part V, HCII 2013, LNCS 8008, pp. 3–9, 2013.

magnetic resonance imaging (fMRI), and optical imaging. Among them, MEG, PET, fMRI, and optical imaging are still technically demanding and expensive. Furthermore, PET, fMRI, and optical imaging, which depend on blood flow, have long time constants and thus are not suitable for rapid communication. For now, therefore, scalp-recorded EEG is the main interest due to its advantages of relatively low cost, convenient operation and non-invasiveness [9].

BCI using scalp-recorded EEG up to present, however, have been primarily used in laboratory and medical fields owing to some limitations: 1) wet electrodes - most recording systems use wet electrodes, in which electrolytic gel is required to reduce electrode skin interface impedance. Using wet electrodes is uncomfortable to wear. 2) multi-channels - the recording system with multi-channels requires a long preparation time, thus are unsuitable for real-life applications. Moreover, a large amount of channels make the system quite big, complex and expensive.

On the other hand, some companies have released practical BCI headsets such as EPOC (Emotiv), neural impulse actuator (OCZ Technology), and mindset (Neurosky) [10-13]. These headsets have brought the field of BCI out of its infancy (clinical trials of invasive BCI technologies) into a phase of relative maturity through many demonstrated prototypes in gaming, PC applications, automotive, and robotics industries.

In this paper, to solve the limitations of conventional interfaces and offer a practical solution to disabled person, a real-time bio-signal interface that includes one dry sensor, wireless transmission, and signal processing is developed. Then, a virtual keyboard system using a bio-signal interface is developed. In addition, we evaluate how well do subjects control the developed system. 6 subjects (2 able-bodied subjects, 2 subjects with severe motor disabilities due to spinal cord injury, 2 subjects with amyotrophic lateral sclerosis) participated in preliminary usability test. From the results, the usefulness of the system is verified.

2   The Virtual Keyboard System

Fig. 1 shows the virtual keyboard system. It consists of bio-signal interface, training and feedback program, connecting module, and virtual keyboard.

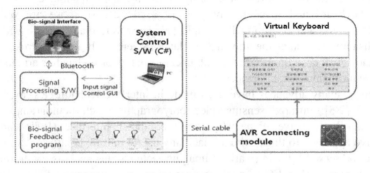

Fig. 1. The configuration of the developed system

2.1   Bio-signal Interface

In developed process, we considered the following issues: 1) for convenient use, one dry sensor technology and wireless transmission are needed. 2) To obtain stable performance, recorded signal should be intensive to interference. 3) It is difficult for people with severe motor disabilities to adjust the contact of sensors without the help from others when the poor contact is occurred. To solve the problem, the structure that sensors can be firmly attached to the forehead of the user is required. 4) To apply the developed system to a large number of users with diverse range of disabilities, detection of various signals including brain wave is needed. Then we co-developed the bio-signal interface with Neurosky. Fig. 2 shows the developed bio-signal interface. It includes the sensor that touch the forehead, reference points located on ear-clip and on-board chip that processes all of the data.

The developed interface collects bio-signals that are intended by subject. Then obtained signal are classified into 5 different control signals: attention (the level of mental "focus" of "attention"), meditation (the level of mental "calmness" or "relaxation"), eye blink, double blink, and jaw clench. Each control signal can be used like a on/off switch using the threshold set by the subject.

Fig. 2. The developed bio-signal interface

2.2   Training and Feedback Program

We developed the training and feedback program like as Fig. 3. The visual feedback is provided to the subjects to monitor performance results. In addition, during training, the level of thresholds can be adjusted based on the subject's condition. If the level of each control signal is greater than the threshold set by the user, the light bulb will turn on.

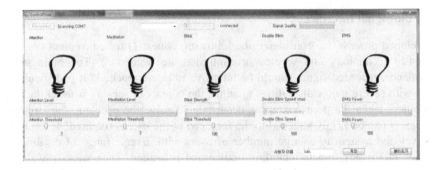

Fig. 3. The training and feedback program

2.3   Connecting Module

The connecting module uses 8-bit MCU to control the virtual keyboard according to PS/2 keyboard protocol. It sends 'Tab' key to PC whenever the light bulb turns on.

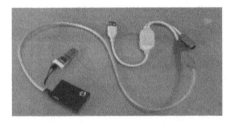

Fig. 4. The connecting module

2.4   Virtual Keyboard

A virtual keyboard presents an image of a keyboard on the computer screen and the subject selects the keys on the screen image. The name of virtual keyboard used in this paper is 'Clickey'. In scanning mode, lights scan letters and symbols displayed on computer screen. The subjects use the bio-signal interface to make selection.

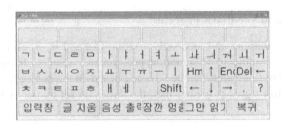

Fig. 5. Clickey

3   Preliminary Usability Test

3.1   Subjects

6 subjects (2 able-bodied subjects, 2 subjects with severe motor disabilities due to spinal cord injury, 2 subjects with amyotrophic lateral sclerosis) participated in preliminary usability test. None of them had ever utilized the developed system. They assisted to their informed consent to participate in the test.

3.2   Test Setup

The objective of the test is to assess the possibility of practical use and obtain a wide variety of opinions on improvement. So, we lend the developed system to participants for 10 days and perform the usability test three times (1st day, 5th day, and 10th day). Fig. 6 shows ALS subject using the developed system.

Fig. 6. ALS subject using the developed system

The following are the procedure of the usability test.

1. The headband is oriented so that the forehead sensor is facing the subject and ear-clip is on the left. Then, the headband is placed towards subject's head and the band is wrapped around head with a secure fit.
2. According to the subject's properties, the thresholds of control signals are adjusted.
3. In case of the first test, training with feedback can be continued for up to 30 minutes.
4. In each session, the subject is instructed to activate control signals 5 times in random order. Subject's task is to turn on the light bulb of target control signal. In case of an incorrect activation, subjects have to continue the given instruction. Each subject perform three sessions on the same day with 5-minutes breaks between the sessions to minimize fatigue.
5. After 10-minutes break, the subject selects the kind of control signal. Then the subjects is instructed to write 25 letters by using virtual keyboard.

6. We conduct a survey about the appearance of the system, performance degree, level of satisfaction, intention of purchase. Each item is scored by the subject according to a ten-point scale.

3.3   Results

The Table 1 shows the results of test. From the results, we are known that the developed system could be a practical solution because all the participants completed the test more than twice and they showed a high understanding of the developed system. Besides, there is little difference between able-bodied and disabled subjects. On the other hand, intra-subject variability (the degree of variability between control signals for each participant) and inter-subject variability (the degree of variability among participants in each control signal) were slight high. This high intra- and inter-subject variability denoted that they had different properties according to control signals and they activated the control signal on their own way. Therefore, it is very important that every person has to know which control signal is right for him or her by using training program.

Especially, according to the survey on intention of purchase, subjects with amyotrophic lateral sclerosis showed higher interest in that system. In addition, we obtained a wide variety of opinions on improvement (feelings of wearing, accuracy, and so on).

Table 1. The results of test

		Able-bodied subjects		SCI subjects		ALS subjects	
		Sub 1	Sub 2	Sub 1	Sub 2	Sub 1	Sub 2
Appearance of the system	1st	7	7	2	7	5	9
	5th	7	8	4	7	5	8
	10th	7	7	5	8	6	8
Performance degree	1st	5	8	4	8	8	6
	5th	3	8	7	7	6	9
	10th	8	8	3	8	6	9.5
Level of satisfaction	1st	5	8	4	8	8	8
	5ht	5	8	7	6	5	9
	10th	6	8	4	7	5	9
Intention of purchase	1st	8	7	5	8	9	8
	5th	8	7	8	7	10	9
	10th	8	7	7	8	10	9
Mission time (25 letters)	1st	29:30	23:51	16:54	16:25	11:06	-
	5th	12:21	14:59	12:39	-	21:03	18:53
	10th	16:36	09:29	09:17	26:36	-	15:22

4   Conclusions

In this paper, the virtual keyboard system is proposed in order to solve the limitations of conventional AAC and offer a practical solution for disabled person. It consists of bio-signal interface, training and feedback program, connecting module, and virtual keyboard. To apply the developed interface to users with diverse range of disabilities, it can measure bio-signals and obtained signals are classified into 5 different control signal. In addition, to help subjects modulating the control signal activation, training and feedback program is developed.

The developed system was validated by 6 subjects (2 able-bodied subjects, 2 subjects with severe motor disabilities due to spinal cord injury, 2 subjects with amyotrophic lateral sclerosis) in preliminary usability test. In the future, we will improve the developed system based on the results of survey.

Acknowledgments. This research was supported by a grant (code #12-A-01, #13-A-01) from Korea National Rehabilitation Research Institute.

References

1. Cook, A.M., Hussey, S.M.: Assistive Technologies: principles and practices, 3rd edn. Mosby (2008)
2. Tai, K., Blain, S., Chau, T.: A review of emerging access technologies for individuals with severe motor impairments. Assit. Technol., 204–209 (2008)
3. Schalk, G., et al.: BCI 2000: A general-purpose brain-computer interface (BCI) system. IEEE Trans. Biomed. Eng. 51, 1034–1043 (2004)
4. Lin, C.-T., et al.: Development of wireless brain computer interface with embedded multitask scheduling and its application on real-time driver's drowsiness detection and warning. IEEE Trans. Biomed. Eng. 55, 1034–1043 (2008)
5. Kauhanen, L., Jylanki, P., et al.: EEG-based brain-computer interface for tetraplegics. Comput. Intell. Neurosci., 1–11 (2007)
6. Barreto, A.B., et al.: A practical EMG-based human-computer interface for users with motor disabilities. Journal of Rehabilitation Research and Development 37, 53–64 (2000)
7. Lturrate, I., et al.: A noninvasive brain-actuated wheelchair based on a P300 neurophysiological protocol and automated navigation. IEEE Trans. Robot. 25, 614–627 (2009)
8. Muller-Putz, G.R., Pfurtscheller, G.: Control of an electrical prosthesis with an SSVEP-based BCI. IEEE Trans. Biomed. Eng. 55, 361–364 (2008)
9. Graimann, B., et al.: Brain-Computer Interfaces, ch. 8. Springer, Heidelberg (2010)
10. Wolpaw, J.R., et al.: Brain-computer interfaces for communication and control. Clin. Neurophysiol. 113, 767–791 (2002)
11. Emotiv. EPOC neuroheadset,
    `http://emotiv.com` `http://www.emotiv.com/apps/epoc/299`
12. OCZ Technology. NIA game controller,
    `http://ocztechnology.com`,
    `http://www.ocztechnology.com/nia-game-controller.html`
13. Neurosky, mindset, `http://neurosky.com`,
    `http://neurosky.com/Products/MindSet.aspx`

Unifying Conceptual and Spatial Relationships between Objects in HCI

David Blezinger[1], Ava Fatah gen. Schieck[1], and Christoph Hölscher[2]

[1] Bartlett School of Graduate Studies
Central House, 14 Upper Woburn Place, London WC1H 0NN, UK
[2] Center for Cognitive Science, Institute for Computer Science and Social Research
Friedrichstr. 50, 79098 Freiburg, Germany
mail@david-blezinger.de

Abstract. To design interfaces which occupy a continuous space of interaction, the conceptual model of an interface needs to be transferred to a spatial model. To find mappings between conceptual and spatial structure which are natural to people, an experiment is undertaken in which participants organize objects in a semi-circle of shelves around their body. It is analyzed how conceptual relationships between objects such as categorial relationships and sequential relationships within task performance are represented in spatial configurations of objects as chosen by the participants. In these configurations, a strong correlation between conceptual and spatial relationships is observed between objects.

Keywords: HCI frameworks, spatial interface, conceptual model, information architecture, navigation, object-based, task-based, spatial configuration, spatial cognition, embodied interaction, categories, visual identity.

1 Introduction

Before the information architecture of a Human-Computer Interface is implemented in wireframes and visual designs, conceptual models are used to develop relationships between objects [1][2][3]. Within conceptual maps, objects are spatially related to each other, but this spatiality is not transferred to the visual layout of the final interface.

Although the necessity to incorporate space into interface design is emphasized in the literature [4][5] there is currently no framework that allows a mapping of conceptual relationships between objects to spatial relationships between the same objects. Instead, objects are laid out in consecutive windows which are not spatially related to each other. This divides the space through which the user navigates rather than providing a continuous space that she encounters in her natural, non-digital environment.

Interaction within a continuous space is not restricted to a certain form of interface. It can equally be applied to screen interfaces, tangible interfaces and other forms of spatial interfaces. More important than the technical or material features of the

M. Kurosu (Ed.): Human-Computer Interaction, Part V, HCII 2013, LNCS 8008, pp. 10–18, 2013.

interface is the way in which its objects relate to each other. Although a spatial interface can be an interface that takes up 3D space around the human body, it can also be a screen based interface at smaller scale in which objects of use are related to each other in one continuous visual space.

To inform a layout of objects in a spatial interface it needs to be known what types of relationships between these objects are relevant to the user's perception, cognition and interaction, and how these relationships can be mapped to spatial relationships between the same objects.

We take categorial relationships between objects and relationships of sequential order of objects within task performance as two main factors which determine the concepualization of object relationships in people's cognition (potentially related to [6][7]), and which are likely to strongly influence the way in which participants of our experiment will choose to organize objects spatially.

We use a conception of tasks that include only such tasks which consist of movement between objects as it would occur in navigating a window-based interface, but in a continuous space. We exclude tasks that involve the manipulation of objects, to create a more uniform experimental setup.

Both categorial relationships between objects and relationships of sequence in which objects follow each other within task performance exist independently of the spatial layout in which objects are laid out and tasks are performed. Thus, these relationships can be treated as conceptual relationships between objects that can be designed within a conceptual model before implementing the interface spatially (for category effects on spatial search see [8]). To enable such an implementation, principles by which conceptual relationships between objects can be mapped to spatial relationships between the same objects need to be found.

By spatial relationships, we mean relationships between objects as relevant to individual perception, cognition and interaction (for embodied interaction with objects see [9][10]). Thus, our concept of spatial relationships is tied not only to relationships between objects but also to the relationship between the human body and these objects. Thus, spatial relationships between objects as described here are relationships as perceived and conceptualized in cognition [11], and are thus tied to the viewpoint of the perceiving, thinking and interacting individual.

We represent the human body through its standing point and reach of arms, which leads to a sphere of possible movement trajectories around the human body [12]. Tasks that include navigation towards several objects in a certain sequence can thus be represented as trajectories of movement between spatially distributed objects in relationship to the human body. In our experimental setup, we choose an angular distribution of objects around the participant's standing point to match this sphere of performative body movement, thus enabling to integrate the representation of body movement and the representation of the spatial environment in which the experiment is performed. This allows an analysis of the influence of movement, perception and cognition on object relationships within an environmental representation.

The spatial framework model [13] states that the position of objects in one's spatial environment are memorized in relationship to the axes of the human body. We aim to relate our experimental setup to the left/right axis and to the top/bottom axis, integrating standing point and movement of the participant with object related cognition.

2   The Experiment

The design of the experimental setup was inspired by an analysis of task performance within a kitchen environment. In a kitchen, objects of daily use are stored in cupboards and shelves. To perform a task such as making coffee, a set of objects is taken from their storage position to a working surface where the task is performed. Each time groceries are bought at the store, they need to be stored in the shelves and cupboards. The objects are stored in relationship to objects which are already present in space, and when no relationship can be found, a new location needs to be found, or the overall configuration of objects needs to be changed to accommodate the new objects.

The experimental setup is designed to contain both activities, and to potentially reveal a mutual influence between the two. First, the participant needs to organise a set of objects in shelves, as she would do it with objects of her daily life. Second, she performs a set of four tasks with these objects, and third, she organises the objects in the shelves again, after the experience of the tasks.

We adapted the environment and the way tasks are performed (Fig.1), to suit our method of analysis and to allow the findings to inform the design of HCI. As described above, the spatial environment in our experimental setup forms a near semi-circle around the participant's standing point, thus equaling the distance at which the participant can position objects with her hands. By choosing this layout and by placing our shelves at a convenient height , we strive to minimize the influence of ergonomic factors on the spatial positions of objects that the participant chooses. Cognitive rather than ergonomic factors are in the focus of the study.

As mentioned above, we define tasks as sequences of object use and thus sequences of movement between objects. Instead of placing the objects on the

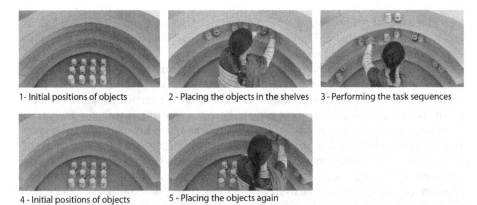

1- Initial positions of objects   2 - Placing the objects in the shelves   3 - Performing the task sequences

4 - Initial positions of objects   5 - Placing the objects again

Fig. 1. The order in which each participant performs the experiment

working surface as it would occur in a kitchen environment, tasks in our experimental environment only require grabbing the objects at their spatial location within the shelves and moving them from the back to the front of the shelves. We assume that this action will reinforce the memory of the spatial position of an object. To further reinforce the memory of the spatial configuration of objects in relationship to task performance, the participant needs to perform the set of tasks twice.

The objects are chosen to form categorial relationships of variable strengths. Common categories (fruits, sweeteners, glasses) are chosen which are likely to be shared across participants, even from different cultural backgrounds. To put the focus on cognitive / spatial relationships between objects, the influence of affordances (shape, size) [14] on the organization of objects is eliminated by choosing an equal size, shape and weight for all objects. The graphic identity of each object consists of a picture of the visual appearance of the actual object that the uni-size object represents, and of a color band. In the first object set which we will call „match", the color is chosen to match each object and thus its category (e.g. a red color is chosen for raspberries and a green color for limes).

We expect that the categorial relationships will be strongly represented in the spatial configuration participants choose when organizing the objects in relation to each other. Participants are likely to group objects of the same category together. We expect that the repeated performance of sequential movements (tasks) on a set of spatially distributed objects will have an influence on the conceptualization of relationships between objects and may thus lead to a re-configuration of objects when spatially organizing the objects again, after task performance.

To test whether the influence of color or categorial information of objects on their spatial organization is stronger, we introduce a second object set in which the color does not match the object. Instead, the pictures of the objects are displayed in greyscale, and the color is chosen randomly, not matching the true color of each object. We call this object set „no match". The participants are divided into two groups. Group A performs the experiment with object set „no match", group B with object set „match". Each group contains 16 participants, each of which takes about 20 minutes to perform the experiment.

2.1   Analysis of Object Distributions

To find a simple yet suitable computational and visual representation that enables an analysis of the results, we map the angular distribution of objects within the shelves onto a plane. This enables a maintenance of angular information while adapting the representation to a more versatile 2D plane. The preservation of metric information would be possible using this representation, but as spatial-configurational relationships between objects rather than metric information are at the focus of the experiment, a metrically correct mapping of the experimental setup was not required.

Each object distribution within the shelves, as chosen by the participant, is transferred to an excel spreadsheet which is saved as a csv file for further computational analysis and visualisation (Fig. 2). Each column of the spreadsheet represents a part of the shelf which has the width of one object. There are three rows per layer of shelves, to represent the numerical identity of each object, and additional information about the grouping of objects.

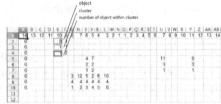

Object distribution and clustering   Computational representation of object distribution

Fig. 2. Transferring the spatial distribution of objects to a spreadsheet

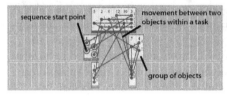
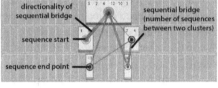

Visualisation of the trajectories of task sequences   Visualisation of movement between clusters

Fig. 3. Visualisation of movement between objects and between groups of objects

Based on the data entered in the spreadsheet, the spatial distribution of objects and the sequential movement between the objects within the tasks are visualized (Fig.3).

3   Results and Discussion

In our analysis of the experimental findings, we focus on spatial relationships between objects and how these relate to the conceptual relationships between the same objects (categorial relationships and task-based, sequential relationships). The large variation in geometric patterns across participants shows that individual choice plays a large part in spatially organizing objects. However, it seems that there are factors which constrain the variation in pattern generation and lead to the observation of repetition in the rules by which the patterns are formed. Different types of spatial organisation can be differentiated by the size of object groups, by the extension at which the objects are distributed in the shelves, by the simplicity of sequential pathways between object groups and by the change of group size and group content.

3.1   Grouping of Objects

When placing objects in shelves, participants tend to group objects together. In our spreadsheet, we gathered information about which objects are grouped together in each of the two object distributions. This enables an analysis of object grouping across all participants of each group. We analyse the frequency with which pairs of objects remain grouped together in both object distributions, and compare the results of both participant groups to each other. As described above, the colour of each object matches the categorial identity of the object in group B, while the colours are chosen randomly in group A.

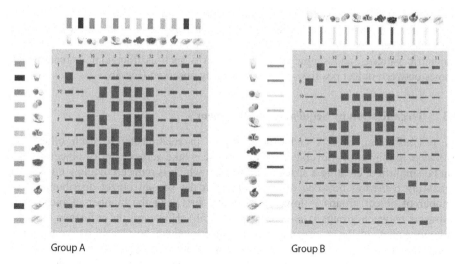

Group A   Group B

Fig. 4. Relative frequency of object pairs which remain grouped in both object distributions

In both participant groups, objects of the same category remain grouped most frequently (Fig.4), which indicates that categorial information has a strong influence on the grouping of objects while colour does not serve as a dominant principle of organising objects into groups. When categorial information is available, no influence of colour on the grouping of objects can be observed. However, within groups the objects were frequently organised in a colour gradient in both participant groups.

3.2 Patterns of Spatial Positions

In some cases, it appears that objects are organized by spatial positions and patterns among these spatial positions independent of the actual objects stored at these positions. It can be observed that some participants choose almost the same pattern

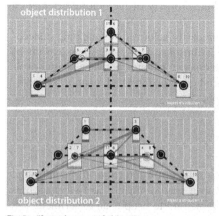

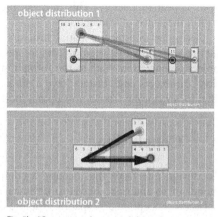

Fig. 5a: "Spread groups of objects"   Fig. 5b: "Concentrated groups of objects"

Fig. 5. Examples of different types of spatial organization

between spatial positions at which groups of objects are stored when they place the objects for the second time while the objects themselves are exchanged between different spatial positions. (Fig.5a) Sometimes, symmetry is used to order objects into geometric patterns, and simple geometric patterns such as triangular or rectangular shapes between object positions can be frequently observed (Fig.5b).

3.3   Organisation by Sequences of Movement

By re-configuring objects and object groups in relationship to each other, the pathways that a set of movement (task) sequences take across a set of object groups can be simplified, creating a space of movement which is reduced to a low number of remaining pathways between object groups. In the second object distribution, the number of object groups is lower than in the first, and the number of movement pathways between groups is also reduced (Fig.5b). However, the influence of tasks and movement on the re-organisation of objects cannot be clearly defined, as the grouping of objects acts as a concurring or complementary organisational principle.

The results deliver a set of organisational principles that could be applied in the design of Human-Computer Interfaces. Because these principles are derived from an experiment in which participants were asked to organize objects as they would usually do it, the organizational principles derived from the experiment are likely to be natural to human perception, cognition and interaction.

Categorially related objects can be grouped together, colours can be used to organise objects in gradients, geometric patterns of spatial positions can serve to organise objects within an interface, and the space of task sequences can be simplified by re-configuring the objects.

The results suggest that geometric patterns of spatial positions of objects or groups of objects can serve as features to organize objects spatially and to give orientation in interaction, independently of the content that is stored at each spatial position. Thus, spatial positions can serve as „containers" for objects, and patterns between them can facilitate the memorization of a spatial system of containers. Strong geometric patterns such as triangles between spatial positions or symmetrically ordered spatial positions appear repeatedly. This may be due to their property of facilitating orientation within the space of interaction.

Giving an interface a spatial structure can be partly achieved by mapping conceptual relationships between objects to spatial relationships. However, some aspects of object organization, such as the formation of geometric patterns between spatial positions of objects and clusters have a purely spatial component which cannot be generated by a mapping from a conceptual model alone.

4   Conclusion

To design interfaces that occupy a continuous space of interaction, the conceptual model of an interface needs to be transferred to a spatial model. To find mappings between conceptual and spatial structure that are natural to people, an experiment was undertaken in which participants organise objects in a semi-circle of shelves around their body.

The experimental findings may provide principles which can inform a mapping from a conceptual model to a spatial interface structure and which are natural to human perception, cognition and interaction. Thus an implementation of the observed principles into the design of Human-Computer interfaces could potentially make an interaction more natural to the users.

In common window-based interfaces, the desktop is used by many people to store their most relevant files and folders. The findings of this study, supported by further research, could enable an implementation of organisational relationships between objects into the spatial and visual organisation within the desktop. Hierarchical and associative relationships between objects which are otherwise hidden in folder structures could be spatially and visually present within a continuous layer of space. The functionality of storing and organising frequently used objects could also be transferred from a classical screen-based desktop to a three-dimensional interface which occupies the space around the user's body. Further research based on this study may already be conducted in relation to a concrete application in the design of human-computer interfaces.

There are limitations to the design of interfaces that occupy a continuous space, however. In the interfaces we interact with every day, we encounter a vast number of objects all of which we can potentially interact with. A logic of consecutive windows supports the interaction with large numbers of objects whereas a continuous space that needs to accommodate all objects comes to its boundaries as the number of objects is increased.

Thus, a selection needs to be made as to which objects a continuous space of interaction should contain. Such a selection could be enabled through configurability of the interface, or through automatic adaptation. The number of objects a space can contain while still being intelligible is likely to depend on its size and also on the patterns by which the objects are spatially organised.

Acknowledgments. We would like to thank Nadia Berthouze, Sam Griffiths, Sean Hanna and Martha Tsigkari for the feedback and support in the course of this study.

References

1. Beck, A., Janssen, C., Weisbecker, A., Ziegler, J.: Integrating Object-Oriented Analysis and Graphical User Interface Design. Fraunhofer-Institut für Arbeitswirtschaft und Organisation (1993)
2. Allen, J., Chudley, J.: Smashing UX Design. John Wiley & Sons, Chichester (2012)
3. John, B.E., Kieras, D.E.: Using GOMS for User Interface Design and Evaluation: Which Technique? ACM Transactions on Computer-Human Interaction 3(4), 287–319 (1996)
4. Hornecker, E., Buur, J.: Getting a grip on Tangible Interaction: A Framework on Physical Space and Social Interaction. In: CHI 2006, Montreal, Québec, Canada, April 22-28 (2006)
5. Ullmer, B., Ishii, H.: Emerging Frameworks for tangible user interfaces. IBM Systems Journal 39(3-4), 915–931 (2000)
6. Burgess, C.: From simple associations to the building blocks of language. Behavior Research Methods. Instruments & Computers 30(2), 188–198 (1998)

7. Lund, K., Burgess, C.: Producing high-dimensional semantic spaces from lexical co-occurrence. Behavior Research Methods. Instruments & Computers 28(2), 203–208 (1996)
8. Kalff, C., Strube, G.: Everyday navigation in Real and Virtual Environments Informed by Semantic Knowledge. In: Carlson, L., Hoelscher, C., Shipley, T.F. (eds.) Proceedings of the 33rd Annual Conference of the Cognitive Science Society. Cognitive Science Society, Austin (2011)
9. Fatah gen. Schieck, A., Moutinhou, A.: ArCHI - Engaging with museum objects spatially through whole body movement. In: Academic MindTrek 2012: International Conference on Media of the Future, Tampere, Finland (2012)
10. Sharlin, E., Watson, B., Kitamura, Y., Kishino, F., Itoh, Y.: On tangible user interfaces, humans and spatiality. Personal and Ubiquitous Computing 8(5), 338–346 (2004)
11. Robbins, P., Aydede, M.: The Cambridge Handbook of Situated Cognition. Cambridge University Press, Cambridge (2009)
12. Laban, R.: Choreutics. Jarrold and Sons Limited, Norwich (1966)
13. Tversky, B., Morrison, J.B., Franklin, N., Bryant, D.J.: Three Spaces of Spatial Cognition. Professional Geographer 51(4), 516–524 (1999)
14. Vingerhoets, G., Vandamme, K., Vercammen, A.: Conceptual and physical object qualities contribute differently to motor affordances. Brain and Cognition 69, 481–489 (2008)
15. Beach, K.: Becoming a Bartender: The Role of External Memory Cues in a Work-directed Educational Activity. Applied Cognitive Psychology 7, 191–204 (1993)
16. Fatah gen. Schieck, A.: Embodied, mediated and performative: Exploring the architectural education in the digital age. In: Voyatzaki, M., Spiridonidis, C. (eds.) Rethinking the Human in Technology-driven Architecture. Transactions on Architectural Education, vol. (55) (2012)
17. Gibson, J.J.: The ecological approach to visual perception. Psychology Press, New York (1986)
18. Moggridge, B.: Designing Interactions. The MIT Press, Cambridge (2006)
19. Norman, D.A.: Affordance, Conventions, and Design. Interactions (May-June 1999)
20. Dourish, P.: Where the Action is. The MIT Press, Cambridge (2001)
21. Hardiess, G., Gillner, S., Mallot, A.: Head and eye movements and the role of memory limitations in a visual search paradigm. Journal of Vision 8(1), 7, 1–13 (2008)
22. Hardiess, G., Basten, K., Mallot, H.A.: Acquisition vs. Memorization Trade-Offs Are Modulated by Walking Distance and Pattern Complexity in a Large-Scale Copying Paradigm. PLoS ONE 6(4) (2011)
23. Hillier, B.: The Social Logic of Space. Cambridge University Press, New York (1984)
24. Derry, S.J.: Cognitive Schema Theory in the Constructivist Debate. Educational Psychologist 31(3/4), 163–174 (1996)
25. Hirtle, S.C., Jonides, J.: Evidence of hierarchies in cognitive maps. Memory & Cognition 13(3), 208–217 (1985)
26. Jameson, A.: Adaptive Interfaces and Agents. DFKI, German Research Center for Artificial Intelligence (2008)
27. Kuhn, G.: Die "Frankfurter Küche". Bonn: Wohnkultur und kommunale Wohnungspolitik in Frankfurt am Main 1880-1930, 142–176 (1998)
28. Penn, A.: Space Syntax and Spatial Cognition. Environment and Behaviour 35(1), 30–65 (2003)
29. Pirolli, P., Card, S.K.: Information foraging. Psychological Review 106, 643–675 (1999)
30. Tversky, B.: Cognitive Maps, Cognitive Collages, and Spatial Mental Models. In: Campari, I., Frank, A.U. (eds.) COSIT 1993. LNCS, vol. 716, pp. 14–24. Springer, Heidelberg (1993)
31. Zacks, J.M., Tversky, B.: Event Structure in Perception and Conception. Psychological Bulletin 127(1), 3–21 (2001)

Context-Aware Multimodal Sharing of Emotions

Maurizio Caon[1,2], Leonardo Angelini[1], Yong Yue[2],
Omar Abou Khaled[1], and Elena Mugellini[1]

[1] University of Applied Sciences of Western Switzerland, Fribourg
[2] University of Bedfordshire, Luton
{Maurizio.Caon,Leonardo.Angelini,Omar.AbouKhaled,
Elena.Mugellini}@hefr.ch, Yong.Yue@beds.ac.uk

Abstract. Computer mediated interaction often lacks of expressivity, in particular for emotion communication. Therefore, we present a concept for context-aware multimodal sharing of emotions for human-to-computer-to-human interaction in social networks. The multimodal inputs and outputs of this system are distributed in a smart environment in order to grant a more immersive and natural interaction experience. The context information is used to improve the opportuneness and the quality of feedback. We implemented an evaluation scenario and we conducted an observation study during some events with the participants. We reported our considerations at the end of this paper.

Keywords: affective computing, multimodal interaction, computer mediated communication, social sharing of emotions.

1   Introduction

The communication and information technology changed the way people communicate. In particular, Internet played a key role in this communication revolution that led to the concept of human-to-computer-to-human interaction (HCHI) as explained by Clubb in [1]. All the paralanguage or normal face-to-face nonverbal communication is missing in HCHI. In particular, the need of expressing emotions through messages became important. Indeed, there has been an attempt in messaging to integrate this kind of non-verbal communication that knew a large success: the emoticons [9]. After they have firstly been introduced by Scott E. Fahlman in 1982, they evolved and spread all over the world. In particular, some studies highlighted that people need to express emotions in popular communication systems as Facebook and Twitter, which are recently referred as "Social Awareness Streams" (SAS) [2]. In fact, shared emotional states have been observed creating engagement and participation among the users [2]. This phenomenon is due to the natural need of the human being of social sharing of emotion. In fact, Rime et al. showed that most emotional experiences are shared with others shortly after they occurred [10]. They pointed out that social sharing represents an integral part of emotional experiences. Since currently the technology is ubiquitous and the access to Internet is indispensable [12], people share their emotions in the SAS in order to openly communicate about the emotional

M. Kurosu (Ed.): Human-Computer Interaction, Part V, HCII 2013, LNCS 8008, pp. 19–28, 2013.
© Springer-Verlag Berlin Heidelberg 2013

circumstances and their feelings and reactions. In this paper, we present a multimodal system that allows people to socially share text, images and their emotions. This system exploits multimodality since the use of natural means of communication facilitates the human interaction with the machine [3]. This statement is particularly valid when context information is taken into account to choose best modality in an opportune manner. The main contribution of this paper consists of introducing the concept of multimodal social interaction in affective computing for the HCHI. In particular, we wanted to explore the communication in SAS.

The rest of this paper is structured as follows: Section 2 analyzes the literature of the Affective Computing domain with focus on computer mediated communication; Section 3 explains the concept of the proposed system whose architecture is described in Section 4; Section 5 shows the implemented evaluation scenario and the discussion; finally, Section 6 reports the conclusion and the future work.

2   Related Work

Many psychologists argue that it is impossible for a person to have a thought or perform an action without engaging, at least unconsciously, his or her emotional systems [13]. This is true also for the interaction with computers. In fact, recent research in psychology and technology assesses that emotions play a key role also in every computer-related activity [14]. From these findings, a new domain emerged and is called affective computing. This name comes from Rosalind W. Picard, one of the first pioneers of this new and exciting domain (the fascinating story of the birth of affective computing has been narrated in the introduction of the first issue of the IEEE Transactions on Affective Computing [34]). She defined the affective computing as the "computing that relates to, arises from or deliberately influences emotions" [15]. This definition gives the idea of how wide this domain is; in fact, affective computing "can mean recognizing user affect, adapting to the user's affective state, generating 'affective' behavior by the machine, modeling user's affective states, or generating affective states within an agent's cognitive architecture" [11]. The first issue tackled by the researchers in affective computing was the emotion detection and recognition. Indeed, the main human communication modalities have been thoroughly studied in order to extract the emotional information (e.g., body gestures and motion [16], facial expressions [17] and speech recognition [18]). The aforementioned communication means are also the most important in the human judgment of behavioral cues [19]. In contrast, computers can capture other signals that the human sensory system completely ignores. For instance, specific sensors on the user can capture physiological signals that have been demonstrated being reliable for the emotion recognition (e.g., electromyography, electrocardiography, electrodermal activity, blood volume pulse, peripheral temperature and respiration as in [20], and electroencephalography as in [21]).

As showed, the scientific community intensively studied the techniques for the emotion recognition but this was only the beginning. Afterwards, the researchers in human-computer interaction broadened the spectrum of this domain exploring how the emotional information can be exploited to enhance the user experience. For example, the user's emotional state can be used to recommend the opportune video [22] or the best song [23] or for marketing purposes [29]. Another interesting application of affective computing can be found in [24] where the authors developed an emotion-aware game-based system for adaptive learning; they observed an enhancement of motivation and satisfaction of the students. Affective computing can also help the rehabilitation process, as proposed in [25], or just help to make the user happier [28].

The aforementioned applications refer to the role of emotion in the interaction between a user and the computer in order to make it more natural, but another important aspect of affective computing is related to HCHI. In fact, the *affective interactions* are not limited to the human-human interaction and the relation between the user and the system, but they refer also to computer mediated communication [26]. Nowadays, it is easy to find people who spend more time interacting with a computer than with other humans in face-to-face encounters [27] where affect not only creates richer interaction, but also helps to disambiguate meaning, allowing for more effective communication [14]. Hence, a current and important trend of affective computing focused on the transmission of emotions in order to fill the gap due to the lack of expressivity in HCHI, where all the paralanguage and non-verbal communication are excluded. In fact, some works focused on introducing a remote system for the communication of touch in order to create empathy between two users who were distant; an example is the "Huggy Pajama", which remotely reproduced the feeling of a hug [30]. The "keep in touch" project followed a similar concept, but it associated the haptic feedback to the voice for an enhanced experience of connectedness [31]. In particular, multimodal interfaces are considered suitable for the communication of emotions. A perfect example of this kind of application is the system for tele-home health care presented in [32]. This system sensed the user's emotional and affective states in order to remotely monitor him/her and opportunely respond to this information by a multimodal anthropomorphic interface agent. As the above-mentioned system, our work takes advantage of the rich expressivity of multimodal interfaces but applied to HCHI. In particular, our system has been designed to provide a feedback representing the emotional states of other members of a social group, exactly as the Emotishare project [33] but with context-aware multimodal representation of emotion in the ambient.

3 Concept

In the previous sections we showed how the scientific community highlighted the importance of emotional information communication in HCHI. With this concept, we address the representation of the users' emotional states in a social group. This system aims at allowing the users to share a message, a picture and their emotional state in an SAS, for instance Twitter. The SASs do not allow a variety of modalities and usually have also a limited number of characters for messages; therefore, the expressivity of emotions in SASs is limited. In order to enhance the user experience, the system provides a multimodal feedback of the shared content to all the followers. The feedback is displayed in the whole environment using different modalities in order to grant a fully immersive experience. In particular, the system should be able to handle the sound (i.e., the vocal synthesis of the message), to display images and to represent emotions. The emotions can be displayed in personal devices (i.e., as pictures) or in the ambient; the emotion state is displayed in the ambient via Aphrodite [7]. Aphrodite is a robotic painting that depicts Venus' head from Botticelli's "the Birth of Venus". This painting comes alive as the magic paintings in Harry Potters' saga; in fact, Aphrodite interacts with the user since she is able to mime the human facial expressions and to reproduce sounds using her artificial voice in order to communicate a certain number of emotional states. We adopted this approach since some researches demonstrated that anthropomorphic artificial agents are very effective for emotion communication [35].

Since communicating using different interaction modalities can be considered as an innate characteristic of human beings, we provided multiple input modalities. The user can communicate his/her emotional state via text, adding an emoticon, using a picture and performing a facial expression. The system manages several distributed devices (as smartphones, PCs, interactive surfaces et cetera) and can recognize the user's facial expression.

The system coordinates inputs and outputs according to the context. The contextual information is used to choose the opportune means of communication in order to improve the user experience.

4   System Architecture

The system presented in this paper has been developed adopting two frameworks: Inter-Face and NAIF (as depicted in Fig. 1).

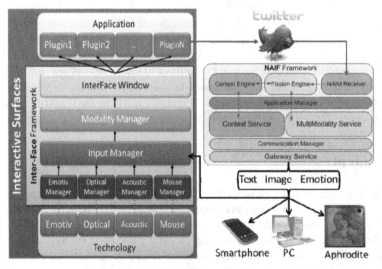

Fig. 1. System architecture integrating the structures of the Inter-Face and NAIF frameworks

The Inter-Face framework has been conceived in order to guarantee the compatibility among multiple heterogeneous technologies by providing a runtime environment for the dynamic adaptation of an application to a surface. In fact, Inter-Face allows transforming any object into an interactive surface [4]. Moreover, it integrates the drivers for the management of the Emotiv EPOC neuroheadset [5]. The NAIF framework aims at interconnecting heterogeneous devices in smart environments. These devices can offer resources to display information to the user and can contribute to the gathering of contextual information. Context information is managed by NAIF [6] and can be used to deliver opportune feedback to the user using the best available modalities. Moreover, each device can send communication intents to the framework, which automatically performs the context-aware multimodal fission of the message content.

Our system allows the user to interact with several surfaces distributed in the environment (e.g., tables, walls, TVs, smartphones et cetera); then, he/she can share a text, an image and his/her personal emotional state. In fact, the user can select the image in the specific application and type the text through touch interaction on the surfaces; in addition, he/she can decide to press the button for the emotion detection via the EPOC neuroheadset. The Emotiv EPOC senses the facial expression through the surface electromyography (sEMG) and then the recognized user's emotional state is added to the message. Afterwards, the user presses the "tweet" button to share his/her message on Twitter. Inter-Face supplies another plugin dedicated to the reception of messages "tweeted" by the people "followed" by the user. These plugins are showed in Fig. 2.

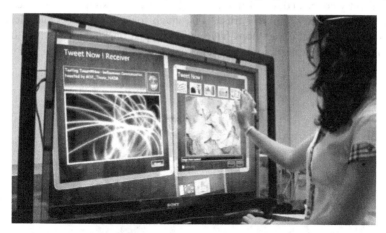

Fig. 2. A user wearing the EPOC neuroheadset and interacting with a touch-enabled TV with Inter-Face running the Twitter application

5   Evaluation Scenario and Discussion

The system presented in the previous section has been implemented in a smart environment, i.e., a smart living room (as depicted in Fig. 3). The NAIF gateway runs on a dedicated server and interconnects all the devices present in the environment. In this scenario, the user can send tweets using his/her smartphone, or laptop, or interacting with the smart surfaces distributed in the environment as the TV and the table.

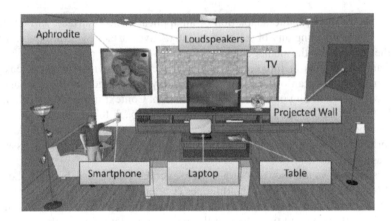

Fig. 3. Study of the evaluation scenario: all the input and output devices are highlighted

The user's emotional state is captured whenever he/she wears the Emotiv EPOC neuroheadset; in fact, the emotion is directly connected to the user's facial expression that is recognized through the sEMG signals. Messages are automatically composed using a specific protocol as described in [6]. The set of available emotions are 5 and they are directly connected to 5 different facial expressions: smile for happiness ":-)", frown for sadness ":-(", wink for trust ";-)", laugh for ecstasy ":-D" and clench for anger ":-@". The facial expressions have been mapped on Plutchik's wheel of emotions [36]. Fig. 4 shows the emotional states represented by Aphrodite.

Fig. 4. Aphrodite can mime several facial expressions that are mapped on the Plutchik's wheel of emotions: a) is the neutral expression with no associated emotion; b) represents the smile for happiness; c) is frown for sadness; d) is frown and glum for anger; e) is wink for trust; f) is laugh for ecstasy.

NAIF takes advantage of several output generators: text messages can be communicated either via Text-To-Speech (TTS) in loudspeakers or via TV speakers;

emotions are shared via emoticons in the displays or via Aphrodite; images are shown in the available displays. In this scenario, the Twitter receiver is implemented in different devices: besides the aforementioned Inter-Face plugin (in the TV and the interactive table), other applications are running in Microsoft Windows (in the laptop and the projected wall).

Output modalities are chosen according to context information. For instance, if the luminosity of the room is high, the system chooses displaying information on the TV instead of the projected wall in order to grant the best quality of feedback (this scenario is shown in Fig. 5).

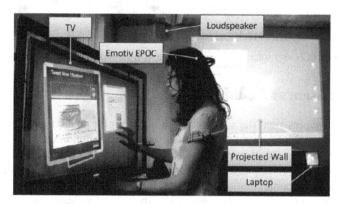

Fig. 5. A user wearing the EPOC neuroheadset and interacting with the touch-enabled TV using the Twitter application

The luminosity information is employed also for the reasoning about emotion sharing: in low light conditions, the robotic painting may not be visible, thus the emotion is displayed in the projected wall or the TV or the table (as depicted in Fig. 6). Similarly, if the room is noisy, audio feedback is avoided. In the current prototype, luminosity and noise information is retrieved through the popular Phidgets sensors [8]. Another context factor is based on the availability of the devices. Only the connected and functioning devices are taken into account during the multimodal fission and if they are not busy performing other tasks. Moreover, messages can be addressed to a specified user: "tweet" messages containing the @User tag can be delivered privately to the specified user according to his/her position. In fact, the user can identify him/herself near to a personal device using an RFID tag. Thus, all these messages are delivered only to the specific user's device in a discrete manner.

The system has been presented during a public event called "Socialize" [39] and other private demonstrations. During these events, we conducted an observation study on the evaluation scenario with the participants. The attenders of the "Socialize" event were young people aged between 20 and 35 years and enthusiast for social network technologies. Although the users showed an initial skepticism, they demonstrated a large interest about our system. The participants were very engaged while experiencing the interaction scenario; they appreciated the possibility of choice for the modality of input (even if the Emotiv EPOC resulted being very cumbersome) and of output. Finally, the users were particularly fascinated by Aphrodite and her capability of reproducing human emotions, who further engaged them.

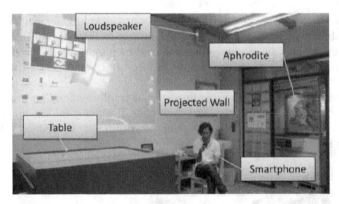

Fig. 6. A user interacting with the system via smartphone application

6   Conclusion and Future Work

In this paper, we presented the architecture of a system for context-aware multimodal HCHI. This system focuses on the computer mediated communication of emotional states via an SAS, i.e., Twitter, with pervasive multimodal feedback in the ambient in order to provide a more immersive user experience. An integrated multimodal fission engine evaluates the context information (e.g., luminosity level, noise level, user identification et cetera) to opportunely enhance the quality of feedback. The user can interact on several devices distributed in the smart environment through gestures and facial expression or traditional graphical user interfaces.

New input modalities will be integrated in the future development of this system. For example, adding vision-based face expression recognition during the interaction with devices that have embedded cameras (e.g., smartphone, laptop, TV et cetera). Recently, more and more novel products provide new ways for the emotion detection as the Affectiva Q-Sensor wristband [37] and the Interaxon headset [38]. Therefore, exploiting the flexibility of the adopted frameworks, these new devices will be integrated in order to gather more information about the users' emotional state and to provide a less cumbersome interface. Moreover, NAIF framework could benefit of the information of the sensed user's emotional state to further improve the quality and the opportuneness of feedback. The future version of this system will include also a virtual avatar representing Aphrodite for ubiquitous representation of emotions.

References

1. Clubb, O.L.: Human-to-Computer-to-Human Interactions (HCHI) of the communications revolution. Interactions 14(2), 35 (2007)
2. Kivran-Swaine, F., Naaman, M.: Network properties and social sharing of emotions in social awareness streams. In: Proceedings of the ACM 2011 Conference on Computer Supported Cooperative Work, p. 379 (July 2011)
3. Dumas, B., Lalanne, D., Oviatt, S.: Multimodal interfaces: A survey of principles, models and frameworks. In: Lalanne, D., Kohlas, J. (eds.) Human Machine Interaction. LNCS, vol. 5440, pp. 3–26. Springer, Heidelberg (2009)

4. Mugellini, E., Abou Khaled, O., Pierroz, S., Carrino, S., Chabbi Drissi, H.: Generic Framework for Transforming Everyday Objects into Interactive Surfaces. In: Jacko, J.A. (ed.) HCI International 2009, Part III. LNCS, vol. 5612, pp. 473–482. Springer, Heidelberg (2009)

5. Emotiv EPOC neuroheadset, http://www.emotiv.com/apps/epoc/299/ (last visited: March 2013)

6. Perroud, D., Angelini, L., Khaled, O.A., Mugellini, E.: Context-Based Generation of Multimodal Feedbacks for Natural Interaction in Smart Environments. In: Proceedings of the Second International Conference on Ambient Computing, Applications, Services and Technologies, pp. 19–25 (2012)

7. Aphrodite project, https://project.eia-fr.ch/aphrodite/Pages/Introduction.aspx (last visited: March 2013)

8. Phidgets sensors, http://www.phidgets.com/ (last visited: March 2013)

9. Hwang, H.C., Matsumoto, D.: Nonverbal Behaviors and Cross-Cultural Communication in the New Era. Language and Intercultural Communication in the New Era, 116 (2013)

10. Rimé, B., Finkenauer, C., Luminet, O., Zech, E., Philippot, P.: Social Sharing of Emotion: New Evidence and New Questions. European Review of Social Psychology 9, 145–189 (1998)

11. Hudlicka, E.: To feel or not to feel: The role of affect in human–computer interaction. International Journal of Human-Computer Studies 59, 1–32 (2003)

12. Hoffman, D.L., Novak, T.P., Venkatesh, A.: Has the Internet become indispensable? Communications of the ACM 47, 37–42 (2004)

13. Picard, R.W.: Does HAL cry digital tears? Emotions and computers. In: Hal's Legacy: 2001's Computer as Dream and Reality, pp. 279–303 (1997)

14. Brave, S., Nass, C.: Emotion in human–computer interaction. In: The Human–Computer Interaction Handbook, pp. 81–93 (2003)

15. Picard, R.W.: Affective Computing. The MIT Press (1997)

16. Kleinsmith, A., Bianchi-Berthouze, N.: Affective Body Expression Perception and Recognition: A Survey. IEEE Transactions on Affective Computing, 1 (2012)

17. Sandbach, G., Zafeiriou, S., Pantic, M., Yin, L.: Static and dynamic 3D facial expression recognition: A comprehensive survey. Image and Vision Computing 30, 683–697 (2012)

18. Anagnostopoulos, C.N., Iliou, T., Giannoukos, I.: Features and classifiers for emotion recognition from speech: a survey from 2000 to 2011. Artificial Intelligence Review, 1–23 (2012)

19. Pantic, M., Sebe, N., Cohn, J.F., Huang, T.: Affective multimodal human-computer interaction. In: Proceedings of the 13th Annual ACM International Conference on Multimedia, p. 669 (2005)

20. Canento, F., Fred, A., Silva, H., Gamboa, H., Lourenco, A.: Multimodal biosignal sensor data handling for emotion recognition. In: Proceedings of 2011 IEEE SENSORS, pp. 647–650 (2011)

21. Schaaff, K., Schultz, T.: Towards emotion recognition from electroencephalographic signals. In: Proceedings of the 3rd International Conference on Affective Computing and Intelligent Interaction and Workshops, pp. 1–6 (2009)

22. Benini, S., Canini, L., Leonardi, R.: A Connotative Space for Supporting Movie Affective Recommendation. IEEE Transactions on Multimedia 13, 1356–1370 (2011)

23. Dornbush, S., Fisher, K., McKay, K., Prikhodko, A., Segall, Z.: XPOD - A Human Activity and Emotion Aware Mobile Music Player. In: Proceedings of the 2nd Asia Pacific Conference on Mobile Technology, Applications and Systems, pp. 1–6 (2005)

24. Tsai, T.W., Lo, H.Y., Chen, K.S.: An affective computing approach to develop the game-based adaptive learning material for the elementary students. In: Proceedings of the 2012 Joint International Conference on Human-Centered Computer Environments, p. 8 (2012)
25. Mihelj, M., Novak, D., Munih, M.: Emotion-aware system for upper extremity rehabilitation. In: Proceedings of Virtual Rehabilitation International Conference, pp. 160–165 (2009)
26. Paiva, A.: Affective interactions: towards a new generation of computer interfaces. In: Paiva, A.M. (ed.) Affective Interactions. LNCS (LNAI), vol. 1814, pp. 1–8. Springer, Heidelberg (2000)
27. Picard, R.W., Cosier, G.: Affective intelligence — the missing link? BT Technology Journal (1997)
28. Tsujita, H., Rekimoto, J.: Smiling makes us happier. In: Proceedings of the 13th International Conference on Ubiquitous Computing (2011)
29. McDuff, D., Kaliouby, R.E., Picard, R.W.: Crowdsourcing Facial Responses to Online Videos. IEEE Transactions on Affective Computing 3, 456–468 (2012)
30. Teh, J.K.S., Cheok, A.D., Peiris, R.L., Choi, Y., Thuong, V., Lai, S.: Huggy Pajama. In: Proceedings of the 7th International Conference on Interaction Design and Children, p. 250 (2008)
31. Wang, R., Quek, F., Tatar, D., Teh, K.S., Cheok, A.: Keep in touch. In: Proceedings of the 2012 ACM Annual Conference on Human Factors in Computing Systems, p. 139 (2012)
32. Lisetti, C., Nasoz, F., LeRouge, C., Ozyer, O., Alvarez, K.: Developing multimodal intelligent affective interfaces for tele-home health care. International Journal of Human-Computer Studies 59, 245–255 (2003)
33. Willis, M.J., Jones, C.M.: Emotishare: emotion sharing on mobile devices. In: Proceedings of the 26th Annual BCS Interaction Specialist Group Conference on People and Computers (2012)
34. Picard, R.W.: Affective Computing: From Laughter to IEEE. IEEE Transactions on Affective Computing 1, 11–17 (2010)
35. Moser, E., Derntl, B., Robinson, S., Fink, B., Gur, R.C., Grammer, K.: Amygdala activation at 3T in response to human and avatar facial expressions of emotions. Journal of Neuroscience Methods 161, 126–133 (2007)
36. Plutchik, R.: The Nature of Emotions. American Scientist 89(4), 344–350 (2001)
37. Affectiva Q-Sensor, http://www.affectiva.com/q-sensor/ (last visited: March 2013)
38. Interaxon headset, http://www.interaxon.ca/ (last visited: March 2013)
39. Socialize event at Fribourg, http://www.socialize-network.com/29-novembre-2012/ (last visited: March 2013)

Supportive User Interfaces for MOCOCO (Mobile, Contextualized and Collaborative) Applications

Bertrand David, René Chalon, and Florent Delomier

Université de Lyon, CNRS,
Ecole Centrale de Lyon, LIRIS, UMR5205,
36 avenue Guy de Collongue, F-69134 Ecully Cedex, France
{Bertrand.David,Rene.Chalon,Florent.Delomier}@ec-lyon.fr

Abstract. Enhancing interaction with supplementary Supportive User Interfaces: Meta-UIs, Mega-UIs, Extra-UIs, Supra-UIs, etc. is a relatively new challenge for HCI. In this paper, we describe our view of supportive user Interfaces for AmI applications taking into account Mobility, Collaboration and Contextualization. We describe proposed formalisms and their working conditions: initially created for designers in the design stage; we consider that they can now also be used by final-users for dynamic adjustment of working conditions.

Keywords: Interactive and collaborative model architectures, formalisms, Ambient Intelligence, pervasive and ubiquitous computing, tangible UI.

1   Introduction

The Supportive User Interface (SUI) issue is a relatively new CHI research field, as illustrated by the first workshop devoted to this issue in 2011 [8]. However, several approaches to this issue were proposed already a long time ago without calling them Supportive UIs, just as Mr. Jourdan wrote prose, … We have a relatively long experience in this approach, as since 1994 we have proposed a formalism the aim of which was to support designers' and final-users' manipulations to allow dynamic changes during execution of interactive applications, first purely interactive, then collaborative, and now mobile, collaborative and contextualized.

2   State of the Art

In the workshop mentioned above [8] several approaches were recalled such as Meta-UIs, Mega-UIs, Extra-UIs, Supra-UIs, and others presented as SEUI (Self Explanatory UI), ISATINE and its SUI as well as supportive UIs derived from Collaborative User Interfaces.

3   Our Proposals

To present our general approach and detailed contributions, we start with a brief reminder of the historical evolution of UI design. The first step was software architecture geared

M. Kurosu (Ed.): Human-Computer Interaction, Part V, HCII 2013, LNCS 8008, pp. 29–38, 2013.

towards the need to separate Presentation, Control and Application behavior. Architecture models accounted for this need first by hierarchical models such as the SEEHEIM model or ARCH model. Then a reactive agent model emerged with mainly a PAC model. In a MDA (Model-driven architecture) approach, several models and formalisms such as ours (see hereafter) were proposed. The CAMELEON reference framework [1] proposed three layered approaches with CUI, SCUI and FUI. We based our work in relation to this state of the art.

3.1 AMF Architecture Principles Model

Our first contribution for SUIs (supportive user interfaces) is in relation to our proposal of the AMF model [10]. The AMF model is structured on the basis of agents, in a way that can be compared to the PAC model [3]. The main difference between PAC and AMF lies in the generalization of the number of facets: 3 in PAC, multiple in AMF. This choice was justified by the need to acquire a clear, clean expression of interaction control. This means that our goal was to express in the control facet only different relationships between other facets, the aim of which is to express other aspects of behavior. If a new behavior such as explanation, trace collection, and so on, occurs, and if it is considered as interesting and repeatable in other applications, a new facet is created. In this way, control can be and is expressed by a graphic representation relating different facet inputs and outputs by clearly expressed administrators (Fig.1). The main goal of this graphic expression of control is to allow the designer to model this behavior in a coherent and comprehensible manner.

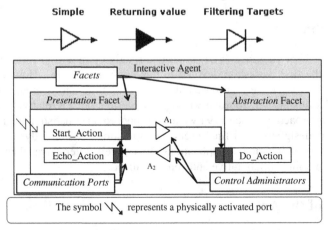

Fig. 1. AMF administrators and their use in control modeling [9]

An interesting consequence of this choice is the possibility of showing this model to the final-user and allowing him to modify the control behavior of his application by using a contextual editor. In this way, the user observing the model can modify it in order to change the behavior of his application. If he considers that the Echo_Action is not meaningful for the user, he can remove the relationship between Do_Action and Echo_Action and the corresponding Control Administrator. Consequently, the manipulation of AMF graphic representations is a Supportive User Interface that can be assimilated to Meta-UI or Mega UI.

3.2   AMF-C Extensions

In the continuity with the AMF model approach, in 1999 the AMF-C model [10] was proposed as an extension of the AMF model for cooperative applications. In this context, the same idea of clean, clear control, expressed graphically, is used. In a distributed collaborative environment, we proposed either to replicate each AMF agent on each workstation, or to fragment each AMF agent and locate only the presentation facet on each workstation. One of the possible implementations is the "replicated mode". In this case, each AMF agent is replicated on each user workstation, and a synchronization mechanism is used at network level to propagate manipulation echoes to other collaborating actors. In this way, with an augmented set of administrators it is possible, just as for AMF, to design the control graphically. In this case, the control expresses not only individual interactions, but also, and in particular, collaborative behavior.

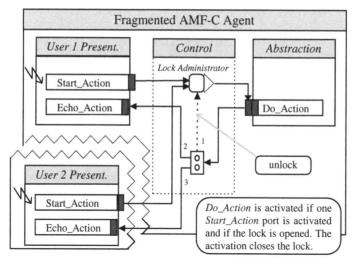

Fig. 2. AMF-C control modeling, expressing awareness.

As regards AMF, our view is that this visual programming is both appropriate and easy to use by the final-user for modifying several behavioral aspects of a collaborative system or an application using a contextual editor. We can mention, as an example of dynamic adjustment of an application behavior, the decision of awareness of a set of manipulations. This means that echoes of operations and corresponding manipulations or results are not propagated to other contributors, as shown in Fig. 2. All adjustments relating to control can be formulated in this way and at the time of execution using a contextual editor.

3.3   IRVO Model

Another model and formalism we proposed, called IRVO (Interaction with Real and Virtual Objects) is devoted to organizing interactions in augmented reality [2]. This approach is based on the idea of using a graphic expression to model the design of this kind of application. However, just as for AMF and AMF-C, we propose using IRVO as a formalism that can be given to the final-user to allow him to modify the composition of an interaction choice, i.e. create a new interaction configuration based on available real and virtual objects. In this way, several augmented reality configurations can be studied (Fig. 3). In this figure and next we present different modeling related to Lea(r)nIt a serious game oriented to Lean Manufacturing mastering which is working in real augmented environment [7]. Le(a)rnIT is an 8-player game during which each student plays an operator role in an industrial production line to understand the complexity of its dynamicity and how to improve it. Raw materials and processed materials are moved between the player's tables by a warehouseman handling a cart. After each simulated working sequence, the teacher and students debrief their working experience in order to find improvements to apply to the production line.

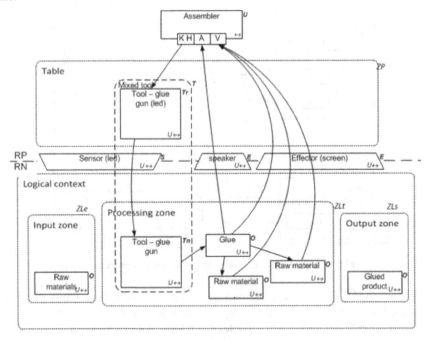

Fig. 3. IRVO modeling of a working space of Lea(r)nIt serious game

3.4   ORCHESTRA Model

ORCHESTRA [5,6] is a formalism, the aim of which is to express orchestration of collaborative applications. This formalism works at a symbolic level, and is able to express in a music-like form, the behavior of a collaborative application between its contributors, their tasks structured in a workflow (states and transitions), and the relevant manipulated artifacts (objects and tools) in a given context. Small working periods can be organized appropriately to express workflow, with repetitions, optional sections, linking up between periods, and so on. This graphic description is decorated by working choices related to cooperative styles, coordination styles, etc.

Different key signatures express collaboration properties such as synchronous or asynchronous collaborations, collaboration modes, and styles of coordination (computational 🖳 or social ☺, implicit ● or explicit ■■):

@ - **Asynchronous** with infinite answer delay

@@ - Asynchronous with limited answer delay corresponding to "**on call**" participation

& - Synchronous "**in-meeting**" cooperation

&& - Synchronous "**in-depth**" cooperation

In synchronous collaboration, two different participations must be distinguished:

- **instantaneous**, short-term collaboration, also known as implicit and expressed by ● i.e. vote activity,
- **long-term** participation, long-term collaboration, also known as explicit and expressed by ■■ i.e. sketching activity.

In short-term activity, as a vote, an implicit collaboration is appropriate (short exclusive access to the shared space), while in a long-term activity, such as sketching, explicit participation must be requested and authorised (long-term access to the shared space) either by **social coordination** (☺), i.e. one of the human contributors is in charge of this coordination, or a **computational** (🖳) contributor i.e. the computer fulfils it. We graphically formulate instantaneous collaboration by a **dot** over concerned chords, while for long-term collaboration we use a horizontal line ■■ and a symbol expressing social or computational coordination (☺, 🖳) i.e. coordination performed by one of the contributors or by interaction (asking for, receiving and returning exclusive access right to shared space).

An important notion in CSCW is **awareness**. Its goal is to allow different contributors to know (or not) what has been done by another contributor. It is important to decide statically (by the designer) or dynamically by the contributor himself the scope of information propagation to other contributors. Statically, we propose expressing awareness in ORCHESTRA formalism. Special marks are proposed:

- 👥 for no awareness,
- 👥 for partial awareness (for specific contributors),
- 👥 for overall awareness (for all contributors).

The main goal of this description is to allow in a comprehensible manner the design of collaborative application behavior. This description or this model is initially devoted to the designer; however, we consider that it can be used also to support dynamic remodeling of application behavior, which can be proposed to users, or at least to experienced users. The following example (Fig. 4) shows how to change the behavior of a part of a collaborative application. This specification can be modified either at the "cords" description level, i.e. to add a new activity or tool to be used, or at decoration level, i.e. allowing change in awareness level.

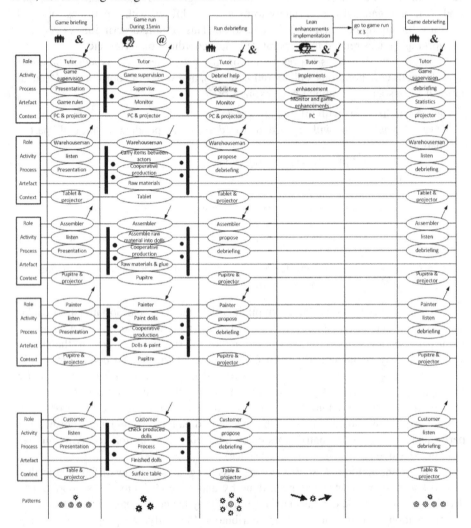

Fig. 4. ORCHESTRA modeling of Lea(r)nIt serious game

3.5   ORCHESTRA+ Approach

The ORCHESTRA mechanism is symbolic-level oriented; we thus propose to take into account progressively more precise considerations related to the user interface finalization process. We use an example to explain thus. In collaboration on interactive mono or multi-touch tables, and mono or multi-users, it is important to be able to express and manage distribution of users and actions. For this reason we are currently working on an extension of ORCHESTRA formalism, called ORCHESTRA+, the goal of which is to progress in concretization from the conceptual user interface (CUI) to the final user interface (FUI) in the CAMELEON reference framework [1]. We can explain this using an example. For artifact manipulation, it appears important to create logical zones in which these artifacts (objects and tools) "can be alive", i.e. used, stored, manipulated, etc. This first structuring is more precise than the symbolic description of ORCHESTRA, but not enough to be considered as final, and to be used in CUI or FUI.

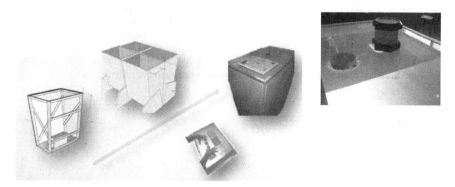

Fig. 5. Tables and tablets as collaboration support and tangible interfaces

To continue this process, we propose two complementary notions and mechanisms: **layers** and **physical zones**. The notion of layer is used to facilitate the expression of access to the artifacts. For a user or a category of users, it is possible to decide that a particular layer is either not visible, or only visible or updatable. The layer appears to be an appropriate mechanism for this access problem. The notion of physical zone is important to allow physical distribution of users' workspaces. In this way, we can dynamically determine the distribution and sharing of physical workspaces. Let us now describe an interesting situation. In a collaborative application, which is a serious game, we use several tablets and tables to work. In this context, we found it very useful to be able to dynamically determine the working conditions for each contributor.

Fig. 6. Work on different workspaces with tangible interfaces

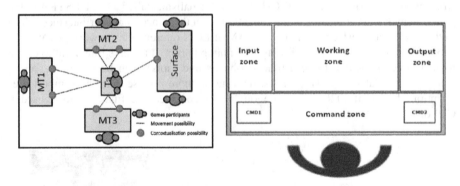

Fig. 7. a) A possible flow of materials between different workspaces (manipulation tables, distribution tablet and Surface based shop), b) A workspace with organization of different zones

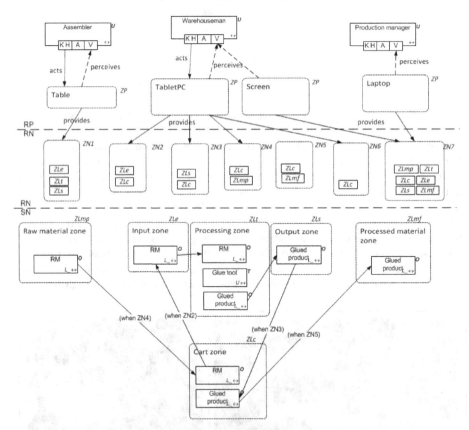

Fig. 8. IRVO Physical, numerical and logical locations modeling of Lea(r)nIt serious game

We can allow each user one tablet or table (Fig. 5), or combine all the tables to create a larger one. Moreover, we can associate different logical working zones with physical zones and thus give users the possibility to act on a zone of a table and also on another zone of another table. We can thus create or adjust interesting working conditions (Fig. 6,7a,b). As we indicated, we can dynamically modify this distribution. We therefore created an appropriate graphic representation, suitable for the designer's work, but also and, in particular, we propose it to the final user.

The user could thus change working conditions dynamically, and allow users to work together in a table consisting of all tables, or to distribute tables and working zones according to the goal of a future working phase (Fig. 8).

4   Conclusion

In this paper we explained our view of Supportive User Interfaces (SUI). Our choice was to propose a variety of formalisms for agent-based user interface modeling (AMF), for cooperative application modeling (AMF-C), for augmented reality interaction (IRVO), and for cooperative application orchestration (ORCHESTRA). All these formalisms are initially design – developer oriented, and completed by tools and editors. As these formalisms are easy to understand, we decided to propose them also to final users, at least experienced final users, who can use a contextual editor to modify the proposed description and thus dynamically change the working application. Through their use during execution of applications, they become supportive user interfaces.

In the last case (ORCHESTRA+), we integrated 3 formalisms to describe (design) a collaborative contextual mobile augmented reality-based serious game called Lea(r)nIt. We used ORCHESTRA to describe the game and its workflow, IRVO to define augmented reality artifacts (tools and objects) working on tables or tablets, and ORCHESTRA+ to distribute work zones on workspaces of different contributors. SUI thus proposed allows us to modify dynamically the workflow of the game and to modify artifacts (tools and objects) by deciding their real or virtual being. It is also possible to remodel topological and geometrical distribution of working zones on different tables, as well as the distribution of artifacts (tool end object).

The least but not the last problem to be taken into account relates to orchestration of interactive applications in the context of their integration in a more important system. In this situation, it is important to be able not only to orchestrate these applications from their layout point of view but also from their functional rules point of view, i.e. which operation each user can execute [4].

References

1. Calvary, G., Coutaz, J., Thevenin, D.: A Unifying Reference Framework for Multi-Target User Interfaces. Interacting with Computer 15(3), 289–308 (2003)
2. Chalon, R., David, B.T.: IRVO: an Interaction Model for designing Collaborative Mixed Reality Systems. In: HCI International 2005, Las Vegas, USA, July 22-27 (2005)
3. Coutaz, J.: PAC, on Object Oriented Model for Dialog Design. In: Interact 1987 (1987)

4. David, B., Chalon, R.: Orchestration Modeling of Interactive Systems. In: Jacko, J.A. (ed.) HCI International 2009, Part I. LNCS, vol. 5610, pp. 796–805. Springer, Heidelberg (2009)
5. David, B., Chalon, R., Delotte, O., Masserey, G., Imbert, M.: ORCHESTRA: formalism to express mobile cooperative applications. In: Dimitriadis, Y.A., Zigurs, I., Gómez-Sánchez, E. (eds.) CRIWG 2006. LNCS, vol. 4154, pp. 163–178. Springer, Heidelberg (2006) 978-3-540-39591-1 ISSN 0302-9743
6. David, B., Chalon, R., Delotte, O., Masserey, G.: ORCHESTRA: formalism to express static and dynamic model of mobile collaborative activities and associated patterns. In: Jacko, J. (ed.) Human-Computer Interaction, Part I, HCII 2007. LNCS, vol. 4550, pp. 1082–1091. Springer, Heidelberg (2007)
7. Delomier, F., David, B., Benazeth, C., Chalon, R.: Situated and collocated Learning Games. In: Conference EC-GBL 2012, European Conference of Game Based Learning, Cork, Ireland (September 2012)
8. Proceedings of the 1er International Workshop on Supportive User Interfaces : SUI 2011 at the 3rd ACM SIGCHI Symposium on Engineering Interactive Computing Systems, Pisa, Italy (June 13, 2011)
9. Tarpin-Bernard, F., David, B.T.: AMF a new design pattern for complex interactive software? In: International HCI 1997, Design of Computing Systems, San Francisco, August 24-29, vol. 21 B, pp. 351–354. Elsevier (1997) ISBN 0444 82183X
10. Tarpin-Bernard, F., Samaan, K., David, B.: Achieving Usability of Adaptable Software: The AMF-based Approach. In: Seffah, A., Vanderdonckt, J., Desmarais, M. (eds.) Human-Centered Software Engineering: Software Engineering Models, Patterns and Architectures for HCI. Springer HCI Series, pp. 237–254 (2008)
11. Tarpin-Bernard, F., David, B.T., Primet, P.: Frameworks and Patterns for Synchronous Groupware: AMF-C Approach. In: Chatty, S., Dewan, P. (eds.) Proceedings of the IFIP Tc2/Tc13 Wg2.7/Wg13.4 Seventh Working Conference on Engineering For Human-Computer Interaction, September 14 - 18. IFIP Conference Proceedings, vol. 150, pp. 225–241. Kluwer B.V, Deventer (1998)

RFID Mesh Network as an Infrastructure for Location Based Services for the Blind

Hugo Fernandes[1], Jose Faria[2], Paulo Martins[1], Hugo Paredes[1], and João Barroso[1]

[1] INESC TEC (formerly INESC Porto) and University of Trás-os-Montes e Alto Douro,
Vila Real, Portugal
[2] University of Trás-os-Montes e Alto Douro, Vila Real, Portugal
{hugof,pmartins,hparedes,jbarroso}@utad.pt, jfaria@utad.pt

Abstract. People with visual impairments face serious challenges while moving from one place to another. This is a difficult challenge that involves obstacle avoidance, staying on street walks, finding doors, knowing the current location and keeping on track through the desired course, until the destination is reached. While assistive technology has contributed to the improvement of the quality of life of people with disabilities, people with visual impairment still face enormous limitations in terms of their mobility. There is still an enormous lack of availability of information that can be used to assist them, as well as a lack of sufficient precision in terms of the estimation of the user's location. This paper proposes an infrastructure to assist the estimation of the user's location with high precision using Radio Frequency Identification, providing seamless availability of location based services for the blind, whether indoor or outdoor.

Keywords: Computer-augmented environments, blind, navigation, rfid.

1 Introduction

The task of moving from one place to another is a difficult challenge that involves obstacle avoidance, staying on street walks, finding doors, knowing the current location and keeping on track through the desired path, until the destination is reached. While assistive technology has contributed to the improvement of the quality of life of people with disabilities, with major advances in recent years, people with visual impairment still face enormous limitations in terms of their mobility. Most navigation systems are designed to be used by users without any major disability and are based on information systems which are mainly focused on road navigation (outdoor) and commercial and tourist destinations.

In recent years, several approaches have been made to create systems that allow seamless tracking and navigation both in indoor and outdoor environments. However there is still an enormous lack of availability of information that can be used to assist the navigation of users with visual impairments (or other kinds of impairment), as well as a lack of sufficient precision in terms of the estimation of the user's location. All these factors combined, maintain a situation of large disparity between the availability of such technology among users who suffer from physical limitations and those who do not suffer such limitations.

M. Kurosu (Ed.): Human-Computer Interaction, Part V, HCII 2013, LNCS 8008, pp. 39–45, 2013.

In this paper, an infrastructure is proposed to assist the estimation of the user's location with high precision to provide for the seamless availability of location based services for the blind, whether indoor or outdoor. In Section2, an overview of currently available, related, technologies is presented. In Section 3, a radio frequency identification (RFID) mesh network is proposed, which is the focus of this work. In Section 4, some considerations are presented about the features and limitations of the presented infrastructure, as final remarks.

2   Related Work

Location and navigation systems have become very important and widely available in recent years as a base for finding the quickest or optimal route to a specific destination or simply to retrieve contextual information about the environment and nearby points-of-interest (POI). Most of these systems use the Global Positioning System (GPS) and only work well in outdoor environment, since GPS signals cannot easily penetrate and/or are greatly degraded inside of buildings.

To address the task of finding the user location in indoor environments several techniques and technologies have been used such as sonar, radio signal triangulation, radio signal (beacon) emitters, or signal fingerprinting. All these technologies can be, and have been, used to develop systems that help enhancing the personal space range of blind or visually impaired users [1].

Another technology widely used in this context is Radio-Frequency Identification (RFID). RFID tags are built-in with electronic components that store an identification code that can be read by an RFID tag reader. In recent years some teams [2][3][4] have developed navigation systems based on this technology. In the case of outdoor environments, some hybrid systems have been proposed that use GPS as the main information source and use RFID for correction and minimization of the location error, like the prototype developed in the SmartVision project by the team at the University of Trás-os-Montes e Alto Douro (UTAD)[5]. The Blavigator project [6] aims to create a small, cheap and portable application as an extension of the work done in the SmartVision project. The new prototype is built with the same modular structure as in SmartVision (**Error! Reference source not found.**).

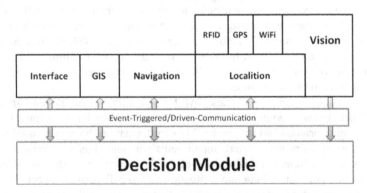

Fig. 1. Modular structure of the Blavigator prototype

The Location Module is responsible for calculating the user's current location based on the inputs given by GPS, Wi-Fi and RFID technologies. The Navigation module is responsible for using the current location to provide guiding instructions and geographic contextual information found in the GIS Module through the Interface Module. The Computer Vision Module performs object and obstacle recognition and also helps to keep the user in safe paths [7]. The Decision Module acts as a mediator and message forwarder between all these modules.

In terms of the use of RFID technology to get the user location, the prototype uses an electronic white cane that senses tags on the floor placed on a topology that consists of connected lines and clusters [8]. This set of connected lines and clusters compose a network of safe paths and points of interest (**Error! Reference source not found.**). The software is able to run offline on a mobile device using a local representation of the data present in the Geographic Information System (GIS Module) [9], namely the tag identifiers, the corresponding geographic coordinates and the tag ownership (lines and clusters).

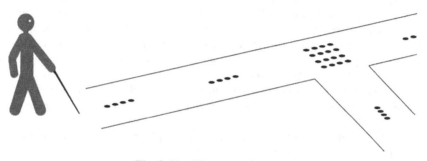

Fig. 2. Line/Cluster topology

The ability to run offline is an advantage when compared to previous works, since older systems relied on a central server to calculate the desired routes and provide information about user location, leaving the user somewhat 'alone' when there was no network coverage. Another advantage of this type of setup is that the user can be certain that he will stay on safe routes and locations, as long as he carefully follows the lines and clusters. However, when compared to normally sighted people, blind users still face some limitations in terms of where they can go while using the system.

3   Mesh Network

To address these limitations we propose a change in the topology from a network of connected lines and clusters to a mesh network of RFID tags (Fig. 3). This topology creates a layer that covers large areas and increases the places that the user can safely navigate.

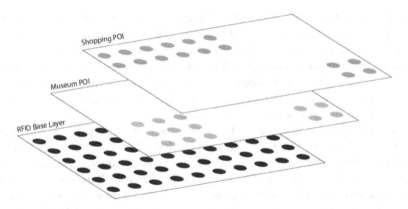

Fig. 3. Mesh network of RFID tags and example layers of points-of-interest

The base layer is the actual infrastructure of individual RFID tags and their geographic location in terms of absolute coordinates. On top of that, it is possible to create an infinite range of layers that can be used to provide location-based services to be shared by both blind and normally sighted users. By clustering some of the tags it is possible to create points-of-interest that can be categorized (layered) according to the needs of the actual location owner or promoter, much like traditional navigation systems, and the path between these POIs can be dynamically calculated using routing algorithms and the RFID base layer.

The contextual information stored can range from the cultural and intellectual promotion of places or events, such as navigation in museums, to commercial advertisements in shopping areas. At the same time, it is possible to extend the coverage area of the system, reducing the gap between the places where blind and normally sighted people can safely navigate.

3.1 RFID Base Layer

The base layer proposed here is a mesh/grid of connected RFID tags. In terms of hardware, the tags are placed individually on the floor. The connections are established in the GIS database. Each tag has its individual information stored in the GIS, but also has a cluster representation, where information about its connecting tags (or exit points) is stored (Fig. 4).

This information can be used to trace routes between the user location and any desired POI and to provide guiding instructions, in case of assisted navigation.

The accuracy is defined by the precision of the tag reader. Most small readers of passive RFID tags work well in a range of a few centimetres and each tag has its individual geographic coordinates stored in the GIS, related to the tag identifier. This reading distance is defined by the RFID reader manufacturer and is considered to be the maximum deviation of the user from its actual location, according to the tag read.

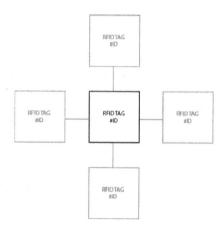

Fig. 4. Tag clustering allowing for the application of routing algorithms

3.2   Points-of-Interest

As an infrastructure for location based services, the mesh network must allow for the identification of places and services, according to user's needs and/or promoter's requirements. These geographic features can range from cultural venues to commercial services. They can also be used to distinguish between danger zones and traffic connections.

To categorize all these features, the RFID tags can be clustered according to the areas covered by the POIS to be represented. Each cluster is defined by the tags by which it is composed, as well as by the geographic coordinates associated with its location. If the POI covers a large area, this information can be used to assure that the user is able to know that the POI has been reached, without having to be physically present near the geographic coordinate that defines the place. In other cases, like the existence of a zebra crossing, this information can be used to inform the user as to when he/she entered the crossing and when he/she left the crossing with high precision. This would be very difficult by just knowing the coordinated of the zebra crossing alone, in traditional navigation systems with low accuracy.

All the POI can be grouped in layers according to the features to be represented and the user can filter the information about which he wants to be warned about. This fact can be used to extend the assistance of the proposed system to all kinds of disabilities which pose limitations to safe navigation, not being restricted to visual disabilities. Navigation between POIs or from the user's current location to a specific POI can be assisted using the RFID base layer and routing algorithms. While navigating, the user can also know about nearby POIs as contextual information, much like normally sighted people can while roaming to a specific destination. This contextual information can describe the environment, warn about danger zones or even be used for commercial purposes. As the GIS can be updated through time, so will the navigation systems that are synced to it, and all the information about the environment.

4   Final Remarks

This paper proposes an RFID mesh/grid as an infrastructure for location based services for the blind. This infrastructure is being assembled at the moment in the campus
of the University of Trás-os-Montes e Alto Douro (UTAD), Portugal, as part of the
development of the Blavigator project. Previous work, cited in Section 2 of this paper
as led the team to the conclusion that the line/cluster topology, while keeping the user
on safe paths, lacked some of the versatility that normally sighted users had while
navigating. Moreover, as the geographic location of the points of interest varies over
time and the tags are embedded in the floor, the mesh topology has arisen as a natural
evolution. Although a test scenario is currently being assembled on campus, to test
and refine this proposal, some remarks can already be considered at this time.

As the reading distance is defined by the RFID reader manufacturer and is it considered to be the maximum deviation of the user from its actual location, according to
the tag read, the distance between tags is of the upmost importance. It defines the
special resolution in terms of location accuracy. Using passive RFID tags, most
manufacturers assure a reading range of a few centimetres (aprox. 10-15cm maximum). While this setup assures for a good accuracy, and passive RFID tags are cheap,
the overall cost of this setup can be somewhat expensive. Using self-powered active
RFID tags, the range can be greatly increased and the costs can be dramatically reduced. But this setup decreases in accuracy. While the first scenario can be used in
museums and shopping areas, with the financial support of entities which intent to
provide its user's with an added value using this infrastructure, the second scenario
can be more appealing to provide alerts that don't require much resolution.

Another scenario can be considered if active RFID tags are mixed among the mesh
of passive tags, decreasing the number of passive tags required. In this setup, active
tags could be used to alert the user of POI that will not change often through time and
represent danger areas. As these locations will hardly be used by people with disabilities, the spatial resolution is of less importance.

While the test scenario is still being assembled, and no tests have been made to test
this proposal, previous experience with the line/cluster topology indicates that the
results of this implementation will be promising. As the geographic location of the
points of interest varies over time and the tags are embedded in the floor, the mesh
topology has arisen as a natural evolution. Following work will made after the test
scenario is assembled and results will be published as a follow up to this paper.

Acknowledgements. The work presented in this paper is funded by the Portuguese
Foundation for Science and Technology (FCT), through the project
RIPD/ADA/109690/2009 – Blavigator: A cheap and reliable navigation aid for the
blind.

References

1. Strumillo, P.: Electronic interfaces aiding the visually impaired in environmental access, mobility and navigation. In: 3rd Conference on Human System Interactions (HSI), pp. 17–24 (2010)
2. Chumkamon, S., Tuvaphanthaphiphat, P., Keeratiwintakorn, P.: A Blind Navigation System Using RFID for Indoor Environments. In: 5th International Conference on In Electrical Engineering/Electronics, Computer, Telecommunications and Information Technology, Thailand, vol. 2, pp. 765–768 (2008)
3. Willis, S., Helal, S.: RFID Information Grid for Blind Navigational and Wayfinding. In: Proceedings of the 9th IEEE International Symposium on Wearable Computers, Osaka, pp. 34–37 (2005)
4. DAtri, E., Medaglia, C., Panizzi, E., DAtri, A.: A system to aid blind people in the mobility: A usability test and its results. In: Proceedings of the Second International Conference on Systems, Martinique, p. 35 (2007)
5. Fernandes, H., du Buf, J., Rodrigues, J.M.F., Barroso, J., Paredes, H., Farrajota, M., José, J.: The SmartVision Navigation Prototype for Blind Users. Journal of Digital Content Technology and its Applications 5(5), 351–361 (2011)
6. Fernandes, H., Adão, T., Magalhães, L., Paredes, H., Barroso, J.: Navigation Module of Blavigator Prototype. In: Proceedings of the World Automation Congress 2012, Puerto Vallarta (2012)
7. Costa, P., Fernandes, H., Vasconcelos, V., Coelho, P., Barroso, J., Hadjileontiadis, L.: Landmarks detection to assist the navigation of visually impaired people. In: Jacko, J.A. (ed.) Human-Computer Interaction, Part III, HCII 2011. LNCS, vol. 6763, pp. 293–300. Springer, Heidelberg (2011)
8. Fernandes, H., Faria, J., Lopes, S., Martins, P., Barroso, J.: Electronic white cane for blind people navigation assistance. In: Proceedings of the World Automation Congress 2010, Kobe (2010)
9. Fernandes, H., Conceição, N., Paredes, H., Pereira, A., Araújo, P., Barroso, J.: Providing accessibility to blind people using GIS. UAIS - Universal Access in the Information Society 11(4), 399–407 (2012)

An Ontology-Based Interaction Concept
for Social-Aware Applications

Alexandra Funke, Sören Brunk, Romina Kühn, and Thomas Schlegel

TU Dresden – Junior Professorship in Software Engineering of Ubiquitous Systems,
Dresden, Germany
{alexandra.funke,soeren.brunk,romina.kuehn,
thomas.schlegel}@tu-dresden.de

Abstract. With the usage of mobile devices becoming more and more ubiquitous, access to social networks such as Facebook and Twitter from those devices is increasing at a fast rate. Many different social networking applications for mobile devices exist but most of them only enable access to one social network. As users are often registered in multiple social networks, they have to use different applications for mobile access. Furthermore, most applications do not consider the users' social context to aid them with their intentions. This paper presents our idea to model the user's social context and intentions in social networks within an ontology. Based on this ontology we describe an interaction concept that allows publishing information in different social networks in a flexible way. We implemented a prototype to show how our findings can be presented. To conclude, we highlight some possibilities for the future of ontology-based social-aware applications.

Keywords: interaction, ontology, semantic modeling, social-aware, social media.

1   Introduction

The number of people using social networks is still rising at a fast pace. Furthermore many people are using several different social networking services simultaneously due to various reasons. One of the reasons is the different focus and functionality offered by various social networks. Some are specialized in offering functionality, such as storage and sharing of photos, videos or other media while others emphasize the aspect of networking and communication.

Another reason is that often different contacts are members of different social networks. In order to reach or follow them all, accounts in multiple social networks are necessary. However, having accounts in multiple social networks raises new issues. Each social network offers its own interface to access it. Thus users have to log on to several websites or use several applications. A lot of functionality is also redundant or overlaps so that people have to decide to use just one social network for a specific intention or to repeat things to reach everyone. Besides being tedious work, duplicating postings may lead to redundant information overflow for people participating in multiple social networks.

M. Kurosu (Ed.): Human-Computer Interaction, Part V, HCII 2013, LNCS 8008, pp. 46–55, 2013.
© Springer-Verlag Berlin Heidelberg 2013

A fast increase of people with mobile devices [1] is pushing the success of social networks even more. Increasingly, people are using those devices to access social networks from everywhere. Besides mobility, there are other advantages as well in using mobile devices for social networking. For example, most devices have an integrated camera allowing sharing of photos or videos immediately. Furthermore, sensors such as GPS can be used to easily share location information. However, there are also challenges, mostly related to hardware restrictions such as small displays and keyboards that make the use of mobile devices more difficult. Besides having mobile websites, many social network providers offer native applications for the most popular mobile platforms. Those applications are also restricted to their own social network. In order to use the functionality of multiple social networks, users have to install and use one application for each network they want to use. Reading and sharing content within multiple social networks is even more difficult on mobile devices due to the mentioned hardware restrictions. Currently, just a few mobile applications allow using the functionality of multiple social networks from one interface. Existing ones are mainly restricted to show a combined social stream. Another interesting aspect that is missing in most social networking applications is that they do not take advantage of the social context to aid users in reaching their goals. The social context consists of the user's contacts, relationships, preferences and memberships in social networks. In addition, the user's intentions influence the usage of the social network. On the one hand, they are part of the social context and represent the reasons to share any kind of information. On the other hand, they focus on aspects such as who wants the user to be contacted or which kind of information is intended for whom.

To illustrate our approach, this paper is structured as follows: in Section 1 we introduce some related work that deals with ontologies for applications in the field of social media. Some applications available for different mobile operating systems are considered and compared in Section 2. These applications integrate a variety of social networks and social media to support the use of many different networks. A definition for "social-awareness" is given in Section 3. This section also presents our ontology, the interaction concept for social-aware applications and the prototype implementation. The paper concludes with a summary and discussion of our concept and future work.

2   Related Work

"An ontology is a specification of a conceptualization." [2] It offers a flexible way to model the knowledge of a domain.

Currently, there is a lack of research in the area of ontology-based approaches for meta-social networks or the modeling of user goals for social media. A meta-social network is an application or a network that enables the user to send messages and status updates in many different social networks at the same time. Different aspects are included in this field. On the one hand, the development of ontologies for context-aware systems plays an important role. On the other hand, approaches that focus on the user needs and different kinds of social network services and their integration in a meta-social network are essential requirements for our interaction concept.

Context is defined as all information about the interaction between user and application. [3] Neto et al. [4] have developed a context model that is based on a domain-independent ontology that providing classes, properties and relations. Lower ontologies can import them easily for specific domains. Tietze and Schlegel [5] describe a possibility to use social networks and meta-social networks in the domain of public transportation. Starting with a classification of social network services and the specification of functionalities, meta-social networks are presented. In their work, an example is given describing how the passengers of public transportation could inform others about things like delays or cancellations using a meta-social network application. We use some of their specific aspects but apply them to our more general approach by integrating their model of social network capabilities into our ontology.

2.1 Mobile Applications for Social Media

There is a small number of mobile applications supporting the use of multiple social networks at the same time. The iOS application Sociable[1] integrates four social networks: Facebook, Twitter, LinkedIn and Tumblr. It enables the user to send status updates to all selected networks at the same time. An update can contain text, images or the user's location. It is not possible to read data from the social networks, such as streams, messages, or profiles. Furthermore, there is no option to specify the recipients of the message or to send the contents based on the goals of the user.

Seesmic[2] is another app, which is designed for social information exchange. It is available for different operating systems. Seesmic integrates Facebook and Twitter. Users can browse through their streams, friend lists or profile pages. Text, images or locations can be sent to both networks at the same time. It is possible to select a single contact or all contacts as recipients of a message. Seesmic only enables to specify a single recipient of a message by going to a friend's profile and posting a message on his or her stream. A further disadvantage is that the application does not save any settings of previous interactions.

The applications shown above, Sociable and Seesmic, do not support functionalities such as selecting specific persons or groups as recipients for a message. Neither can updates be sent via additional communication channels such as e-mail. Only contacts from the social networks can receive messages. It is not possible to select other people from the contacts list of the smartphone. Users of multiple networks cannot specify one particular network as preference for different types of content (e.g. to share images) using the apps mentioned above. Another important point is that for every new message, users have to explicitly define the settings for sending, as there is no way to save the settings and reuse them later. Based on those facts, we have developed a concept that considers these aspects and tries to integrate them in a user-friendly way. The focus of our work lies on the members of social networks and their needs, the so-called users' intentions.

---

[1] http://andysmart.org/work/sociable/
[2] https://seesmic.com/

2.2   Semantic Modeling for Social Media

FOAF (Friend of a Friend)[3] is one of the first semantic models to grasp social interconnections between people. Persons, their activities and relationships to other people or objects are modeled in this ontology. Moreover, FOAF can also describe documents and organizations. This data is implemented machine-readable in RDF (Resource Description Framework) [6] and OWL (Web Ontology Language). [7] FOAF profiles can be used to find out, for example, who knows whom and if two persons have the same friends. To model these relationships, every profile gets a unique identifier. The base information of people are name, e-mail address and photos. Furthermore, blog, interests and publications can be specified. The focus of our work lies on the goals and intentions of the users, which cannot be modeled with FOAF. Nevertheless, FOAF can be used as a basis for our developed ontology.

3   Social-Aware Applications

Users' contacts, relationships, preferences and memberships in different networks are social information of the user and part of the user context. Therefore, we call it the social context of the user. An application that is sensitive to this data is social-aware. This means that such an app collects social information and responds through adapting the visualization or interaction. Similar to context-sensitive applications that are aware of the location or environment of the user, such an application is aware of the user's social context. If the application knows the contacts of the user and the frequency of communication with them, it can predict a possible target audience and appropriate functionalities for sharing content with users. Furthermore, information, such as the classification of contacts in social groups (e.g. mother and father are part of the family group), and the location in combination with friends (e.g. university and fellow students) can also play a major role in social-aware applications. Boldrini at al. [8] state that social-aware applications can learn automatically, collect social information and use it to predict future movements of the user. They developed a content sharing service that tracks the social context and relationships to optimize the distribution of data to interested users. In our work, the focus lies on the social contacts within social networks and the distribution of personal data over various communities and functionalities.

3.1   An Ontology for Social-Aware Applications

With an ontology it is possible to represent arbitrary concepts and their relations within a formal model. OWL ontologies are described in terms of classes of individuals as well as the properties of those individuals. Properties can connect different individuals or they can relate data attributes to an individual. Relations are described in a formal way with strictly defined semantics. In doing so, it is possible to apply inference rules to infer implicit facts from existing ones.

---

[3] http://www.foaf-project.org/

Our main motivation for using an ontology is the possibility to model the social context of the user in a flexible way and independently from one concrete technology or social network. This is a key factor in order to integrate the functionality of multiple social networks and to actually use the social context to aid users. Another advantage is the possibility to integrate and reuse other existing ontologies. In our work, we include several other ontologies describing related concepts, such as FOAF, to model relations among people.

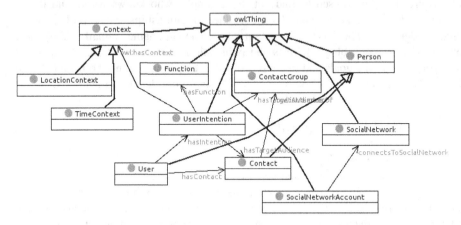

Fig. 1. Simplified view of the main concepts within our ontology

In the following paragraphs, we describe some of the main concepts of our ontology. Figure 1 gives an overview of these concepts.

The use of a social network is always related to some intentions of its users. For example, they might want to check if there are any status updates of their friends or upload a photo for their family. Based on those intentions, they interact with the user interface of their social network to achieve their goal. Their intentions are translated into some actions within the available functionality. Our main focus is to grasp what users intend to do within their social networks. This information is modeled within the *UserIntention* class. The purpose of this class is to describe a user's intentions and actions in social networks in an abstract and social-network independent way. Intentions are described in a formal way as precise as possible. This enables us to map an intention to possibly multiple specific networks, the concrete functionality within those networks as well as to recipients and groups of recipients in different networks.

An example showing some individuals of our ontology as well as their relations is shown in Figure 2.

An intention contains the desired functionality of a social network, such as posting a message or uploading a photo. Desired functionality is modeled in the *Function* class. Combining other actions into new ones can compose more complex actions. This gives us the flexibility necessary to support things like uploading a photo and posting it together with a text message. An intention also has some options, usually

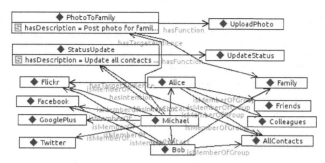

Fig. 2. An example part of ontology instances describing intentions

dependent on the function. For example, an option would be whether the current location should be shared together with a message and the importance of that message. In order to match the desired functionality to the features supported by social networks, we require a compatible model of the network capabilities. For that, we include an existing ontology describing the capabilities of different social networks. [5] Modeling the capabilities enables us to do things such as automatically selecting an appropriate social network for certain input and functionality requirements. Another part of the intention is the target audience serving as list of recipients. This functionality is important to allow users to select who can see their content and many social networks support it. A target audience can be linked to an intention. It is modeled independently of concrete social networks but can be mapped to concrete contacts as described below. A special target audience can also be the public, which is mapped to the corresponding settings in concrete social networks as well. Some social networks or publication channels such as blogs do not allow content to be published to a restricted set of recipients. This can be modeled within our ontology to avoid publishing content meant only for a private audience. An intention can also contain a human readable description allowing easier reuse.

The User class covers information regarding data of individual users and the social networks they use. It includes the users' accounts in social networks as well as their preferences. Preferred social networks in general or for specific functions can be described here. For example, users might prefer a specialized social network such as Flickr as their primary network for uploading photos. This information can be used to post photos to Flickr and automatically create links to it in other social networks.

As social networks are primarily about communication, a user's contacts or friends are one of the core concepts within our ontology. The *Contact* class and the properties connected to it describe a user's contacts and the relation with them. As contacts often have accounts in multiple social networks, it is important to have a mapping of those accounts to one identity. The *Contact* class has a property that maps accounts in different social networks to one person instance. This mapping could be done automatically, e.g., based on the contact's name or e-mail address. The user could also manually choose which contact in different social networks is the same person. The *ContactGroup* class allows for a flexible creation of groups. Similar to contacts, it allows a mapping to groups of a social network or to contacts directly if groups are not supported. For example, a *ContactGroup* "Family" can be mapped to the corresponding groups in different social networks and could include a family member who has only an e-mail address. Preferences of contacts, e.g., regarding a preferred social network, can be modeled within our ontology as well. Together with the users' preferences and the mapping of contacts to their accounts in different social networks, it can be used to avoid posting the same information multiple times to the same people in multiple social networks but instead make sure that it reaches each contact just once through their preferred channel. The *Contact* and *ContactGroup* classes enable a straightforward mapping of their instances to an address book structure for integration into existing applications.

3.2   An Ontology-Based Interaction Concept

As shown above, an ontology can serve as a flexible model to represent a user's social and physical context and to integrate multiple social networks. However, an ontology on its own does not make a social-aware application as it is merely a way to create a model of the users' intentions and their context. We need to take advantage of this model to help users communicate in an efficient and user friendly way. Therefore, we have developed an interaction concept that uses our ontology to aid the users in their utilization of social communication.

Similar to the ontology, our interaction concept is built around the user's intention. As shown in the previous section, intentions in the ontology are modeled in an abstract way and are independent of a specific network. We can use this fact to move from a social network centric view to a task or intention based view. Our concept includes an integrated view on the social networks used by one person also known as a meta-social network. While most meta-social networks or applications focus on having an integrated stream of content from multiple social networks, they are rather limited when it comes to the creation and distribution of content over multiple social media channels as shown in 2.1. Our goal is to decouple the creation and sharing of content in social networks from one single provider and to help users to express their intentions to the system by using the social context.

A typical interaction flow within our concept would look like this: The first interaction step is always to choose the desired functionality and to select or create the content to be shared. This can be done in two ways. The first way is to select the functionality first and then the content. For example, users might choose to update their status and then provide text using an input field. The second way to choose desired functionality is to determine it automatically based on the type of provided content. For example, it is possible to take a photo and immediately share it through the application. The next step is to select options such as showing the current location with a message. These options mostly depend on the chosen functionality and content type. After that, the target audience can be selected. The selection allows choosing from any combination of contacts and groups as stored in the ontology. As described in the previous chapter, the identities of the same person in different social networks are all connected to one contact, which allows us to select a contact only once, independent of their social network accounts/identities. The same is true for groups as they are mapped to each social network as well. Special groups, such as "public", can be chosen, too. The selection can be done using the local address book if it can be synchronized with the ontology.

A selection of social networks is done automatically based on the chosen functionality, the target audience and preference settings. Obviously, only networks supporting the desired functionality can be used. Based on the target audience, only networks that have at least one recipient are chosen. This selection is further refined if one person has an account in multiple networks. To avoid duplicates for the recipient, only one channel is chosen per contact if possible. This is in turn based on the contact's preferred network or, if that information is not available, based on the user's preferences. As an optional step, the user can then refine that selection manually. If the content cannot be delivered to all recipients of the target audience due to a deselection, a warning will show up. As the last step, sharing of content can be triggered and it will be distributed through the selected channels.

The intention will be saved automatically for possible reuse later to increase efficiency. Together with the intention, context information is available, such as the current location. Optionally, users can assign a description to the stored intention if they intend to use it again with new content. A simple approach would be to show a list of most recent intentions as suggestion to the user, or the most frequently used. However, it is also possible to utilize the social context contained in the intentions in a more sophisticated way. That is, for example, the information which contacts or groups have been used how often recently, which networks were used, what functionality and so on. This information can be used not only to suggest a previously used intention but may also help to create a new intention by suggesting likely values based on history. For example, by recommending most recent contacts or often used functionality. Furthermore, social context can be combined with the current selection when creating an intention. After a user chose to share a photo, the target audience used most frequently will be suggested together with photo sharing.

Using the social context could help making the interaction of users with their social networks more efficient and user friendly. However, for user acceptance, it is important not to decide too many things automatically but leave control with users. For many interaction steps, we therefore compute the most likely values based on ontology, history and context and show them as an unobtrusive suggestion leaving the choice to the user whether to accept or to ignore the recommendation.

3.3   Implementation

In order to evaluate our concept, we have created a prototype implementation for mobile devices using the Android platform. The application enables the user to send status updates and messages or to share pictures in the social network Facebook, the micro blogging service Twitter and the photo hosting service Flickr using a single common interface. Thus, users can type a text, choose a picture from the gallery of their smartphone and send this to certain people. As recipients, single persons can be selected as well as groups of persons, e.g. the group "Family". A special feature of this app is that contacts may not only be chosen from the social networks but also from the address book of the mobile device. After all parameters are set the message is sent as specified.

Flickr operates as primary network for sharing photos. Thus, if this network is selected, the picture will be uploaded to that platform and the posts for Facebook and Twitter only contain a link to this page. Otherwise, the images are uploaded directly to the target platform of the posting. When the user sets all preferences and sends a status update, the settings are saved as an intention. Furthermore, the application can be started via the intent to share photos. The user only has to choose a picture in the gallery, select "Share" and our meta-social network application. This integrates a task based approach with the platform. Some of the functionality presented in our theoretical concept cannot be implemented because of restrictions of the social networks' APIs. For example, it is not possible to send direct messages via the Facebook API. According to Facebook, this restriction is necessary in order to avoid spam. The procedure for sending a message is simple. The application starts and the user can type in a text. Afterwards, users may select the group via the group button and shares the picture through the preferred streams, e.g., upload the picture in a Flickr album and send a link to the members of the group "Family" via Facebook, Twitter and maybe via e-mail. For users, it is possible to personalize the message and the recipients of the information when using our application. Furthermore, they can choose people from the contacts list, not only from the communities. One of the most important facts is that the intentions are saved and can be reused for later messages.

4   Conclusions and Future Work

In this paper we have shown how social context and intentions of users in social networks can be modeled within an ontology. Based on that, we described an interaction concept and showed how it can be implemented in a prototypical application. As a preliminary work, we presented several applications that are currently available for smartphones and showed the limitations of those apps. Furthermore, we discussed related work that deals with ontologies for context sensitive applications, the modeling of user intentions and the development of meta-social networks.

We have shown some advantages of using an ontology to model the social context such as flexibility and the possibility to integrate existing models. The functionalities, contacts and settings were described in an abstract way to capture the social context independently of concrete social networks and technologies. Thus, the ontology provides easy modeling and mapping of users and functionalities from different social networks. The user-centered interaction concept based on this utilizes the social context in order to help users to connect to each other and share information in a fast and simple way. Users can not only share content with all their contacts easily, they can also save intentions through their social context for later reuse. The implemented prototype puts one possibility of our interaction concept into practice. It integrates different networks with different functionalities in a single application using a common interface.

Future work could contain the implementation of not only social context but also other context information to enhance functionality. For example, the user intentions suggested could be based on temporal or physical context in addition to social context and including the location context could support suggesting intentions based on this information. An intention that was created at a certain location can be recommended for reuse. In general, context ontologies can be used for that purpose. In some social networks, it is possible to tag different people on photos. If an application enables this feature, it can suggest the tagged people as recipients of the corresponding photo. Furthermore people who are closely related to them may be suggested. It is also conceivable that the ontology tracks content that is shared via different social networks. Hence, comments which were written as reply to this message could be associated with each other.

Therefore, we think our approach explores interesting aspects of social-aware applications and leads to intriguing questions for further research in the fields of ontology-based interaction concepts and of social-aware applications for ubiquitous social media.

References

1. Google Inc.: Mobile Internet & Smartphone Adoption, New Insights into Consumer Usage of Mobile Devices. the Shift to Smartphones & the Emergence of Tablets (2011)
2. Gruber, T.R.: A Translation Approach to Portable Ontology Specifications. Knowledge Acquisition 5(2), 199–220 (1993)
3. Dey, A.K.: Understanding and using context. Human-Computer Interaction Institute (2001)
4. de Freitas Bulcao Neto, R., da Graca Campos Pimentel, M.: Toward a domain-independent semantic model for context-aware computing. In: Third Latin American Web Congress, Buenos Aires, Argentina (2005)
5. Tietze, K., Schlegel, T.: On modeling a social networking service description. In: GeNeMe Gemeinschaft in Neuen Medien: Virtual Enterprises, Communities & Social Networks, Dresden, Germany (2011)
6. Lassila, O., Swick, R.R.: Resource Description Framework (RDF) Model and Syntax Specification. W3C (1999)
7. W3C OWL Working Group, OWL 2 Web Ontology Language Document Overview, W3C (2009)
8. Boldrini, C., Conti, M., Delmastro, F., Passarella, A.: Context- and social-aware middleware opportunistic networks. Journal of Network and Computer Applications, 525–541 (2010)

Sensor-Based Adaptation of User Interface
on Android Phones

Tor-Morten Grønli[1], Gheorghita Ghinea[1,2], and Jarle Hansen[2]

[1]Norwegian School of Information Technology
Oslo 0185, Norway
[2]Brunel University
Uxbridge, UB8 3PH, United Kingdom
tmg@nith.no, george.ghinea@brunel.ac.uk,
hansjar@jarlehansen.net

Abstract. The notion of context-aware computing is generally the ability for the devices to adapt their behavior to the surrounding environment, ultimately enhancing usability. Sensors are an important source of information input in any real world context and several previous research contributions look into this topic. In our research, we combine sensor-generated context information received both from the phone itself and information retrieved from cloud-based servers. All data is integrated to create a context-aware mobile device, where we implemented a new customized home screen application for Android enabled devices. Thus, we are also able to remotely configure the mobile devices independent of the device types. This creates a new concept of context-awareness and embraces the user in ways previously unavailable.

Keywords: sensor, interface adaptation, Android.

1 Introduction

Context-aware computing builds on a combination of various technologies such as computers, mobile devices, human, sensors, cloud, web, Internet and software services. All such technologies and services are interconnected in order to communicate with each other and exchange of useful information such as location, weather information, traffic conditions, road direction, and health and safety.

1.1 Sensors

Sensors are an important source of information input in any real world context and several previous research contributions look into this topic. For instance, Parviainen et al. (2006) approached this area from a meeting room scenario. They found several uses for a sound localization system, such as: automatic translation to another language, retrieval of specific topics, and summarization of meetings in a human-readable form. In their work, they find sensors a viable source of information, but also

M. Kurosu (Ed.): Human-Computer Interaction, Part V, HCII 2013, LNCS 8008, pp. 56–61, 2013.
© Springer-Verlag Berlin Heidelberg 2013

acknowledge there is still work to do, like improving integration. Modern smart phones have a number of built in sensors, all of which can usually be accessed as local services through well-defined APIs. In our work, by taking the aspect of sensors and context-awareness and integrating it in a mobile application we reduce the workload for the end users. Thus, in combination with a cloud-based server application we are able to remotely configure the mobile devices independent of the device types, creating a new concept of context-awareness.

1.2   Cloud Computing

Cloud computing has received considerable attention in the software industry. It is a buzzword frequently used for all sorts of services, ranging from hosted virtual machines to simple web based email applications. So what is cloud computing? We have used the definition created by NIST [6] (National Institute of Standards and Technology, United States): *"Cloud computing is a model for enabling ubiquitous, convenient, on-demand network access to a shared pool of configurable computing resources ..."*. Large IT companies like Microsoft, Google and IBM, all have initiatives relating to cloud computing [7] which have spawned a number of emerging research themes, among which we mention: *cloud system design* [6], *benchmarking of the cloud* [5] and *provider response time comparisons*. Mei et al. [5] have pointed out 4 main research areas in cloud computing that they find particularly interesting, namely *Pluggable computing entities, data access transparency, adaptive behaviour of cloud applications and automatic discovery of application quality.*

Our work focuses on *data access transparency*, where clients transparently will push and pull for data from the cloud, and *adaptive behavior of cloud applications*. We adapted the behavior of the Google App Engine server application based on context information sent from the users' devices thus integrating context and cloud on a secure mobile platform [1].

Accordingly, in our research, we combine sensor-generated context information received both from the phone itself and information retrieved from cloud-based servers. All data is integrated to create a context-aware mobile device, where we implemented a new customized home screen application for Android enabled devices. Thus, we are also able to remotely configure the mobile devices independent of the device types.

2   Design

We developed a proof-of-concept client application on an Android device (HTC Nexus One). The HTC manufactured Nexus One represents one of the first worldwide available commercial Android phones. The Nexus One features dynamic voice suppression and has a 3.7-inch AMOLED touch sensitive display supporting 16M colours with a WVGA screen resolution of 800 x 480 pixels. It runs the Google Android operating system on a Qualcomm 1 GHz Snapdragon processor and features 512 MB standard memory, 512 MB internal flash ROM and 4 GB internal storage. Furthermore, the Nexus One also has 5 sensors available: accelerometer, magnetic field, orientation, proximity and light.

The developed application utilized context-aware information from the device in the form of time, location and sensors. Additionally it utilized context-aware information from the cloud-integrated backend to acquire dynamic interface content, contacts and calendar data. At launch, the application would look as illustrated in Fig. 1. This interface would change depending on the user's given context. The applications available would be adapted and customized to match the current computed user context and thereby unobtrusively alter the user experience.

Fig. 1. Home Screen Interface

2.1   Meta-Tagging

To make it possible for users to tag their appointments and contacts with context information we added special meta-tags. By adding a type tag, for example *$[type = work]* or *$[type= leisure],* we were able to know if the user had a business meeting or a leisure activity. We then filtered the contacts based on this information. If the tag *$[type=work]* was added, this lets the application know that the user is in a work setting and it will automatically adapt the contacts based on this input. In a work context only work related contacts would be shown. To add and edit these tags we used the web-interface of Google contacts and calendar.

2.2   Sensors as Input Data

The sensors on the mobile device were also used as input to the application. We used the API available on the Android platform and through the *SensorManager* base class called were able to access all of the built-in sensors (eg. accelerometer, light) on the mobile device. We started out by just showing the input values from the sensors in our pilot study. When we expanded the application after the initial tests we wanted to use the sensor input to further enhance the user experience. We ended up with using two features directly in the application, the accelerometer and the light sensor. The accelerometer was used to register if the device was shaking, which probably meant that the user was on the move, for example running or walking fast. In these cases, the user interface was automatically changed to a much simpler view which has bigger buttons and is easier to use when on the move.

The second sensor we used in our experiment was the light sensor. By constantly registering the lighting levels in the room we adjusted the background color of the application (Fig. 2). We changed the background color of the application very careful-ly, as it would be very annoying for the users if color changes were happening often and were drastic. Accordingly, color was gradually faded when the lighting values measured from the environment changed.

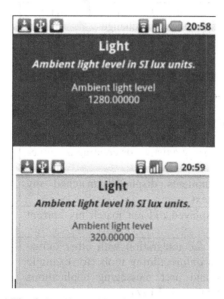

Fig. 2. Interface adaptation to ambient light

3   Evaluation Results

The developed prototype was evaluated in two phases. In the first, a pilot test was performed with a total of 12 users. These users were of mixed age, gender and com-puter expertise. The results from this phase were fed back into the development loop, as well as helped remove some unclear questions in the questionnaire. In the second phase, the main evaluation, another 40 people participated. Out of the 40 participants in the main evaluation, two did not complete the questionnaire afterwards and were therefore removed making the total number of participants 38 in the main evaluation. All 50 participants were aged between 20 and 55 years old, had previous knowledge of mobile phones and mobile communication, but had not previously used the type of application employed in our experiment. None of the pilot test users participated in the main evaluation.

Table 1. Evaluation results

	Statement	Mean	Std. Dev.
User Interface			
1.	It is easy to see the available functions	3.50	0.51
2.	The features of the application are hard to use	1.89	0.73
3.	The adaptability of the application is a feature I approve	3.45	0.55
Sensor Integration			
4.	The background color in the application changes when the lighting in the room changes	3.55	0.65
5.	When moving around, a simplified user interface is not presented	2.11	1.06
6.	I find sensor integration annoying and would disable it on my device	1.84	0.72
Context Awareness			
7.	The close integration with Google services is an inconvenience, I am not able to use the system without changing my existing or creating a new e-mail account at Google	1.76	0.88
8.	Calendar appointments displayed matched my current user context	3.58	0.55
9.	The contacts displayed did not match my current user context	1.29	0.52
10.	I would like to see integration with other online services such as online editing tools (for example Google Docs) and user messaging applications (like Twitter and Google Buzz)	3.29	0.73

The questionnaire that was employed in the second phase had three different parts, dealing with the user interface, sensor integration, and context-awareness, respectively, and the evaluation results are summarized in Table 1. Edwards [4] argued that such tailoring of data and sharing of contextual information would improve user interaction and eliminate manual tasks. Results from the user evaluation support this. Users find it both attractive as well as have positive attitudes towards automation of tasks such as push updates of information by tailoring the interface.

4   Conclusions and Future Work

Sensors are an important source of information input in any real world context. The work presented in this paper has shown that it is feasible to implement sensor-based context-aware integration through a suitable interplay between on-device context-aware information, such as sensors, and cloud-based context-aware information such as calendar data, contacts and applications, building upon suggestions for further research on adaptive cloud behavior as identified in [2,3].

By taking advantage of the rich hardware available on modern smartphones, the developed application is able to have tighter and more comprehensively integrated sensors in the solution. From user evaluation one can learn that although sensor integration as a source for context-awareness is well received, there is still research to do. In particular this has to do with the fact to what extent what thresholds should be used for sensor activation and deactivation. We have shown that it is feasible to implement sensors and extend their context-aware influence by having them cooperate with cloud-based services. Future interesting work could investigate to what extent what thresholds should be used for sensor activation and deactivation and explore if there are differences in people's perceptions of different sensors.

References

1. Binnig, C., Kossmann, D., Kraska, T., Loesing, S.: How is the weather tomorrow?: towards a benchmark for the cloud. In: Proceedings of the Second International Workshop on Testing Database Systems. ACM, Providence (2009)
2. Christensen, J.H.: Using RESTful web-services and cloud computing to create next generation mobile applications. In: Proceedings of the 24th ACM SIGPLAN Conference Companion on Object Oriented Programming Systems Languages and Applications. ACM, Orlando (2009)
3. Dey, A., Abowd, G.D.: Towards a Better Understanding of Context and Context-Awareness. In: 1st International Symposium on Handheld and Ubiquitous Computing (1999)
4. Edwards, W.K.: Putting computing in context: An infrastructure to support extensible context-enhanced collaborative applications. ACM Transactions on Computer-Human Interaction (TOCHI) 12, 446–474 (2005)
5. Mei, L., Chan, W.K., Tse, T.H.: A Tale of Clouds: Paradigm Comparisons and Some Thoughts on Research Issues. In: Proceedings of the 2008 IEEE Asia-Pacific Services Computing Conference, pp. 464–469. IEEE Computer Society (2008)
6. Mell, P., Grance, T.: The NIST Definition of Cloud Computing. National Institute of Standards and Technology, Special Publication 800-145 (2011)
7. Parviainen, M., Pirinen, T., Pertilä, P.: A Speaker Localization System for Lecture Room Environment. In: Renals, S., Bengio, S., Fiscus, J.G. (eds.) MLMI 2006. LNCS, vol. 4299, pp. 225–235. Springer, Heidelberg (2006)

Perception and BDI Reasoning Based Agent Model for Human Behavior Simulation in Complex System

Jaekoo Joo

Systems and Management Engineering, Inje University
197 Inje-ro, Gimhae-si 621-749, Republic of Korea
jjoo@inje.ac.kr

Abstract. Modeling of human behaviors in systems engineering has been regarded as an extremely complex problem due to the ambiguity and difficulty of representing human decision processes. Unlike modeling of traditional physical systems, from which active humans are assumed to be excluded, HECS has some peculiar characteristics which can be summarized as follows: 1) Environments and human itself are nondeterministic and dynamic that there are many different ways in which they dynamically evolve. 2) Human perceives a set of perceptual information taken locally from surrounding environments and other humans in the environment, which will guide human actions toward his or her goal achievement. In order to overcome the challenges due to the above characteristics, we present an human agent model for mimicking perception-based rational human behaviors in complex systems by combining the ecological concepts of affordance- and the Belief-Desire-Intention (BDI) theory. Illustrative models of fire evacuation simulation are developed to show how the proposed framework can be applied. The proposed agent model is expected to realize their potential and enhance the simulation fidelity in analyzing and predicting human behaviors in HECS.

Keywords: Human Behavior, Affordance theory, BDI theory, Agent-based Simulation, Social Interaction.

1 Introduction

Both cognitive and rational reasoning aspects of human behaviors must be accommodated in developing common framework for modeling and simulation of Human-Environment Complex System (HECS) due to the critical role of humans in systems operation and dynamics. However, modeling of human behaviors in systems engineering has been regarded as an extremely complex problem due to the ambiguity and difficulty of representing the nondeterministic and dynamic nature of human decision processes, which makes the research difficult and slow. Unlike modeling of traditional physical systems, from which active humans are assumed to be excluded, HECS has some peculiar characteristics which can be summarized into two. First, environments and human him/herself are nondeterministic and dynamic that there are many different ways in which they dynamically evolve. The characteristics mainly

M. Kurosu (Ed.): Human-Computer Interaction, Part V, HCII 2013, LNCS 8008, pp. 62–71, 2013.

stem from dynamic properties of environmental information and its using during human decision making processes. The nondeterministic characteristics of human itself means that human can take a number of different actions even though they are located in the same environmental situation. Furthermore, his/her physical and emotional conditions that can be easily changed according to outside situation may affect decision making processes. These nondetermistic aspects make human do not take an action just by following action route predetermined off-line. Second, human perceives a set of perceptual information taken locally from surrounding environments and use it with a series of complex rational reasoning processes, which will guide human actions toward his or her goal achievement. The perceptual information is classified into environs static data, environs dynamic data, and social interaction data with other humans. At any instant of time, there may exist various assigned objectives that the human is asked to accomplish. In order to achieve a goal(s), a human anticipates perceptually-available outcomes and opportunities to take an action using the perceptual information. In order to overcome the challenges due to the above characteristics, we present a generic modeling framework for mimicking perception-based rational human behaviors by combining the ecological concepts of affordance- and the Belief-Desire-Intention (BDI) theory. According to Gibson, the perceived information regarding sets of affordance-effectivity taken from their surrounding environs is used for a human to make decisions to take action [1]. Rao and Georgeff argued the necessity of three attitudes of Belief, Desire and Intention (BDI), representing respectively, the information, motivational, and deliberative states of human agent [2]. While the perceptual property of affordance provides a basic idea of how a perception guides human actions, the mental attitudes of BDI are critical for achieving adequate or optimal performance when deliberation is subject to resource bounds [3]. In our previous work, we developed and verified an agent-based formal simulation framework of affordance-based human behaviors in emergency evacuation situations [4]. In the work, however, our perspective on human behavior was limited to individual decision making with only human perceptions rather than more complex problem domains involving rational reasoning and human interactions. In this paper, we will extend the framework by accommodating the rational reasoning aspects of human behavior in HECS. To this end, the perceptual information is firstly classified to clarify their influence on human action decisions in the context of the developed framework. Second, the functionalities and processes of perception and BDI reasoning of human agent are discussed. Finally, an exemplary scenario is developed and illustrative models of fire evacuation simulation are developed to show how the proposed framework can be applied. The proposed human behavioral modeling framework is expected to enhance the modeling fidelity and simulation credibility for human-included complex systems.

2   Background

2.1   Affordance Theory and Perception-Based Action

The perception-based action was initiated by Gibson who regarded a human action as a consequence of direct perception of affordance (action opportunity provided by the environment) and effectivity (an individual's ability to take a specific action) [1]. Thereafter, Turvey defines affordance as a real property of the animal-environment system (AES) that is perceived directly toward the execution of a potential action [5]. Turvey bases the definition of affordance in terms of properties that represent a potential state and are not currently realized (called dispositional properties or dispositions). Dispositions occur in pairs in which a property of the environment (i.e., walk on – ability for the person) is complemented by a property of the animal's capability known as an effectivity (i.e., to walk on the stairs' surface). So the terms of affordance and effectivity can be combined together so that they incur a different property (i.e., climb the stairs) to be activated. For example, in case of a person-climbing-stairs system (W), a person (Z) can walk (q), stairs (X) can support something (p), and they together yield climbing property (r). This formal definition of affordance, effectivity, and juxtaposition function can be mapped to the precondition set of state transition function and provides a foundation that the concept of an affordance can be combined with software engineering and systems theory. Kim et al. have suggested an affordance-based descriptive formalism for complex human-involved systems using finite state automata [6]. In their work, an environmental system is defined as a set of nodes and arcs that describe discrete states of the system and the transitions between states, respectively.

2.2   BDI Theory and BDI Agents

Rao and Georgeff argued the necessity of three attitudes of Belief, Desire and Intention (BDI) [2]. The core concepts of the BDI paradigm allow use of a programming language to describe human reasoning and actions in everyday life. Because of this straightforward representation, the BDI paradigm can map extracted human knowledge into its framework relatively easily. Raubal suggests a perceptual way-finding model that integrates simulated environmental states and agent beliefs within a "Sense-Plan-Act" framework [7]. Shendarkar et al. propose the use of an agent-based simulation modeling paradigm to construct a crowd simulation [8]. However, they do not use the concept of direct perception, which produces immediate human actions with reference to dynamic environments.

2.3   Human Behavior Modeling and Simulation

Traditionally human agents in a system had been modeled as a part of physical resources and assumed to be passive elements taking actions and making decisions based upon pre-programed/rule-based logics in modeling and simulation problems. However, complex cognitive processes corresponding to human decision behaviors cannot be easily inferred using a logical rule-based model, a statistical model, or an analytical predictive model. In agent-based modeling, a flexible set of attributes is assigned to each person, so that an intelligent agent mimics the abstract characteristics of a human. Evacuation models such as Egress, Building Exodus, Simulex, Exit, and Wayout can be used to simulate the evacuation efficiency of buildings [9]. Building Exodus and Simulex, widely used as commercial software for evacuation simulations, assume the presence of a rational agent able to assess the optimal escape route and avoid static physical obstructions [10-11]. However, none of them is grounded on both the ecological concept of affordance and a formal system that enables individual decision making based on human perceptions of dynamic environmental elements and rational reasoning for the simulation.

3   Agent-Based Simulation of Human-Environment Complex System

In this section, we briefly explain overall framework of our approach to agent-based simulation of human behaviors. In software engineering, an agent is defined as a computer system situated in an environment and capable of autonomous action to meet system objectives [12]. An agent in a simulation model implies a nature for each entity and expresses the complex interactions with other agents in the environment so that the simple agent rule can generate complex system behaviors. For an agent-based simulation of human behaviors, there should be two kinds of agent model: human agent model and environmental agent model. While a human agent model is represented by goals, perception abilities, a decision making algorithm, and action capabilities, an environmental agent model maps the dynamics of environmental elements onto the system model. Several attributes and characteristics of each agent are defined to reflect the diversity of the humans and environmental elements in the system.

3.1   System Architecture for Agent-Based Simulation of Human-Environment Complex System

By accommodating and reflecting the above characteristics of agent-based simulation of human behaviors, system architecture of the agent-based human behavior simulator is developed as depicted in Figure 1. The simulation model is composed of three major parts: 1) Human agent, 2) Environmental agent, and 3) FSA-based state transition map. Human agent represents each human in the system. It perceives environs data which is classified dynamic environs data from environmental agent and static environs data from other static elements in the environment such as building structure and sign information. Human agent interacts with other human agents in the system and receives social interaction data as a result of the interaction. Both with the perceived environmental and social information, human agent makes action decisions by a series of cognitive and rational reasoning processes using a series of algorithms. Environmental agent maps the dynamics of environmental elements in the system such as fire, smoke, and flood onto the simulation model. It acts according to its peculiar physical law of state transition dynamics. It should be noted that only fire is considered as environmental agent in this paper. FSA-based state transition map is a formal automata model of HECS describing the whole state map including a goal state, which can be transited by human actions and environmental dynamics in the system. The FSA model provides dynamic (temporal and spatial) situations and the preconditions of possible transitions for agents in the system. While the FSA-based state transition map itself is a descriptive model for representing a system, the agent models generate each event to drive the FSA model according to the dynamically changing situation.

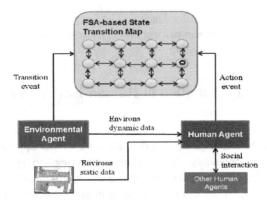

Fig. 1. System architecture of the proposed agent-based simulator

4   Perception and BDI Reasoning-Based Human Agent Model

In this section, we explore the logical cognition and reasoning processes of human action from the point of environmental perception to decision making and execution in order to design a human agent model for the simulation as illustrated in figure 2. The human agent model is designed to generate an action decision through three phases: perception, reasoning and cognition, action decision making. Each of the phases is illustrated below.

4.1   Perception Phase

When human is placed in a situation forcing to make urgent action decisions like emergent evacuation, he/she will try to grasp the situation by sensing current information on outside environment. If there are other humans who are close enough to interact with in the system, the human may consider the social interaction data in making action decisions. Furthermore, human should check his/her internal physical and psychological states such as emotion and cultural tendency on the current system before taking an action. The external perceptual data obtained from outside affect the outcome of internal perception.

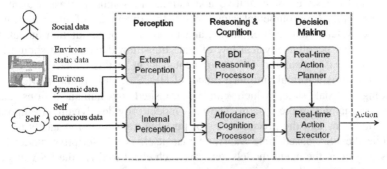

Fig. 2. Functional structure of perception and BDI reasoning-based human agent model

4.2   Reasoning and Cognition Phase

As soon as receiving the external and internal perceptual information, human starts to process them through complex mental processes of reasoning and cognition to make a series of action decisions which is believed to be able to lead to his/her goal achievement. In this paper, while perception is defined as a mental process of receiving external and internal information, reasoning and cognition is defined as some sorts of mental processes of computation (including screening) and fabrication of the information. There are two different modules in this phase: BDI reasoning processor and Affordance cognition processor. The outputs of these reasoning and cognition processors are used for next decision making phase.

BDI Reasoning Processor

The BDI reasoning processor generates a set of reasoning information including beliefs, desires, and intention by receiving external perceptual information. Its functional structure is illustrated in figure 3. Beliefs are not just perceptual information simply given from environments, but information the agent has about the environmental world through some computation and reasoning processes as a result of receiving perceptual data. They are synthesized information showing action alternatives and their results that a human agent could currently choose. Therefore, they represent the structure of possible action alternatives within his/her PB or partial map of the environmental world (Belief-accessible possible worlds). They are stored at long-term memory. Beliefs generator in figure 3 transforms the environs and social data into beliefs. Beliefs are lately used not only for desire generation and deliberation, but for real-time action planning. Desires are state of affairs that the agent would wish to bring about. They contain the information about the objectives to be accomplished, the priorities and payoffs associated with the various objectives. Based on the beliefs and initial intentions, the human agent decides desires via Desire generator. For example, if there are several exits or intermediate positions to escape from fires in a building and a human agent select some of them as goals of its evacuation movement, they can be decided as desires in the simulation. Intentions means desires the agent has committed to achieve. The human agent filters the desires and selects some of them to commit via Deliberator. They play a critical role in practical reasoning by reducing options during action decision making processes.

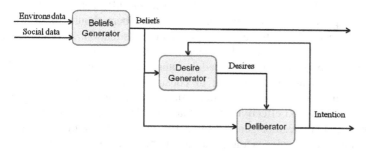

Fig. 3. Functional structure of BDI reasoning processor

Affordance Cognition Processor

Based on not only the perceived external information taken from surrounding environs, but self conscious data obtained from internal perception of human, Affordance cognition processor judges whether there exists any affordance-effectivity sets which satisfying the action possibilities to achieve the selected intentions. While beliefs are stored at long-term memory and obtained through computation or reasoning processes, affordance (a perceptual property of the environment that provides an action opportunity offered to human) is a quick response as a result of receiving perceptual information without any reasoning processes and it is stored at short-term memory. Affordance and effectivity (an individual's ability to take a specific action) can be mapped to the precondition of specific action which is triggered by a juxtaposition function [7]. Affordance is lately used not only for real-time planning, but for action decision executing.

4.3   Decision Making Phase

Real-Time Action Planner

The real-time action planner generates a feasible sequence of actions reaching from the current toward the goal state for a human agent. In order to do that, the planner refers to not only beliefs and intentions generated by BDI reasoning processor, but also sets of affordances and effectivities produced by the affordance cognition processor. Intention is used as goal state. Beliefs provide the planner a set of important information (for example, the layout of an environmental space) to determine an optimal route from the current to goal state among possible alternative routes. The dynamic data of affordance and effectivities is used to screen the possible alternative routes out by eliminating the alternative routes which do not satisfy the action condition of affordance-effectivities. If an unexpected or undesired situation occurs in the system while it makes transitions from one state to another on the state map of the FSA-based state transition map. it causes a transition that leads to deviation from the active plan. If it happens, the planner immediately recalculates the plan in order to cope with the dynamic change of the environment.

Action Decision Executor

Action decision executor, an event generator dedicated for human agent, generates each action event according to pre-defined human action plans by receiving perceived affordances and effectivities from human agents. Since the affordance-based FSA model is just a descriptive model for representing a system, there should be an event generator to drive it according to the dynamically changing situation. Action events are generated based on the following four steps: 1) An external state transition triggers the action decision generator to reschedule future events for the simulation. 2) The Action decision executor takes a plan to consider human's willingness and preference, and evaluate the chance of the next action that is planned. 3) As soon as the condition of affordance and effectivity for actualizing the next action on the plan is met, the action decision executor generates event to make a transition to the next state. 4) Finally, the events are generated based on the plan, which will make a transition to the next state as a result.

5   Illustration: Warehouse Fire Evacuation Problem

5.1   Scenario

To verify the applicability of the perception and BDI reasoning-based agent model for human behavior simulation, the Warehouse Fire Evacuation (WFE) problem [5] is applied. In this considered WFE Problem, a fire breaks out in a warehouse in which two human operators are working. The warehouse area is equally divided in a rectangular grid of 0.8 0.8 m2 which is used for either storage or passageway. In this storage area, goods are stacked up so high that operators cannot observe what happens over the storing lots. Fires are firstly broken out at three different locations in the warehouse and are fast propagated to neighboring lots in a certain amount of speed. As soon as an operator perceives the fire, he/she shall have to find a possible and safe route to an exit along passageway by considering perceived surrounding situations in order to escape from the fires. When he/she tries to move to a next passageway lot, if the lot is already occupied with a fire, he/she cannot access to the lot and have to find another passageway that offers an affordance to move.

5.2   Model Description

The human behaviors in this evacuation problem can be interpreted as a typical example of the perception and BDI reasoning-based human agent model of human behavior simulation. Dynamics of the warehouse can be implemented by the floor layout of the warehouse and a dynamic interaction between human agents and propagating fire agents in the warehouse. In most cases of evacuation problems, human obtains environs data and social interaction data from perception. Furthermore, human behaviors are greatly affected by physical states due to injury or psychological states such as normal or panic. In case of the WFE example, the external perceptual information includes positions of fire and goods and their distance from current position. The social interaction data include the positions of other workers in the floor. There are two kinds of internal human data: physical and psychological. Human action abilities can be leveled according to physical states such as severely wounded, wounded, and normal. These kinds of perceptual information are used to generate beliefs, desires, and intentions. In case of WFE problem, beliefs contain location and distance of each cell from current position of human agent. Beliefs are made by using not only his or her prior knowledge about the floor layout in the warehouse, but perceived information on surrounding situations within the PB. Since the beliefs are determined through dynamic perception of the floor according to movement of each human agent, they have relativistic values. Beliefs are used as a basis for generating plans to the way to the exits. The goal of each evacuee is to get out of the warehouse via one of the exits while avoiding a spreading fire. Based on the beliefs on exits or intermediate position to the exits, BDI reasoning processor determines desires and then filters them to generate intentions. Cells occupied by goods or fires are assigned with very high values to ensure that there is no affordance of move-ability to an evacuee and he or she will never attempt to occupy them.

To perform this behavior, each human agent perceives whether its adjacent lots within PB provide it with the affordance, "is-move-able" for the human agent, or not. In this warehouse evacuation problem, the affordance is "move-ability for an evacuee to an adjacent lot," and the accompanying effectivity for a human agent is "capability to move to an adjacent lot." Based on the beliefs, intentions, and affordance-effectivity pairs, each evacuee will make a plan for his or her movements along passageways to the exit. The affordance-effectivity pairs are also used as trigger for generating action event by human decision executor.

6   Conclusion

In this research, a modeling framework for designing human agents that is able to mimic perception-based rational human behaviors in human-environmental complex systems is proposed by combining the ecological concepts of affordance- and the Belief-Desire-Intention (BDI) theory. The proposed modeling framework deal with not only perceptual aspects of human behaviors, but can accommodate more complex problem domains involving rational reasoning and human interactions and communications. The perceptual information is firstly classified in order to clarify its influence on human action decisions. The functionalities and processes of the proposed human agent are discussed in detail. An exemplary scenario is developed and illustrative models of fire evacuation simulation are developed to show how the proposed framework can be applied. However, the aspects of emotion and culture are not deeply explored how they will affect reasoning and decision making process of human behaviors. While social factors such as interactions and communications within and between groups of people are groundings of the simulation models, they will be considered (e.g. social psychology, emotions, cultures, and knowledge levels) in the future in order to enhance swarm intelligence and behaviors into account. The proposed framework is expected to realize their potential and enhance the simulation fidelity in analyzing and predicting human behaviors in HECS. Also, implementation and validation of the simulation will should be followed via human experiments in a suitable task environment (e.g. virtual reality).

Acknowledgement. This research was supported by Basic Science Research Program through the National Research Foundation of Korea(NRF) funded by the Ministry of Education, Science and Technology(20120938).

References

1. Gibson, J.J.: The Ecological Approach to Visual Perception. Houghton Mifflin, Boston (1979)
2. Rao, A.S., Georgeff, M.P.: BDI Agents: From Theory to Practice, Tech. Rep. 56, Australian artificial intelligence institute, Melbourne, Austrailia (1995)
3. Bratman, M. E.: Intentions, plans and practical reason. Cambridge University Press (1987)

4. Joo, J., Kim, N., Wysk, R.A., Rothrock, L., Son, Y., Oh, Y., Lee, S.: Agent-based simulation of affordance-based human behaviors in emergency evacuation. Simulation Modeling Practice and Theory 32 (2013)
5. Turvey, M.T.: Affordances and Prospective Control: An Outline of the Ontology. Ecological Psychology 4(3), 173–187 (1992)
6. Kim, N., Joo, J., Rothrock, L., Wysk, R.: An Affordance-Based Formalism for Modeling Human-involvement in Complex Systems for Prospective Control. In: Proceedings of the 2010 Winter Simulation Conference, Baltimore, MD, pp. 811–823 (2010)
7. Raubal, M.: Human wayfinding in unfamiliar buildings: a simulation with a cognizing agent. Cognitive Processing (2-3) 363-388 (2001)
8. Shendarkar, A., Vasudevan, K., Lee, S., Son, Y.: Crowd simulation for emergency response using BDI agents based on immersive virtual reality. Simulation Modelling Practice and Theory 16, 1415–1429 (2008)
9. Shi, J., Ren, A., Chen, C.: Agent-based evacuation model of large public buildings under fire conditions. Automation in Construction 18, 338–347 (2009)
10. Virtual Environment V 5.9. Simulex User Guide, Integrated Environmental Solutions Ltd.
11. Galea, E.R., Lawrence, P. J., Gwynne, S., Filippidis, L., Blackshields, D., and Cooney, D: Building EXODUS V 4.06, User Guide and Technical Manual (2006)
12. Zhang, Z., Zhang, C.: Agent-Based Hybrid Intelligent Systems: An Agent-Based Framework for Complex Problem Solving. Springer, New York (2004)

Long-Term Study of a Software Keyboard That Places Keys at Positions of Fingers and Their Surroundings

Yuki Kuno, Buntarou Shizuki, and Jiro Tanaka

University of Tsukuba, Japan
{kuno,shizuki,jiro}@iplab.cs.tsukuba.ac.jp

Abstract. In this paper, we present a software keyboard called Leyboard that enables users to type faster. Leyboard makes typing easier by placing keys at the positions of fingers and their surroundings. To this end, Leyboard automatically adjusts its key positions and sizes to users' hands. This design allows users to type faster and more accurately than using ordinary software keyboards, the keys of which are unperceptive. We have implemented a prototype and have performed a long-term user study. The study has proved the usefulness of Leyboard and its pros and cons.

Keywords: Touch screen, text entry, software keyboard, long-term study.

1 Introduction

QWERTY software keyboards are available for text entry on devices with touch screens. Although their key layout is the same as those of physical QWERTY keyboards, we find it difficult to place our fingers on target keys since we cannot obtain any physical feedback.

On the other hand, software keyboard can easily change their key positions and sizes so that they can fit each user. Utilizing this advantage may be able to compensate for the lack of physical feedback and to improve input speed.

In this paper, we describe a software keyboard that can automatically adjust its key positions and sizes to the user's hands. Longitudinal experiments have proven the usefulness of the proposed keyboard.

The novel features of our software keyboard are as follows:

- We place keys at the touch points of all fingers and their surroundings, so that key layout suits the position of all fingers.
- We move all keys for the thumb to prevent the hand from breaking its posture, while users press a key with pressing another key (thumb based sliding).
- We place some keys around the position of thumbs and enable them to input by swiping them with thumbs (thumb swipe input).
- We combine thumb based sliding and thumb swipe input, so that multiple keys can be input without breaking the posture of the users' hand.

M. Kurosu (Ed.): Human-Computer Interaction, Part V, HCII 2013, LNCS 8008, pp. 72–81, 2013.

2   Related Work

LiquidKeyboard [1], CATKey [2], Personalized Input [3], and the study of Gunawardana et al. [4] adopt a similar approach to our research, which adjusts shapes and places of keys to users' hands. These keyboards can input letters but not some keys available on ordinary keyboards, including numbers and some symbols. In contrast, we use the combination of thumb based sliding and thumb swipe input to input such keys.

The study of McAdam et al. [5,6] and SLAP widgets [7] take an approach to provide users tactile feedback, which is different from ours. In contrast, we reduce the difficulty of input arising from the lack of feedback by fitting the place of keys to each user.

Gestyboard [8] and Bimanual Gesture Keyboard [9] use gesture input in text entry. In our method, gesture input is used only as a modifier, not for text entry itself.

3   Leyboard

We designed a prototype of the software keyboard that places keys at positions of fingers and their surroundings [10]. In this section, we describe the essential features, way to input, and design of this prototype. We named this prototype "Leyboard". We named it Leyboard by replacing 'K' with 'L', since we expect this will be a more advanced than an ordinary Keyboard.

3.1   Key Placing on Positions of Fingers and Their Surroundings

The keys of Leyboard are placed to be based on the touch points of users' fingers. Each layout of keys is determined from Voronoi diagrams. Leyboard places keys on the basis of each finger's position and QWERTY layout. A, S, D, F, Space, Enter, J, K, L, and semi-colon keys are placed at each finger's position. Hereafter, we call these keys home position keys and the rest non-home position keys.

Calibration. In this study, calibration is the name of the procedure for determining the layout of Leyboard. Leyboard checks the correspondence of touch points and users' fingers, when users place all their fingers (i.e., ten fingers) on the touch screen. Then, Leyboard calculates gradients of hands from position of the index and little finger of each hand.

Leyboard places home position keys at each finger's position, and places non-home position keys around them. Positions of non-home position keys are concyclic, where they are rotated in accordance with the gradients of hands, with the home position key at the center. We considered that keys can be input easily as making their distance from home position key the same as that of other keys. On the other hand, those distances are different on ordinary software keyboards.

The calibration is completed when users release their fingers from the touch screen. Now the keys of Leyboard are placed at the positions of users' fingers and their surroundings. Thus the key layout suits the position of users' fingers, which enables users to input the keys they intend to input easily.

Determining Key Area. We make Leyboard to input the key the coordinates of which are the closest to the users' touch point. This involves making the area of the keys as large as possible to enable users to press keys easily. We have made Leyboard determine the area of keys by Voronoi diagrams. We used Fortune's algorithm [11] to draw lines of Voronoi diagrams.

Placing Keys Around Thumbs. On physical keyboards, thumbs are used to press the Space key. On Leyboard, not only Space but also many keys are placed around the position of thumbs. Keys that change the key set, which we describe below, are also included in those keys. Therefore, the key layout of Leyboard is strictly different from the QWERTY layout. The point is that all keys are placed at the positions of users' fingers or their surroundings in this design. Therefore, users' do not need to move their hands as widely as they do with ordinary QWERTY layout keyboards. We consider this enable users to keep their fingers at the position of home position keys while they input, thus reducing error inputs.

We made Leyboard able to provide three key sets (Fig. 1 to Fig. 3). With Leyboard, users change the key sets as necessary, while they type. We placed keys to change the key set around thumbs as described earlier. Key sets were provided because we cannot put all necessary keys for text entry in one state (i.e. one key set) in the design of Leyboard, where all keys are at the position of users' fingers or their surroundings. While users press keys to change key sets, which is shown by circle in Fig. 2 and Fig. 3, the key set changes into the one shown in these figures. When users released their finger from the key for changing key sets, the key set returns to the alphabet set (Fig. 1). As a result, Leyboard is able to input 102 keys for total in these three key sets.

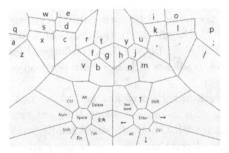

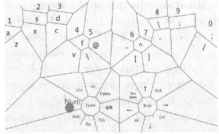

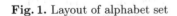

Fig. 1. Layout of alphabet set   **Fig. 2.** Layout of numbers and symbols set

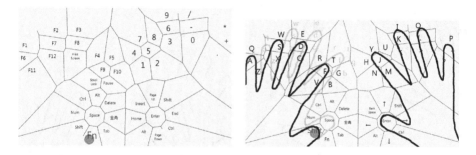

Fig. 3. Layout of functions and numerical keypad set   **Fig. 4.** Example of thumb based sliding

3.2 Thumb Based Sliding

We developed a technique called "thumb based sliding". Two keys must be pressed to input some characters (e.g., 'f' and Shift to input capital 'F'). The thumb based sliding technique makes such simultaneous input easy with our key layout. Fig. 4 shows an example of thumb based sliding. Assume that the left thumb presses Shift. Then Leyboard moves all keys for the left hand like Fig. 4. This design allows users to press keys while pressing another key without breaking the posture of the hand, which enables users to input smoothly and therefore, quickly.

3.3 Thumb Swipe Input

Keys at thumbs and their surroundings are able to be input with "thumb swipe input". Users can input these keys by swiping their fingers from one key to the next. Note that keys to change the key set are only functional while users press these keys. Thumb swipe input is used when users need to input modifier keys such as Shift while the key set is changed.

3.4 Combination of Thumb Based Sliding and Thumb Swipe Input

Leyboard enables users to input many types of keys without breaking the postures of their hands by combining thumb based sliding and thumb swipe input. Fig. 5 shows an example of the combination of thumb based sliding and thumb swipe input. Note the rectangle at the upper right of the left index finger. The key displayed in the rectangle changes while the user swipes his or her left thumb. Here, the keys are constantly at the user's fingers or their surroundings.

3.5 Sound Feedback

We give users sound feedback when they input a key or change the key set. Leyboard makes clicking sound to notify users that events described above has occurred.

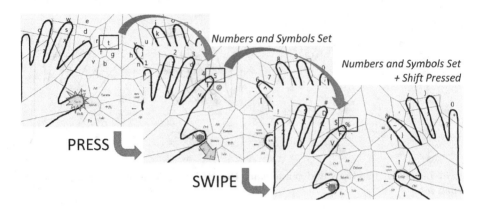

Fig. 5. Example of combination of thumb based sliding and thumb swipe input

4   Developing Environment

We chose C# as a programming language for developing Leyboard. We implemented Leyboard as a WPF application, which supports multi touch by using API of .NET Framework 4/WPF4.

5   Long-Term User Study

We conducted a long-term user study to compare an ordinary software keyboard and Leyboard. We chose the software keyboard regularly installed in Windows 7. Hereafter, we call this software keyboard the Windows 7 keyboard. The study lasted about one year, which is a considerable term.

5.1   Environment

We used Acer's ICONIA-F54E to operate Leyboard in the study. Fig. 6 shows our experiment environment where we used ICONIA-F54E. ICONIA-F54E has a 14-inch touch screen, 1366×768 pixels resolution (WXGA), and can detect up to 10 touch points.

5.2   Participant and Tasks

The participant was one of the authors. We chose tasks of inputting English pangrams. The pangrams contained capital letters and some also included symbols. Characters in one pangram range from 31 to 63. Hereafter, we call inputting 10 different pangrams a set. The participant had three sets of tasks for each software keyboard on each day. We conducted this evaluation from February 15, 2012 to February 14, 2013, for 364 days (cutting 2 days on which no task occurred). This is equivalent of 1092 sets. For the first seven days, the participant performed tasks first with the Windows 7 keyboard and then Leyboard. These orders were reversed in a seven-day cycle.

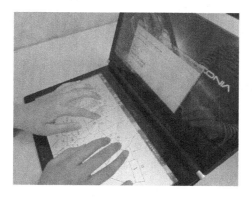

Fig. 6. Experiment environment

5.3   Results

We calculated an input rate in words per minute (wpm), since completion time itself is not a metric of performance of software keyboards. This is because pangrams inputted by the participant differed in every sets. Wpm is a unit of presenting words inputted in a minute that is defined by Gentner [12]. It is calculated as follows:

$$\frac{1}{5}\frac{\text{Incoming keystrokes except mistakes (times/set)}}{\text{Input time (minutes/set)}}$$

Fig. 7 shows the input rate in wpm on each software keyboard, with fitted curves, which are approximated to logarithmic curves.

Although the difference is not yet large, Leyboard still outperforms ordinary software keyboards according to the fitted curve. The maximum input rate of each keyboard was 54.8 wpm for the Windows 7 keyboard and 56.4 wpm for Leyboard. The average input rate of each keyboard was 39.7 for the Windows 7 keyboard and 41.6 for Leyboard. Leyboard has a higher input rate than the Windows 7 keyboard in both values.

Fig. 8 shows the error rate on each software keyboard. The average error rate of each keyboard was 6.07 % for the Windows 7 keyboard and 6.40 % for Leyboard. The value of Leyboard is slightly higher than that of the Windows 7 keyboard.

When a long-term study is conducted, values of evaluation follow the power law. The power approximation curves of values become almost straight lines in a double logarithmic chart. Fig. 9 shows the double logarithmic chart of time for inputting 100 characters on average. As the power approximation curves of Fig. 9 are nearly straight lines, values in the evaluation are considered reasonable and proper.

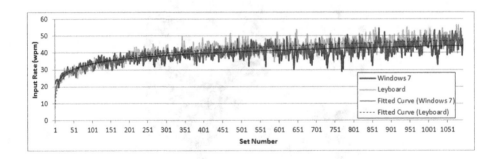

Fig. 7. Input rate (wpm) on each software keyboard

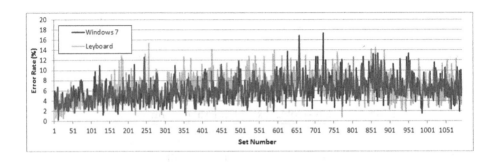

Fig. 8. Error rate on each software keyboard

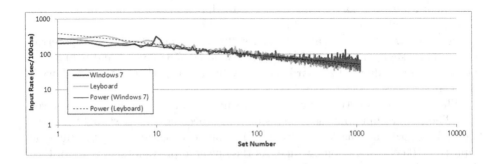

Fig. 9. Input rate (seconds per 100 characters) on each software keyboard

6   Discussion

We have conducted paired t-test to verify that wpm and error rate averages of both keyboards significantly differ. As a result, the input rate of Leyboard is significantly higher than that of the Windows 7 keyboard ($t = -14.6591, p = 2.2e - 16 < 0.01$); but its error rate is also significantly high ($t = -3.708, p = 0.0002 < 0.01$).

Furthermore, we have analyzed the content of the errors. We collected logs of all key inputs during the study. Therefore, we referred to the log and determined every mistaken or omitted input using the following algorithm:

1. Errors are determined in prefix search.
2. The algorithm compares the text and input by characters. The correct input will be skipped.
3. If the character and the input were different, focus on the next character. If it matches the input, the input is considered an omission. Otherwise, it is considered a mistake.
4. If there are any errors, ignore every input from the next onwards until the input come to the correct. This is to avoid the slippage of input that arises from the omission.

For example, if the text was "puppy" and the input was "pupy", the algorithm would find the omission of 'p'. For another example, if the text was "lazy" and the input was "kazy", the algorithm would find the mistake of 'l'. Note that the participant cannot go farther on the tasks as long as he or she inputs wrong characters and eventually has to input the correct character. As a result, the actual inputs with an error would have inserted certain character strings compared with the text. Thus, the actual input of the first example becomes "pupypy" (pup(y)py), and the second becomes "klazy" ((k)lazy).

There were 17,215 errors on the Windows 7 keyboard and 17,490 on Leyboard. The Windows 7 keyboard had 12,878 mistakes and 4,337 omissions. Leyboard had 11,148 mistakes and 6,342 omissions. Even though Leyboard had more errors than the Windows 7 keyboard, Leyboard seems to have had fewer mistakes and more omissions. To describe the tendency of errors on both keyboards, we draw graphs of error frequencies. Fig. 10 shows the mistakes, and Fig. 11 shows the omissions. Here we cut the characters that have errors below a certain number (100 for mistakes and 50 for omissions) to make these graphs conspicuous.

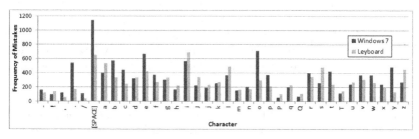

Fig. 10. Breakdown of mistakes

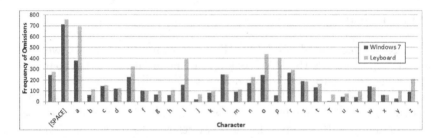

Fig. 11. Breakdown of omissions

In mistakes, the Windows 7 keyboard mistook period, space, 'o' and 'y' more frequently than Leyboard. Period was mostly mistaken as comma (316 times), Space as 'n' (605), 'o' as 'p' (368) and 'y' as 'u' (276). They are all keys placed horizontally except Space and 'n'. On the other hand, Leyboard mistook 's' and 'z' more frequently than the Windows 7 keyboard. 's' was mostly mistaken as 'x' (246) and 'z' as 'a' (221). They are both keys placed vertically.

In omissions, the Windows 7 keyboard did not exceed Leyboard for any specific key. On the other hand, Leyboard omitted 'a', 'i', 'o' and 'p' more frequently than the Windows 7 keyboard. Especially omitted were 'a' in "and" (37), 'i' in "quiz" (151), 'o' in "of" (40), and 'p' in "nymph" (66). It seems that Leyboard is poor at inputting keys with little fingers, such as 'a' and 'p'. Actually, little fingers were hardly used on the Windows 7 keyboard in the study; annular fingers were used instead. This is because keys for little fingers on the QWERTY layout are hard to reach for little fingers on a software keyboard. This is not a specific case because at the time we conducted the user study on the early version of Leyboard, there were users who did not use little fingers while inputting on the Windows 7 keyboard for the same reason. It is also difficulty to input adjacent keys in the top row continuously. This seems to be because we placed non-home position keys con-cyclically, which strains the key layout, especially of the top row.

7   Conclusion and Future Work

In this paper, we have presented Leyboard, a software keyboard that enables faster typing than ordinary software keyboards. Leyboard places home position keys of the QWERTY layout at the touch point of each finger and non-home position keys at their surroundings. Leyboard enables many keys to be pressed with a small amount of hand movement by combining thumb based sliding and thumb swipe input.

A long-term user study found that Leyboard exceeds input rate of the regularly installed Windows 7 software keyboard. However, its error rate is also high. This is because the current version of Leyboard tends to make users input wrong keys placed vertically, is poor at inputting with little fingers, and has difficulty inputting adjacent keys in the top row continuously. Therefore, our future work is to make Leyboard input these for sure. Also, we are considering having user studies with more participants and comparing Leyboard with physical keyboards, including ergonomic products.

References

1. Sax, C., Lau, H., Lawrence, E.: LiquidKeyboard: An ergonomic, adaptive QWERTY keyboard for touchscreens and surfaces. In: Proceedings of the Fifth International Conference on Digital Society, ICDS 2011, XPS, pp. 117–122 (2011)

2. Go, K., Endo, Y.: CATKey: Customizable and adaptable touchscreen keyboard with bubble cursor-like visual feedback. In: Baranauskas, C., Abascal, J., Barbosa, S.D.J. (eds.) INTERACT 2007. LNCS, vol. 4662, pp. 493–496. Springer, Heidelberg (2007)

3. Findlater, L., Wobbrock, J.: Personalized input: improving ten-finger touchscreen typing through automatic adaptation. In: Proceedings of the 2012 ACM Annual Conference on Human Factors in Computing Systems, CHI 2012, pp. 815–824 (2012)

4. Gunawardana, A., Paek, T., Meek, C.: Usability guided key-target resizing for soft keyboards. In: Proceedings of the 15th International Conference on Intelligent User Interfaces, IUI 2010, pp. 111–118. ACM, New York (2010)

5. McAdam, C., Brewster, S.: Distal tactile feedback for text entry on tabletop computers. In: Proceedings of the 23rd British HCI Group Annual Conference on People and Computers, BCS-HCI 2009, pp. 504–511 (2009)

6. McAdam, C., Brewster, S.: Mobile phones as a tactile display for tabletop typing. In: Proceedings of the ACM International Conference on Interactive Tabletops and Surfaces, ITS 2011, pp. 276–277 (2011)

7. Weiss, M., Wagner, J., Jansen, Y., Jennings, R., Khoshabeh, R., Hollan, J.D., Borchers, J.: Slap widgets: bridging the gap between virtual and physical controls on tabletops. In: Proceedings of the 27th International Conference on Human Factors in Computing Systems, CHI 2009, pp. 481–490 (2009)

8. Coskun, T., Artinger, E., Pirrilli, L., Korhammer, D., Benzina, A., Grill, C., Dippon, A., Klinker, G.: Gestyboard: A 10-finger-system and gesture based text input system for multi-touchscreens with no need for tactile feedback. In: Proceedings of the 10th Asia-Pacific Conference on Computer-Human Interaction, APCHI 2012, pp. 701–702 (2012)

9. Bi, X., Chelba, C., Ouyang, T., Partridge, K., Zhai, S.: Bimanual gesture keyboard. In: Proceedings of the 25th Annual ACM Symposium on User Interface Software and Technology, UIST 2012, pp. 137–146 (2012)

10. Kuno, Y., Shizuki, B., Tanaka, J.: Leyboard: A software keyboard that places keys at positions of fingers and their surroundings. In: Proceedings of the 10th Asia-Pacific Conference on Computer-Human Interaction, APCHI 2012, pp. 723–724 (2012)

11. Fortune, S.: A sweepline algorithm for voronoi diagrams. In: Proceedings of the Second Annual Symposium on Computational Geometry, SCG 1986, pp. 313–322. ACM, New York (1986)

12. Gentner, D.R.: Keystroke timing in transcription typing. In: Cooper, W.E. (ed.) Cognitive Aspects of Skilled Typewriting, pp. 95–120. Springer (1983)

Fast Dynamic Channel Allocation Algorithm for TD-HSPA System

Haidong Li, Hai-Lin Liu, and Xueyi Liang

School of Applied Mathematics, Guangdong University of Technology,
Guangzhou 510520, China
hlliu@gdut.edu.cn

Abstract. In order to make full use of channel, a new dynamic channel allocation algorithm for TD-HSPA system is proposed. The proposed algorithm gives priority to consider the time slot distribution in uplink channels. This paper uses low order modulation coding in uplink channels, but uses high order modulation coding in downlink channels. The transmission rate of uplink and downlink are asymmetric. In his paper, we propose a criterion sharing channel for each other through main and auxiliary frequency when the voice channel is idle. As a result, the system capacity is increased 50% larger than the past. Simulation results show that the proposed algorithm can decrease the call blocking ratio and dropping packet rate of data service, improve the channel utilization efficiency, and increase the number of data users dramatically.

Keywords: TD-HSPA, asymmetric transmission, frequency sharing, dynamic channel allocation.

1   Introduction

The fast DCA (Dynamic Channel Allocation) algorithm, which can dynamically adjust the channel resource, allocates channel resources to the user according to telecommunication network. The resource of wireless channel includes carrier, slot, and spreading codes in TD-HSPA (Time Division-High Speed Packet Access) system. The DCA algorithm searches the channels which have smaller interference and which can provide stable service to supply the users. The TD-HSPA system has three kinds of key techniques which include AMC (Adaptive Modulation and Coding), HARQ (Hybrid Automatic Repeat reQuest) and FPS (Fast Packet Scheduling), which can better support asymmetric data transmission[1-6]. The existing fast dynamic channel allocation algorithms mainly include MRG MB DCA, PCR MB DCA algorithm[7-11], etc. Literature [8] makes data business be able to use idle speech channel through a movable boundary method, which reduces the data service loss rate of packet and the average waiting time. The paper [9] sets a different priority for voice users and reserves parts of channels for high precedence voice users based on the paper [8] to reduce call blocking rate of handover voice users. However the paper [9] does not consider the data slots allocated in uplink time slots, allocated and transmitted method in the case

M. Kurosu (Ed.): Human-Computer Interaction, Part V, HCII 2013, LNCS 8008, pp. 82–91, 2013.

of asymmetric transmission, and frequency sharing (Main and auxiliary carrier can borrow channel resources from each other).

This paper proposes a new algorithm for the three key techniques in TD-HSPA. Firstly, we give priority to the uplink time slot and allocate a data slot in the uplink time slot for data transmission. Secondly, we transmit different rate of data service when uplink time slots and downlink time slots are asymmetric. we use low order modulation to transmit low speed data service in uplink time slots and use high order modulation to transmit high speed data service in downlink time slots for achieving the maximization of channel utilization efficiency. Thirdly, we propose a criterion based on BRU number when main and auxiliary frequency borrow channel resources from each other. We can fast determine data channel which can borrow idle voice channel of main frequency or auxiliary frequency. More users get system service by way of frequency sharing that main and auxiliary frequency can borrow channel resources from each other. We will ensure more users to access the system through the switch which is from 3G to 2G when the volume of service is very large. The proposed algorithm is called TD-HSPA MB (movable boundary) DCA algorithm in this paper. Simulation results show that the proposed algorithm can decrease the call blocking ratio and increase the performance of data service and the number of data users dramatically.

2   Resource Allocation

I In the time slot of the main and auxiliary frequency, the channel with SF (Spreading Factor) 16 can be looked upon as one BRU, twhich means one time slot is composed by sixteen BRU (Basic Resource Unit). In TD-HSPA system, resource allocation mainly involves the allocation of the BRU made up of carriers, time slots and channel codes. Transmission rate of voice service, HSUPAHigh Speed Uplink Packet Accessdata service, stream data service, browser data service and interactive (background) data service are 12.2kbit/s, 32kbit/s, 64kbit/s, 128kbit/s and 384kbit/s,respectively. And the corresponding numbers of BRU are 2, 4, 8, 16 and 48[12].

According to the changes of wireless channel, TD-HSPA uses AMC technique to select the appropriate modulation and coding mode, which makes users use network in most effective way. In uplink time slot, base station sends data to the user by using low-level modulation and low-rate channel coding, which ensures the quality of communication. Difference from the way of uplink time slot, the user receives data from base station by using higher order modulation and high-rate channel coding so that they have a high transmission rate in downlink time slot. TD-HSPA uses HARQ technology to ensure the reliability of data transmission. TD-HSPA FPS is transferred to the base station from the radio network controller. Furthermore, scheduling signaling can be transmitted directly between base stations and mobile terminals, so it reduces the scheduling time. FPS makes the resource allocation of base station be more flexible and rapid, and increases the cell throughput.

3   Fast Dynamic Channel Allocation Algorithms

In the view of the negative impact caused by the dropped calls or data traffic delay, it is important to reduce the rate of dropped calls and speed up data transmission. This can bring bilateral benefits for telecom operators and users.

To make use of system resources better, TD-HSPA MB DCA algorithm is proposed. It reduces call blocking rate and improves the performance of data service.

3.1   Chanel Modulation and Coding

In this paper, we should allocate two time slots for uplink channel and four time slots for downlink channel. Furthermore, in the uplink time slot, one time slot is allocated for voice service and the other is for data service. In the uplink and downlink time slot, we use the QPSK (Quadrature Phase Shift Keying) and 16QAM (Quadrature Amplitude Modulation) modulation, respectively. Base station transmits data with low speed 32kbit/s in uplink time slot. Meanwhile, uplink time slot resource is shared, which makes the uplink code channel be not restricted and maximize the total capacity of the channel. Base station transmits data with high speed in the downlink time slot, and AMC adjusts the modulation mode according to instantaneous changing of downlink channel. When current line-channel environment is very good, we use 16QAM modulation mode. However we use QPSK modulation when channel environment is poor or close to the edge of the cell. The ratio of uplink and downlink time slot is 2:4, so the transmission rates of uplink and downlink are different. To use time slots efficiently, data is transmitted with low speed in the uplink time slot and high speed data service in the downlink time slot.

Using PCR MB DCA algorithm [9], it allocates all two uplink time slots to voice service. PCR MB DCA algorithm model is shown in Figure 1.

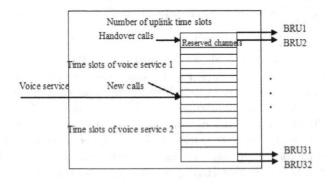

Fig. 1. PCR MB DCA algorithm model

Compared with PCR MB DCA algorithm using symmetrical transmission in uplink and downlink time slots, TD-HSPA MB DCA algorithm uses asymmetric transmission. Shown in Figure 2 and Figure 3TS is used to represent one time slot. And the arrow of towards up denotes the uplink time slot. Similarly, the arrow of towards down represents the downlink time slot), the utilization rate of downlink time slot for PCR MB DCA algorithm is half of the TD-HPSA MB DCA algorithm.

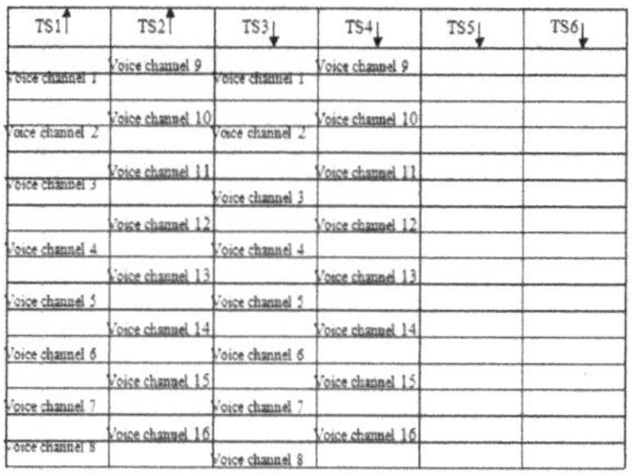

Fig. 2. Channel allocation model for PCR MB DCA algorithm

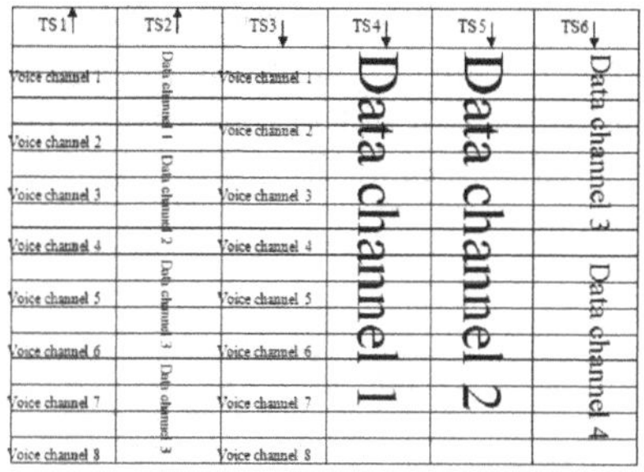

Fig. 3. Channel allocation model for TD-HSPA MB DCA algorithm

3.2 Borrowing Criterion of Main, Auxiliary Frequency Channel Resource

In TD-HSPA MB DCA algorithm, when the channel of main frequency is not enough to support data service, it uses free voice channel of auxiliary frequency. In time slots of main frequency, data service uses free voice channel. When there is no free voice channels in main frequency the algorithm borrows free voice channels from auxiliary frequency. Since voice service has a high priority, voice service can occupy the channel used by data services when new voice calls arrive.

In this paper, cross-frequency network is selected as a networking mode. Every cell has nine frequencies, which include three main frequencies and six auxiliary frequencies. We take a main frequency and an auxiliary frequency for example. In uplink time slots of main and auxiliary frequency, they both have 32 BRU. Voice service requires two BRU while data service (32kbit/s) requires 4 BRU. In the ith main frequency, when voice user x_i and data users y_i satisfy $0 \leq x_i \leq 8, 4 \leq y_i \leq 8$ and $2x_i + 4y_i \leq 32$, data service borrows the free voice channels from ith frequency of main frequency. Similarly, when voice user x_i and data users y_i satisfy $0 \leq x_i \leq 8, 8 \leq y_i \leq 12$ and $2x_i + 4y_i \leq 48$, data service borrows the free voice channels from ith frequency of auxiliary frequency.

When the free voice channels are borrowed by data service, they are divided into groups as follows. Data service (32kbit/s) requires 4 BRU, so we view 4 BRU as a unit. And the BRU resources of other data services are an integer multiples of the unit.

One group means that it can transmit data service with 32kbit/s. Within time slots of main frequency, if the data buffer queue is not empty and groups from 1to 4 are free, they are used to transmit data service of 128kbit/s TV or two-way data service of 64kbit/s stream service of 64kbit / s or four-way data service of 32kbit/s HSUPA service. Similarly, if the 1th, 2th and 3th groups are free, they are used to transmit corresponding data. By this method, the probability of borrowing free voice channels is increased greatly.

If data channels of main frequency are not enough, it can borrow free voice channels from auxiliary frequency. Moreover, if voice channels of auxiliary frequency are still not enough, the TD-HSPA system switches to 2G to ensure the data transmission. The Model of TD-HSPA MB DCA algorithm is shown in Figure 4.

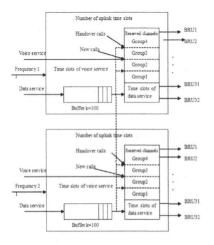

Fig. 4. Model of TD-HSPA MB DCA algorithm

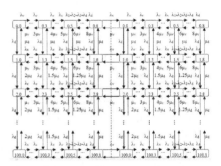

Fig. 5. Markov queuing model of TD-HSPA MB DCA algorithm

3.3   Performance Analysis for TD-HSPA MB DCA Algorithm

Suppose that packet arrival rate, the average time of transmission and the intensity of data traffic are λ, $\frac{1}{\mu_d}$ and $a_d = \frac{\lambda_d}{\mu_d}$ respectively. Let the total arrival rate of voice service be λ_ν containing λ_n (new arrival rate of voice calls) and λ_h (arrival rate of handover voice calls). Set the arrival rate of VIP high-speed handover voice calls, VIPlow-speed handover voice calls, ordinary high-speed handover voice calls and ordinary low-speed handover voice calls to be $\lambda_1, \lambda_2, \lambda_3$, and λ_4 .The average duration of voice calls is $\frac{1}{\mu_\nu}$, and the traffic intensity of voice calls is $a_\nu = \frac{\lambda_\nu}{\mu_\nu}$.

In main frequency, voice service requires two BRU, and each time slot contains sixteen BRU. The number c of voice traffic channel is set to be 8. Data service using free voice channels borrows the 1th, 2th, 3th and 4th group each of which represents 0.25 time slot. The number of data slot d can be 1, 1.25, 1.5, 1.75 and 2 when the queue length of data buffer k is 100.

Under the premise of the criteria for borrowing main and auxiliary frequency resources and the performance analysis of TD-HSPA MB DCA algorithm, we create the Markov queuing model in figure 4[13, 14]. In this figure, the left part of the dotted line represents the Markov queuing model of main frequency A, and the right part represents the Markov queuing model of auxiliary frequency B. This figure shows that Markov queuing model of main frequency A and auxiliary frequency B borrow resources from each others. In this figure, each small rectangle node (m, n) denotes one system state, where m represents the number of packets in the buffer and n represents the number of voice calls serviced. Figure 4 shows that free voice channels can be borrowed by data service of main and auxiliary frequency from each other[13, 14].

3.4   The Formula of TD-HSPA MB DCA Algorithm

When the uplink time slots have two time slots, one slot is allocated to voice service and the other for data service. The utilization rate of uplink channel is

$$\eta = \frac{a_v(1 - p_c)/8 + a_d(1 - p_{pd})}{2}. \tag{1}$$

When the voice call channels is fully occupied, the voice call blocking ratio is

$$\eta = \frac{\frac{a_v^c}{c!}}{\sum_{k=0}^{c} \frac{a_v^k}{k!}}. \tag{2}$$

When the data buffer is full, the new arrival data service will be refused. The packet loss rate of data services is

$$P_{pd} = \sum_{n=0}^{8} p(100, n). \tag{3}$$

When data channels and voice channels are utilized fully, the new arrival data service will be queued in the buffer. Queue length L of packet and average waiting time W of packet are

$$L = \sum_{m=0}^{100} \sum_{n=0}^{8} mp(m, n), \quad W = \frac{L}{\lambda(1 - P_{pd})}. \tag{4}$$

Reselection from TD-HSPA to GSM is a switching process in different systems. In the mobile communication network, the measurement system can be opened or closed by setting the parameters, which controls cell reselection among disparate systems [15].

4   Simulation Results

In this paper, we compare TD-HSPA MB DCA algorithms with PCR MB DCA algorithm from several aspects such as number of data users, channel utilization efficiency, call blocking rate, loss rate of data packet and average waiting time of data packet. The parameters of computer simulation are set as follows. Given a main frequency and a auxiliary frequency in a cell, there are six time slots with the time slot rate 0.5 of uplink and downlink in one frequency. A time slot is composed by sixteen BRU. Voice service requires two BRU and the data service with 32kbit/s, 64kbit/s and 128kbit/s requires the corresponding numbers of BRU are 4, 8 and 16. The arrival rate of voice service is set to be 0, 30, 60, 90, 120 and 150 call/h. And about 40% to 50% of voice service involves handover service [16]. Arrival rate of voice handover service is 0, 10, 20, 40, 50 and 60 call/h, and arrival rate of data service is 0, 12, 24, 36, 48, 60 packet/s [17]. Average waiting time of voice call is 90 seconds, while average transmission time of data packet is 0.1068 seconds, 0. 0534 seconds, 0.0267 seconds. The voice time slot ratio of uplink and downlink service is 1:1, while the ratio of data service is 1:3. A main and an auxiliary frequency can borrow four BRU from each other, and handover rate from 3G service to 2G service is 5% [15, 18]. Arrival of voice calls and data packets obeys the Poisson distribution, while the duration time of the voice calls and transmission time of data packet obey the negative exponential distribution.

Using the above parameters, the simulation results PCR MB DCA and TD-HSPA MB DCA algorithm are shown from figure 6 to figure 12.

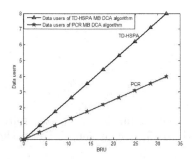

Fig. 6. Available data service users of TD-HSPA MB DCA algorithm and PCR MB DCA algorithm

In figure 6, we can see that the data user number of TD-HSPA MB DCA algorithm is two times of PCR MB DCA algorithm. Because uplink channel coding of TD-HSPA uses low-level modulation, it can transmit low-rate data service and make the uplink code channel be not restricted. Furthermore, uplink channel resource can be shared. Hence, it increased the uplink capacity greatly

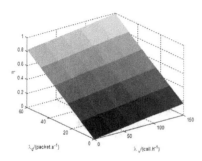

Fig. 7. The channel utilization efficiency of PCR MB DCA algorithm

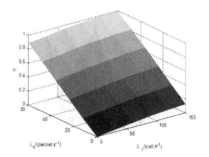

Fig. 8. The channel utilization efficiency of TD-HSPA MB DCA algorith

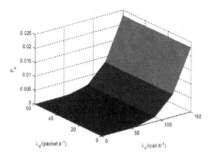

Fig. 9. The voice call blocking rate of PCR MB DCA algorithm

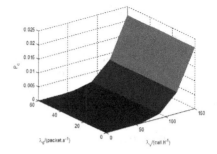

Fig. 10. The voice call blocking rate of TD-HSPA MB DCA algorithm

and service for more users with adding a small amount of system signaling overhead, which can be accepted by the system. Comparing figure 7 with figure 8, when the arrival rate of data and voice service is high and low, respectively, the channel utilization efficiency of PCR MB DCA algorithm is 0.83, but that of TD-HSPA MB DCA algorithm is 0.87, increased by 4.8%. With high arrival rate of data service and voice service, the channel utilization rate of PCR MB DCA algorithm is 0.980, while that of TD-HSPA MB DCA algorithm is 0.995, increased by 1.53%. In TD-HSPA MB DCA algorithm, the channels of main and auxiliary frequency can be borrowed from each other, and downlink time slots of high order modulation and the channel coding of high speed can utilize the time slot resources effectively, so the total channel utilization rate is increased. Similarly, comparing figure 9 with figure 10, call blocking rate of voice services of TD-HSPA MB DCA algorithm is lower by 5% than that of PCR MB DCA algorithm, which is caused by voice handover from TD-HSPA system to GSM system. Compared with the packet loss rate of PCR MB DCA algorithm in figure 11, the packet loss rate of TD-HSPA MB DCA algorithm is reduced by 5%. The reason is that, in uplink time slots of TD-HSPA MB DCA algorithm, it uses the low order modulation mode and low speed channel coding for ensuring the

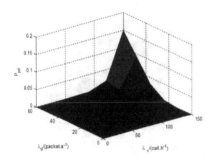
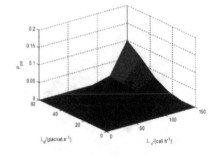

Fig. 11. The channel utilization efficiency of TD-HSPA MB DCA algorithm

Fig. 12. Data packet loss rate of TD-
HSPA MB DCA algorithm

communication quality, and it uses AMC technology to ensure the reliability of
the transmission in the downlink time slots of TD-HSPA MB DCA algorithm.
FPS of TD-HSPA system makes scheduling period become smaller base station
allocate resource more flexibly and rapidly. In summary, TD-HSPA MB DCA
algorithm is effective to the cells with asymmetry time slot of uplink and downlink and many voice users, especially for the cells possessing a large number of
data users. Meanwhile, it increased the system capacity significantly.

5   Conclusions

In this paper, TD-HSPA MB DCA algorithms are proposed. By asymmetric
transmission rate of uplink and downlink, a reasonable arrangement for time
slots and channels is obtained in uplink and downlink. For increasing the number of users and the system capacity, the main and auxiliary frequency borrow
channel resources from each other. The proposed algorithm can decrease the
call blocking ratio and the dropping packet rate of data service, and improve
the channel utilization efficiency. Furthermore, TD-HSPA MB DCA algorithms
provides some references for WCDMA-HSPA and TD-LTE systems.

Acknowledgment. This work was supported in part by the Natural Science
Foundation of Guangdong Province (S2012010008813), and in part by the projects
of Science and technology of Guangdong Province (2012B091100033),and in part
by the projects of science and technology of the department of education of
Guangdong province (2012KJCX0042), and in part by Zhongshan projects of
science and technology (20114A223).

References

1. Peng, M., Wang, W., Chen, H.-H.: TD-SCDMA evolution. IEEE Vehicular Technology Magazine 5(2), 28–41 (2010)
2. Peng, M., Zhang, X., Wang, W., et al.: Performance of Dual-Polarized MIMO for
TD-HSPA evolution systems. IEEE Systems Journal 5(3), 406–416 (2011)

3. Zhang, X., Wang, W., Li, Y., et al.: Performance analysis for dual polarization antenna schemes in TD-HSPA system. In: IEEE Personal Indoor and Mobile Radio Communications, September 26-30, pp. 2052–2056 (2010)
4. Peng, M., Wang, W.: A framework for investigating radio resource management algorithms in TD-SCDMA systems. IEEE Radio Communications, S12–S18 (June 2005)
5. Wang, L.-C., Huang, S.-Y., Tseng, Y.-C.: Interference Analysis and Resource Allocation for TDD-CDMA Systems to Support Asymmetric Services by Using Directional Antennas. IEEE Transactions on Vehicula Technology 54(3), 1056–1068 (2005)
6. Chang, K., Lee, K., Kim, D.: Optimal timeslot and channel allocation considering fairness for multicell CDMA/TDD Systems 33(11), 3203–3218 (2006)
7. Zhao, Y.Z.: Study on the Dynamic Channel Allocation Technology in TD-SCDMA System. Jilin University (2007)
8. Shi, W., Zhao, Y., Zhao, S.: A novel fast dynamic channel allocation scheme based on TD-SCDMA system. Journal of Jilin University (Engineering and Technology Edition) 38(4), 955–959 (2008)
9. Shi, W., et al.: Fast dynamic channel allocation algorithm based on priority channel reservation. Journal Communications 30(7), 59–66 (2009)
10. Chen, L., Yoshida, H., Murata, H.: A dynamic channel assignment algorithm for asymmetric traffic in voice/data integrated TDMA/TDD mobile radio. In: International Conference on Information, Communications and Signal Processing (ICICS), pp. 215–219 (1997)
11. Xu, J., Yao, F.: Dynamic channel allocation strategy between uplink and downlink for TD-SCDMA system. Journal of PLA University of Science and Technology (Natural Science Edition) 10(5) (October 2009)
12. 3GPP TSG UTRA TDD: Radio Transmission and Reception. 3GPP TS 25.105.2001
13. Sun, R., Li, J.: Basis Theory of Queuing. Publishing House of Science, Beijing (2002)
14. Shi, W., Zhang, L., Hu, K.: Communication Network Principle and Applications. Publishing House of Electronics Industry, Beijing (2008)
15. Zhao, S., et al.: Network deployment, operation and optimization practice in TD-SCDMA system. Publishing House of Electronics Industry, Beijing (2010)
16. Ismail, M., Aripin, N.: Downlink Soft Handover Performance for Different Cell Selection Schemes in WCDMA System. IEEE (2005)
17. Guo, S.: Radio resource management of TD-SCDMA system. ZTE Communications 5(12), 36–39 (2006)
18. Zhang, X., et al.: HSUPA/HSPA Network technology. Posts & Telecom Press, Beijing (2008)

Evaluating Intelligibility Usage and Usefulness in a Context-Aware Application

Brian Y. Lim and Anind K. Dey

Carnegie Mellon University, 5000 Forbes Ave., Pittsburgh, PA 15213
{byl,anind}@cs.cmu.edu

Abstract. Intelligibility has been proposed to help end-users understand context-aware applications with their complex inference and implicit sensing. Usable explanations can be generated and designed to improve user understanding. However, will users want to use these intelligibility features? How much intelligibility will they use, and will this be sufficient to improve their understanding? We present a quasi-field experiment of how participants used the intelligibility features of a context-aware application. We investigated how many explanations they viewed, how that affected their understanding of the application's behavior, and suggestions they had for improving its behavior. We discuss what constitutes successful intelligibility usage, and provide recommendations for designing intelligibility to promote its effective use.

Keywords: Context-Awareness, Intelligibility, Explanations, User Study.

1   Introduction

Context-aware applications use implicit sensing and complex inference to automatically and calmly adapt for users [2]. End-users may not be aware of what these applications know, and struggle to understand and trust their behaviors [11]. To counter this, context-aware applications should be intelligible by providing explanations of their behavior [4]. Indeed, there have already been several context-aware applications that support some level of intelligibility (e.g., [1, 14, 15]). These systems support a limited set of explanations users can ask for: What, Certainty, Inputs, Why, and Why Not. However, Lim & Dey [5] found that users ask a wider range of questions of context-aware applications, and that different explanations have different impacts on user understanding. To support this wider range of explanations, Lim & Dey [7] designed Laкsa, which provides explanations to 8 question types for several context types. While that work provides a crucial step for designing intelligibility to be more usable and interpretable, it stopped short of evaluating the impact of intelligibility on users. Lim & Dey [8] investigated the impact of intelligibility on understanding and impression, but this was studied with questionnaires and 'paper' prototypes rather than an interactive prototype. Furthermore, intelligibility was shown "always on" to participants, so they were biased to look at the explanations. This leaves open the research questions: even if intelligibility can improve user understanding and trust, will users

M. Kurosu (Ed.): Human-Computer Interaction, Part V, HCII 2013, LNCS 8008, pp. 92–101, 2013.

want to use it, and, if so, how much? Moreover, given how much they do use, how much will that improve their understanding of context-aware applications?

Related work has explored the impact of explanations on end-users as they used context-aware systems. Tullio et al. [14] evaluated an intelligible interruption door display over six weeks, and found that users were able to "attribute concepts of machine learning to their system," but had difficulty remembering relevant features. Cheverst et al. [1] deployed the Intelligent Office System that provided explanation visualizations of rules and confidence. However, regarding explanations, their evaluation focused on eliciting user preference about visualization format, and not on their impact. Vermeulen et al. [15] conducted a pilot user study of PervasiveCrystal in a simulated museum with five participants, who "were able to use the questions interface to find the cause of events" of three tasks. We add to this body of work evaluating intelligible context-aware systems by explicitly measuring intelligibility usage in a high-fidelity prototype that provides over 9 explanation types (e.g., Certainty, Why, Why Not, What If) for three context types (Availability, Place, Sound). We also investigate the impact of this usage on user understanding of the application's inference. Our contributions show intelligibility is useful by investigating:

1. How much participants *use* intelligibility in a real context-aware application,
2. Their *opinion* of the *usefulness* of the explanations to understand application behavior and situations, and
3. How *useful* their use of intelligibility is on understanding and handling of these situations.

The rest of the paper is organized as follows: we describe an intelligible context-aware prototype which we developed for this study. Next, we elaborate on the quasi-field experiment we conducted, where participants engaged in "everyday" scenarios *in-situ* using the prototype. Following this, we present results of how participants used intelligibility and how that improved their understanding of application inference. Finally, we discuss design implications due usage patterns and constraints, and how to encourage users to use more intelligibility to further improve their understanding.

2   Laкsa 2 Prototype

Mobile phones allow people to keep in touch with others and be easily reachable. However, there are inappropriate times to receive calls, as they can be socially disruptive (e.g., in meetings and movie theatres), or they interrupt productive work. We have developed Laкsa, a mobile application which can automatically change the phone's ringer mode (e.g., [12]). It senses and infers the following contexts:

- **Availability:** *Available, Semi-Available, Unavailable* — is inferred from rules regarding the following three factors.
- **Place:** *Office, Café, Library, etc.* — is inferred by sensing latitude and longitude and matching to a pre-specified named place. The user's sensed location is modeled as a radial Gaussian, with decreasing likelihood further away from the latitude

and longitude coordinates (as in [7]). Each Place is inferred with different certainty based on how much the user's estimated location area "overlaps" with the circular area of the named place: more overlap leads to higher certainty.

- **Sound:** *Talking, Music, and Ambient Noise* — is inferred using a naïve Bayes classifier on features extracted from the phone microphone. Features extracted are similar to [9]: e.g., mean of power, low-energy frame rate, spectral flux, and bandwidth. These are renamed to lay terms that end-users can understand. The machine learning was implemented with an Android port of Weka [16].
- **Schedule:** *Personal, Work, Unscheduled* or *Other Event.*

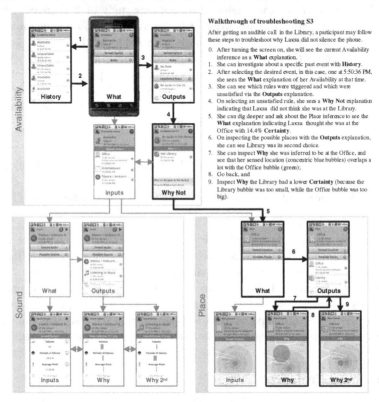

Walkthrough of troubleshooting S3

After getting an audible call in the Library, a participant may follow these steps to troubleshoot why Laksa did not silence the phone.

0. After turning the screen on, she will see the current Availability inference as a **What** explanation.
1. She can investigate about a specific past event with **History**.
2. After selecting the desired event, in this case, one at 5:50:36 PM, she sees the **What** explanation of her Availability at that time.
3. She can see which rules were triggered and which were unsatisfied via the **Outputs** explanation.
4. On selecting an unsatisfied rule, she sees a **Why Not** explanation indicating that Laksa did not think she was at the Library.
5. She can dig deeper and ask about the Place inference to see the **What** explanation indicating Laksa thought she was at the Office with 14.4% **Certainty**.
6. On inspecting the possible places with the **Outputs** explanation, she can see Library was its second choice.
7. She can inspect **Why** she was inferred to be at the Office, and see that her sensed location (concentric blue bubbles) overlaps a lot with the Office bubble (green);
8. Go back, and
9. Inspect **Why** the Library had a lower **Certainty** (because the Library bubble was too small, while the Office bubble was too big).

Fig. 1. Screenshots of Laksa showing several explanation types of the upper-tier context Availability, and lower-tier contexts Sound and Place. Arrows between each screenshot shows how a user can transition from one explanation to another. The bold trace indicates how one may view explanations in 9 steps to troubleshoot Scenario 3 after the phone rang in the Library. The Intelligibility UI was adapted from [7, 8] using the "bubbles" metaphor to explain how Place is inferred, and the "weights of evidence" bar charts to explain feature votes for different Sounds.

Laksa is *intelligible* [5] to help users to understand what it knows and how it makes inferences. Using the Intelligibility Toolkit [6], it provides explanations to questions:

1. **What** is the inference for the context? With how much **Certainty**? **When** was this value inferred?

2. **History**: what was the inference at time *H*?
3. **Inputs**: what details affect this context? (Factors, related details, etc.)
4. **Outputs**: what values can this context be inferred as? With how much **Certainties** are these values inferred?
5. **Why** was this value inferred?
6. **Why Not (Alt)**: why wasn't this inferred as *Y*, instead?
7. **What if** the factors are different, what would this inference be? (Requires user manipulation; only provided for Availability)
8. **Description**: meaning of the context terms and values.
9. **Situation** of what was happening to affect the inference to provide a ground truth of what was being inferred, e.g., playing an audio clip of what was heard.

We developed Laкsa for Android 2.2, and deployed it on the Motorola Droid for the user study. Sensing and inferencing were performed using background services on the phone every 30 seconds. Fig. 1 shows several screenshots of the Laкsa prototype. Users can transition from one explanation *page view* to another by clicking on buttons, option menu items, and flinging (swiping).

3   Scenario-Driven Quasi-Field Study

We explored intelligibility usage with a controlled scenario-driven user study to (i) present participants with critical incidences, and (ii) observe and measure their subsequent behaviors. We employed a quasi-field design (similar to 13] where each participant was brought to the necessary places to engage in various activities. The experimenter enacted critical incidences (e.g., by calling the participant's phone), and presented a printed flash card describing what was happening. The participant could interact with Laкsa as much or as little as she wished. After each scenario, the experimenter interviewed the participant asking about her opinion of the situation and the application, her understanding of how Laкsa made inferences, and how she may improve its behavior.

3.1   Scenarios

We employed four scenarios to span three situational dimensions: (i) Exploration / Verification (S1) of Laкsa's functionality and explanations; (ii) Fault Finding (S2, S3) to diagnose Laкsa's inappropriate behavior; and (iii) Preemptive Exploration (S4) where participants investigated a potential future situation.

S1: Talking in the Office. The participant learned about and freely explores Laкsa's core features and explanations as Laкsa infers the Office location and Talking sounds.

S2: Missed Call while Reading and Listening to Music. The participant read a news article of her choosing from www.cnn.com while she listened to a song (a mostly vocal version of Sound of Silence) through speakers. Meanwhile, she missed multiple calls from a coworker because Laкsa inferred the Music as Talking and automatically silenced its ringer. At the end of the song, the phone finally rang audibly. The participant learned that her coworker was frustrated from trying to call her repeatedly.

S3: Phone Interruption in the Library. The participant walked to a nearby library to search for a specific book to read. Meanwhile, a coworker called her phone, but Laκsa misinferred the participant's Place as still in the Office instead of Library, allowing the phone to ring audibly in the quiet library.

S4: Preemptively Checking Availability in Café. The participant received a flash card describing that she frequents a nearby café, and should check whether she will be able to receive calls there. She was not prompted what to do to achieve this objective.

3.2   Measures and Data Preparation

We measured how useful intelligibility was for the participants in terms of how much they used, and how that impacted their understanding of Laκsa and its issues.

Usage of Intelligibility. We logged when participants viewed each explanation page in the UI. For each scenario, we measured which explanation types each participant viewed, how many (**# Explanation Types**), when they were viewed, for how long (**Duration**), and how often (**View Count**). We built a network graph for each participant scenario to illustrate the *sequence diagram* of how he used intelligibility (e.g., Fig. 1). As a measure of a usage pattern of intelligibility, we compute the **Context Ratio** of how many explanation types of deeper contexts (Place and Sound) were viewed compared to that of the shallower context (Availability). Next, we use these metrics to investigate their influence on user understanding.

User Understanding and Suggestions for Control. We coded transcripts into units of *beliefs* to characterize participant mental models about their understanding, using a coding scheme counting whether the participant indicated knowledge of the inferred **value** (e.g., Sound=Music), **alternative values** (e.g., P01S2 *"Talking (evidence=85.4) very close to music (84.?). Could have gone any way."*), inference **certainty** (e.g., P02S3: *"It was 9.3% certain I was at the Office"*), **inputs** (e.g., Pitch, Periods of Silence; *"the blue bubble was directly over the Library building."*), inference **model** (e.g., P17S3: *"...since the library bubble was very small then it calculated the probability was very low."*), **technical** details (e.g., P18S3: *"It seems to be based on its Wi-Fi connection, and ... because it said networking and it gave the location badly and we're deep inside a bunch of concrete and metal, so the GPS shouldn't be working right now."*), and **situation justification** (e.g., P02S2: *"The music was much more mellow, and they were really singing"*). We calculate an **Understanding Score** for each participant scenario by adding all 7 codes for both Place and Sound (Max=14).

Another measure of how well participants understood Laκsa is how many effective *control suggestions* they provided to overcome any issues or problems in the scenarios. We calculated a weighted **Control Score** with a coding scheme counting whether the participant suggested **availability rules** (e.g., delete rule "Someone's Talking"), changing settings for inferring **Place** or **Sound**, and whether to change their own **behavior** (e.g., lowering the music volume). This score represents the number and effectiveness of suggestions provided for the scenario. Partially effective suggestions with compromising side-effects are given only half a score.

Perception of Application and Explanations. For each scenario, we asked participants their perception of Laкsa's **Behavior Appropriateness** (7-point Likert scale) and if they agreed or disagreed that the explanations were helpful (**Explanation Helpfulness**; 7-point).

4   Results

We recruited 18 participants (11 females) with ages 19 to 65 (Median=26) years. 9 participants were graduate students, and three were undergraduates. P01, P16, and P17 were students in a computer-related field. P18 was a web programmer, while the others spanned a wide range of areas (e.g., actor, pianist, field interviewer, hospital administrator, chemical engineering, retiree). We engaged each participant for 1h 44min on average (range: 1h 29m to 1h 58m). Each participant was compensated $20.

Although participants experienced the same scenarios, due to conducting the experiment in the field, there was some variability in what Laкsa sensed and the resulting explanations. For example, location accuracy depended on where the participant walked to, weather, and other environmental factors; when the participant walks to the café in S4, she may hear background music, or be near people who are talking.

For S4, participants exhibited two distinct behaviors to explore the hypothetical situation: (i) they either just sat where they were and tried to use the What If explanation facility (S4-if, 10 cases), or (ii) walked to the café to test Laкsa *in-situ* (S4-situ, 11 cases; some participants did both). We treat these as distinct scenarios.

Perception of Application Behavior and Explanations. As expected, participants perceived Laкsa's behavior as inappropriate for S2 and S3 ($M_{S2}=-2.1$, $M_{S3}=-2.4$), but appropriate for S4 ($M_{S4-if}=2.0$, $M_{S4-situ}=2.4$): $F_{3,25}=3.90$, p<.05; contrast test: p<.01. Participants generally found the explanations helpful (M=1.5), though explanations were less helpful in S2 ($M_{S2}=0.6$) than in S4 ($M_{S4-if}=2.4$, $M_{S4-situ}=2.5$); Tukey HSD test: p<.05.

Intelligibility Usage. Combining usage logs across S2 to S4, we determined participants' overall usage of intelligibility (see Table 1), and their usage for each explanation type (see Table 2). Most participants actively looked at many Explanation Types (Median=8), many times (View Count Median=21), for about 3 minutes per scenario. They also tended to look more at deeper contexts (Place or Sound) than just Availability (Context Ratio Median=1.4). Usage ranged from very engaged (View Count Max=65, Scenario Duration Max=12.5min), to conservative, e.g., min 2 views (P08S4-if), 1 explanation type (P14S4-situ), scenario duration <1 min (P08S4-if).

Table 2 illustrates which explanation types were more popular, i.e., higher view count, and how much time participants spent looking at each explanation type.

Table 1. Summary statistics of intelligibility usage per participant scenario (S2-S4)

Per Scenario	Mean	SD	Std. Err.	Min	Median	Max
View Count	24	15	2	2	21	65
# Explanation Types	7.9	3.4	0.4	1	8	19
Context Ratio	1.8	1.8	0.2	0	1.4	8
Total Duration (s)	205	136	18	52	196	749

Table 2. Usage of explanation types: total view count of explanation types for all participant scenarios, and median durations for respective views (for Total View Count > 15)

	Total View Count			Median Duration (s)		
	Avail.	Place	Sound	Avail.	Place	Sound
What + Certainty	232	114	69	5.7	3.1	3.2
History	130	5	3	.5	-	-
Outputs + Certainty	84	102	33	6.4	5.6	4.9
Inputs	217	45	60	7.1	6.4	6.0
Why	26	16	44	3.5	9.9	6.1
Why Alt	21	35	41	4.7	9.6	4.9
What If	31	-	-	24.8	-	-
Definition	5	7	28	-	-	6.1
Situation	-	-	15			

User Understanding and Control Suggestions. For each scenario, participants articulated 0 to 8 correct beliefs about Laкsa's behavior (Median=4). 41% of the beliefs were about the awareness of the inferred Value for Place and Sound, 28% about a broader understanding of the inference (Alternative Values and Certainty), 15% about the Inputs state and Model mechanism, and 2.5% about deeper Technical details. 14% of the beliefs were drawn from the Situation to justify Laкsa's behavior.

Participants provided 0 to 6 correct Control Suggestions (Median=2) for each scenario, and had an average Control Score of 2.10 (Std Err=0.29). This is significantly greater than 1 (i.e., H_0: Score>1, p<.01). Participants made effective and partial Control Suggestions about: Availability Rules (29%), Settings (27% Place, 8% Sound), Behavior Change (36%).

The extent and pattern of intelligibility usage did affect how well participants understood Laкsa, as we shall see next.

Impact of Intelligibility Usage on Understanding. We chose View Count and Context Ratio as factors of intelligibility usage. We split View Count into discrete intervals of 10 counts; we split Context Ratio into two groups Shallower (N=34) and Deeper (N=20), where participants saw twice as many explanations about Place or Sound than Availability (ratio ≥2). Fig. 2 summarize these results showing that higher and deeper use of explanations lead to higher Understanding and Control scores.

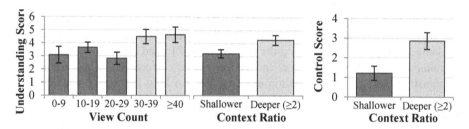

Fig. 2. (Left) Understanding Score is higher when explanation views ≥ 30 than less ($p<.05$) and **(Right)** Understanding and Control Scores are higher (both $p<.05$) when participants ask more explanations about Deeper contexts (Place and Sound) than Availability.

5   Discussion and Recommendations

Our results show how participants were willing to use intelligibility, and how quickly or deeply they used it. This satisfies our hypothesis that more Intelligibility Usage (View Count and Context Ratio) improves Understanding. These have implications on how intelligibility should be provided to facilitate its more effective use.

5.1   Usage and Usefulness of Intelligibility

Our results show that intelligibility was *useful* for participants to (i) engage with intelligibility (some participants deeply so), (ii) rate explanations as helpful, and (iii) better understand application behavior. We next discuss how they used intelligibility, and how certain usage patterns were more effective in improving user understanding.

Diverse Usage of Explanation Types. Participants used a diverse range of explanation types and in diverse ways. What and Inputs were conduits to other explanations for participants to learn deeper reasons. However, although some explanation types were used less than others, some were viewed for longer durations (*e.g.*, Place Why / Alt). Furthermore, as with [7], the sequence diagrams of our participants revealed various usage styles (e.g., quick comparison between Why and Why Alt reasons, diving into a deeper context after going straight to Availability Inputs).

Unlike what was found in [4], our participants felt that the What If explanation was easy to use and liked it (e.g., P11S1: *"[Using] it was just more fun ... I like to think of hypothetical things, but it also gives me a sense of what the phone is capable of, and helps to develop trust when you know what to expect"*). In fact, for S4, 10 participants chose to ask What If instead of immediately walking to the café. However, this fascination with What If can also give users false trust since it obscures potential pitfalls in sensing. Participants who used What If in S4 may not realize how noisy the café may be or that the Place inference was not particularly good there. P11 did not bother to explore Laĸsa's inference in-situ because *"technology is supposed to make your life easier; you shouldn't have to waste time to make sure it works right."* Perhaps providing warnings that sensing can fluctuate due to environmental conditions may help users be more careful when using What If.

Occasionally, participants forgot what had recently happened, e.g., for S2, P07 thought he was talking to the experimenter at the time Laкsa inferred Sound as Talking. Had he played the recorded audio of that time (Situation), he would have learned that only singing was heard. Using the played audio, P15 and P16 were able to identify guitar sounds when Sound was finally recognized correctly as Music. Hence, in combination with History, Situation explanations can help jog a user's memory of what was happening, independent of the application's inference. This helps them form Situation Justifications for the application behavior. How may we also provide Situation explanations for contexts other than Sound? For Place, perhaps by showing a photograph at the location (if one was taken at the same time). For Motion recognition, perhaps by animating an *interpreted* diagram of how the phone was moving (derived from accelerometer data).

While earlier research into intelligibility sought to prioritize providing some explanation types over others (e.g., [4, 5]), along with [8], our findings suggest instead to provide a diversity of explanation types will be helpful to support different learning and troubleshooting strategies users have.

Deeper Usage of Intelligibility. Our quantitative results indicate that viewing more explanations, especially about deeper contexts can lead to deeper understanding, and more effective control suggestions for improving the application behavior. So, to promote user understanding, we need to encourage users to dig for more explanations, and to dig deeper. Perhaps, if the user starts asking questions, the application can hypothesize faults, and highlight which factors are probably causing them. These guesses could come from a knowledge base of typical faults [7], or be triggered when inferences Certainty becomes too low (e.g., <80% [8]).

5.2   Constraints for Intelligibility

While the upper bounds of our participants' usage of intelligibility may give an indication of engagement, the lower bound may portend the limits to which some users are willing to use intelligibility. Therefore, we derive some time and view constraints for intelligibility. Participants only spent about 3-10 seconds viewing each explanation, so each explanation page needs to be correctly and effectively interpreted within that short duration. Perhaps, if an explanation cannot be understood within that duration, it should be split into multiple parts where the user can ask for more *on demand*. Furthermore, our quicker participants spared only about 1-3 minutes exploring explanations for each incident. This may be even shorter without the experimenter demand effect when users explore intelligibility outside of a user study. Hence, question asking should be streamlined to facilitate multiple views (~20) within about 2 minutes before the user gives up. In our scenarios, we have focused on investigating the usage of intelligibility about incidences *in-situ* and in the moment. However, users may postpone investigating an incident until they have more time. Under those circumstances, we expect usage amounts and duration to be higher.

6 Conclusion

We have presented a quasi-field study measuring how participants used an intelligible context-aware application in scenarios representing real-world, "everyday" situations. We found that viewing more explanations, especially more about deeper contexts can further improve user understanding of application inference. We provided implications for promoting more effective intelligibility usage, time constraints within which users are willing to view intelligibility, and discuss how much intelligibility should be provided to sufficiently improve user understanding of context-aware applications.

Acknowledgements. This work was funded by the National Science Foundation under grant 0746428, Intel Research, and A*STAR, Singapore.

References

1. Cheverst, K., et al.: Exploring issues of user model transparency and proactive behavior in an office environment control system. UMUAI 2005 15(3-4), 235–273 (2005)
2. Dey, A.K., Abowd, G.D., Salber, D.: A conceptual framework and a toolkit for supporting the rapid prototyping of context-aware applications. HCI Journal 16(2-4), 97–166 (2001)
3. Hilbert, D.M., Redmiles, D.F.: Extracting usability information from user interface events. ACM Computing Survey 32(4), 384–421 (2000)
4. Lim, B.Y., et al.: Why and why not explanations improve the intelligibility of context-aware intelligent systems. In: CHI 2009, pp. 2119–2128 (2009)
5. Lim, B.Y., Dey, A.K.: Assessing Demand for Intelligibility in Context-Aware Applications. In: Ubicomp 2009, pp. 195–204 (2009)
6. Lim, B.Y., Dey, A.K.: Toolkit to Support Intelligibility in Context-Aware Applications. In: Ubicomp 2010, pp. 13–22 (2010)
7. Lim, B.Y., Dey, A.K.: Design of an Intelligible Mobile Context-Aware Application. In: MobileHCI 2011 (2011) (to appear)
8. Lim, B.Y., Dey, A.K.: Investigating Intelligibility for Uncertain Context-Aware Applications. In: Ubicomp 2011 (2011) (to appear)
9. Lu, H., et al.: SoundSense: scalable sound sensing for people-centric applications on mobile phones. In: MobiSys 2009, pp. 165–178 (2009)
10. Milewski, A.E., Smith, T.M.: Providing Presence Cues to Telephone Users. In: CSCW 2000, pp. 89–96 (2000)
11. Muir, B.: Trust in automation: Part i. theoretical issues in the study of trust and human intervention in automated systems. Ergonomics 37(11), 1905–1922 (1994)
12. Rosenthal, S., Dey, A.K., Veloso, M.: Using Decision-Theoretic Experience Sampling to Build Personalized Mobile Phone Interruption Models. In: Lyons, K., Hightower, J., Huang, E.M. (eds.) Pervasive 2011. LNCS, vol. 6696, pp. 170–187. Springer, Heidelberg (2011)
13. Roto, V., et al.: Examining Mobile Phone Use in the Wild with Quasi-Experimentation. Helsinky Institute for Information Technology (HIIT), Technical Report (2004)
14. Tullio, J., et al.: How it works: A field study of non-technical users interacting with an intelligent system. In: CHI 2007, pp. 31–40 (2007)
15. Vermeulen, J., et al.: PervasiveCrystal: Asking and Answering Why and Why Not Questions about Pervasive Computing Applications. IE 10, 271–276 (2010)
16. Weka for Android, https://github.com/rjmarsan/Weka-for-Android (retrieved August 26, 2011)

Strangers and Friends

Adapting the Conversational Style of an Artificial Agent

Nikita Mattar and Ipke Wachsmuth

Artificial Intelligence Group, Bielefeld University
Universitätsstr. 25, 33615 Bielefeld, Germany
{nmattar,ipke}@techfak.uni-bielefeld.de

Abstract. We demonstrate how an artificial agent's conversational style can be adapted to different interlocutors by using a model of Person Memory. While other approaches so far rely on adapting an agent's behavior according to one particular factor like personality or relationship, we show how to enable an agent to take diverse factors into account at once by exploiting social categories. This way, our agent is able to adapt its conversational style individually to reflect interpersonal relationships during conversation.

Keywords: embodied conversational agents, conversational style, social categories, personality, relationships, situational context.

1 Motivation

More than two decades ago, Wahlster and Kobsa [19] argued that systems employing long-term models of users are not in the focus of interest, one reason being privacy concerns. That the systems did not need to account for recurring interactions since these did not happen that often, was considered another important reason.

With nowadays technology, these assumptions do not seem to hold anymore. Smartphones bring technology to our fingertips that is consulted on a daily basis and that is able to have access to voluntarily provided private data of their owners. The availability of such technologies fosters the need for more natural interaction metaphors. For instance, Apple's Siri allows users to use natural language to query information, delegate tasks, and to have minimal task-related conversations (e.g., adding a new appointment in the calendar). Embodied conversational agents (ECAs) go even further, exploiting factors like personality characteristics or interpersonal relationships of the interlocutors [1]. Still closer to human-like partners, in terms of length of interaction and closeness of relationship, are companion agents as envisioned by Yorick Wilks [20]. If his vision came true, encounters with more than one human are inevitable. Such agents would need facilities to keep track of their relationships with different individuals and to adapt accordingly.

However, most of the current approaches do not pay enough attention to the interaction between the factors that play a role in human-human interaction.

M. Kurosu (Ed.): Human-Computer Interaction, Part V, HCII 2013, LNCS 8008, pp. 102–111, 2013.

Research findings in psychology indicate that the factors are interwoven and cannot be considered in separate that easily. Therefore, we stress the importance of providing artificial agents, that are to engage in long-term interactions, with means to represent individual information about different persons and the conversational style that is appropriate for the agent to use when interacting with a certain interlocutor. Access to this kind of information should allow an agent to regard the interaction in its individual interpersonal context.

The paper is structured as follows. In Sect. 2, work on conversational systems taking personality factors and interpersonal relationships into account is reviewed. Section 3 is dedicated to the factors that influence conversational style. In Sect. 4, our approach of exploiting a Person Memory to control the conversational behavior of a virtual agent is presented. The paper concludes with a brief summary and hints about future work.

2   Related Work

The possibility of equipping a technical system with a human-like embodiment changed the intended use from mere tools to systems, in which the relationship to the user influences the success of the system. For example pedagogical agents have to interact over a longer period of time with an individual learner where embodiment fosters the development of a relationship. In turn, the relationship has important effects on the learner [7].

Even in single-encounter interactions, the relationship plays an important role. Bickmore and Cassell [1] demonstrated how small talk can be used to affect the (trust) relationship between an agent and its interlocutor in a task-oriented interaction. The authors found that their real estate agent REA when engaging in small talk was more preferred by extroverts than by introverts [1]. In more recent work, Bickmore and Schulman describe their approach on building and influencing a relationship [2]. Here, dialogue acts are grouped into four relationship categories (stranger/professional, more than professional relationship, casual friend, close friend). While their system is able to infer the current relationship, it is not able to change the relationship into a more intimate direction. As one reason why their system fails to do so, they regard individual differences in the behavior tolerated as appropriate for a given relationship.

While the former approaches are concerned with the effects of the type of relationship on the interaction, Mairesse and Walker [11] examined how the personality of a system and its interactants influence their conversation and demonstrate how utterances of a dialogue system can be adapted according to different personalities. They show that the utterances produced by their system can reliably be assigned different personalities.

The systems described above have in common that they attempt to improve human-computer interaction by adapting the system's conversational behavior in respect to relationship or personality factors. In social psychology, factors that affect the conversational behavior are referred to under the term **conversational style**, as discussed in the following.

3   Influences on Conversational Style

According to Tannen [17], conversational style is not to be understood as something humans can choose to use during conversation or not. Stylistic strategies make up the conversational style of a person. Tannen states that the repertoire of strategies are determined by the individual's context (e.g., a narrow geographical region or a culture), and are habitual and rather learned automatically. This way conversational style of a person can serve as an indicator for a person's personality [17].

In social psychology, **personality traits** are an important tool to describe differences in human behavior. The Five Factor Model (FFM), one of the most prominent models to describe personality [15], comprises of the so called *Big Five* dimensions *Openness, Conscientiousness, Extraversion, Agreeableness*, and *Neuroticism*. It has been widely used in ECAs to model the personality of the agent, the interlocutor, or even both. Some of the traits are directly linked to conversational style. For instance, extroverted people tend to be more talkative and socially engaging, whereas introverts are more shy and reserved, especially during initial encounters. Conscientiousness, being an indicator of self-control, has effects on disclosure of personal information [15]. Mairesse [10] summarizes further effects of the Big Five personality traits on linguistic features. Furthermore, according to the similarity-attraction hypothesis, similarity in personality traits can serve as an indicator about how relationships develop. The more similar two persons are in a certain personality trait, the more likely they develop a closer relationship [16], [15].

Different models have been developed to describe **relationships**. In Bickmore and Picard's review of work on human-human relationships [4], the following five types of models are listed: *dyadic, provision, economic, dimensional*, and *stage models*. According to them, dyadic models are the most prominent in recent social psychology research. Here, a relationship is made up of the interpersonal interaction and the **situational context** of the interlocutors [21]. Since the behavior of one partner depends on the behavior of the other, a person with a certain personality does not necessarily behave in the same way in every interaction [16], [14], [21]. In dimensional models, which are frequently used in ECA design [8], diverse dimensions have been used to describe a relationship (see e.g., [16], [3]). Some of these dimensions, like dominance, friendliness, distance, or intimacy, are used to assess the level of emotional appraisal [8], use of certain topics, use of structural elements, irony, humor, and the tendency to be more agreeing or confronting during conversation [18], [6], [17].

In summary, several factors determine the conversational behavior of persons. Some of these factors depend on the individual relationship and the direct interaction of the partners. In human-human interaction, most of these factors are not under the direct control of the individuals. Nevertheless, personality traits, individual relationships and the situational context have influence on the conversational style. To enable an artificial agent to adapt its conversational style to different types of interlocutors (in particular, strangers or friends) our idea is to equip an agent with a Person Memory.

4   Exploiting an ECA's Person Memory to Adapt Conversational Style

Our model of Person Memory [13] enables the embodied conversational agent Max [9] to remember and retrieve information about the diverse interlocutors he interacts with. Furthermore, Max is able to exploit existing information to react to different kinds of persons. The model consists of two parts (see Fig. 1): The first part is used to represent information about groups of individuals in the form of *social categories*. The second part is used to represent information about individuals like *biographical facts, preferences,* and *events.*

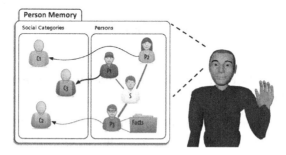

Fig. 1. Max and his model of Person Memory [13]. Social categories and individual information are associated to certain persons.

The Person Memory is embedded into the static knowledge layer of the agent's deliberative component that was introduced in [9] (cf. Fig. 2). In the deliberative component we differentiate dynamic knowledge, that is maintained during an ongoing conversation, and more static knowledge, that consists of long lasting information. In addition to the representation of the agent itself (denoted with S in Fig. 1), a representation of a person is *activated* in Person Memory when an interaction with a certain individual starts. The Person Memory transfers information relevant to the current interaction into the situation and discourse model as it has direct access to the dynamic knowledge layer (see Fig. 2). During conversation, different components use this information to influence the selection of appropriate dialogue plans.

In earlier work [13], we used information from the Person Memory to determine suitable subjects for initial conversations with strangers (cf. Fig. 3). A further extension of the deliberative component [12] allows to control the types of **dialogue sequences** Max uses during conversation (e.g., *"Question/Answer"* vs. more complex sequences like *"Question/Counter/Probe/Reply"*). The process of social categorization can now be extended to hold information on how to influence the conversational style according to different personality traits, relationships and further factors, like social roles, for instance by influencing the agents choice of such *dialogue sequences.*

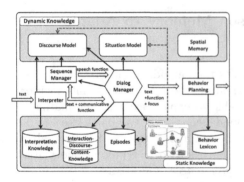

Agent: I have never met you here before, Paula.
       Are you waiting for someone?
Paula: Yeah, I am waiting for a friend of mine.
Agent: Do you study computer science? Most of
       the people I meet here do.
Paula: No. I am a student of sport sciences.
Agent: Oh, nice to finally meet someone from
       another discipline.
Agent: So do you live here in Bielefeld?
Paula: Yes, I moved here two years ago.
Agent: Oh, ok.
Agent: Did you see the last match of Bielefeld's
       soccer club?
Paula: Yes, that was a great game.

Fig. 2. Deliberative component of our ECA Max ([9], extended). The Person Memory has direct access to the dynamic knowledge layer.

Fig. 3. Excerpt of a conversation between our agent and a previously unknown person [13]

4.1 Representing Personality Traits, Relationships, and Social Roles in Person Memory

Building on our previous work, a certain type can be assigned to a category, to distinguish between them in the Person Memory. Three types – *Relational, Personality, Generic* – for representing categories that describe relationships, personality traits, and generic social categories, respectively, are used (see Fig. 4). Categories of the types *Relational* and the subtypes of *Personality* are considered mutually exclusive. That is, every person in the Person Memory can only be assigned one relational and one personality category for each personality dimension at a given time. However, the same person can belong to several generic categories at once. Note that categories like *boss, co-worker*, etc., that could also be considered relational categories, are treated as social roles and therefore belong to the generic categories.

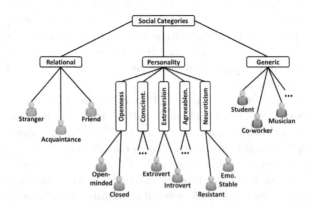

Fig. 4. Types of social categories in our Person Memory

In addition to the information we previously associated with social categories [13], information that has impact on the conversational style of the agent can now be provided. Table 1 depicts a sample category C. Slots with the name *sequence* are used to represent information about the **choice of dialogue sequences** [12]. In this case, the optional argument $[Arg]$ is used to specify a probability $\in [0, 1]$ for the agent to choose a certain sequence when its interlocutor belongs to the category C. In this example three types of dialogue sequences can be used by the agent during conversation: *short, medium,* and *long.* The sum of the probabilities is normalized to 1. It is possible to only provide probabilities for some of the sequences. The difference between the sum of the probabilities provided and 1 is distributed equally among the remaining sequences.

The **use of topics** is implemented following Breuing and Wachsmuth's [5] approach. The *topic*-slots hold information about the probability for the agent to use topics from a certain topic category (immediate, external, communication).

The Situational Context. Among the *active* representations of the agent and its interlocutor (see above), the situational context of the interaction is constituted by a set of "salient" social categories: a relational category, categories of the different personality dimensions, and further generic categories. Note that a person is always assigned to a relational category, whereas personality categories, and other generic categories are optional. Whenever an interaction starts, the Person Memory accesses the situational model of the dynamic knowledge layer (see Fig. 2). On the one hand, the situational model includes descriptions, like name, location, and type of the current situation (cf. Table 2) that can trigger certain categories (e.g., second and third slot in Table 1). On the other hand, weights $w_g, w_p, w_r \in [0, 1]$, with $w_g + w_p + w_r = 1$, for the three types of categories can be provided. Again, the difference of the sum of the probabilities of the provided categories and 1 is distributed equally among the remaining categories.

Combining Several Categories. Since several categories can contain information influencing the agent's conversational style, there needs to be a way of combining this information. To give an idea, the final probabilities for the *choice of dialogue sequences* are calculated as follows.

Let $S = \{short, medium, long\}$ be the set of available sequences, C_g, C_p, C_r three categories of the types *Generic, Personality, Relation,* respectively, and T the current situation with weights w_g, w_p, w_r. Since the sum of these weights is normalized to 1, they can be used to represent the prior probabilities $P(c_j) = w_j$, with $w_j \in \{w_g, w_p, w_r\}$, of a category $c_j \in \{C_g, C_p, C_r\}$. The law of total probabilities can be applied to determine $P(s_i)$:

$$P(s_i) = \sum_{j \in \{g,p,r\}} P(s_i|c_j)P(c_j) \ . \tag{1}$$

The final probability values $P(s_i)$ are transferred into the discourse model as described above. In this case, the sequence manager (see Fig. 2) uses these probabilities to determine which dialogue sequence the agent uses next.

Table 1. Excerpts of a social category representing information influencing the conversational style

Slot	Value	[Arg]
cat_type	generic	
trigger	situation_type	casual
trigger	situation_location	work
sequence	short	0.75
sequence	medium	0.15
sequence	long	0.1
topic	immediate	0.6
topic	external	0.25
topic	communication	0.15
...	...	...

Table 2. Two different situations

Slot	Value	[Arg]
situation_name	freetime	
situation_type	casual	
situation_location	home	
cat_weight	relational	0.7
cat_weight	personality	0.3

Slot	Value	[Arg]
situation_name	lunchbreak	
situation_type	casual	
situation_location	work	
cat_weight	generic	0.5

4.2   Encounters with Strangers and Friends

With the extensions of the Person Memory introduced above, our ECA Max is now able to adapt his conversational style taking several factors into account.

In Fig. 5, four encounters $E1_a$, $E1_b$, $E2_a$, $E2_b$, with two persons P_a, P_b, taking place in different situations $S1$, $S2$ are depicted. Situations $S1$ and $S2$ correspond to the situations depicted in Table 2. Person P_a is considered to belong to the category *stranger*, P_b to the category *friend*. In addition, both are assigned to the personality category *introvert* and a generic category *co-worker*. Next to the categories, the probability models for the *choice of dialogue sequences* $p_{C_j}(S)$ and for the *use of topics* $p_{C_j}(T)$ are given. The first, second, and third value denote the probabilities for *short*, *medium*, and *long* dialogue sequences, and *immediate*, *external*, and *communication* topic categories, respectively. The generic category only influences the *use of topics*, since no probabilities for the *choice of dialogue sequences* are given. The final probability models $p(S)$ and $p(T)$ for each encounter are calculated as described in Sect. 4.1.

The weights of situation $S1$ let the relational category of the interlocutor dominate the conversational style in encounters **E1**. Whereas in encounters **E2**, the generic category *co-worker* has the strongest influence on the conversational style of the agent. In both encounters $E1_a$ and $E2_a$ with person P_a, the dialogue engine of the agent mainly produces dialogue sequences of the types *short* (e.g., *Question/Answer*). However, the agent is more likely to select topics from the communication topic category in encounter $E1_a$ than in $E2_a$. In the latter, the agent focuses on topics dealing with the immediate situation. In encounters $E1_b$ and $E2_b$ with person P_b, the difference in the conversational style of the agent becomes more evident. In encounter $E1_b$, the likelihood for the dialogue engine to select dialogue sequences of type *long* is much higher than in the more formal encounter $E2_b$. Here the dialogue engine mainly uses *short* dialogue sequences,

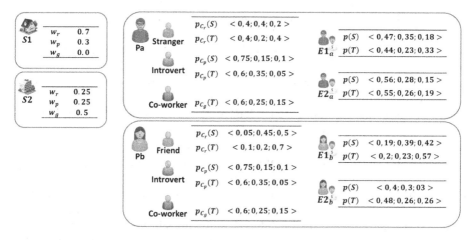

Fig. 5. Four encounters $E1_a$, $E1_b$, $E2_a$, $E2_b$, with two different persons P_a, P_b, taking place in different situations $S1$, $S2$

again. Furthermore, the agent focuses on private topics from the communication topic category in encounter $E1_b$, whereas he predominantly uses topics of the immediate situation in $E2_b$.

4.3 Discussion

As demonstrated in Sect. 4.2, the further exploitation of social categories enables the agent Max to integrate information like personality, relationship, and social roles with the ongoing situation. Therefore, Max is able to consider the individual interpersonal context the interaction takes place in. In encounters $E1$ in Sect. 4.2, the dialogue engine is able to take the private situation $S1$ into account. The agent focuses on the relationship and uses more private topics. Whereas, in encounters $E2$ the more formal situation is reflected. Whatever the relationship status between the interlocutors might be, discussing topics concerning the immediate situation is more appropriate than talking about private matters in a work context.

5 Conclusion and Future Work

In this paper, we demonstrated how information of a Person Memory can be used to adapt the conversational style of the agent Max towards different persons. While we showed how to manipulate two particular features, review of related work in Sect. 2, and the discussion of conversational style in Sect. 3, revealed more ways how conversational style can be influenced.

Currently, the dialogue engine of Max is able to differentiate between three relational categories (stranger, acquaintance, friend), two categories along the extraversion personality dimension (extrovert, introvert), and several generic social categories. However, while the dialogue engine is able to assign generic categories automatically during conversation (cf. [13]), right now mechanisms to assign the relational categories and to assess the interlocutors personality are in work to come.

The dialogues produced by our agent differ in the topics being talked about, and total length due to the selected types of dialogue sequences. A formal evaluation of these varying dialogues is planned as future work. In that, it will be examined if differences in the conversational style of Max can be correctly assigned to different person/situation constellations.

Besides an evaluation of the dialogues produced by Max, as a next step, we will focus on how to integrate and exploit strategies to change the relationship between an agent and its interlocutors into a certain direction (e.g., from stranger to friend). Building on the work presented, our agent will be enabled to choose among strategies appropriate for different individuals in different situations. To be noted, finally, the approach is not restricted to influencing conversational style and could be extended to trigger non-verbal behaviors like gesturing, use of eye contact, or proxemics and can therefore be exploited in diverse human-agent interaction scenarios.

References

1. Bickmore, T., Cassell, J.: Relational Agents: A Model and Implementation of Building User Trust. In: Proceedings of the SIGCHI Conference on Human Factors in Computing Systems, CHI 2001, pp. 396–403. ACM, New York (2001)
2. Bickmore, T., Schulman, D.: Empirical Validation of an Accommodation Theory-Based Model of User-Agent Relationship. In: Nakano, Y., Neff, M., Paiva, A., Walker, M. (eds.) IVA 2012. LNCS, vol. 7502, pp. 390–403. Springer, Heidelberg (2012)
3. Bickmore, T.W.: Relational Agents: Effecting Change through Human-Computer Relationships. Ph.D. thesis. Massachusetts Institute of Technology (2003)
4. Bickmore, T.W., Picard, R.W.: Establishing and Maintaining Long-Term Human-Computer Relationships. ACM Transactions on Computer-Human Interaction (ToCHI) 12(2), 293–327 (2005)
5. Breuing, A., Wachsmuth, I.: Let's Talk Topically with Artificial Agents! Providing Agents with Humanlike Topic Awareness in Everyday Dialog Situations. In: ICAART 2012 - Proceedings of the 4th International Conference on Agents and Artificial Intelligence, pp. 62–71. SciTePress (2012)
6. Eggins, S., Slade, D.: Analysing Casual Conversation. Cassell (1997)
7. Gulz, A., Haake, M., Silvervarg, A., Sjödén, B., Veletsianos, G.: Building a Social Conversational Pedagogical Agent: Design Challenges and Methodological approaches. In: Perez-Marin, D., Pascual-Nieto, I. (eds.) Conversational Agents and Natural Language Interaction: Techniques and Effective Practices, ch. 6, pp. 128–155. IGI Global (2011)
8. Kasap, Z., Ben Moussa, M., Chaudhuri, P., Magnenat-Thalmann, N.: Making Them Remember–Emotional Virtual Characters with Memory. IEEE Computer Graphics and Applications 29(2), 20–29 (2009)

9. Kopp, S., Gesellensetter, L., Krämer, N.C., Wachsmuth, I.: A Conversational Agent as Museum Guide – Design and Evaluation of a Real-World Application. In: Panayiotopoulos, T., Gratch, J., Aylett, R.S., Ballin, D., Olivier, P., Rist, T. (eds.) IVA 2005. LNCS (LNAI), vol. 3661, pp. 329–343. Springer, Heidelberg (2005)

10. Mairesse, F.: Learning to Adapt in Dialogue Systems: Data-driven Models for Personality Recognition and Generation. Ph.D. thesis, University of Sheffield, Department of Computer Science (2008)

11. Mairesse, F., Walker, M.A.: Towards Personality-Based User Adaptation: Psychologically Informed Stylistic Language Generation. User Modeling and User-Adapted Interaction 20, 227–278 (2010)

12. Mattar, N., Wachsmuth, I.: Small Talk Is More than Chit-Chat. In: Glimm, B., Krüger, A. (eds.) KI 2012. LNCS, vol. 7526, pp. 119–130. Springer, Heidelberg (2012)

13. Mattar, N., Wachsmuth, I.: Who Are You? On the Acquisition of Information about People for an Agent that Remembers. In: ICAART 2012 - Proceedings of the 4th International Conference on Agents and Artificial Intelligence, pp. 98–105. SciTePress (2012)

14. Reis, H.T., Capobianco, A., Tsai, F.F.: Finding the Person in Personal Relationships. Journal of Personality 70(6), 813–850 (2002)

15. Selfhout, M., Burk, W., Branje, S., Denissen, J., van Aken, M., Meeus, W.: Emerging Late Adolescent Friendship Networks and Big Five Personality Traits: A Social Network Approach. Journal of Personality 78(2), 509–538 (2010)

16. Svennevig, J.: Getting Acquainted in Conversation: A Study of Initial Interactions. Pragmatics & Beyond. John Benjamins Publishing Company (1999)

17. Tannen, D.: Conversational Style: Analyzing Talk among Friends. Oxford University Press, Oxford (2005)

18. Ventola, E.: The structure of casual conversation in english. Journal of Pragmatics 3(3-4), 267–298 (1979)

19. Wahlster, W., Kobsa, A.: User models in dialog systems. In: Kobsa, A., Wahlster, W. (eds.) User Models in Dialog Systems, pp. 4–34. Springer, Berlin (1989)

20. Wilks, Y.: Artificial Companions as a new kind of interface to the future Internet. Research Report 13, Oxford Internet Institute/University of Sheffield (2006)

21. Zayas, V., Shoda, Y., Ayduk, O.N.: Personality in Context: An Interpersonal Systems Perspective. Journal of Personality 70(6), 851–900 (2002)

suGATALOG: Fashion Coordination System That Supports Users to Choose Everyday Fashion with Clothed Pictures

Ayaka Sato[1], Keita Watanabe[2], Michiaki Yasumura[3], and Jun Rekimoto[1,4]

[1] The University of Tokyo, 7-3-1 Hongo, Bunkyo, Tokyo 113-0033, Japan
[2] JST ERATO, Igarashi Design Project
[3] Keio University
[4] Sony Computer Science Laboratories Inc.
ayakasato@acm.org, watanabe@gmail.com, yasumura@sfc.keio.ac.jp,
rekimoto@acm.org

Abstract. When deciding what to wear, we normally have to consider several things, such as color and combination of clothes, as well as situations that might change every day, including the weather, what to do, where to go, and whom to meet with. Trying on many possible combinations can be very tedious; thus, computer support would be helpful. Therefore, we propose suGATALOG, a fashion coordination system that allows users to choose and coordinate clothes from their wardrobe. Previous studies have proposed systems using computer images of clothes to allow users to inspect their clothing ensemble. Our system uses pictures of users actually wearing the clothes to give a more realistic impression. suGATALOG compares several combinations by swapping top and bottom images. In this paper, we describe the system architecture and its user interface, as well as an evaluation experiment and a long-term trial test to verify the usefulness of the system.

Keywords: Fashion coordinate, Clothes, Life-log.

1 Introduction

When going out, a typical activity is to decide which clothes to wear. When selecting clothes, we have to consider several things, including the type of clothes we want to wear, the colors, and their combinations, as well as other situations, such as the reason for going out, the weather, where to go, and whom to meet with. We normally decide by looking through our wardrobe or actually trying them on until we are satisfied. It is not a trivial task and is often time consuming. After putting on the clothes, we often realize that the complete combination or an aspect of the clothes, such as the length and shape, is not as we expected; therefore, we have to try other combinations. Such trials might continue until we are satisfied or until we run out of time. Moreover, after trying on several combinations, we might still be unable to decide. For these reasons, we are inclined to wear similar and "safe" combinations that we already know work well.

M. Kurosu (Ed.): Human-Computer Interaction, Part V, HCII 2013, LNCS 8008, pp. 112–121, 2013.

Based on our survey about the everyday attire of 54 people (including male and female, teenagers to hexagenerians), 83% tend to wear the same combinations of clothes often, 80% have clothes that they have hardly worn or have never worn, and 76% think the clothes they already own could be better coordinated. These results indicate that they have difficulty in coordinating the clothes that they have, and imply that computer-assistance clothing coordination would be helpful.

In this research, we propose suGATALOG, a system that coordinates the user's own clothes using their "wearing image." We define a "wearing image" as an image of what the user is actually wearing. In this system, a user first takes a picture of his/her attire with a camera attached to a mirror and splits this picture into two—a top clothes part and a bottom clothes part. Then the user coordinates his/her clothes using these separated pictures. This paper outlines a prototype system and reports the results of an evaluation experiment.

2   Related Work

Interest and demand in fashion coordination is increasing, and many services and products on the market support this activity. Systems using pictures showing only clothes [5, 6] can support the selection of daily clothes with using pictures of owing clothes. However, because these systems only present pictures of clothes, users still have difficulty imagining the final image when the clothes are worn. Systems using computer graphics allows users to interactively control a 3D model of themselves at home using a depth camera [3], or combine clothes according to the user's movements [2, 4]. For this system to be used at home, the user's body must be scanned in 3D, and it is unrealistic to prepare 3D models of all clothes that a user already owns. Systems using user's body enables users to change the color of the worn clothes in the digital mirror [1], or enables online social fashion comparisons in physical stores based on multi-camera perception [8]. These systems are useful as a shopping support system at apparel shops. However, it is not suitable for home use, because the required equipment is currently still complicated and takes up a large space.

3   suGATALOG

suGATALOG is a system that allows users to choose and coordinate clothes from their wardrobe using pictures taken in front of a full-length mirror in their room. Users can take a picture before going out or when checking their own reflection in the mirror. In this section, we will describe the design and implementation of our system.

Our system has three main advantages over the previous systems:

- This system gives a realistic impression by showing pictures of the clothes that a user is actually wearing.
- By recording past combinations automatically, it helps a user to find a combination that the user has seldom worn or never tried.
- It does not require expensive or large equipment; therefore, it is easy to use at home.

3.1   Approach

We focused our attention on the differences in usability between pictures showing only clothes and pictures showing worn clothes. Fig. 1 shows a comparison of images showing only clothes (Fig. 1-A, B, C) and images showing worn clothes (Fig. 1-A', B', C'). In A and A', the shape of the sweater is different, and the absence of a face, arms, and legs makes it difficult to imagine whether the clothes will fit the face, hairstyle, and the body. In B and B', the width of the skirt is different. In C and C', the width of the dress is different, and the relative length of the dress is difficult to judge without trying it on. Our approach is to coordinate clothes using images like A', B' and C' where the clothes are worn, so that a realistic image showing the clothes fitted on the user's own body and also showing the user's own face. We use this approach to coordinate clothes in a "wearing image" without changing clothes in reality.

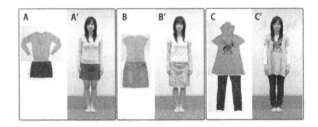

Fig. 1. Comparing pictures of clothes only (A, B, C) with pictures of clothes worn (A', B', C')

3.2   Brief Overview of the System

suGATALOG consists of a web camera attached to a full-length mirror and a web application. A user takes a picture with the camera in front of the mirror, and the picture is uploaded and shown on the web application. The picture is split into two parts: a top part and a bottom part. The user can coordinate clothes using these pictures on the application without actually wearing the different clothes.

3.3   Functions

In this section, we describe the functions of the web application.

Shooting Function. The Shooting function takes a picture of the user. When the user clicks on the camera icon on the bottom left of the screen (Fig. 2), a video from the web camera pops up with a guideline indicating where the user should stand. The user stands in front of the camera according to the guideline shown on the screen, and a picture is taken after 5 or 10 s (selectable).

Calendar Function. The Calendar function displays a calendar showing pictures that were previously taken and uploaded on specific days (Fig. 2). Users can view their fashion ensembles for each day of the month, which enables them to review what they have worn previously.

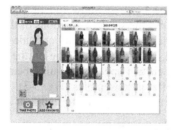

Fig. 2. An example screen of the calendar and the fitting room

Fitting Room Function. The Fitting Room function simulates ensembles using the pictures on the calendar. In Fig. 2, we named the left part of the screen as "fitting room." Each day of the calendar is divided in two parts at the center. If the user clicks on the upper part, that part of the fitting room is replaced by the clicked picture, and the same with the lower part. The picture is divided at the vertical center but the length of tops might differ, so users can adjust the length by using a slider. In this way, we coordinate different clothes by superimposing an adjusted tops picture over a bottoms picture. Fig. 3 shows the work sequence of simulating ensembles.

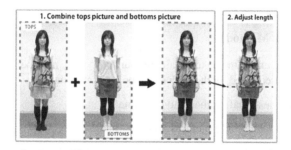

Fig. 3. To simulate an ensemble in suGATALOG, (1) select top and bottom pictures from the calendar and superimpose them, and (2) adjust the length of the top

Favorite Function. The Favorite function saves and compares several ensembles. If the user clicks on the "ADD FAVORITE" button at the bottom of the fitting room portion of the screen, the ensemble shown in the fitting room is saved as a favorite. This function enables users to save their favorite ensembles and to compare several ensembles side-by-side.

Auto-coordinating Function. The Auto-coordinating function coordinates several ensembles at a time. Generally, when selecting a combination of clothes, we select either tops or bottoms first, and then select the other. In a similar way, clicking on either the tops picture or the bottoms picture in the fitting room, several combinations for the selected half will be shown. This function helps users to find new combinations easily, which requires time and effort without the system, and to find unexpected combinations, which they had never tried previously or which they thought would not match.

Fashion Show Function. The Fashion Show function presents the ensembles in a fashion show style. Fig. 5 shows the screen transitions of this function. Two randomly selected pictures from the calendar appear from both sides of the fashion show stage (Fig. 5-1) and move forward, like models walking on the stage. When they come to the front center, an animated curtain appears and disappears (Fig. 5-2), and their tops are switched, creating two different ensembles (Fig. 5-3). This function enables users to review what they have worn in the past and to find new combinations.

Fig. 4. Fashion show function

4   Experiment

In this section, we describe the method, result, and observations of an evaluation experiment.

4.1   Experiment Method

We conducted an experiment to evaluate our proposed method for coordinating clothes. We compared three methods that can be used at home.

Method 1: Choose actual clothes and coordinate

Method 2: Use clothes-only pictures and coordinate on PC

Method 3: Use clothes-worn pictures and coordinate on PC (proposed method)

Method 1 is most typically used at home, Method 2 was proposed in a related work, and Method 3 is what we propose in this paper. In the experiment, 14 subjects (6 males and 8 females, with ages between 20 and 39) participated; they were all students who usually select clothes every day. Each subject brought five tops and five bottoms and created twenty-five combinations using these methods. We decided to use the subjects' own clothes so that they could coordinate clothes in their usual style. Participants first evaluated the realism of the ensemble image from each method, and then ranked the three methods.

Experiment 1: Comparing the realism of the ensemble. We conducted a survey with the questionnaire below on each of the 25 ensembles to verify how realistic subjects think the ensembles from each method are.

Q.1 Would you go out wearing this ensemble? [Yes / No]

Q.2 How much confidence do you have in your answer for Q.1? [Higher confidence 5 / 4 / 3 / 2 / 1 Lower confidence]

We asked these questions because asking "Do you think this simulation image looks real?" is too direct. By answering "Would you go out wearing this ensemble?," the subject would care about how others see him/her; therefore, he/she can answer the question in a normal situation. Regardless of the answer to Q.1, a higher value for Q.2 indicates that the ensemble is more realistic. We describe the experimental procedure for each method below.

Method 1: Coordinate by Viewing Actual Clothes

First, we labeled clothes which the subject brought in (from 1 to 5 for tops and from A to E for bottoms), and hang them on the hanger rack. Then, the subject answers the questions for each of the 25 ensembles by selecting and overlapping the clothes on the hanger rack.

Method 2: Coordinate on PC Using Clothes-Only Pictures

First, we take individual pictures of each of the subject's 10 pieces of clothing, and then scan the pictures into the PC and add them to the application shown in Fig. 6-1. The operating instructions were given beforehand. Then, the subjects answered the questions for all 25 ensembles in the same order as Method 1.

Method 3: Coordinate on PC Using Pictures of Clothes That a Subject Is Wearing (Our Proposed Method)

First, we take pictures of the subject wearing a combination of 1A (top: 1, bottoms: A), 2B, 3C, 4D, 5E. As with method 2, scan the pictures into the PC and add them to the application as shown in Fig. 6-2. The operating instructions were given beforehand. Then, the subjects answered the questions for all 25 ensembles in the same order as Method 1 and 2.

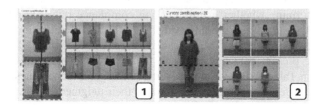

Fig. 5. Operation screen used in Method 2 (left) and Method 3 (right)

Experiment 2: Ranking of three methods. After Experiment 1, the subjects prioritized the three methods based on the strength of their convictions and gave the reasons for the rankings.

4.2   Results

Table 1 shows the results of Q.1, the degree of confidence that the subject would actually go out wearing the specified ensemble. We removed the combinations 1A, 2B, 3C, 4D, and 5E from each method, because the subjects have already tried them on in the experiment. The higher rate of confidence indicates that the subject is more convinced of the reality of when wearing that particular combination of clothes.

Method 3, which used pictures of worn clothes, received an average score of 4.0, which is the highest average value of confidence for all combinations. Colored squares in Table 1 show the highest confidence rate for the three methods for each test subject, and Method 3 was ranked highest among the three methods by the most number of people (a total of 11). Table 2 shows the results of Experiment 2, the ranking of the strongest conviction on the Q.1 answer. The best ranked was Method 3, which uses pictures of worn clothes on the PC, the second place was Method 1, which uses real clothes, and the third place was Method 2, which uses only-clothes pictures on the PC.

Table 1. Result of Experiment 1—comparison of degrees of confidence. M is male subject and F is female subject

Subjects Method	M1	M2	M3	M4	M5	M6	F1	F2	F3	F4	F5	F6	F7	F8	Average	Male Average	Female Average
Method 1	3.5	3.7	3.9	3.3	2.9	4.1	4.2	4.2	3.4	3.8	4.0	4.0	3.9	2.9	3.7	3.6	3.8
Method 2	3.7	4.2	3.8	3.4	3.4	3.9	4.2	3.8	3.2	3.9	4.1	3.2	3.9	3.0	3.7	3.7	3.8
Method 3	4.2	3.9	3.9	3.1	3.9	4.4	4.5	4.3	3.8	4.9	3.8	4.2	3.9	3.4	4.0	3.9	4.1

Table 2. Result of Experiment 1—comparison of degrees of confidence

	Method 1	Method 2	Method 3
1st	1	1	12
2nd	11	2	2
3rd	2	11	0

4.3   Observations

As shown in Table 3, we carried out an analysis of variance of a two-way layout with repetition to determine whether a difference in confidence rates existed among the three methods using the R. We confirmed a significant difference among the three methods with a 0.1% level of significance. We also confirmed that significant differences were present between male and female participants with a 5% level of significance from Table 3. Based on the average of the confidence rates of males and females in Table 1, females are more confident in their judgment.

The total average of confidence rate was 3.77 and the standard deviation was 1.03. To verify the effect of the operating order of the three methods, performed a two-way layout analysis of variance on the method and the order. As a result, the p-value was 0.69 and the level of significance was 5%. Therefore, we consider that the order of the methods had no effect on the results.

Table 3. Analysis of variance (ANOVA) table

	DOF	Variation	Unbiased variance	F-value	p-value
Method difference	2	18.37	9.1869	8.9502	0.0001426
Gender difference	1	4.54	4.5433	4.4262	0.0356892
Residual error	836	858.11	1.0264		

Method 1 was the second best in giving the users confidence regarding their clothing choice before going out. However, it received the lowest average of confidence among all three methods. In this method, subjects can actually pick up the clothes to see the colors and texture and to feel the fabrics. On the other hand, they are too close to their image in the mirror to see the combination objectively. This method took longer than 100 seconds per ensemble on average, and this method requires large body movements compared to other methods.

Method 2 received the least confidence in judgment. With this method, several ensembles can be compared side-by-side on the PC, and this enables the user to see them objectively. From the interview, we concluded that pictures showing only the clothes made it difficult to imagine wearing selections. Further, one of the interviewees said that all combinations seemed to fit together because of the flatness of the clothes. These comments indicate that the simulation using clothes-only pictures enables the user to coordinate colors, but it makes it difficult to imagine wearing the ensembles.

Method 3 received the highest result in confidence. The biggest reason could be that it produces a stereoscopic effect when the clothes are on the user's own body. In addition, the ensemble can be seen objectively as in method 2. Users discover that some combinations surprisingly fit and they choose to wear the new ensembles. Most people determined that this technique gave them the highest degree of confidence about going out wearing the selected ensemble. However, tops with lower hems could cause part of the top to remain with the bottom part of the photo, thereby causing confusion for the user. In fact, if the photo showed the hem of another top in the bottom portion, the evaluation was lower than a photo where the photo does not show extraneous fabric.

5 Trial Use

suGATALOG is designed for long-term use at home. Here, we initiated a long-term trial use of the system. Two males and two females participated in this trial for three weeks. Each participant registered for the web application and set up a web camera on their room mirror to take pictures while standing at the same location. They used the web application from their own PC's web browser. We interviewed each participant after the trial, and we report the comments from the interview below. The Fashion Show function was not yet available at the time of this trial.

Regarding the Shooting function, three participants said that taking pictures was troublesome. Our system requires users to use a web browser on a PC, and this may have been inconvenient for participants to turn on the PC and run the application. The only participant who did not think it was inconvenient ran his computer all the time. Having a PC that always runs is important for the usability of this method. Regarding the Calendar function, all participants said that they started to think about fashion more often. This may be because the Calendar function allows them to view what they wore every day. One male said that he realized he was wearing similar colors even when he was wearing different clothes; therefore, he started to consciously select different colors. In fact, after wearing different gray shirts for three days, he wore a yellow shirt, and he wore similar colors less frequently after those three days.

Regarding the Fitting Room function, one female participant said that she used to think her choices were not good enough after she had left the house, but she became more confident about her choices with this system, probably because she could view her ensemble choices objectively. Being able to mix and match with one click, she could coordinate until she was satisfied without changing clothes in real life. On the other hand, one male participant said that he could not choose one combination from various options by himself. Regarding the Favorite function, we found out that only one participant used it for more than three times, based on trial data. This function is used to compare several ensembles, but participants were using the system mainly in the morning when they are in a hurry. Regarding the Auto-coordinating function, one female participant found this function useful for comparing several ensembles when she has a specific piece of clothing that she wants to wear. After using this function, another female participant started to wear clothes that she rarely wore. In fact, she wore those clothes twice more during the trial. One male participant actually wore some combinations he never thought of before. These comments indicate that our system triggers participants to wear combinations they have never tried.

6   Discussion

6.1   The Process of Recording Pictures

We intended our system to record candid pictures in usual poses, such as when checking one's self in a mirror so that the recording process would not be an additional task to do every day. However, the results from the trial suggested that it was a bothersome task for most of the participants. Our future work is to make this recording process simpler and easier; for example, taking pictures automatically when the wardrobe door is closed [7], or attaching sensors and a camera on the front door and taking pictures before the user walks out. Attaching a camera on the front door enables the system to coordinate the clothes with shoes as well.

6.2   Application Ideas

All the pictures were uploaded on the web; therefore, they can be accessed from anywhere. This feature can be useful when shopping at apparel stores to coordinate previously purchased clothes with clothes at the store. This can also be used by shop assistants to recommend clothes that would work with customers' existing clothes. Picture images also include hairstyle, make-up, and body shape, which can be used at hair salons to cut the same hairstyle or apply the same makeup style. It can also be applied to support a weight-loss diet using the differences in body shape during a span of time.

7   Conclusion

We proposed and implemented suGATALOG, a fashion coordination system that allows users to choose and coordinate their clothes. Our system uses pictures of clothes that the user is actually wearing, and this enables users to try various combinations easily without actually changing. We conducted an evaluation experiment to verify the usefulness of the system by comparing three methods of coordinating: using actual clothes, using clothes-only pictures on the PC, and our proposed method of using pictures of clothes that the user is actually wearing on the PC. The result revealed that our proposed method for coordinating clothes was superior to other methods in achieving realism.

References

1. Cheng, C.-M., Chung, M.-F., Yu, M.-Y., Ouhyoung, M., Chu, H.-H., Chuang, Y.-Y.: Chromirror: a real-time interactive mirror for chromatic and color-harmonic dressing. In: CHI 2008 Extended Abstracts on Human Factors in Computing Systems, pp. 2787–2792 (April 2008)
2. Cordier, F., Magnenat-Thalmann, N.: A data-driven approach for real-time clothes simulation. In: Proceedings of 12th Pacific Conference on the Computer Graphics and Applications, pp. 257–266 (May 2004)
3. Hauswiesner, S., Straka, M., Reitmayr, G.: Free viewpoint virtual try-on with commodity depth cameras. In: Proceedings of the 10th International Conference on Virtual Reality Continuum and Its Applications in Industry, pp. 23–30 (December 2011)
4. Hoshino, J., Saito, F.: Building virtual fashion simulator by merging cg and humans in video sequences. Transactions of Information Processing Society of Japan, 1182–1193 (2000)
5. SONY.   MyStylist,   http://www.jp.playstation.com/scej/title/my/stylist (access date February 2013)
6. Tsujita, H., Tsukada, K., Kambara, K., Siio, I.: Complete fashion coordinator: a support system for capturing and selecting daily clothes with social networks. In: Proceedings of the International Conference on Advanced Visual Interfaces, pp. 127–132 (May 2010)
7. Tsujita, H., Tsukada, K., Siio, I.: A wardrobe to support creating a picture database of clothes. In: Adjunct Proceedings of the 6th International Conference on Pervasive Computing, pp. 49–52 (May 2008)
8. Zhang, W., Matsumoto, T., Liu, J., Chu, M., Begole, B.: An intelligent fitting room using multi-camera perception. In: Proceedings of the 13th International Conference on Intelligent User Interfaces, pp. 60–69 (January 2008)

Interacting with a Context-Aware Personal Information Sharing System

Simon Scerri[1], Andreas Schuller[2], Ismael Rivera[1], Judie Attard[1], Jeremy Debattista[1],
Massimo Valla[3], Fabian Hermann[2], and Siegfried Handschuh[1]

[1] Digital Enterprise Research Institute, National University of Ireland, Galway, Ireland
[2] Fraunhofer-Institute for Industrial Engineering, Stuttgart, Germany
[3] Telecom Italia Labs, Torino, Italia
firstname.surname@deri.org,
firstname.lastname@iao.fraunhofer.de,
massimo.valla@telecomitalia.it

Abstract. The di.me userware is a decentralised personal information sharing system with a difference: extracted information and observed personal activities are exploited to automatically recognise personal situations, provide privacy-related warnings, and recommend and/or automate user actions. To enable reasoning, personal information from multiple devices and online sources is integrated and transformed to a machine-interpretable format. Aside from distributed personal information monitoring, an intuitive user interface also enables the i) manual customisation of advanced context-driven services and ii) their semi-automatic adaptation across interactive notifications. In this paper we outline how average users interact with the current user interface, and our plans to improve it.

1 Introduction

The development of a context-aware, proactive information management system relies on the following three core system requirements:

i) the extraction, resolution and integration of various types of distributed data in the personal information sphere [10][1];
ii) the realtime processing of personal data and event streams in order to detect situations benefitting from warnings/suggestions/automation;
iii) the design of an intelligent user interface that is able to non-intrusively but effectively support people with their day-to-day tasks.

Each of the above corresponds to realised components of the di.me userware's architecture, shown in Fig. 1. The first is embodied within the Personal Information Model (PIM): an ontology-based representation of the unified personal information sphere. A wide-variety of information is extracted from personal devices (including computers, gadgets and smartphones) and online sources (including social networks) [9]. In particular, the Personal Information Crawler targets retreivable resources, profiles and social

---

[1] This term refers to all of a user's digital information, including heterogenous personal identities and resources, stored on various devices as well as online accounts and services.

M. Kurosu (Ed.): Human-Computer Interaction, Part V, HCII 2013, LNCS 8008, pp. 122–131, 2013.

interactions belonging to both a person and their contacts, whereas the Context Extractor targets realtime personal and contact activities by way of various device (e.g. GPS positioning, environment, device usage) and virtual sensors (e.g. location, people tagged nearby). The extracted information is semantically lifted on top of a set of integrated ontologies[2] covering various domains. Following their extraction, entities are processed in order to perform cross-source resolution and integration, thus enabling di.me to be used as an entry-point for distributed personal information management.

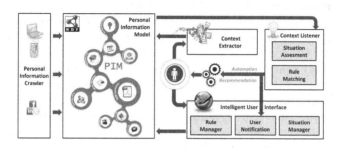

Fig. 1. The main concepts of our approach

Apart from providing improved personal information management, gathered PIM knowledge (including not only static information but also context data streams) is exploited by the *Context Listener* (Fig. 1) to pro-actively assist users with their knowledge-intensive tasks. The context listener provides two services: situation assesment and rule matching. The objective of the former is to automatically recognise recurring situations that have been marked by the user, whereas the objective of rule matching is to trigger context-driven rules, as defined by the user. These two services are enabled by a number of underlying techniques that can reason over the PIM knowledge.

The two above services depend on the user's interaction with di.me. To enable situation assesment, a person is required to initially mark situations of interest (which can subsequently be manually adjusted at all times). In contrast, the customisation of context-driven rules is targeted for manual setup by the user. Thus, an intelligent *User Interface* (UI) is a necessary di.me architecture component (Fig. 1), and the very success and usability of the userware relies on its ability to hide the complexity of the semantic processing from the users, while allowing them to take full advantage of its potential [5]. This places the di.me UI squarely within the category of intelligent UIs, i.e., "human-machine interfaces that aim to improve the efficiency, effectiveness, and naturalness of human-machine interaction by representing, reasoning, and acting on models of the user, domain, task, discourse, and media" [7], where, in di.me, the said models are provided by the ontologies. As a complex system built upon comprehensive knowledge models, the di.me userware needs an appropriate way of interacting with the user. Therefore, as a second focus of this paper we will describe how the chosen UI design can, and to what extent, simplify the complexity of the existing models.

---

[2] The ontologies will be published under OSCAF: http://oscaf.sourceforge.net/

A survey of intelligent UIs categorises the observed functionality under five broad aims [4]. di.me's objectives fall under two of these categories. The first and primary purpose of the di.me UI is to "take over tasks from the user", reducing their workload by attempting to recognise their intent and propose or automise actions, thus allowing them to focus on other things. Specifically, the intelligent UI interacts with users to learn when automation/notification is most desired. In addition, the UI is also responsible for providing suggestions/warnings, or automating user tasks straightaway, as instructed by the user and recognised by the context-recognition system. The ultimate objective here is then to proactively suggest or even perform tasks for them according to their real life situation. A secondary aim of the di.me UI is to "support new and complex functionality", in order to ease the adoption and uptake of the di.me userware. As explained above, the di.me UI serves as an interface between the user's (mental model) and the system's (ontology representation) view of their personal information model. In addition, the implemented UI strives to make the user fully aware of the di.me userware's abilities by means of various interactive features, such as, context-dependent explanations accompanying intelligent suggestions, history-based functionality previews, etc.

In the remainder of this paper we describe the situation assesement and rule matching techniques, focusing on the user-device interactions across the intelligent UI. Before concluding, we provide an outline of future tasks and a comparison to related work.

2   Providing Context-Aware Intelligence

The Context Listener looks out for context changes registered in the PIM, resulting from either user (e.g. movement, device usage) or system (e.g. new file created, new contact added) activities, and corresponding to a respective PIM update (e.g. new location registered, new resource created). This information is then combined with other personal knowledge available, in order to detect i) recurring situations stored by the user, and ii) fire context-driven rules defined by the user. In this section, we give an overview of the techniques for situation assesement and context-driven rule matching, and demonstrate how the user's interaction with their UI can provide the desired intelligence.

2.1   Adaptive Situation Assesement

Context is any kind of information that can be used to characterise the situation of a person, as an entity [3]. Thus, to recognise recurring situations, di.me continuously monitors and interprets personal activities by exploiting the following three sources: Personal Devices, Services and Social Networks. Modern user devices (computer, smart phones, gadgets, tablets, etc.) provide a vast amount of personal activity context information. User attention monitoring constantly tracks what the user is doing on the device (e.g., foreground applications, web browsing activities, IM presence, etc.), whereas device-embedded sensor data is also monitored in the background to gather additional context (e.g geographical positioning, environmental conditions, network connectivity, etc.). Personal services registered on devices are monitored for additional context (e.g. weather, calendaring, emailing services). Finally, social networks are also targeted for context information, with both semi-structured (e.g. check-ins, tagging nearby people) and unstructured (e.g. live post text containing references to nearby people, locations and current activities) being considered for a more complete snapshot of user activity context. This snapshot is constantly updated by the Context Extractor (Fig. 1) and stored within the user's PIM, as an instance of the purposely-engineered DCON Context Ontology[3], which we refer to as the *Live Context* [8].

Whenever they desire, users of the di.me system can mark personal situations (e.g. "Business Meeting", "Driving to Airport") through a specific button in the main UI dashboard. This will popup the new situation window, as shown in the (di.me-mobile UI, Fig. 2-a). Although a user can simply save the situation without looking at further options, the UI has been designed in a way that encourages them to provide additional information about the situations that they have just saved. This information is subjective and highly-dependant on the user's personal understanding of each individual situation. As shown in the example, people can remove irrelevant context items (e.g. the fact that the weather is Sunny), and assign different weights to relevant items (e.g. Jane is more linked to this situation than John is). The extremities of the 'weight bars' double as exclusion and necessity flags (e.g. if the user is in location DERI, then this situation can never recur). All these concepts are supported by different elements in the DCON ontology, which is also used to store situation instances. The shown UI design attempts to simplify the complexity of the DCON semantics so that non-technical people can fully-grasp them, and interact with the system to provide better situation representations.

To recognise a situation's recurrence, the Context Listener continuously compares its representation to the constantly-changing live context. Since both consist of structured RDF[4] graphs, a graph matching technique performs value matching at both DCON context element and attribute level. Therefore, even if two context elements (instances) co-occur in both graphs (e.g. Jane Doe is nearby in both the situation and the live context), context attributes (e.g. Jane Doe's actual detected distance) are compared to adjust the level of similarity accordingly. Similarly, if two context elements do not co-occur but have the same type (e.g. the situation graph refers to Jane Doe, whereas the live context graphs refers to John Doe), a partial match can still be detected based on the context attributes (e.g. they are both very near, or belong to a same group). A similarity function then determines which situations could be recurring, by taking into account all element-matching scores between each candidate situation and the live context, in combination with any user-defined element weights.

Users are notified of a situation that is matched with a high-enough confidence level (Fig. 2-b). For convenience, the situation matching bar always shows the three highest-matched situations at all times, even when initiated by the user. Through the bar, users can interact with the situation assesment results by dismissing or confirming the match. Live Context snapshots in the former case are stored as negative instances of the situation, whereas confirmed matches are stored as positive instances. Eventually, a situation (a DCON instance) will refer to multiple context logs. This way, the user's interaction with the UI results in the automatic increase/decrease of specific context element weights, resulting in a semi-automatic adaptive situation assesment service.

---

[3] http://www.semanticdesktop.org/ontologies/dcon/
[4] http://www.w3.org/RDF/

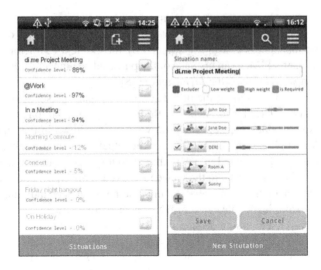

Fig. 2. Situation Notification (a) and Management (b)

2.2   Context-Driven Rule Matching

Context-driven rules (e.g., 'If Email from X received', *then:* 'Forward to Y', 'If in Situation "Business Meeting"', *then:* 'Change PhoneMode to Silent', 'If sharing items with untrusted contacts', *then:* 'Notify') can be created and managed through the *Rule Manager*. Inspired by approaches such as [6], user-defined rules in di.me are modelled on the Event-Condition-Action rule pattern concept: $if E[c_1, .., c_m] \implies [a_1, .., a_n]$ where the *event E* represents a rule that consists of a combination of conditions, trigerring one or more resulting *actions a*.

Rule conditions correspond to the wide-variety of concepts and information items that a typical person deals with through their devices (e.g., files, emails, contacts, locations, networks). Thus, rules are constructed out of items that can be stored in the PIM, based on concepts in the underlying ontologies and their attributes (e.g., an e-mail with a specific sender, a person with a low trust value, a meeting with certain people, etc.). To help the user with constructing personalised rules, the Rule Manager displays a selection of both i) ontology concepts (e.g., Person) and ii) known instances of these concepts (e.g., 'John Doe'). Whereas the latter identifies a specific instance that is known in the PIM, the former is equivalent to stating that any instance of the chosen concept will match the condition (e.g., any Person). The Rule Manager also allows users to specify *constraints* on top of conditions, based on the applicable ontology attributes. For example, a Person concept or a specific Person instance can both be constrained as follows: [Person].isAdded/.isRemoved (from the PIM), .isAddedToGroup, .isMentionedInMicropost, .approachesArea/.leavesArea (as detected by context sensors), .hasTrustValueChanged, .hasDataAccessGranted/.hasDataAccessWithdrawn (to any PIM resource).

Fig. 3-a shows three rules that have been created through the di.me-web UI Rule Manager. The latter enables users to create, modify and discard rules, as well as enable or disable them (tickbox on the left). Rules are constructed by selecting and constraining

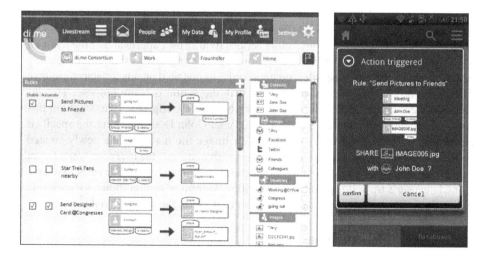

Fig. 3. Creating Context-driven Rules in the Rule Manager (a) and Rule Match Notifications (b)

PIM concepts and/or instances, as shown on the right hand side (e.g. contacts, groups, contacts, stored situations, images) . The first rule consists of three adjoined conditions: a specific user-created Situation, a Contact and an Image. The latter two conditions have been constrained to specify that the Contact needs to be nearby (isNearby) and a member of a specific group ('Friends'), and the Image needs to just have been created (isCreated). Therefore, the full rule can be described as: "When in Situation 'going out' *and* in the vicinity of any Contact(s) belonging to a Group 'Friends' *and* a new Image is created *then:* Share the Image with that Contact(s)".

Rule *events* trigger different types of actions such as *Notify, Post, Share, Change, Tag* and *Send.* A number of items are applicable to each of these actions (e.g. you can post a status message, link or photo; send emails and instant messages; change a person's trust level, a file's privacy level or your online presence message; share files and profiles etc.). Since the current di.me prototype only implements the sharing action, Fig. 3-a only demonstrates rules with this type of action.

Saved rules are translated to instances of the DRMO Rule Management Ontology[5]. The Context Listener monitors context-changing events in order to compare them against the defined rules. Context-changing events include: the creation of new PIM items (e.g. new e-mail received, new contact saved), the recurrence of known user activity context elements (e.g. user goes near a known location, connects to known wifi), and the re-activation of entire situations (as highlighted in Section 2.1). In order to match events to rule conditions, DRMO rules are mapped into SPARQL queries[6]. Stream event data is then matched against SPARQL triple patterns, pushing the rule onto a queue if a rule condition is matched. The context listener monitors future events, such that if all conditions are satisfied, it triggers the corresponding action(s) [2].

---

[5] http://www.semanticdesktop.org/ontologies/drmo/
[6] http://www.w3.org/TR/rdf-sparql-query/

The behaviour of the UI once a rule has been matched depends on the triggered action(s). Whereas notifications will simply result in dashboard system messages (including warnings and suggestions), other actions require user approval. For the featured sharing action, an interactive dialog box shows up, offering users a shortcut for sending matched items (images, other files, contact cards) to matched agents (contacts/groups). The dialog box includes an explanation of why the action is being propose. This includes presenting the user with matched variables for the trigerred rule, which in this case correspond to a matched Situation, a Contact (John Doe) matching the specified constraints (isNearby, Group:Friends), and an Image file matching the newly-created constraint (isNew). For this type of privacy-sensitive actions, the userware waits for an explicit go-ahead before performing the intended action (Fig. 3-b). For rules that have gained the user's trust, it is possible to override ('Automate' option shown for each rule in Fig. 3) the action confirmation box in order to perform the action(s) rightaway.

3   Future Enhancements

Through additional user-studies we intend to identify the best UI scheme to increase user interaction with the intelligent functionality provided by di.me. In particular, we will investigate non-intrusive techniques for i) avoiding 'cold start' userware problems by automatically detecting and suggesting candidate new situations and rules, rather than fully rely on the user for their initial setup; ii) providing improved justifications for system suggestions, warnings, etc.; and ii) enabling richer user feedback for the intelligent features, in order to speed-up the userware's learning process and user adaptivity.

The Situation Assesment technique will be enhanced to present candidate new situations to the user. For the purpose, the context matcher will closley monitor the changing context and look out for significant/sudden changes. The existing UI will therefore be extended to include new situation suggestions in the di.me dashboard, potentially sparing users the cognitive effort required to remember to mark new situations. At the moment, it is only possible for situations to be saved while they are still active. Since this is not always realistic, a context history visualisation will enable users to save important situation at a later time, by scrolling through a timeline and visually identifying it. Technically this is already possible due to automatic system logging, which also takes regular snapshots of a user's context for persistence.

Situation match notifications will be improved in order to provide justification as to why di.me detects that the situation has been reactivated. This could take the form of notification similar to the rule matching dialog box (Fig. 3-b), where instead of matched system/user events, the UI will display a list of context elements (e.g. people, wi-fi connections, time periods, running applications, locations, etc.) that are common to both the live context and the matched situation, together with their assigned weights. A further extension of this UI will also give the user an option to directly influence the situation assesment technique. At the moment, following a high match score, the user is able to either confirm (tick) or dismiss (untick) the detected situation (Fig. 2-a). As the proposed situation match notification extension will contain a list of detected context elements, the user could interact by removing irrelevant ones, or adjust their weight. Thus, the automatic weighting algorithm will be supplemented by direct user input, speeding up the learning process for the representation of the abstract situations.

The Rule Manager will be extended to support more complex user-defined rules, as supported by the DRMO ontology. Although the examples shown in Fig. 3 only employ the use of the *and* operator between both conditions and constraints, DRMO rules enable the use of additional logical operators. These include simple operators such as *or*, *and* and *negation*, as well as time-oriented operators like *succeededBy* and *preceededBy*. Supporting these operators in the Rule Manager will enable rules of type: "When in Situation 'Working at the office' *succeededBy* Situation 'Driving home from work' *then:* Change PhoneMode to Normal". Similarly, a more advanced UI will also support literal value constraints in addition to object attribute constraints. Here, users will be able to specify constraints on string or numeric values, through the use of simple operators such as $==, >=, <=, <, >$, as well as more complex ones like *contains* and *similar*. This will also enable rules of type: "New e-mail sent from 'John Doe' *and not* (subject *contains* 'di.me') *then:* Forward to 'Jane Doe' ".

Fig. 4. History-based Rule Previews

A special feature being investigated regards the exploitation of the existent automatic system logging, in order to enable history-based rule previews. Here, users will be able to how a rule that they have just defined will function, based on their recent history. Thus, they will get a feel of how the rule will function in the future, by being shown all recent occasions which would have triggered the same rule. Fig. 4 shows instances of when the first rule shown in Fig. 3 would have been triggered in recent days, in this case showing which image would have been shared with whom, and why. This feature will easen the user's comprehension of created context-driven rules.

4 Related Work

Nokia Situations[7] is a context-aware smartphone application that adjusts one's phone settings based on the detected situation. Similar to di.me, Nokia Situations consist of user-configurable profiles (e.g. lunch, sleeping, watching television, etc.) that include a number of context parameters (e.g. location, time of day, time of week, wifi connectivity, etc.). Although the UI is able to let users interact with the concept of personal situations, the variety of context parameters supported by di.me through the DCON ontology remains unrivalled. Unlike di.me, the application does not adapt to the user by auto-adjusting situation representation during use. Nokia Situations also allow users to set different system actions (restricted to mobile phone configuration) based on a detected situation. In contrast, di.me can also fire rules based on other perceived system events (e.g. receiving an e-mail, adding a contact to a group).

Few of the comparable context-driven rule matching efforts enable users to personalise and manage their own rules. Beltran et. al. developed SECE [1] to enable user-defined rules for recommendations, based on open linked data services. Although rules are defined in an English-like 'rule language', the developers acknowledge the need for a UI that enables users having no technical skills to also create rules. In addition, di.me's adoption of ontologies as a standard knowledge representation format, as opposed to a system-dependant rule language, means that di.me rules can be processed, or re-used in external RDF-compliant systems.

Other efforts that are directly comparable to our personalisable rule manager UI are the on{x}[8] and IFTTT[9] services. on{x} also defines rules in terms of triggers and actions. Triggers are limited to the phone's sensors and abilities (e.g. GPS location, weather, wifi connections, battery). Like in di.me, a user can also select from various actions (e.g. notify, send text messages, invoke a Web service). Rules are created through the Website and pushed on to the user's phone, whereby they can only be disabled or deleted. Feature-wise, IFTTT is very similar to on{x}, providing a list of 'Channels' (e.g. Facebook app, Email service), triggers (e.g. new photo on Facebook, new email) and actions (e.g. post status message on Facebook, send an email) for setting rules through its interface. Both on{x} and IFTTT rules allow for the creation or the download of existing rules. The latter are termed 'recipes', and are a core feature of both services. Likewise, di.me offers a limited number of pre-defined 'generic' rules, for initial consumption by all users. The major contrast between on{x} and IFTTT is that whereas on{x} rule customisation requires Javascript proficiency, users can interact with IFTTT rules through an intuitive UI. Similarly, the di.me rule manager UI enables average users to visually select from a wide variety of events and their conditions, with the UI to DRMO instance translation being performed by the intelligent userware. In comparison to both on{x} and IFTTT, the di.me UI allows for more triggers, including detected personal situations, the creation/modification/deletion of known types of information elements (including documents, emails, persons, calendar events, etc.), and context changes detected by additional devices and sensors (e.g. online check-ins).

---

[7] http://www.pastillilabs.com/Situations

[8] https://www.onx.ms

[9] https://ifttt.com/

5   Conclusion

In this paper we introduce the main techniques behind a context-aware, intelligent information system that is able to automatically recognise recurring situations, and provide context-aware warnings, suggestions and automate system actions. In order to enable these functions, personal information is aggregated from personal devices and online sources, and combined in a personal knowledge repository with activities that are observed by various device and virtual sensors. The resulting machine-interpretable personal information is then utilised by an intelligent UI to enable the average device user to configure and manage i) personal situation detection and ii) personalisable context-driven rules. We also describe how the UI-user interactions are exploited to adapt the di.me userware to the user, by gradually improve the underlying techniques based on user feedback. Through descriptions of additional experimental features that are planned for future releases, we give an idea of the di.me userware's potential as a novel, intelligent personal information management system.

Acknowledgments. This work is supported in part by the European Commission under the Seventh Framework Program FP7/2007-2013 (*digital.me* – ICT-257787) and in part by Science Foundation Ireland under Grant No. SFI/08/CE/I1380 (*Líon-2*).

References

1. Beltran, V., Arabshian, K., Schulzrinne, H.: Ontology-based user-defined rules and context-aware service composition system. In: García-Castro, R., Fensel, D., Antoniou, G. (eds.) ESWC 2011. LNCS, vol. 7117, pp. 139–155. Springer, Heidelberg (2012)
2. Debattista, J., Scerri, S., Rivera, I., Handschuh, S.: Ontology-based rules for recommender systems. In: Proceedings of the International Workshop on Semantic Technologies meet Recommender Systems & Big Data (SeRSy12) at ISWC 2012 (November 2012)
3. Dey, A.K.: Understanding and using context. Technical report, Future Computing Environments Group, Georgia Institute of Technology, Atlanta, GA, USA (2001)
4. Ehlert, P.: Intelligent user interfaces: Introduction and survey. Technical report, Data and Knowledge Systems Group, Faculty of Information Technology and Systems, Delft University of Technology, Delft, Netherlands (2003)
5. Hermann, F., Schuller, A., Scerri, S., Thiel, S.: The digital.me user interface: an interaction concept for the management of personal information and identities. In: Proceedings of the 15th International Conference on Human-Computer Interaction(HCI 2013) (2013)
6. May, W., Alferes, J.J., Amador, R.: An ontology- and resources-based approach to evolution and reactivity in the semantic web. In: Meersman, R., Tari, Z. (eds.) OTM 2005. LNCS, vol. 3761, pp. 1553–1570. Springer, Heidelberg (2005)
7. Maybury, M.T., Wahlster, W. (eds.): Readings in intelligent user interfaces. Morgan Kaufmann Publishers Inc., San Francisco (1998)
8. Scerri, S., Attard, J., Rivera, I., Valla, M., Handschuh, S.: Dcon: Interoperable context representation for pervasive environments. In: In Proceedings of the Activity Context Representation Workshop at AAAI 2012 (2012)
9. Scerri, S., Cortis, K., Rivera, I., Fr, C.: Deliverable d03.03 data mining and semantic matching engine. Technical report, di.me Consortium (2012)
10. Scerri, S., Gimenez, R., Herman, F., Bourimi, M., Thiel, S.: digital.me—towards an integrated Personal Information Sphere. In: Workshop on Social Networks, Interoperability and Privacy by W3C, W3C Workshop, Berlin (June 2011)

Design and Evaluation of Eco-feedback Interfaces to Support Location-Based Services for Individual Energy Awareness and Conservation

Yang Ting Shen[1], Po Chun Chen[2], and Tay Sheng Jeng[1]

[1] Department of Architecture, National Cheng Kung University, Tainan, Taiwan
[2] National Cheng Kung University, Tainan, Taiwan
{bowbowyangting,bootwooo}@gmail.com,
tsjeng@mail.ncku.edu.tw

Abstract. The Eco-feedback technology has widely applied to the energy conservation. Eco-feedback technology is usually represented as any kind of interactive device or interface targeted at revealing energy consumption in order to promote users' energy awareness and then trigger more ecologically responsible behaviors. In this paper, the primary goal is to help the individual user understand his comparative energy consumption through the Eco-feedback energy visualization. The energy information we provide is the comparison between the historical average energy consumption and the instant energy consumption. Based on the instant comparative energy consumption, the user can intuitively understand the current energy consumption is higher or lower than usualness. We develop the location-based individual energy consumption feedback system named EME (Energy MEter). Integrated with the concepts of historical comparison and incentives, three kinds of eco-feedback interface prototypes including the Dichotomy type, the Accumulation type, and the Numeral type are designed and deployed in practical fields. The user study both from quantitative and qualitative surveys is conducted in order to find out the potential interface which links user and energy consumption data better.

Keywords: Eco-feedback, Energy awareness, Energy conservation, Comparative energy consumption.

1 Introduction

Reducing energy consumption by increasing uses' awareness is recognized as one of the current major global issues. Recent interest in use of persuasive technologies has produced a range of interactive interventions designed to change behaviors [8]. The effective persuasive technology design usually integrates with HCI (Human Computer Interaction/Interface) to provide users well-illustrated information in order to convince them. One particularly popular form of HCI research is the design and study of eco-feedback, which is defined as technologies that provide feedback on individual or group behaviors with a goal of reducing energy consumption [11][14]. Eco-feedback technology is usually represented as any kind of interactive device or interface

M. Kurosu (Ed.): Human-Computer Interaction, Part V, HCII 2013, LNCS 8008, pp. 132–140, 2013.

targeted at revealing energy consumption in order to promote users' energy awareness and then trigger more ecologically responsible behaviors [17]. Eco-feedback is based on the working hypothesis that most people lack awareness and understanding about how their everyday behaviors affect energy conservation [9]. Once the energy information is revealed, the cognitive reflection derived from awareness will trigger further energy saving behaviors.

Comparison is one of the most common methods to achieve persuasive behavior change via Eco-feedback. There are two main approaches of Eco-feedback comparison design: normative comparison (social comparison) and historical comparison (self-comparison) [7]. Normative comparison means people share and evaluate their information through comparison with others [6]. Normative comparison facilitates competition tapping into user's intrinsic drive for cognition and extrinsic need for social status. As the development of social media such as Google+, Facebook, or Twitter etc., massive real-time normative comparison has become possible and presented significant effectiveness in community-based energy conservation. In contrast to normative comparison, historical comparison is defined as the act of an individual comparing himself at two or more different point in time [1]. In the field of HCI, historical comparison is most often depicted in charts over a certain period of time aimed at satisfying user's need for self-evaluation and learning [16]. The comparison of individual's achievements in the past with current performance is effective in motivating action, especially when assuming the previous consumption levels are lower than the present [2]. According to different conditions and scenario, both of those two comparison methods have their particular contributions to the energy conservation. For example, the narmative comparison may take advantage of persuasive effectiveness from intimate groups or community due to the peer pressure. However, in consideration of single user or cold start situation, historical comparison may provide a possible solution because it only needs to record and illustrate individual information. In addition, in regard to the individual energy usage information provided by eco-feedback system, the data format relative to users' cognition needs to be considered and represented carefully. The statistic researches indicate that energy illustration with components of *historical comparison* and *incentives* drive higher engagement and thus reduction in energy consumption [3][12]. Historical comparison is defined as the ability of users to view current and pass consumptions; moreover, the needs of incentives which motivate effective awareness and behaviors concretize the feedback about the system status derived from behavior change. In this context, the primary interest in this paper is in helping the individual user understand his *comparative energy consumption*. The novel energy information we provide is the comparison between the historical average energy consumption and the instant energy consumption. Based on the comparative energy consumption, the user can intuitively understand the current energy consumption is higher or lower than usualness. We develop the location-based individual energy consumption feedback system named EME (Energy MEter). Integrated with the concepts of historical comparison and incentives, three kinds of eco-feedback interface prototypes including the dichotomy type, the numeral type, and the accumulation type are designed and deployed in practical fields. The user study both from quantitative and qualitative surveys is conducted in order to find out the potential interface which links user and energy consumption data better.

2   System Architecture

The Eco-feedback system called EME is developed for the individual energy awareness and conservation.   The EME system are consisted of two parts including the EME interface and the Livindex Cloud service. The EME interface is designed for visualizing the energy information generated from the Livindex Cloud service. We will describe it in the next section integrated with experiment design. Here we discuss the system architecture of Livindex Cloud service which works as an energy monitor system for energy data harvesting and processing. Compared with general household energy projects which monitor collective energy consumption from all family members, the EME system focuses on the individual energy consumption because we want to motivate individual user for responsible behavior change. Therefore, the service provied by the Livindex Cloud is designed for the historical comparison (self comparison) of personal electricty consumption based on the user's location. The framework of Livindex Cloud service includes three procedures (Figure 1):

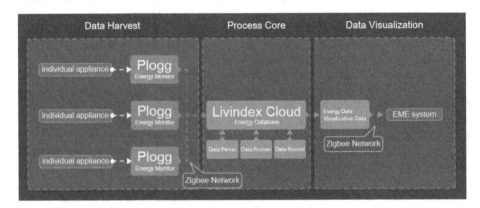

Fig. 1. The Livindex Cloud service framework

1. Data Harvest: several energy monitors called smart outlet are disposed in the individual working space to record personal energy consumption history. Smart outlets harvest individual energy data from distributed energy devices and deliver them to the server via RF signals(ZigBee) per minute.
2. Process Core: the server receives and procsses collected raw energy data into comparative energy consumption data. The process begin with data parser that written in Python script. CSV parser library and URL library work as the transmitter to parse and post energy data into our MSQL database. The raw data recorded in the database are further processed and compiled to the average energy consumption. The final result called instant *comparative energy consumption data* are generated from the difference between the average energy consumption and the instant energy consumption.
3. Data Visualization: the server transmits the instant comparative consumption data to the EME interface for further visualization. In this research, the instant comparative consumption data are compiled to three kinds of data format in order

to correspond to our three EME interface designs. The data transmition between the Livindex Cloud service and distributed EME interfaces are also based on the RF signals(ZigBee).

3   Interface Design

The EME interface aims to simplify complex energy data harvested by EME system and deliver the visualized comparative energy consumption information to the target user. The interface design is based on the Eco-feedback technology which reveals the user's energy information in order to promote users' energy awareness and then trigger more ecologically responsible behaviors [17]. Hence, the effectiveness of Eco-feedback awareness and comprehension is the primary design criterion to EME interface. We design three kinds of tangible interfaces including dichotomy type, numeral type, and accumulation type to support individual Eco-feedback information. Those three interfaces share the same tangible shape but present different information visualization types. The experimental studies are conducted to assess their performance of energy information representation.

3.1   Location-Based Energy Harvest

The EME interface serves an individual user as the ambient device in his working space. We set up several experimental environments distributed in campus to simulate the individual working spaces. For the collection of individual energy usage, especially electricity usage, we plug the power log which can harvest the power usage in the individual main power socket. All of the personal power usage collected from extension cords are transmitted via RF signal (ZigBee) and recorded in the cloud database call Livindex Cloud Service. The Livindex Cloud Service is running a MySQL which is updated by all individual power logs every 5 minutes. The collected data are allocated into individual fields according to log ID. After about two week collection, we figure out the individual average power consumption for each user. Then we compare the instant individual power consumption with the individual average power consumption (Figure 2 upper). Two kinds of comparative data are generated: one is below the average and how low against the average; another is beyond the average and how high against the average. Those two kinds of comparison data are visualized by tangible interfaces as eco-feedback information. Instead of the limitation of screen-based interface such as GUI, we try to distribute individual energy information into location-based services.

3.2   Three Interface Types

In order to support the user-centric and location-based services, we design the tangible interface which can be carried by individual user and plugged into the socket for power supply and system activation, and therefore the services can be triggered according to the user's location. Once plugged into the socket, the tangible interface starts to support instant comparison information from Livindex database by RF signals. Compared with the screen-based interface, the critical issue of our Eco-feedback interface is how to simplify complex energy data harvested by EME system and deliver the visualized comparative energy consumption to the target user. According some commercial products [4][5]   and academic researches[9] [17], the visualization types can be classified into three kinds of prototypes including dichotomy, accumulation, and numeral types. This conclusion inspires us to develop three kinds of EME interfaces based on Eco-feedback technology (Figure 2 lower) to see which one has the better performance for energy awareness and comprehension:

— *Dichotomy Type Interface*

The dichotomy interface is implemented with a Green and a Red LED to illustrate only the below status (green LED on) and the beyond status (red LED on).

— *Accumulation Type Interface*

The accumulation interface in implemented with two LED bars to illustrate the abstract below proportion (ten grids in blue) and the abstract beyond proportion (ten grids in red).

— *Numeral Type Interface*

The numeral interface is implemented with a seven-segment display to illustrate the below value (negative number) and beyond value (positive number).

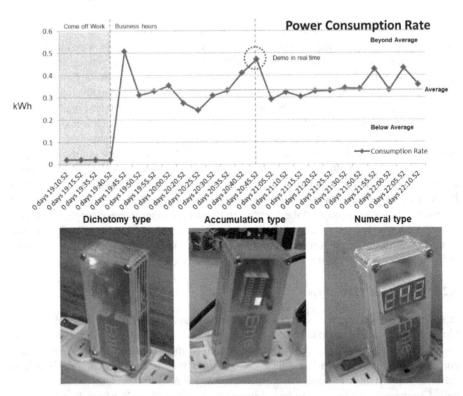

Fig. 2. Upper: the real-time energy consumption rate (per 5 min) and average energy consumption. Lower: Three kinds of tangible interface: Dichotomy, Accumulation, and Numeral types.

The instant comparative energy consumption information from the Livindex Cloud Service are compiled into three kinds of data formats and transmitted to the corresponding interfaces. The individual user can perceive personal energy information via the EME interface. In order to assess the performance of each EME interface, we set up the experimental environment to run the same energy consumption script but different interfaces in turn for an individual user. We will discuss the evaluation and result in the next section.

4   Evaluation and Result

The aim we develop three kinds of EME interfaces is to study the potential visualization design which links the user and his comparative energy consumption information better. Eight subjects are participated in our experiments and following evaluations. The subjects represented a mix of design practitioners and academic researchers. The semi-construction evaluation including quantitative and qualitative survey is conducted to assess three EME interfaces individually and the primary result will focus on the comparison between them.

In the beginning of pilot study, the NASA Task Load Index (TLX) [10] with six criteria (mental, physical, temporal, performance, effort, frustration) is applied to survey each subject's perception of task demands for the usability score. The results of three interfaces are illustrated in Figure 3 left.   Compared with the dichotomy type and the numeral type interface, the accumulation type interface performs equally in all criteria. In addition, the radar diagram shape is similar with the ideal radar diagram illustrated in Figure 3 right. This result may explain why most of subjects indicate their preference to the accumulation type interface. However, it doesn't mean the accumulation type interface is the best choice to fit all criteria. For instance, the dichotomy type interface performs well in all criteria except the frustration criterion. Therefore, the further heuristic evaluation is conducted to collect qualitative feedbacks from subjects.

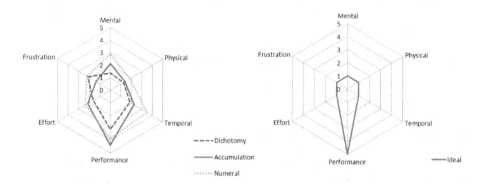

Fig. 3. Left: The stacked radar diagram presents three kinds of interface scores based on NASA 6 TLX criteria. Right: The ideal interface score based in NASA 6 TLX criteria.

Our goal in this heuristic evaluation is to obtain feedback on a set of heuristics that is based on Nielsen's [15] and related researches [13], but modified to be more applicable to ambient displays and Eco-feedback interfaces. We conclude ten heuristic criteria for our Eco-feedback interface evaluation including 1.sufficient information design, 2.visibility of system state, 3.aesthetic & pleasing design, 4.peripherality of display, 5.flexibility & efficiency of use, 6.consistent & intuitive mapping, 7.match between system & real world, 8.easy transition to more in-depth information, 9.recognition rather than recall, and 10.display better than its replacement. We use those eight criteria to design our questionnaires and assess them to the same eight subjects who participated in our pilot study before. Each participant is asked to provide a relevance rating on a scale of 1 to 5 (5 being highest) and comments about each of the heuristics. Participants were also encouraged to suggest additional heuristics at the end. We list the statistical and qualitative results of three interfaces individually below:

Dichotomy Type. (Score: 3.025)
Subjects can quickly understand the rough information by glancing the light signals. However, it is hard for subjects to acquire further detail information. For instance, subjects can't assess how much higher or lower energy consumption than the routine. Sometimes it may cause confusion about the degree of energy consumption associated with behaviors.

Accumulation Type. (Score: 4.30)
The abstract bar LED interface which transforms the degree of energy consumption into the percentage give the user two levels information for different situations. For the roughly understanding level, subjects can rapidly grasp the approximate consumption information by the volume of LED bar. In addition, for further detail demand, subjects can count the grids of LED bar to understand more accurate information.

Numeral Type. (Score: 3.85)
The interface presents the comparative energy consumption in numeral kWatt format directly. Most subjects indicate that the memory demand is quite high because they need to recall not recognize the visualized information. However, subjects also admit that the numeral information may support most accurate detail and become useful if subjects get familiar with the data format after long-time training.

5   Discussion

The results from our evaluation may conclude into two primary findings: 1.the need of multiple levels display for diverse usage scenarios and 2.the further incentives beyond comparative information for motivating the behavior change. Firstly, about the eco-feedback design, three tangible interfaces have different values depended on users' skill levels. For instance, amateurs may perform better in the dichotomy interface. However, experts may benefit by the numeral or accumulation type due to the detail awareness. Furthermore, the learning effect may also need to be considered. In addition to the users' skill levels, the need of information also shows high connection to the users' situations. Compared with Dichotomy (3.025) and Numeral (3.85) type interfaces, users grade the Accumulation type interface with high score (4.30). The primary reason is that the visualization with ten grids LED bar demonstrates the flexibility of perception levels. For the roughly understanding level, users can rapidly grasp the approximate consumption information by the volume of LED bar. In addition, for further detail demand, users can count the grids of LED bar to understand more accurate information. Therefore, the need of different levels visualization to deal with users' learning effect and complex use scenarios is necessary.

Furthermore, we find those three tangible interfaces can support incentives to attract users' attention but NOT necessarily trigger consequent behaviors for energy saving. Some users say they won't take any action even the interface indicates their comparative energy consumption is higher than the average. This disable situation may derive from three critical reasons:

1. The energy consumption is necessary for daily work. In some situations, the higher energy consumption is due to the necessary energy requirements such as heavy works, not the energy wasting behaviors. Users have difficulty to reduce the energy consumption if the energy saving behavior affects their works.
2. There is no supporting knowledge for users to develop the energy saving strategy. Most of users appreciate our Eco-feedback interface design to reveal their energy consumption information. However, they also indicate that the interfaces doesn't provide further strategies to suggest them the potential solutions for energy saving.
3. There is no enough incentive to motivate further behaviors. The evaluation demonstrates that comparative energy consumption information indeed plays a significant role for energy awareness and comprehension.

However, knowing the consumption situation doesn't have the positive causal relationship to trigger behaviors. For instance, some users say they won't change their behaviors if the energy consumption can't have the dramatic drop after their efforts. They expect to see the amplified effect or accumulated record to strengthen or go beyond the relationship between energy conservation and behaviors. Therefore, the incentive which represents the gains and losses derived from behaviors may bring more effectiveness. It suggests us some further development related to "Rewards and Penalizations" mechanism in the future. For example, we can design a kind of virtual credit to provide the metaphor or the link between the users' behaviors and the "Rewards and Penalizations" mechanism. The users' behaviors may not directly link to the energy visualization but affect the individual credits. Hence the accumulated influence from every behavior can be recorded and represented. Furthermore, the compensation mechanism such as riding a bike to generate energy can be involved to compensate for the necessary energy consumption in daily works. In conclusion, the credit concept may reveal the opportunity to design everyday things linking to the energy conservation.

References

1. Albert, S.: Temporal comparison theory. Psychological Review (1977)
2. Becker, L.: Joint effect of feedback and goal setting on performance: A field study of residential energy conservation. Journal of Applied Psychology (1978)

3. Darby, S.: The effectiveness of feedback on energy consumption: A review for DEFRA of the literature on metering, billing and direct displays. Environmental Change Institute, University of Oxford (2006)
4. DIY Kyoto, http://www.diykyoto.com/wattson/how-wattson-works (retrieved July 3, 2008)
5. Eco-Eye.com, http://www.eco-eye.com/ (retrieved July 3, 2008)
6. Festinger, L.: A Theory of Social Comparison Processes. Human Relations 7, 117–140 (1954)
7. Fischer, C.: Feedback on household electricity consumption: a tool for saving energy? Energy Efficiency 1, 79–104 (2008)
8. Fogg, B.: Persuasive technology: using computers to change what we think and do. Ubiquity, 89–120 (2002)
9. Froehlich, J., Findlater, L., Landay, J.: The design of eco-feedback technology. In: Proceedings of the SIGCHI Conference on Human Factors in Computing Systems (CHI 2010), pp. 1999–2008 (2010)
10. Hart, S., Staveland, L.: Development of NASA-TLX (Task Load Index): Results of empirical and theoretical research. Human Mental Workload (1988)
11. Holmes, T.: Eco-visualization: combining art and technology to reduce energy consumption. In: Proceedings of the 6th ACM SIGCHI Conference on Creativity & Cognition, pp. 153–162 (2007)
12. Jain, R.K., Taylor, J.E., Peschiera, G.: Assessing eco-feedback interface usage and design to drive energy efficiency in buildings. Energy and Buildings 48, 8–17 (2012)
13. Mankoff, J., Dey, A.K., Hsieh, G., Kientz, J., Lederer, S., Ames, M.: Heuristic evaluation of ambient displays. In: Proceedings of the Conference on Human Factors in Computing Systems, CHI 2003, p. 169 (2003)
14. McCalley, L., Midden, G.: Computer based systems in household appliances: the study of eco-feedback as a tool for increasing conservation behavior. In: Proceedings of the Third Asian Pacific Computer and Human Interaction (APCHI 1998), p. 344 (1998)
15. Nielsen, J., Molich, R.: Heuristic evaluation of user interfaces. In: Proceedings of the SIGCHI Conference on Human Factors in Computing Systems, pp. 249–256 (1990)
16. Petkov, P., Köbler, F., Foth, M., Krcmar, H.: Motivating domestic energy conservation through comparative, community-based feedback in mobile and social media. In: Proceedings of the 5th International Conference on Communities and Technologies, pp. 21–30 (2011)
17. Pierce, J., Odom, W., Blevis, E.: Energy aware dwelling: a critical survey of interaction design for eco-visualizations. In: Proceedings of the 20th Australasian Conference on Computer-Human Interaction: Designing for Habitus and Habitat, pp. 1–8 (2008)

Fuzzy Logic Approach for Adaptive Systems Design

Makram Soui[1,2], Mourad Abed[1], and Khaled Ghedira[2]

[1] Univ. Lille Nord de France, F-59000 Lille, France
UVHC, LAMIH, F-59313 Valenciennes, France
CNRS, UMR 8201, F-59313 Valenciennes, France
Souii_makram@yahoo.fr,
Mourad.Abed@univ-valenciennes.fr
[2] Institute of Management of Tunis,
41, Rue de la liberté, Cité Bouchoucha, Bardo 2000, Tunisia
khaled.ghedira@isg.rnu.tn

Abstract. Adaptive system is a field in rapid development. Adaptation is an effective solution for reducing complexity when searching information. This article presents how to personalize user interface (UI) using fuzzy logic. Our approach is based on the definition of relations for selection of appropriate and not appropriate of UI components. These relations are based the degree of certainty about the meaning coincidence of metadata elements and user' preferences. The proposed approach has been validated by applying it in e-learning field.

Keywords: Adaptation, Adaptive Systems, Fuzzy logic, Evaluation, User Interface (UI).

1 Introduction

In past few years, there has been a widespread emergence of adaptive systems that go beyond the capabilities of traditional interactive systems. In fact, adaptation deals with the ability of a system to collect user information, to analyse this information, and to adapt the system to users' preferences [1]. In general, we can say that adaptation deals with the capacity of personalization of a user interface (UI) considering some information related to the context (i.e., platform, user and environment characteristics). The adaptation can take into account several aspects (e.g., navigation, structure, functionalities) and it can be performed basically on the UI containers presentation; i.e., layout, colors, sizes, and content; i.e., data, information [2] [3] [4]. Such systems can be used in many different domains; for example: e-learning [5], logistics and transport [6] are domains in which adaptation principles offer new perspectives. In fact, several research studies have been carried out about user modelling, design methods and tools for UI generation. However, the evaluation of such systems is neglected in the HCI (Human-Computer Interaction) literatures. To fill up this lack, it is necessary to envisage new evaluation methods taking into account the evaluation function for assessing the quality of adaptive system. This paper is structured as follows. Section II presents an approach for recommending UI adaptation based on the

M. Kurosu (Ed.): Human-Computer Interaction, Part V, HCII 2013, LNCS 8008, pp. 141–150, 2013.
© Springer-Verlag Berlin Heidelberg 2013

users' preferences. III concludes the paper, with illustrations concerning learning resources and futures perspectives.

2 User-Adaptive Systems

Adaptive system is a field in rapid development. Today, these systems are indispensable to those who want to retrieve appropriate information with less effort at any time and any where. Adaptation is an effective solution for reducing complexity when searching information. In this way, the user feels like the system was developed for him/her. Two categories of adaptation are often distinguished in the literature: adaptability and adaptivity.For [7] "Adaptability is used to refer to self-adaptation that is based on knowledge (regarding the user, the interaction environment, the context of use, etc.) that is available to (or is collected by) the system prior to the commencement of interaction, and leads to adaptations which also precede the commencement of interaction".

But, "Adaptivity refers to self-adaptation that is based on knowledge which is collected and / or maintained by the system during interaction sessions (either directly from the user, or through monitoring / inferencing techniques) and which leads to adaptations that take place while the user is interacting with the system". [8] defines adaptation as the process of modifying systems to work adequately in a given context, which means the system suits perfectly user expectation in a given context. In general, we can say that adaptation deals with the capacity of adaptation of a system considering some information related to the context of use (P: platform, U: user and E: environment). In general, we can say that adaptation deals with the capacity of adaptation of a system considering some information related to the context of use (P: platform, U: user and E: environment).

3 Approach Overview

For us, UI is depicted as a set of components or interactive objects. So, the personnalization of system requires adaptation of UI components presentation to a set user' preferences. Our approach is based on the definition of relations for selection of appropriate and not appropriate of UI components. These relations are based on fuzzy logic that represent the degree of certainty about the meaning coincidence of metadata elements and user' preferences. To understand better our contribution, we will start by giving the definitions of important concepts used in our proposed approach such as fuzzy rule-based systems and metadata for UI description.

3.1 Fuzzy Rule-Based Systems

Fuzzy logic may be viewed as an extension to classical logic systems for dealing with the problem of knowledge representation in an environement of uncertainly and impression. Fuzzy logic as it name suggests, is a form of logic whose underlying modes of reasoning are approximate rather than exact. Its importance arises from the fact that most modes of human reasoning are approximate reasoning [9]. Knowledge representation is enhanced with the use of linguistic variables and their linguistic values that are defined by context-dependent fuzzy sets whose meanings are specified by gradual membership functions. In our work, Fuzzy inference systems are developed for adaptive system using Mamdani-type.

3.2 Metadata for UI Description

Our approach is based on metadata which is commonly used for the reuse of UI components. This metadata is a set of data elements useful for the description of UI components. A metadata element has a name which might not be unique. For that reason a data element has a unique identifier. For the description of a UI components, the data element has to be associated with a specific value which characterizes the UI component. Formally, a metadata (useful to annotate UI components) is described as 5-tuple M= (I, N, IN, V, D) where:

— I: is a set of data elements Identifiers.
— N: is a set of data element Names.
— IN (data elements): is a relation from I to N.
— V: is a set of data element Values.
— D (Descriptive data element): is a relation from IN to V. D defines the value associated to each metadata element for describing UI components.

A user interface is a set of services represented by UI components. We assume that a service could be represented by more than one UI components. We assume also that each UI components is annotated with metadata. Formally, an interface is described as a 4-tuple $C = (S, \Omega, R, \Psi)$ where:

— S: is a set of services proposed by UI components.
— Ω : is a finite set of UI components, representing the services of UI.
— R : is a relation from S to Ω. R determines the UI components representing a service.
— Ψ : is a relation from Ω to P (D), P denotes the set of partitions. Ψ defines the set of metadata describing each UI component.

3.3 Semantic Relations between Values of Data Elements and Users' Preferences

Semantic relations represents the coincidence degree between descriptive data elements and users' preferences. Formally, SRVDL is described as 5-tuple R= (L, A, P, M, C) where:

— L: is a set of level values in natural language describing possible user' preferences in relation to adaptation attributs. For example novice and expert are possible values for experience attribut.
— A (adaptation attributs): is a set of dimensions which can be used for UI adaptation. Each adaptation attribut is related to a set of level values that describe possible user' preferences with respect to the adaptation attributs.
— P (users' preferences): is a relation from L to A. P defines the linguistic terms that are related to a particular adaptation attributs.
— M (Metadata element associated with users' preferences) is a relation from P(D) to P, where P denotes the set of partition.

— C (Coincidence Degree of Metadata elements with users' preferences) is a relation
from M to [0..1]. C specifies the coincidence degree of metadata elements and users'
preferences.

3.4 The Proposed Approach

The proposed method consists in recommending UI adaptation based on the users'
preferences. By exploiting principles proposed by the inference engine of Mamdani
fuzzy rule-based system, the principle of this proposed method will be based on fuzzy
logic that represent the degree of certainty about the meaning coincidence of metadata
elements and user' preferences. Our method has three phases (Fig. 1.): (1) Fuzzification interface. (2) An inference system. (3) A defuzzification interface. The proposed
system use a rule base (RB) which stores the available knowledge about the problem
in the form of fuzzy. A general model of a fuzzy inference system is shown in Fig. 1.

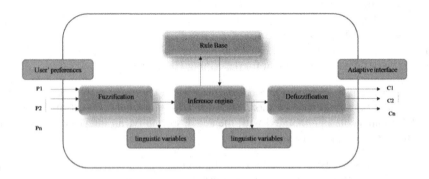

Fig. 1. A general model of a fuzzy inference system

- **Fuzzification Interface:** the fuzzifier maps input numbers (user' preferences) into corresponding fuzzy memberships. This is required in order to activate rules that
are in terms of linguistic variables. A combination of all the input values defines a
user profil. The fuzzifier takes input values and determines the degree to which they
belong to each of the fuzzy sets via membership functions. The linguistic variable
which represents the user' preference is described as a triplet $T = (p, T(p), Up)$ where:

— p: is the name of preference or adaptation attribut for example level of knowledge
— $T(p) = \{E_1, E_2, ..., E_n\}$: is a set of values in natural language describing possible
   user' preferences levels for example {low, medium, high}.
— U_p: universe of discourse of p.

This phase is composed of three steps: (1) Determination of discourse universe for
each user' preference, (2) Definition of fuzzy sets. (3) Definition of fuzzy membership functions.

- **An Inference System:** the inference engine defines mapping from input fuzzy
sets into output fuzzy sets according to the information stored in the RB. Once the
inputs are fuzzified, the corresponding input fuzzy sets are passed to the inference
engine that processes current inputs using the rules retrieved from the rule base.

Each rule is in IF-THEN form:
IF X1 is A1 and ....and Xn is An THEN Y is B

With Xi and Y being imput and output linguistic variables, respectively and with Ai and B being linguistic labels with fuzzy sets associated defining their meaning.

- A Defuzzification Interface: the defuzzifier maps output fuzzy sets obtained from the inference process into a crisp action that constitutes the global output (adaptive interface). There are a number of methods of doing this, and the most common one among them is the centroid or centre of gravity method. The centre of gravity is simply the weighted average of the output membership function. The result is calculated using the formula:

$$\overline{X}(centroid) = \frac{\int_{b}^{a} x\mu(x)dx}{\int_{b}^{a} \mu(x)dx}$$

Where [a, b] is the interval of the aggregated membership function.

-Adapted Degree of UI Components to Users' Preferences. The idea is to select UI components to user' preferences. So, we define a relation for determining the degree of UI component appropriateness to a user' preferences and also a relation for determining the degree of not appropriateness ones to a users' preferences basing on the metadata (m). These relations are defined as follows:

- AdaptedDegree (Adapted degree of UI components (co) to users' preferences (p))
  AdaptedDegree : $P \times \Omega \rightarrow [0..1]$
  AdaptedDegree (p, co) = Max {C (m, p) / m : $P(D)$ and (co, m) $\in \Psi$}.
- NotAdaptedDegree (NotAdapted degree of UI components (co) to users' preferences (pr))
  NotAdaptedDegree : $P \times \Omega \rightarrow [0..1]$
  NotAdaptedDegree (p, co)= 1- AdaptedDegree (p, co)

The relations AppDegree and NotAppDegree are used to determine Adapted and Not Adapted UI components.

4   Case Study: Contextual Adaptation of Learning Resources

The use of information technology and communication has greatly improved the way we read and learn. These advances are revolutionizing our way of learning by adapting access to content and services. A large amount of educational resources is produced continuously on the Web. Given the cost of these resources and the expertise to produce them, it is essential to make them easily accessible, adaptable and reusable. The learner hope to have at his/her disposal only some information, just what he/she is directly interested in. The system offers him/her adaptive interface according to his/her preferences. To achieve this aim, we follow the classical phases of fuzzy logic system. The fuzzification interface takes as inputs the user' preferences, and generates as output an adaptive course with a set of personnlized services.

As an illustration, we consider a fuzzy inference system with two inputs and one output. Let the two inputs represent the number of years of education and the number of years of experience, and let the output of the system be difficulty which describes a complexity level of course. Let p_i is the study level which indicates the number of years of education, $T_{study\ level}$ represent its term set {low, medium, high}, and the universe of discourse be [1-20]). Let p_2 indicate the number of years of experience, the universe of discourse be [0 -30], and the corresponding term set be {*low, medium, high*}. Similarly, c_1 is the difficulty of course which is an output variable characterizing the courses. In order to map input variables p_1 and p_2 to output c_1, it is necessary that we first define the corresponding fuzzy sets. The membership functions for the input and output variables are shown in Fig 2, 3 and 4.

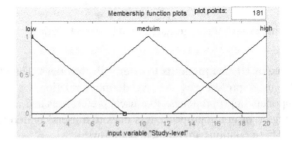

Fig. 2. Fuzzy membership functions for the study level

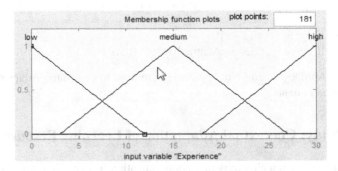

Fig. 3. Fuzzy membership functions for the expeience

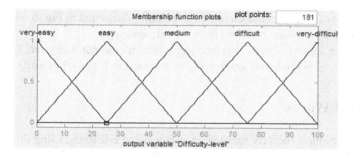

Fig. 4. Fuzzy membership functions for the experience

A fuzzy rule base contains a set of fuzzy rules R (r1, r2,...,rn). For the given example, the rules are stated as

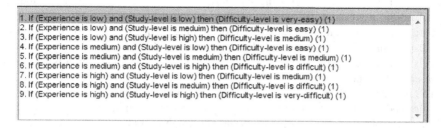

Fig. 5. Examples of rules

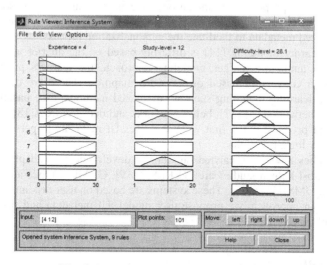

Fig. 6. Operation mode the inference system

The inference engine considered is the classical one employed by Mamdani which considers the minimum t-norm as conjunctive and implication operators and a mode A-FATI defuzzification interface where the aggregation operator G is modeled by the maximum t-conorm, whilst the defuzzification method is the centre of gravity.

The operation of the inference engine is graphically illustrated in Fig. 6 which depicts the membership functions resulting from the inference step.

Finally, the defuzzification interface aggregates the output fuzzy sets by means of the maximum. This process is graphically represented in Fig. 6 (right part).

5   Related Works

Several studies have recently focused on personalized e-learning using different techniques. [10] describes a data mining process based on Moodle Course Management System [11]. (e.g., the number of assignments done by a student, the number of quizzes failed, the number of quizzes passed, the total time spent on assignments, etc.). This approach is based on user activity data. However, this data is not analysed. A different viewpoint on educational data mining or data mining in e-learning is provided by [12], where the following categories are identified: prediction, including classification, regression and density estimation, clustering, relationship mining (including association rule mining, correlation mining, sequential pattern mining and causal data mining), distillation of data for human judgement, and discovery with models. In our approach, we propose an automatic analyse method which is based on the user background (history).

The majority of E-learning systems model the learner as an entity accompanied by a static predefined set of interests without modeling the learning resources. We can cite for example, Smart E-learning environment which is composed of two processes: teacher apprentice for authoring (TAA) and tutor apprentice for delivery (TAD) [13]. In our approach, we propose to define an interface as metadata for UI description that could be used during the adaptation process. In this way, the designer and the evaluator will use shared metadata to facilitate the personalization process.

Automatic learner modeling [14] is differing based on the queries attributes used previously. In automatic learner modeling approach, the learners profile is constructed using a conversion based on keyword mapping. There are different techniques used in learning modeling such as rule based methods, case based reasoning [15], Bayesian networks[16] [17], belief networks and decision trees[18]. To our best knowledge, our proposal is the first work that uses UI metadata to personalize interface using fuzzy logic principle.

Several studies have been carried out for the development of adaptive interface. Some of the most known studies are: TERESA [19], CAMELEON framework [20], SUPPLE [21], DMSL [22] [23]. These systems are based on user task and transformation rules. In our approach, we propose user model, UI metadata and fuzzy rules to personalize interface.

6   Conclusion

In this paper, we presented a rigor and generic approach based on fuzzy logic for the automatic prevision of UI adaptation. Mathematical notation is used for two reasons. First, mathematical notation enhances the rigor/precision of our approach. Second, the high abstract level of mathematical notation is used for defining the approach as a generic solution. The proposed approach exploits semantic relations between data elements and users' preferences to determine adapted UI components appropriate to users' characteristics. Future directions of this research will deal with extending the proposed approach and proposition of metrics for evaluation of adaptation strategies. Since we have tested our approach only in the e-learning context, it would be also interesting to generalize the approach with other fields of application (transport, logistics, etc).

References

1. Kobsa, A.: Personalized hypermedia presentation techniques for improving online customer relationships. The knowledge Engineering Review 16(2), 111–115 (2001)
2. Ledoux T.: Etat de l'art sur l'adaptabilité. Research Report RNTL ARCAD, number Livrable D1.1; Li, Y., Hong, J.I., Landay, J.A.: Design Challenges and Principles for Wizard of Oz Testing of Location-Enhanced Applications. IEEE PervasiveComputing 6, 70–75 (2001)
3. Zimmermann, A., Andreas, L.: LISTEN: a user-adaptive audioaugmented museum guide. In: User Modeling and User-Adapted Interaction, pp. 390–414 (2008)
4. Hervás, R., Bravo, J.: Towards the ubiquitous visualization: Adaptive user-interfaces based on the Semantic Web. Interacting with Computers 23, 40–56 (2011)
5. Essalmi, F., Jemni Ben Ayed, L., Jemni, M., Kinshuk, Graf, S: A fully personalization strategy of E-learning scenarios. International Journal of Computers in Human Behavior 26(4), 581–591(2010)
6. Huang, Y., Bian, L.: A Bayesian network and analytic hierarchy process based personalized recommendations for tourist attractions over the Internet. Expert Systems with Applications 36(1), 933–943 (2009)
7. Paramythis, A., Totter, A., & Stephanidis, C. A modular approach to the evaluation of adaptive user interfaces. Paper Presented at the Proceedings of Workshop at the Eighth International Conference on User Modeling, UM 2001, Freiburg, Germany (2001).
8. Ledoux T.: Etat de l'art sur l'adaptabilité. Research Report RNTL ARCAD, number Livrable D1.1; Li, Y., Hong, J.I., Landay, J.A.: Design Challenges and Principles for Wizard of Oz Testing of Location-Enhanced Applications. IEEE PervasiveComputing 6, 70–75 (2001)
9. Zadeh, L.A.: Fuzzy Sets. Information and Control 8, 338–352 (1965)
10. Amershi, S., Conati, C.: Combining Unsupervised and Supervised Classification to Build User Models for Exploratory Learning Environments. J. Educ. Data Min. 1(1), 18–71 (2009)
11. Moodle: Moodle.org: Open-source community-based tools for learning (2010), http://www.moodle.org
12. Koedinger, K., Baker, R., Cunningham, K., Skogsholm, A., Leber, B., Stamper, J.: A data repository for the EDM community: the PSLC datashop. In: Romero, C., Ventura, S., Pechenizkiy, M., Baker, R.S. (eds.) Handbook of Educational Data Mining. CRC Press, Boca Raton (2010)
13. Shehab, A., Gamalel, D.: Smart E-learning: A greater perspective; from the fourth to the fifth generation E-learning. Egyptian Informatics Journal Elsevier (accepted May 25, 2010)
14. Ebru, O., Gözde, B.: Automatic detection of learning styles for an E-learning system. Computers and Education, 355–367 (2009)

15. Ileh, M.T., Saleh, I.: Case Based Reasoning Approach of Tutoring in E-learning Platform. In: IEEE ICMCS 2009, Maroc, April 2-4 (2009)
16. Dan, Y., XinMeng, C.: Using Bayesian Networks to Implement Adaptivity in Mobile Learning. In: Proceedings of the Second International Conference on Semantics, Knowledge, and Grid, SKG 2006 (2006)
17. García, P., Amandi, A., Schiaffino, S., Campo, M.: Evaluating Bayesian networks' precision for detecting students' learning styles. Computers & Education 49, 794–808 (2007)
18. Ueno, M.: Animated Pedagogical Agent Based on Decision Tree for e-Learning. In: ICALT 2005, pp. 188–192 (2005)
19. Berti, S., Mori, G., Paternò, F., Santoro, C.: TERESA: A Transformation-Based Environment for Designing Multi-Device Interactive Applications. In: Proceedings of CHI 2004, Extended Abstracts on Human Factors in Computing Systems, pp. 793–794. ACM Press, Wien (2004)
20. Calvary, G., Coutaz, J., Thevenin, D., Limbourg, Q., Bouillon, L., Vanderdonckt, J.: A Unifying Reference Framework for Multi-Target User Interfaces. Interacting with Computers 15(3), 289–308 (2003)
21. Gajos, K., Weld, D., Wobbrock, J.: Automatically generating personalized user interfaces with Supple. Artificial Intelligence 174(12-13), 910–950 (2010)
22. Savidis, A., Antona, M., Stephanidis, C.: A Decision-Making Specification Language for Verifiable User-Interface Adaptation Logic. International Journal of Software Engineering and Knowledge Engineering 15(6), 1063–1094 (2005)
23. Antona, M., Savidis, A., Stephanidis, C.: A Process–Oriented Interactive Design Environment for Automatic User Interface Adaptation. International Journal of Human Computer Interaction 20(2), 79–116 (2006)

Part II

Computational Vision in HCI

Semi-supervised Remote Sensing Image Segmentation Using Dynamic Region Merging

Ning He[1,*], Ke Lu[2], Yixue Wang[3], and Yue Gao[4]

[1] Beijing Key Laboratory of Information Service Engineering,
Beijing Union University, Beijing 100101, China
[2] University of Chinese Academy of Sciences, Beijing 100049, China
[3] Shenyang Institute of Engineering, Shenyang 110136, China
[4] School of Computing, National University of Singapore, Singapore
xxthening@buu.edu.cn, luk@ucas.ac.cn, wyx123624@163.com

Abstract. This paper introduces a remote sensing image segmentation approach by using semi-supervised and dynamic region merging. In remote sensing images, the spatial relationship among pixels has been shown to be sparsely represented by a linear combination of a few training samples from a structured dictionary. The sparse vector is recovered by solving a sparsity-constrained optimization problem, and it can directly determine the class label of the test sample. Through a graph-based technique, unlabeled samples are actively selected based on the entropy of the corresponding class label. With an initially segmented image based semi-supervised, in which the many regions to be merged for a meaningful segmentation. By taking the region merging as a labeling problem, image segmentation is performed by iteratively merging the regions according to a statistical test. Experiments on two datasets are used to evaluate the performance of the proposed method. Comparisons with the state-of-the-art methods demonstrate that the proposed method can effectively investigate the spatial relationship among pixels and achieve better remote sensing image segmentation results.

Keywords: Semi-supervised, Remote Sensing Image, Image segmentation, Dynamic region merging.

1   Introduction

Existing works on remote sensing image segmentation mainly focus on either feature dimension reduction or semi-supervised classification. Traditional feature dimension reduction methods, such as Independent Component Analysis and Principal Component Analysis. The discriminative approach to classification circumvents the difficulties in learning the class distributions in high dimensional spaces by inferring the boundaries between classes in the feature space[1,2]. Support vector machines (SVMs) [3] and multinomial logistic regression [4], are among the state-of-the-art discriminative techniques to classification. Due to their ability to deal with large input

---

[*] Corresponding author.

M. Kurosu (Ed.): Human-Computer Interaction, Part V, HCII 2013, LNCS 8008, pp. 153–162, 2013.

spaces efficiently and to produce sparse solutions, SVMs have been successfully used for hyperspectral supervised classification [5-7]. The multinomial logistic regression has the advantage of learning the class distributions themselves. Effective sparse multinomial logistic regression methods are available [8]. These ideas have been applied to hyperspectral image classification [9]. In order to improve the classification accuracy, some methods have integrated spatial information[10,11,12].

In region-based methods, a lot of literature has investigated the use of primitive regions as preprocessing step for image segmentation [13]. The advantages are regions carry on more information in describing the nature of objects, and the number of primitive regions is much fewer than that of pixels in an image. Starting from a set of primitive regions, the segmentation is conducted by progressively merging the similar neighboring regions according to a certain predicate, such that a certain homogeneity criterion is satisfied. In previous works, there are region merging algorithms based on statistical properties [14], graph properties [15]. Most region merging algorithms do not have some desirable global properties, even though some recent works in region merging address the optimization of some global energy terms, such as the number of labels [16] and the area of regions.

In this paper, we introduce a new semi-supervised clustering algorithm which exploits the spatial contextual information. The algorithm implements two main steps: (a) the semi-supervised clustering algorithm [17] to infer the class distributions; and (b) segmentation, by inferring the labels from a posterior distribution built on the learned class distributions and on a multi-level logistic (MLL) prior. The class distributions are modeled with a multinomial logistic regression, where the regressors are learned using both labeled and, through a graph-based technique, unlabeled samples. The spatial contextual information is used both in building the graph accounting for the feature "closeness"and in the MLL prior. The region merging segmentation is computed via a min-cut based integer optimization algorithm. Fig.1 illustrates the flowchart of the proposed method.

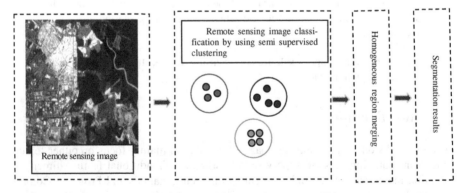

Fig. 1. The flowchart of the proposed remote sensing image classification method by using semi-supervised and region merging

The rest of the paper is organized as follows. Section 2 introduces the proposed remote sensing image classification method by using the spatial information. Experimental results and comparison with the state-of-the-art methods on two datasets are provided in Section 3. Finally, we conclude the paper in Section 4.

2 Remote Sensing Image Segmentation by Using Semi-supervised Classification and Region Merge

In this section, we introduce the proposed remote sensing image segmentation method by using semi-supervised clustering method as shown in Fig.1. First, we introduce the remote sensing image constraint process by semi-supervised clustering. Next, we describe the region merging process on the classified remote sensing image.

2.1 Semi-supervised Image Segmentation

Semi supervised clustering [18] means Grouping of objects such that the objects in a group will be similar to one another and different from the objects in other groups with related to certain constraints or prior information.

The Fig.2 represents the semi supervised clustering model. The three clusters are formed using certain constraints or prior information. Besides the similarity information which is used as color knowledge, the other kind of knowledge is also available by either pair wise (must-link or cannot-link) constraints between data items or class labels for some items. Instead of simply using this knowledge for the external validation of the results of clustering, one can imagine letting it "guide" or "adjust" the clustering process, i.e. provide a limited form of supervision. There are two ways to provide information for semi supervised clustering: search based or similar based.

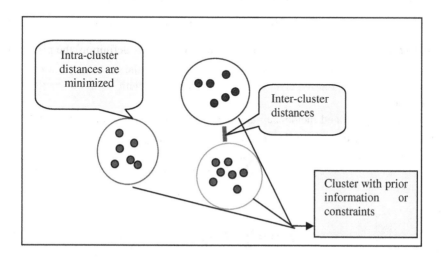

Fig. 2. Semi supervised clustering model

Recently, spectral methods have become increasingly popular for clustering. These algorithms cluster data given in the form of a graph. One spectral approach to semi-supervised clustering is the pixel spatial correlation. In the proposed method, the relationship among pixels in the remote sensing image is formulated in a remote sensing image structure. In this part, we introduce the remote sensing image construction procedure by using pixel spatial correlation. In the constructed remote sensing image $G = \{V, E, W\}$, each vertex denotes one pixel in the remote sensing image $X = \{x_1, x_2, ..., x_n\}$. Therefore, there are n vertices totally in G.

In a remote sensing image structure, each hyperedge connects multiple vertices. To construct the hyperedge, the spatial correlation of pixels are taken into consideration. In this process, each pixel is selected as the centroid and connected to its spatial neighbors, which generates one hyperedge. This hyperedge construction method is under the assumption that spatial connected pixels should have large possibility to have the same labels. As each pixel generates one hyperedge, there is a total of n hyperedges.

Let the selected number of spatial neighbors be K, and there are totally $K + 1$, vertices in one hyperedge. Each hyperedge $e \in E$ is given a weight $w(e) = 1$, which reveals that all hyperedges are with equal influence on the constructed hypergraph structure. Though each hyperedge plays an equal role in the whole hypergraph structure, the pixels connected by one hyperedge may be not close enough in the feature space. Therefore, these pixels may have different weights in the corresponding hyperedge. For a hyperedge $e \in E$, the entry of the incidence matrix H of the hypergraph G is generated by:

$$H(v, e) = \begin{cases} 1 & \text{if } v = v_c \\ \exp(-d^2(v, v_c)/2\sigma^2) & \text{otherwise} \end{cases} \quad (1)$$

where v_c is the centroid pixel, $d(v, v_c)$ is the distance between one v in E and v_c, and σ is the mean distance among all pixels. Under this definition, the pixels in one hyperedge which are similar to the centroid pixel in the feature space can be strongly connected by the hyperedge, and other pixels are with weak connection by the hyperedge.

By using the generated incidence matrix H, the vertex degree of a vertex $v \in V$ and the edge degree of a hyperedge $e \in E$ are generated by:

$$d(v) = \sum_{e \in E} H(v, e) \quad (2)$$

and

$$d(e) = \sum_{v \in V} H(v, e) \quad (3)$$

In the above formulation, D_v and D_e denote the diagonal matrices of the vertex degrees and the hyperedge degrees respectively, and W denotes the diagonal matrix of the hyperedge weights, which is an identity matrix.

2.2   Region Merging

The previous stage only removes redundant regions that do not annoy object semantics. The main purpose of this paper is to represent homogenous objects with few regions. Such homogeneous objects may be extractable more easily than other complex objects, since their components have very similar statistical properties to each other. However, low contrast boundaries between objects may result in merging objects of different semantics. To avoid non-semantic merging, we perform a ternary classification for segmented regions. We determine the class of a region R_i, $C(R_i)$, as follows.

$$C(R_i) = \begin{cases} H, \text{if } \sigma_i^2 \leq VHR\_TH \\ IAH, \text{if } C(R_j) = H, \exists R_j \in \xi(R_i) \\ IH, \text{otherwise} \end{cases} \qquad (3)$$

where VAR_TH is the variance of the largest region. Here, H, IAH and IH are abbreviations of a homogeneous region, inhomogeneous region adjoining a homogeneous region, and inhomogeneous region adjoining only inhomogeneous regions, respectively.

After classification, we examine only regions of class H for merging, and regard regions of class H or IAH as valid merging candidates. That is, we examine R_i and R_j for merging such that $C(R_i) = H$ and $R_j \in \xi(R_i), C(R_j) \neq IH$. This restriction is to prevent two regions of different semantic contents from being merged. Here $\xi_M(R_i)$ is selected by using gradient-based criterion.

Pixel p is defined as a boundary pixel between R_i and R_j, if there are two pixels, p_i and p_j such that $p_i \in R_i$, $p_j \in R_j$, and $p_j \in N_4(p_j)$, where $N_4(p)$ is a set of pixels neighboring p by 4-connectivity. Then, merging candidates $\xi_M(R_i)$ for a given R_i can be determined by considering the weakness of boundary pixels. If R_j satisfies both conditions, $C(R_j) \neq IH$ and at least half of the boundary pixels between two regions have gradient values less than $2\sqrt{VAR\_TH}$, R_j is an element of $\xi_M(R_i)$. Using these merging candidates, we perform the following algorithm until it terminates automatically.

Step 1: Find R_i such that $C(R_i) = H$ and $\xi_M(R_i) \neq \phi$.

Step 2: If there is no such region, the merging procedure is terminated. Otherwise, find merging pair (R_i, R_j) that provides the smallest value of variance after merging, and then merge them.

Step 3: Classify the merged region to H in order to expand this region continuously by merging.

Step 4: Go to step 1.

3 Experiments

In this section, we first describe the testing datasets and then discuss the experimental results and the comparison with the state-of-the-art methods.

3.1 The Testing Datasets

In our experiments, two datasets are employed to evaluate the performance of the proposed method. The first dataset is the Airborne Visible/Infrared Imaging Spectrometer (AVIRIS) image taken over NW Indiana's Indian Pine test site, which has been widely employed. The Indian Pine dataset is with the resolution of 145 ×145 pixels and has 220 spectral bands. 20 bands are removed due to the water absorption bands. There are originally 16 classes in total, ranging in size from 20 to 2455 pixels. Some small classes have been removed and only 9 classes are selected for evaluation. The details information about the selected classes is shown in Table 1.

Table 1. Details of the Indian Pine Dataset

Class	#Pixels	Class	#of pixels	Class	#of Pixels
Soybeans-no till	972	Corn-no till	1428	Grass/pasture	483
Soybeans-min	2455	Corn-min	830	Grass/trees	730
Soybeans-clean till	593	Woods	1265	Hay-windrowed	478
Total	9134				

3.2 Compared Methods

To evaluate the effectiveness of the proposed remote sensing image segmentation approach, the following methods are employed for comparison.

1. Semi-Supervised Graph Based Method [8]. In semi-supervised graph based method, the hyperspectral image classification is formulated as a graph based semisupervised learning procedure. All pixels are denoted by the vertices in the graph

structure, which is able to exploit the wealth of unlabeled samples by the graph learning procedure.

2. Graph-based methods [16], in which each sample spreads its label information to its neighbors until a global stable state is achieved on the whole dataset.

3. Supervised Bayesian approach with active learning [19], which by using supervised Bayesian approach to hyperspectral image segmentation with active learning.

3.3   Experimental Results

In our experiments, the number of labeled training samples for each class varies from 10 to 100, i.e., {10, 20, 30, 50, 100}. To evaluate the hyperspectral image classification performance, the widely used overall accuracy (OA) and the Kappa statistic are employed [9] as the evaluation metrics. In the following experiments, K is set as 12, and $VAR_{max} = 40$.

Experimental comparisons on the testing datasets are shown in Fig. 4 and Fig. 6. In comparison with the state-of-the-art methods, the proposed method outperforms all compared methods in the testing databases. Here we take the experimental results when 10 samples per class are selected as the training data as an example. In the Indian Pine dataset, the proposed method achieves a gain of 1.23%, 3.50%, 0.03%, and 34.38% in terms of the OA measure and a gain of 3.92%, 27.60%, 0.44%, and 30.52% in terms of the Kappa measure compared with semi-supervised, graph-based method, supervised bayesian methods. Experimental results show that the proposed method achieves the best image segmentation performance in most of cases in the testing dataset, which indicates the effectiveness of the proposed method, as shown in Fig. 3.

Figure 4 demonstrates the classification map of the proposed method in the testing dataset with different number of selected training samples per class.

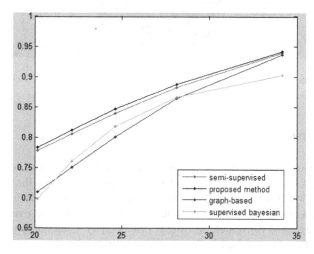

Fig. 3. The segmentation accuracy results of compared methods in the Indian Pine dataset

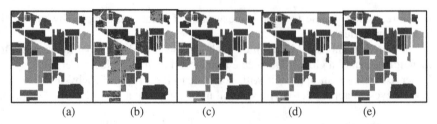

Fig. 4. Segmentation maps of the Indian Pine Sub dataset. (a) Ground truth map with 9 classes (b)-(e) Segmentation maps with 10,20,30 and 50 labeled training samples for each class.

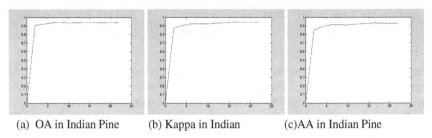

(a) OA in Indian Pine   (b) Kappa in Indian   (c)AA in Indian Pine

Fig. 5. Segmentation performance comparison with different K values by using 10 training sample per class in the Indian Pine dataset

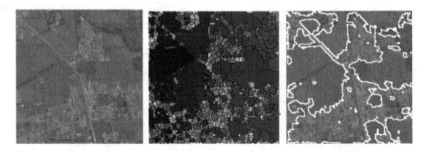

Fig. 6. Segmentation of the proposed method. (a) Ground truth map with 9 classes (b) Semi-supervised classification result (c) Region merging segmentation result.

4   Conclusion

In this paper, we propose a remote sensing image segmentation method by using the semi-supervised classification comprised with region merging. In the proposed method, the relationship among pixels in the remote sensing image is formulated in a semi-supervised clustering. In the constructed remote sensing , each vertex denotes a pixel in the image, and the remote sensing is generated by using the spatial correlation among pixels. Semi-supervised learning on the remote sensing is conducted for remote sensing image classification, and then using the region merging method to segment the classified image. This method employs the spatial information to explore the relationship among pixels, and the high dimensional feature is only used to further enhance the spatial-based correlation in the constructed remote sensing , which is able to avoid the curse of dimensionality.

Experiments on the Indian Pine datasets is performed, and comparisons with the state-of-the-art methods are provided to evaluate the effectiveness of the proposed method. Experimental results indicate that the proposed method can achieve better results in comparison with the state-of-the-art methods for remote sensing image segmentation.

Acknowledgments. This work was supported by the national natural science foundation of China (Grant nos. 61070120, 61103130, 61271435, 61202245); National Program on Key Basic Research Projects (973 programs) (Grant nos. 2010CB731804-1, 2011CB706901-4); Beijing Natural Science Foundation (Grant no. 4112021); Foundation of Beijing Educational Committee (Grant no. KM201111417015); The Project of Construction of Innovative Teams and Teacher Career Development for Universities and Colleges Under Beijing Municipality (no. CIT&TCD20130513).

References

1. Bandos, T., Bruzzone, L., Camps-Valls, G.: Classification of hyperspectral images with regularized linear discriminant analysis. IEEE Transaction on Geoscience and Remote Sensing 47(3), 862–873 (2009)
2. Berge, A., Solberg, A.: Structured gaussian components for hyperspectral image classification. IEEE Transaction on Geoscience and Remote Sensing 44(11), 3386–3396 (2006)
3. Scholkopf, B., Smola, A.: Learning With KernelsSupport Vector Machines, Regularization, Optimization and Beyond, Cambridge, MA. MIT Press Series (2002)
4. Cawley, G.C., Talbot, N.L.C.: Sparse multinomial logistic regression via bayesian L1 regularisation. In: Advances in Neural Information Processing Systems (2007)
5. Camps-Valls, G., Bruzzone, L.: Kernel-based methods for hyperspectral image classification. IEEE Transactions on Geoscience and Remote Sensing 43, 1351–1362 (2005)
6. Fauvel, M., Benediktsson, J.A., Chanussot, J., Sveinsson, J.R.: Spectral and spatial classification of hyperspectral data using SVMs and morphological profiles. IEEE Transactions on Geoscience and Remote Sensing 46(11), 3804–3814 (2008)
7. Plaza, A., Benediktsson, J.A., Boardman, J.W., Brazile, J., Bruzzone, L., Camps-Valls, G., Chanussot, J., Fauvel, M., Gamba, P., Gualtieri, A., Marconcini, M., Tilton, J.C., Trianni, G.: Recent advances in techniques for hyperspectral image processing. Remote Sensing of Environment (2009)
8. Camps-Valls, G., Marsheva, T.B., Zhou, D.: Semi-supervised graph-based hyperspectral image classification. IEEE Transaction on Geoscience and Remote Sensing 45(10), 3044–3054 (2007)
9. Gao, Y., Chua, T.-S.: Hyperspectral image classification by using pixel spatial correlation. In: Li, S., El Saddik, A., Wang, M., Mei, T., Sebe, N., Yan, S., Hong, R., Gurrin, C. (eds.) MMM 2013, Part I. LNCS, vol. 7732, pp. 141–151. Springer, Heidelberg (2013)
10. Gao, Y., Wang, M., Tao, D., Ji, R., Dai, Q.: 3D object retrieval and recognition with hypergraph analysis. IEEE Transactions on Image Processing 21(9), 4290–4303 (2012)

11. Marconcini, M., Camps-Valls, G., Bruzzone, L.: A composite semisupervised svm for classification of hyperspectral images. IEEE Geoscience and Remote Sensing Letters 6(2), 234–238 (2009)
12. Gu, Y., Wang, C., You, D., Zhang, Y., Wang, S., Zhang, Y.: Representative multiple kernel learning for classification in hyperspectral imagery. IEEE Transaction on Geoscience and Remote Sensing 50(7), 2852–2865 (2012)
13. Park, H.S., Ra, J.B.: Homogeneous region merging approach for image segmentation preserving segmentic object contours. In: Proceedings of the International Workshop on Very Low Bitrate Video Coding, Chicago, IL, pp. 149–152 (1998)
14. Fablet, R., Boucher, J.-M., Augustin, J.-M.: Region-based image segmentation using texture statistics and level-set methods. In: Acoustics, Speech and Signal Processing (2006)
15. Nock, R., Nielsen, F.: Statistical region merging. IEEE Transactions on Pattern Analysis and Machine Intelligence 26(11), 1452–1458 (2004)
16. Cevahir, C., Aydin Alatan, A.: Region-based image segmentation via graph cuts. In: 15th IEEE International Conference on Image Processing, pp. 2272–2275 (2008)
17. Li, J., Bioucas-Dias, J.M.: Semisupervised hyperspectral image segmentation using multinomial logistic regression with active learning. IEEE Transactions on Geoscience and Remote sensing 48(11), 4085–4098 (2010)
18. Kulis, B., Dhillon, S.B.I., Mooney, R.: Semi-supervised graph clustering: a kernel approach. In: Proceedings of the 22nd International Conference on Machine Learning, Bonn, Germany (2005)
19. Li, J., Bioucas-dias, J.M., Plaza, A.: Hyperspectral image segmentation using a new bayesian approach with active learning. IEEE Transaction on Geoscience and Remote Sensing 49(10), 3947–3960 (2011)

Correcting Distortion of Views into Aquarium

Yukio Ishihara[1] and Makio Ishihara[2]

[1] Kyushu University, 3-1-1, Maidashi, Higashi-ku, Fukuoka, 812-8582 Japan
iyukio@redoxnavi.med.kyushu-u.ac.jp
http://redoxnavi.kyushu-u.ac.jp/
[2] Fukuoka Institute of Technology, 3-30-1 Wajiro-higashi,
Higashi-ku, Fukuoka, 811-0295 Japan
m-ishihara@fit.ac.jp
http://www.fit.ac.jp/~m-ishihara/Lab/

Abstract. In this paper, we discuss a way to correct light distortion of views into an aquarium. When we see fish in an aquarium, they appear closer also distorted due to light distortion. In order to correct the distortion, the light rays travelling in the aquarium directly towards an observer should hit him/her after emerging from the aquarium. A basic idea is to capture those light rays by a reference camera, then merge the rays as a single view, which is displayed to the observer. An experiment in a real world environment shows that light distortion of a view into an aquarium can be corrected using the multiple reference camera views.

Keywords: distortion correction, aquarium, light distortion.

1 Introduction

Aquariums provide extraordinary experience and opportunities for people to learn about sea creatures. Primary school students might learn what they look, what they eat, how they swim, how they work together to hunt their prey, etc. High school students might learn how food chain works, how sea creatures fight against the water pressure, how fish organize their schools, etc. Researchers might learn how water pollution affects sea environments, what kind of problems the global warming causes in the sea environments, etc. Thus aquariums play an important role in education for a wide range of people.

The structure of aquaria is quite simple. It is an acrylic box with empty space inside, which is filled with sea water. One of problems inevitably focused on is water pressure. Recently, the size of aquaria has become huge to exhibit migratory fish and large fish, also to reduce their stress levels, resulting in high water pressure. For example, the largest aquarium in Japan belongs to Okinawa Churaumi Aquarium. Its size is 35m high x 27m wide x 10m deep. So, the water pressure is tremendous. To resist the water pressure, the acrylic glass of the aquarium is made 60cm thick. The thickness of acrylic glass however causes another problem. That is light distortion.

Do you have any experience of seeing fish distorted while you watch them in an aquarium, especially in a diagonal direction to the acrylic glass? Those

M. Kurosu (Ed.): Human-Computer Interaction, Part V, HCII 2013, LNCS 8008, pp. 163–170, 2013.

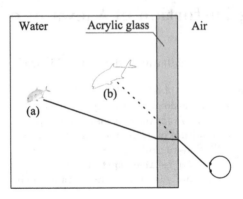

Fig. 1. Light distortion in an aquarium (top view)

fish appear closer also distorted. This is due to light distortion. Technically, this phenomenon stems from light refraction where light rays bend passing through the surface of the acrylic glass. To deal with the light distortion, we use a set of photos taken at specific positions and the image based rendering technique [1].

There have been studies dealing with light distortion for measurement of underwater objects [2]. As mentioned above, photos taken by an underwater camera are affected by light distortion in the same way. It leads to inaccuracy of measurement on the photos. To deal with this problem, correspondence from each pixel on the photos to a single light ray, called ray-map, is created during the calibration process. The map enables accurate measurement of objects on the photos even though the objects appear distorted. In contrast, Sedlazeck et al. study on a way to create photo-realistic underwater images by modeling light refraction, light scattering and light attenuation [3]. So far, no studies are found attempting to construct distortion-free views. 3D structure estimation of underwater objects studied in [4] could construct those views using the extracted geometric information, but it seems inappropriate to create underwater atmosphere by including drifting dust, ascending air bubbles, tiny creatures and the like. Therefore, in this paper, we focus on the image based rendering for distortion-free views.

The rest of the paper is organized as follows. In Section 2 it is explained how light distortion occurs in an aquarium and how different the fish would look. Then a basic idea to correct the distortion is explained in Section 3. In Section 4 it is shown that the idea could correct the distortion in real world environments. Finally we give concluding remarks in Section 5.

2   Light Distortion in Aquarium

Figure 1 illustrates how light distortion occurs in an aquarium. An observer stands in front of the acrylic glass of the aquarium. The observer sees an imaginary fish (b) in the direction represented by the dotted line, which is visually

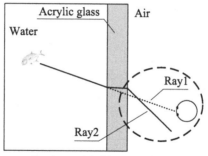

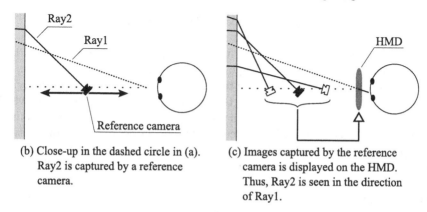

(a) Two key light rays. Ray1 travels without consideration of light refraction while Ray2 is refracted on the two boundaries: water to glass, glass to air.

(b) Close-up in the dashed circle in (a). Ray2 is captured by a reference camera.

(c) Images captured by the reference camera is displayed on the HMD. Thus, Ray2 is seen in the direction of Ray1.

Fig. 2. Our approach to correct light distortion (top view)

created from the real fish (a). The solid line represents the light ray passing through the acrylic glass. The light ray emerging from the fish (a) takes the first contact with a boundary between water and the acrylic glass then the light ray is refracted. It continues to travel inside the acrylic glass and takes the second contact with a boundary between the acrylic glass and air then the light ray is refracted again. After that it travels in the air to the observer's eye. Thus, the observer will be misled into perceiving the fish (b) as the real. The light distortion becomes more significant when the observer stands diagonal to the acrylic glass and sees fish. The light distortion leads to misperceiving of not only the position of the fish but also the size. The next section describes a way to correct the light distortion.

3   Correction of Light Distortion

Figure 2 illustrates how the light distortion is corrected. (a) shows an observer being right in front of the straight line coming from a fish. Ray2 represents the

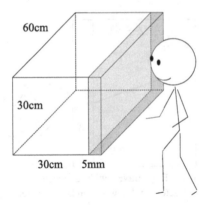

Fig. 3. A simulated glass aquarium

actual light path. Thus, in this case, the observer does not see the fish in that direction represented by Ray1. The point is how Ray2 could be seen by him/her at their position. In this research, we take following two steps to expose Ray2 to the observer. As shown in Figure 2(b), the first step is to gather all light rays which should go straight to the observer from the fish. To gather the light rays, a camera (hereinafter referred to as the reference camera) is placed at a series of positions and captures the rays. Each of the images taken by the reference camera includes a part of color information, all of which is necessary to construct the light distortion-free observer view. That is, in the second step, these images need to be merged as a single view before it is finally displayed on the HMD as shown in Figure 2(c). Consequently, the observer sees the fish in the right direction.

In the next section, we simulate a small-sized aquarium on a computer and show that the light distortion can be corrected.

4   Experiment on Correction of Light Distortion

First of all, the positions and orientations of the reference camera need to be obtained in order to capture the light rays of interest as shown in Figure 2(c). To obtain those, we simulated a small-sized glass aquarium on a computer and traced light rays between the observer and aquarium. Note that in the simulation and following experiment the aquarium used is made of glass instead of acrylic glass. Refractive indices were set to 1.52 for the glass, 1.33 for the water inside. The observer's position was set to 5cm to the aquarium. Figure 3 shows the simulated glass aquarium of 5mm thickness. Figure 4 shows some of the simulated light rays. The dotted and solid lines correspond to Ray1 and Ray2 in Figure 2, respectively. As seen in the figure, becoming closer to the aquarium, more angled light rays need to be captured. To be general, when the observer is at D mm to the aquarium then a ray (or the dotted line which should come to the observer from the direction of θ_o off-center) can be captured at x in the direction of θ_c, which are given by

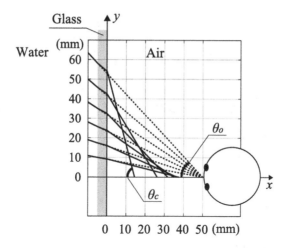

Fig. 4. Simulated light rays between the observer and aquarium (top view)

$$\theta_c = \arcsin\left(\frac{n_3}{n_1}\sin\theta_o\right), \tag{1}$$

$$x = \frac{(G+D)\tan\theta_o - G\tan\theta'}{\tan\theta_c}, \qquad \theta' = \arcsin\left(\frac{n_3}{n_2}\sin\theta_o\right) \tag{2}$$

where n_1, n_2 and n_3 are refractive indices for air, glass and water, respectively. G is the thickness of the glass aquarium.

We have now obtained where and which orientations the reference camera should be placed in to capture the light rays of interest. Next we make sure that a set of the images captured by the reference camera is sufficient to create the light distortion-free observer view. Figure 5 shows a glass aquarium in a real world environment. The aquarium is the same size as the simulated. Water plants and rocks were placed in a way as shown in the figure. Figure 6 shows an experimental setup. The reference camera was placed at 25 to 50mm to the aquarium along the line, also was angled at 50° off-center for 25 to 31mm and by 20° for 32 to 50mm. At each position one reference view was taken, therefore 26 reference views in all were obtained. Another camera representing the observer was placed at 50mm to the aquarium and was angled at 25° off-center. To map pixels on the observer view to those on the reference views, light rays were traced from the observer to the reference camera. This mapping allows the creation of the light distortion-free observer view from the reference views.

Figure 7(a) shows a view into the empty aquarium, which was taken at the observer's position. Figure 7(b) shows a view at the same position but the aquarium is filled with water. As expected, the view is distorted compared to (a). Specifically the objects in (b) appear larger towards the right. Figure 7(d) to (f) represent the reference views. Note that these views are all affected by light distortion. Finally as shown in Figure 7(c), the desired view from the observer

Fig. 5. A glass aquarium in a real world environment. Note that no water in the aquarium to see the right positions of the water plants and rocks inside.

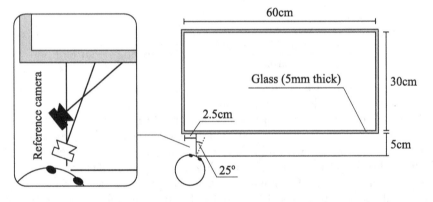

Fig. 6. An experimental setup (top view)

was constructed of the reference views. Pixels surrounded by a line in (d) and (e) contributed to the corresponding part of the view (c). Thus (c) is comprised of a part of each of the reference views. Although 26 reference views were taken, any views at 39 to 50mm positions were not used for (c). Here we calculated the limiting values of θ_c and x when θ_o approaches zero, which are given by

$$\lim_{\theta_o \to 0} \theta_c = 0, \tag{3}$$

$$\lim_{\theta_o \to 0} x = \frac{n_1 n_2 (G + D) - n_1 n_3 G}{n_2 n_3} \approx 38.02. \tag{4}$$

(a) The observer view when no light distortion affects it or no water is filled with.

(d) The view from the reference camera placed at 25mm to the aquarium and angled at 50° off-center.

(b) The observer view when water is filled with, in which the objects appear larger towards the right compared to (a).

(e) The view from the reference camera at 35mm and 20°.

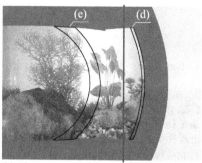

(c) The constructed view from the observer, which is comprised of the reference views represented by (d) to (f).

(f) The view from the reference camera at 45mm and 20°.

Fig. 7. Views from the observer and reference camera in a real world environment. The vertical lines help understand to what extent the view is distorted, also corrected.

Therefore reference views up to 38mm position are sufficient to construct the light distortion-free observer view in this experimental setup. Finally it is obviously shown that the light distortion found in (b) was successfully corrected in (c), but not totally. Some peripheral part was left gray background. The gray background could be drawn if reference views at 1 to 24mm positions were available as well as upward/downward ones. In this experiment, only rightward reference views were used for simplicity. In addition, reference views up to 24mm position were unable to be obtained because of no room to place the reference camera even closer to the aquarium.

5   Conclusions

In this paper, we discussed a way to correct light distortion of views into an aquarium. When we see underwater objects, they appear closer also distorted due to light distortion. In order to correct the distortion, the light rays travelling in the aquarium directly towards an observer should hit him/her after emerging from the aquarium. First we found out where those light rays were heading by a computer simulation. Next those light rays were captured by a reference camera placed at the obtained positions. Finally the reference views were merged as a single view. This experiment in a real world environment confirmed that the light distortion-free observer view was successful constructed of the multiple reference views.

In the future work, we plan to expand the observer view by using not only rightward reference views but leftward/upward/downward ones. Also we seek another way to capture reference views so that it takes less time to complete the observer view. As a result, it will allow moving objects in the aquarium such as swimming fish and swaying water plants.

References

1. Gortler, S.J., Grzeszczuk, R., Szeliski, R., Cohen, M.F.: The lumigraph. In: Proc. of SIGGRAPH 1996, pp. 43–54 (1996)
2. Treibitz, T., Schechner, Y., Kunz, C., Singh, H.: Flat Refractive Geometry. IEEE Trans. Pattern Anal. Mach. Intell. 34(1), 51–65 (2012)
3. Sedlazeck, A., Koch, R.: Simulating Deep Sea Underwater Images Using Physical Models for Light Attenuation, Scattering, and Refraction. In: Proc. of the Vision, Modeling, and Visualization Workshop 2011, pp. 49–56 (2011)
4. Kang, L., Wu, L., Yang, Y.-H.: Two-view underwater structure and motion for cameras under flat refractive interfaces. In: Fitzgibbon, A., Lazebnik, S., Perona, P., Sato, Y., Schmid, C. (eds.) ECCV 2012, Part IV. LNCS, vol. 7575, pp. 303–316. Springer, Heidelberg (2012)

A Dense Stereo Matching Algorithm
with Occlusion and Less or Similar Texture Handling

Hehua Ju and Chao Liang

College of Electronic Information and Control Engineering,
Beijing University of Technology, Beijing 100124, P.R. China
lc_s2009@emails.bjut.edu.cn

Abstract. Due to image noise, illumination and occlusion, to get an accurate and dense disparity with stereo matching is still a challenge. In this paper, a new dense stereo matching algorithm is proposed. The proposed algorithm first use cross-based regions to compute an initial disparity map which can deal with regions with less or similar texture. Secondly, the improved hierarchical belief propagation scheme is employed to optimize the initial disparity map. Then the left-right consistency check and mean-shift algorithm are used to handle occlusions. Finally, a local high-confidence strategy is used to refine the disparity map. Experiments with the Middlebury dataset validate the proposed algorithm.

1   Introduction

Considering the great advances in digital camera technology coupled with the convenience, compactness and relatively low costs, stereo vision is widely used. Stereo vision can reconstruct the 3D geometry of a scene from two or more images of the same scene taken from different views, and then perform localization, mapping and measurement. The core of the stereo vision is the stereo matching, which is to find a unique mapping between the points belonging to two images of the same scene. Generally the camera geometry is known and the images can be rectified, then the stereo matching reduces to a simplified problem, where points in one image can correspond only to points along the same scanline in the other image. However, due to illumination, image noise and occlusion, the stereo matching is still a great challenge.

There exists lots of works on the stereo matching. Scharstein and Szeliski [1] provide an exhaustive comparison of stereo matching algorithms. Generally, these algorithms can be roughly classified into two categories: the local algorithm and the global algorithm. The local algorithms usually utilize local measurements such as image intensity, and aggregate information from multiple around pixels. The simplest method is to minimize the matching error within rectangular windows of fixed size [2]. The accuracy of these methods depends on the size and shape of the window. Some modified algorithms are also proposed, which utilize multiple windows [3, 4] or adaptive windows [5] to minimize the error and give performance improvements. But occlusion and region with less or similar texture are still open problem for local algorithm.

Global algorithms for the stereo matching problem define a global cost function by making some assumptions on the disparity map. The global cost function generally

M. Kurosu (Ed.): Human-Computer Interaction, Part V, HCII 2013, LNCS 8008, pp. 171–177, 2013.

includes terms for local property matching, smoothness terms, and in some cases, penalties terms for occlusions. Then the most efficient minimization scheme is chosen according to the form of the cost function. Finally taking the results of the local algorithm as initial values, the global cost function can be solved. One typical global algorithm is based on dynamic programming techniques [6]. It can obtain good results at occlusion regions, while easily suffers from "streaking" artifacts due to its inconsistency between scanlines. The other popular global algorithms, such as the belief propagation (BP) [7] and Graph Cut [8], can improve the accuracy since they perform the minimization in two dimensions. However since the cost function is a highly nonlinear function, it is hard to obtain the global optimal solution. Additionally, these methods are computationally expensive.

In this paper, a new stereo matching algorithm is presented. The proposed algorithm first use cross-based regions to compute an initial disparity map which can deal with regions with less or similar texture. Secondly, the improved hierarchical belief propagation scheme is employed to optimize the initial result, where a stability factor is introduced into the hierarchical belief propagation scheme to balance the ratio of the data term and smoothness term in BP technique for every pixel. Then, the left-right consistency check and mean-shift are applied to handle occlusion pixels. Finally, a local high-confidence strategy is used to refine the obtained disparity map. Experiments with the Middlebury dataset validate the proposed algorithm.

The paper is organized as follows. Section 2 presents the computation of the initial disparity map with cross-based regions to handling with the less or similar texture region. Section 3 introduces the global optimization for the initial disparity map with the improved hierarchical BP. The proposed occlusion handling and disparity refinement are introduced in Section 4. Experimental results are reported in Section 5. Section 6 concludes this paper.

2   Initial Disparity Map with Cross-Based Regions

For the global stereo matching algorithm, the selection of the initial disparity map generally plays an important role. The accuracy of the local algorithms varies with the size and shape of the matching window. In region low texture region, the matching window should be large enough, while in region with similar texture the matching window should be small enough. So in this paper, we use an adaptive window determined by the cross-based scheme to compute the initial disparity map.

It is reasonable to assume that pixels with similar colors are in the same plane and have the same disparity. According to the cross-based scheme [9], the matching window of a pixel $P = (x_p, y_p)$ can be determined by two orthogonal segmentations: the vertical segmentation $V(P)$ and the horizontal segmentation $H(P)$

$$\begin{cases} H(p) = \left\{ (x, y) \mid x \in \left[x_p - x_p^l, x_p - x_p^r \right], y \in y_p \right\} \\ V(p) = \left\{ (x, y) \mid x \in x_p, y \in \left[y_p - y_p^l, y_p - y_p^r \right] \right\} \end{cases} \tag{1}$$

The details of determine the matching window are as follows. First take P as the start point to compute the vertical length of the matching window. Compare the color difference between the point p_1 and the point $p_2 = (x_p - i, y_p)$ with the threshold (τ) for $i \in [1, r]$, where r is related to the maximum window radius.

$$S(p_1, p_2) = \begin{cases} 1, & \max_{c \in \{R,G,B\}} (|I_c(p_1) - I_c(p_2)| \leq \tau) \\ 0, & otherwise \end{cases} \qquad (2)$$

when $S(p_1 - p_2) = 1$, the length of vertical length of the matching window V_p is added by 1. So the vertical length of the window can be computed by performing this operation in up and down direction. Secondly, determine the matching window by the horizontal segmentation. Take each point on the determined vertical segment as start to perform segmentation in left and right direction along the corresponding horizontal line using the strategy similar to equation (2). The obtained matching window $U(p)$ can be regarded as a set of horizontal segments $H(q)$.

$$U(p) = U_{q \in V(p)} H(q) \qquad (3)$$

With the truncation model

$$e(x, y, d) = \min(\sum_{c \in \{R,G,B\}} |I_c^l(x, y) - I_c^r(x - d, y)|, T) \qquad (4)$$

where T is a threshold for truncation, the cost function can be defined as

$$C(x, y, d) = \frac{1}{\|U_d(x, y)\|} \sum_{(x,y) \in U_d(x,y)} e(x, y, d) \qquad (5)$$

where $\|U_d(x, y)\|$ is the number of pixels in both matching windows and

$$U_d(x, y) = \{(x, y) | (x, y) \in U_l(x, y), (x - d, y) \in U_r(x, y)\} \qquad (6)$$

U_l and U_r is the matching window in left and right image respectively.

To speed up the matching, the orthogonal integral image technique is used to construct the cost function. It needs the following 4 steps to construct the cost function.

- Step 1. Compute the horizontal integral, $S^H(x, y)$, of the line corresponding to an interest point with cost function $e_d(x, y)$.

$$S^H(x, y) = \sum_{0 \leq m \leq x} e_d(m, y) = S^H(x - 1, y) + e_d(x, y) \qquad (7)$$

where $S^H(-1, y) = 0$.

- Step 2. Compute the value of the cost function corresponding each line in the matching window.

$$E_d^H(x, y) = S^H(x + h_e, y) - S^H(x - h_s - 1, y) \qquad (8)$$

where h_e and h_s denotes the end and start of each horizontal segment respectively.

- Step 3. Compute the vertical integral, $S^V(x, y)$

$$S^V(x, y) = \sum_{0 \leq n \leq y} E_d^H(x, n) = S^V(x, y-1) + E_d^H(x, y) \tag{9}$$

- Step 4. Obtain the cost function

$$E_d(x, y) = S^V(x, y + v_e) - S^V(x, y - v_s - 1) \tag{10}$$

Finally, with the winner-take-all strategy, a disparity map can be obtained.

3   Global Optimization with the Improved Hierarchical BP

To refine the results, we apply an improved hierarchical BP algorithm to optimize the initial disparity map. Generally, the cost function $E(p, d)$ for the BP algorithm has the following form [7]

$$E(p, d) = E_d(p, d) + E_s(p, d) \tag{11}$$

where $E_d(p, d)$ is the data term and $E_s(p, d)$ is the smoothness term. Noting that in some regions, the smoothness does not need. So an efficient weighting the rate of the smoothness is introduced, and the cost function of the BP is modified

$$E(p, d) = C_f(p) \cdot E_d(p, d) + \frac{1}{C_f(p) + 0.5} \cdot E_s(p, d) \tag{12}$$

where $C_f(p)$ is the stability factor is defined by the best disparity $C_b(x, y)$ and the second best disparity $C_s(x, y)$

$$C_f(p) = \frac{|C_b(x,y) - C_s(x,y)|}{C_b(x,y)} \tag{13}$$

In this paper, the data term takes the cost function defined in section 2. The smoothness term has the following form

$$E_s^i(p, d) = \sum_{q \in N(p)} M_{q \to p}^i(d) \tag{14}$$

where i is the iteration number, $M_{q \to p}^i(d)$ denotes the message between the pixel p and its neighbor $q \in N(p)$, and the jump cost, $E_j(d_p, d)$, expresses the discontinuity

$$M_{q \to p}^i(d) = \arg\min_{d_q}(E_d(p, d_q) + \sum_{s \in N(q), s \notin p} M_{q \to p}^{i-1}(d_q) + E_j(d_q, d)) \tag{15}$$

$$E_j(d_q, d) = \min(\theta_{bp}, \rho \mid d_q - d \mid) \tag{16}$$

where θ_{bp} is the truncation value to limit the increase of the cost.

4   Disparity Map Refinement

In real applications, occlusion inevitably occurs. The occlusion results in mismatches in the obtained disparity map [1]. To handle this problem, the left-right consistency check is performed to detect the occluded points.

$$D_l(x, y) = D_r(x - D_l(x, y), y) \tag{17}$$

where D_l, D_r is the left and right disparity respectively. Denote the pixel which does no satisfy equation (17) as the occluded points. Once the occluded points are detected, use the mean-shift algorithm [10] to segment the left image into different regions. Then according to segment, assign each occluded points the disparity which corresponds to its segment region.

To further improve the performance, the local high-confidence voting technique [11] can be used to refine the obtained disparity map. At first, smooth the image with the 3×3 median filter. Then for the left image, compute cross-based region image and calculate the histogram of disparity values in each region. Take the largest value of the histogram as the disparity of this region. Finally, filter the final disparity map with a median filter to remove the noise.

5   Experiments

The proposed algorithm is validated with the Middlebury dataset [1]. Parameters in the proposed algorithm are set as follows: the maximum window radius $r = 17$, the threshold $\tau = 20$, the truncation value $T = \text{max\_}d + 5$, $\theta_{bp} = 2 \text{ max\_}d/16$, and $\rho = 1.0$, where $\text{max \_}d$ is the largest disparity of the stereo images pair.

To make comparisons, the VariableCross [9] and the RealTimeBP [7] algorithms are also performed. Results are reported in Table 1. The measurement is computed for three subsets of an image. They are "nonocc": the subset of non-occluded pixels, "all": pixels that are either non-occluded or half-occluded, and "disc": pixels near the occluded areas [1]. It is obviously that the proposed algorithm outperforms the VariableCross and the RealTimeBP. Figure 1 shows results with the proposed algorithm. The proposed algorithm is also applied to an outdoor scene. Figure 2 shows experimental results of one pair of images. We can see for the unstructured outdoor, the proposed algorithm can also obtain good results.

Table 1. Experimental results

Algorithm	Tsukuba			Venus			cones		
	nocc	all	disc	nocc	all	disc	nocc	all	disc
The proposed method	1.24	1.60	6.58	0.35	0.53	3.09	4.64	11.6	10.9
VariableCross	1.99	2.65	6.77	0.62	0.96	3.20	6.28	12.7	12.9
RealTimeBP	1.49	3.40	7.87	0.77	1.90	9.00	4.61	11.6	12.4

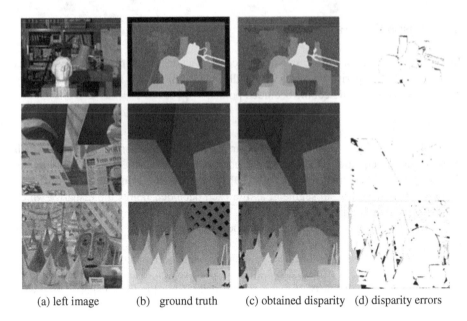

(a) left image   (b) ground truth   (c) obtained disparity   (d) disparity errors

Fig. 1. Experimental results for Middlebury data set

(a) left image   (b) obtained disparity

Fig. 2. Experimental results for a pair of images of outdoor scene

6   Conclusions

In this paper, a new dense stereo matching algorithm is proposed. The proposed algorithm first use cross-based regions to compute an initial disparity map which can deal with regions with less or similar texture. Secondly, the improved hierarchical belief propagation scheme is employed to optimize the initial disparity map. Then the left-right consistency check and mean-shift algorithm are used to handle occlusions. Finally, a local high-confidence strategy is used to refine the disparity map. Experiments with the Middlebury dataset validate the proposed algorithm.

References

1. Scharstein, D., Szeliski, R.: A taxonomy and evaluation of dense two-frame stereo correspondence algorithms. Int'l Journal of Computer Vision 47(1), 7–42 (2002)
2. Okutomi, M., Kanade, T.: A multiple baseline stereo. IEEE Trans. PAMI 15(4), 353–363 (1993)
3. Geiger, D., Ladendorf, B., Yuille, A.: Occlusions and binocular stereo. In: Sandini, G. (ed.) ECCV 1992. LNCS, vol. 588, pp. 425–433. Springer, Heidelberg (1992)
4. Fusiello, A., Roberto, V., Trucco, E.: Efficient stereo with multiple windowing. In: Proceedings of IEEE CVPR, pp. 858–863 (1997)
5. Kanade, T., Okutomi, M.: A stereo matching algorithm with an adaptive window: theory and experiment. IEEE Trans. PAMI 16(9), 920–932 (1994)
6. Salmen, J., Schlipsing, M., Edelbrunner, J., Hegemann, S., Lüke, S.: Real-time stereo vision: Making more out of dynamic programming. In: Jiang, X., Petkov, N. (eds.) CAIP 2009. LNCS, vol. 5702, pp. 1096–1103. Springer, Heidelberg (2009)
7. Yang, Q.X., Wang, L., et al.: Real-time global stereo matching using hierarchical belief propagation. In: Proceedings of BMVC (2006)
8. Boykov, Y., Veksler, O., Zabih, R.: Fast approximate energy minimization via graph cuts. IEEE Trans. PAMI 23(11), 1222–1239 (2001)
9. Zhang, K., Lu, J., Lafruit, G.: Cross-based local stereo matching using orthogonal integral images. IEEE Trans. CSVT 19(7), 1073–1079 (2009)
10. Comaniciu, D., Meer, P.: Mean-shift: A robust approach toward feature space analysis. IEEE Trans. PAMI 24(5), 603–619 (2002)
11. Lu, J., Lafruit, G., Catthoor, F.: Anisotropic local high-confidence voting for accurate stereo correspondence. In: Proceedings of of the SPIE, vol. 6812, pp. 68120J–68120J-12 (January 2008)

Robust Face Recognition System
Using a Reliability Feedback

Shotaro Miwa, Shintaro Watanabe, and Makito Seki

Advanced Technology R&D Center, Mitsubishi Electric Corp., 8-1-1,
Tsukaguchi-Honmachi, Amagasaki City, Hyogo, 661-8661, Japan
Miwa.Shotaro@bc.MitsubishiElectric.co.jp,
Watanabe.Shintaro@dc.MitsubishiElectric.co.jp,
Seki.Makito@dr.MitsubishiElectric.co.jp

Abstract. In the real world there are a variety of lighting conditions, and there exist many directional lights as well as ambient lights. These directional lights cause partial dark and bright regions on faces. Even if auto exposure mode of cameras is used, those uneven pixel intensities are left, and in some cases saturated pixels and black pixels appear. In this paper we propose robust face recognition system using a reliability feedback. The system evaluates the reliability of the input face image using prior distributions of each recognition feature, and if the reliability of the image is not enough for face recognition, it capture multiple images by changing exposure parameters of cameras based on the analysis of saturated pixels and black pixels. As a result the system can cumulates similarity scores of enough amounts of reliable recognition features from multiple face images. By evaluating the system in an office environment, we can achieve three times better EER than the system only with auto exposure control.

Keywords: Face Recognition, Prior Probability, Probabilistic Model.

1 Introduction

Recently cameras are used in many places around us, on streets, in offices, in houses, etc. Some of the key functions of those cameras are authorization or identification using face recognition. As places where cameras are used expand, variations of lighting conditions where cameras are used also expand, and that causes a big problem for face recognition. Generally face recognition degrades its performance under lighting condition and face-pose changes. Until now many face recognition algorithms are proposed [1-8] to tackle this problem. Some use robust features against lighting changes [2,3], and some pick up a stable set of features under pose and lighting changes [4-6], and some use better lighting normalization [7]. But the performance of face recognition depends on the quality of face images, and under worse lighting conditions, more FR(False Rejectance) cases occur.

To solve this issue and improve the performance of face recognition under lighting changes, we propose a face recognition system using a reliability feedback which controls camera parameters based on reliability of the current face image and capture more informative face images to improve the final face recognition performance.

M. Kurosu (Ed.): Human-Computer Interaction, Part V, HCII 2013, LNCS 8008, pp. 178–185, 2013.

This paper is organized as follows. In Section 2 we explain a face recognition combining image-difference features based on a probabilistic model. In Section 3 we explain a face recognition system using a reliability feedback. In Section 4 we report our experimental results, and in Section 5 we conclude our framework.

2   Face Recognition Combining Image-Difference Features Based on a Probabilistic Model

2.1   General Flow of Face Recognition System

General flow of face recognition system is described in Fig.1. Firstly a face detector is applied to an input image. If a face is detected in the image, feature point detectors are applied to the detected face. Then using detected feature point locations the face is cropped into the image with fixed width and height, and a lighting normalization is applied to this cropped image. After these processing the probe image is ready for comparison with a face database. In the comparison with a face database face similarity score between the probe image and each registered gallery image is calculated, and if the score is greater than the operational threshold, the owner of the normalized face is authorized as a registered person.

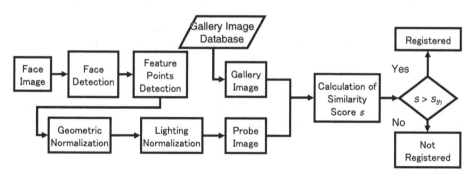

Fig. 1. General Flow of Face Recognition System

2.2   Face Recognition Combining Image-Difference Features Based on a Probabilistic Model

In [6] we proposed face recognition algorithm for access control systems combining image-difference features based on a probabilistic model. In the case of access control systems available cameras are not as good as consumer cameras, and the algorithm needs to be robust against degraded images caused by low cost cameras under lighting changes. The proposed face recognition algorithm is described in Eq. (1) and Eq.(2).

$$S(I_p, I_g) = \sum_{i=1}^{N} h_i(\Delta_i) P(f_{p_i}) P(f_{g_i}) \tag{1}$$

$$h_i(\Delta_i) = \begin{cases} \alpha_i, if \left| f_{pi} - f_{gi} \right| > t_i \\ \beta_i, otherwise \end{cases} \tag{2}$$

In Eq.(1) I_p and I_g show a probe image and a gallery image respectively. $S(I_p, I_g)$ is a similarity function between I_p and I_g. f_{pi} and f_{gi} are values of recognition feature f_i .respectively. $P(f_i)$ is a prior distribution of recognition feature f_i, and N is the total number of recognition features. In Eq.(2) $h_i(\Delta_i)$ is a similarity function of image difference Δ_i. t_i is a threshold, and α_i and β_i are confidence values. Recognition feature f_i, threshold t_i, and confidence values α_i, β_i are all learned by Real AdaBoost [8]. In the following section, face recognition system using a reliability feedback using above face recognition algorithm is proposed.

3   Face Recognition System Using a Reliability Feedback

Recent cameras have auto exposure control functions to get better images under various lighting conditions. But there are many directional lights as well as ambient lights in the real world, and even after the auto exposure combined directional lights produce partial dark and bright regions on faces.

Under these situations it is difficult to extract reliable features of both regions from one image. We propose a face recognition system to capture a sequence of images to get more reliable features by controlling exposure parameters.

3.1   System Flow

The system flow of our proposing face recognition system using a reliability feedback is described in Fig.2. For each probe image the reliability of the image is calculated. If the reliability is less than a threshold, the system continues to capture more faces. To calculate the reliability of the image we use the face recognition algorithm based on a probabilistic model in section 2. It evaluates reliabilities of all features using the prior distributions of each feature. As a result all features are categorized into two groups, reliable features and unreliable features. For faster calculation we use the discretized functions based on prior distributions (Fig.3). If the value of the feature lies inside the range of the prior distribution between $\mu - 4\delta$ and $\mu + 4\delta$, the feature is decided as reliable. If outside the range, the feature is decided as unreliable. The ratio of the number of reliable features to the total number of all features is used as the reliability of the one probe image. To calculate the reliability in a recognition session, we cumulate the all reliable features got in a session until that time. Before continuing to the reliability calculation, if the total number of captured images c is larger than a trial threshold c_{total}, the system jumps to the final similarity calculation.

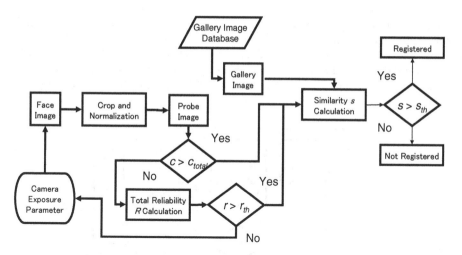

Fig. 2. System Flow of Face Recognition System Using a Reliability Feedback

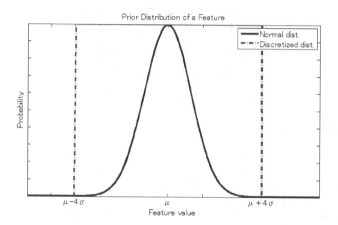

Fig. 3. Discretized Prior Distribution

3.2   Camera Exposure Control

Exposure parameter of the camera is controlled based on the number of black pixels and saturated pixels (Fig.4). The initial exposure control mode is set to auto. But if the reliability r of the first probe image is less than a reliability threshold r_{th}, the exposure control mode is set to manual. In a manual mode exposure parameter is controlled to lower the excessive number of saturated pixels or black pixels allowing some amount of saturated pixels and black pixels.

To set the exposure parameter, the system counts the sum of black pixels n_b which is less than a dark intensity threshold th_b, and the sum of saturated pixels n_s more than a saturated intensity threshold th_s.

If the sum of the saturated pixels n_s is greater than the threshold n_{sth}, we set the exposure control mode to manual and continue to change the exposure to wide to narrow until the reliability is greater than a threshold. If the sum of the black pixels n_b is greater than the threshold n_{bth}, we set the exposure control mode to manual and continue to change the exposure to narrow to wide until the reliability is greater than a threshold.

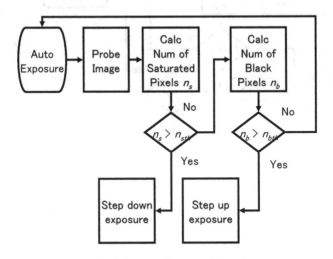

Fig. 4. Camera Exposure Control

3.3   Similarity Score

If the reliability gets greater than a reliability threshold or the number of iterations reaches a trial threshold c_{total}, the system merges multiple reliable features from multiple face images in a session. In Eq.(3) M is the number of the image in a session. G is the all indices of reliable features and g_j is the indices of reliable features newly found in round j. In Eq.(4) similarity score s is the sum of similarity score calculated by reliable features newly found in j-th image in one session.

$$G = \{g_j\}, 1 \le j \le M \qquad (3)$$

$$S(I_p, I_g) = \sum_{i \in g_1} h_i(\Delta_i) + \sum_{i \in g_2} h_i(\Delta_i) + \dots \qquad (4)$$

4   Experimental Result

We evaluated our system in an office environment on a laptop PC with a camera. In the office environment there is a window at the one side of faces and some of all the lights on the ceiling are turned on. Under this condition intensities of pixels are not even, and there exist both dark regions and bright regions on the face (Fig.5).

We compared the system with reliability feedback and the system without reliability feedback using only faces with auto exposure. The total number of subjects is 40. The face detector, feature points detector, and face recognizer are the same ones as used in [6].

Fig. 5. Examples of Evaluated Faces with Auto Exposure

— Face Recognition Performance

Fig.6 shows the comparison of two ROC(Receiver Operating Characteristic) curves.

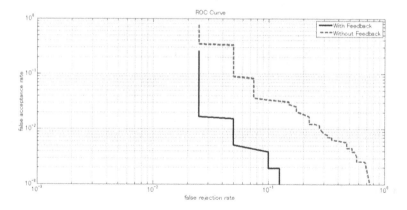

Fig. 6. ROC Curve

Fig.6 shows that EER of the system with a reliability feedback is 2.5% and the system without a reliability feedback is 8.0%, and the system with a achieves three times better performance than without a reliability feedback.

— Analysis of Features

During a recognition session, some new reliable features are sequentially added with changes of the exposure parameter. Fig.7 shows the way of filter calculation where sum of pixels in white regions is subtracted by sum of pixels in black region. Fig.8 shows examples of newly added reliable features. Initially with auto exposure a region at the left side of the face is dark, but by changing exposure parameters, the region gets brighter and new reliable features appear not only in this region and also around the nose bridge regions. As a result those newly found reliable features helps for better face recognition performance.

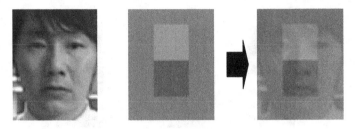

Fig. 7. Filter Calculation

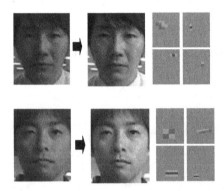

Fig. 8. Newly Added Reliable Features in a Session
(Left: Auto Exposure, Right: Manual Exposure)

5　Conclusion

In this paper the face recognition system using a reliability feedback is proposed to solve the problem of uneven lighting conditions where directional lights cause partial black and saturated regions on faces. Firstly using prior distributions of each feature to decide whether the feature is reliable, or not, the scheme for estimating the reliability of the input image is proposed. Secondly controlling the exposure parameter based on the analysis of saturated pixels and black pixels is proposed. By evaluating the system in an office environment, we confirmed that more reliable features are collected from pixel-intensity improved regions during changes of exposure parameters, and as a result we achieved three times better EER than the system with only auto exposure.

The proposed system can handle the lighting conditions which cause some amounts of saturated pixels and black pixels. But because there are more variety of lighting conditions in the real world, further researches include further evaluations under more lighting conditions.

References

1. Zhao, W., Chellappa, R., Rosenfeld, A., Phillips, P.J.: Face Recognition: A Literature Survey. ACM Computing Surveys, 399–458 (2003)
2. Wiskott, L., Fellous, J.-M., Kruger, N., von der Malsburg, C.: Face recognition by elastic bunch graph matching. IEEE Trans. PAMI 19(7), 775–779 (1997)
3. Ahonen, T., Hadid, A., Pietikainen, M.: Face Description with Local Binary Patterns: Application to Face Recognition. IEEE Trans. PAMI 28(12) (2006)
4. Moghaddam, B.: Principal Manifolds and Probabilistic Subspaces for Visual Recognition. IEEE Trans. PAMI 24(6), 780–788 (2002)
5. Jones, M., Violal, P.: Face Recognition Using Boosted Local Features. In: MERL Tech. Report, TR2003-25 (2003)
6. Miwa, S., Kage, H., Hirai, T., Sumi, K.: Face Recognition for Access Control Systems Combining Image-Difference Features Based on a Probabilistic Model. IEEJ Trans. on Electronics, Information and Systems 131(12), 2165–2171 (2011)
7. Kumar, R., Jones, M., Marks, T.K.: Morphable Reflectance Fields for Enhancing Face Recognition. In: IEEE Conf. on Computer Vision and Pattern Recognition (CVPR), pp. 2606–2613 (2010)
8. Schapire, R., Singer, Y.: Improving boosting algorithms using confidence-rated predictions (1999)

A Developer-Oriented Visual Model
for Upper-Body Gesture Characterization

Simon Ruffieux[1], Denis Lalanne[2], Omar Abou Khaled[1], and Elena Mugellini[1]

[1] University of Applied Sciences of Western Switzerland, Fribourg
{Simon.Ruffieux,Omar.AbouKhaled,Elena.Mugellini}@Hefr.ch
[2] University of Fribourg
Denis.Lalanne@unifr.ch

Abstract. This paper focuses on a facilitated and intuitive representation of up-per-body gestures for developers. The representation is based on the user motion parameters, particularly the rotational and translational components of body segments during a gesture. The developed static representation aims to provide a rapid visualization of the complexity for each body segment involved in the gesture for static representations. The model and algorithms used to produce the representation have been applied to a dataset of 10 representative gestures to illustrate the model.

Keywords: natural interaction, human-computer interaction, multimodality, visualization tools, developer-oriented.

1   Introduction

Recent advances in computer-vision and in low-cost hardware and embedded systems are widening the real-time methods used to recognize body and hands gestures. A few years ago, most systems were restricted to research and game industry; nowadays the miniaturization of hardware such as inertial motion units (IMU) allows embedding them in innovative devices such as watch, clothes, etc. [1-2]; while advances in computer-vision allows efficient recognition of body movements through camera   and depth sensing cameras [3]. These advances and the growing number of different types of sensors available imply more work on the design of gestures for Human-Computer Interaction in order to develop gestures that can be recognized by most sensors individually.

When designing applications for Human-Computer Interaction relying on air gestures, the developers have a large choice of potential gestures from literature and many sensors available on the market. However, they are not always aware of the implications of their preliminary choices. Developers usually design gestures for a specific system with limited considerations for the portability of the gestures to other technologies. However, the recognition rate might largely vary depending on the type of sensors and their location for a same gesture. The representation developed in this work should help providing a facilitated solution to visualize and characterize gestures from a developer point of view by helping them to design gestures and their

M. Kurosu (Ed.): Human-Computer Interaction, Part V, HCII 2013, LNCS 8008, pp. 186–195, 2013.

motion with a particular focus on recognition using specific sensor in the best location or to design specific gestures that can comply all types of sensors by automatically identifying the most significant motion components of gestures. This work focuses on identifying the main motion components involved in a gesture, as retrieved by inertial motion units to infer a characterization for each body segment of the upper-body

This work takes place in the context of the FEOGARM project [4] which goal is to provide a comprehensive framework for facilitating gesture evaluation and recognition methods. This work intends to provide guidelines on how to choose and design gestures depending on the types of sensors considered.

2   Related Work

In the literature, different approaches to classify, characterize and represent gestures have been developed. Classification and taxonomy of the different gestures is a theoretical solution to describe a gesture; discriminating gestures according to their semantic. Textual or visual representations of the motion of gestures is a more practical solution, which uses different methods to describe, illustrate or store the information related to a gesture; such as mathematical definitions, specific file formats or visualization and characterization tools.

Gesture taxonomy and classification has been widely studied by psychologists and computer scientists in the context of human-human interaction [5-6] and in the context of human-computer [7-9] and human-robot interaction [10]. In these researches, several classifications and taxonomies have been proposed. The taxonomy from Pavlovic [9], which divides meaningful gestures from unintentional movements, has been often reused and extended by researchers in the literature. The meaningful gestures are subdivided in different classes according to the presence or not of motion and also according to their semantic meaning. In the work of Aigner & al. [11], they propose an interesting extension of the taxonomy through a schema resuming the information with a visual example for each class of gesture. The meaningful gestures are divided in several sub-categories: pointing gestures, semaphoric gestures (static, dynamic and stroke) are completely unrelated to the meaning and strictly learned; pantomimic gestures represent a specific task being performed, iconic (static and dynamic) gestures are used to convey information about objects or entities such as shape or size or motion path; finally manipulative gestures involve real object manipulation. These classifications are often slightly modified according to the exact subject of application. Such taxonomies are interesting for researchers to share a common language and also important when designing gestures to produce intuitive gestures according to their intended function and taxonomy.

The storage and visual representation of motion has been mostly studied in works related to music and dance, often taking inspiration from the Laban notation providing a similar idea as partitions for music [12–14]. A specific storage format linking music and gesture has been produced in the form of the GDIF format [3]. This format has been developed as a tool for standardizing the way music-related movement data are described, stored and streamed. Storage of the motion information is sensor dependent; the information cannot be stored similarly for video streams or for accelerometers.

In the HCI domain, less research has been developed toward a standardized representation of gestures. Formal definitions have been developed, for example Pavlovic & al. [9] developed the following mathematical definition in the context of a hand gesture: "*Let $h(t) \in S$ be a vector that describes the pose of hands and/or arms and their spatial position within an environment at time t in the parameter space S. A hand gesture is represented by a trajectory in the parameter space S over a suitably defined interval I*". This definition illustrates perfectly what a dynamic gesture is in mathematical terms and is useful when developing algorithms to clarify what to process and recognize; although it can efficiently visually illustrate the translation of gestures involving a single element, it is less suitable to illustrate gestures involving more elements. The visualization of the motion of gestures usually adopts a representation based on the kinematic properties of the human skeleton; a kinematic tree consisting of segments that are linked by joints [15]. This solution is widely used amongst researchers and quite efficient to understand the motion of a particular gesture however it generally requires either video or multiple consecutive pictures to illustrate a dynamic gesture. In works presenting databases of gestures, the gestures are generally illustrated using several classical approaches: with one or more pictures containing arrows to indicate the movement of a subject such as in the work of Song & al. [5], with dashed lines to indicate the final posture of the body such as in the NASA standards[1] or with videos available on a website [2].

In various fields, the characterization of specific features of gestures is used to improve or optimize processes involving motion. In the medical fields, different studies try to characterize medical gestures to improve their efficiency. In [16], they characterized the motion during chest physiotherapy; which can be seen as a repetitive tangible dynamic gesture; they monitored the force and trajectories of the hands of the physiotherapist to infer quality of the medical act to potentially to improve it. In [17], the information retrieved from a Kinect sensor is used to monitor the motion of patients effectuating in-home rehabilitation in order to characterize and improves the gestures. In the musical field, a similar approach has been developed; they use the characterization of the motion of a musician while playing to infer the relation between the sounds produced and the gestures [18].

Recently, in the work by Glomb et al. [1], focusing on the creation of a dataset of hand gestures for HCI, a table illustrating the gestures was partially characterizing the complexity of the gestures by using the most significant motion components of the hand and fingers. The work presented in this paper brings that model further by characterizing automatically each segment of the arm through its main motion components and providing a visual tool to intuitively represent the information.

3   Model

The model developed has several key points: the definition of the terms used to describe the motion components, the segments that have been taken into account, the terms to define the quantity of motion and finally the visualization tool to provide the

---

[1] http://msis.jsc.nasa.gov/

information to the users. Note that the plans and axes described in the following sections reference the definitions from the chapter "Anthropometry and Biomechanics" in Man-Systems Integration standards document from the NASA[2].

The terms that have been chosen to represent the significant component(s) of the motion of a particular segment are "None", "Static", "Translational", "Rotational" and "Complex". The **"None"** component represents the fact that the motion of the segment is not significant for the gesture. This can be inferred by detecting significant variations of the motion of a segment between different occurrences of a same gesture. The **"Static"** component represents the absence of motion of a particular segment during a gesture. If the segment is not static, the gesture would not be recognized. The **"Translational"** component represents linear motion along one of the transverse, vertical or sagittal plane. The **"Rotational"** component represents the rotation of a segment along its axis. The rotations along the two other axes are not considered in the present work as they do not have as much implications for the recognition using visual recognition. This component is mostly represented in the forearm and hand segments for the present work. The **"Complex"** component, also referred as **"Trans&Rot"** in the visual representation, indicates that both translation and rotational motions of the segment occurred during the gesture.

The gestures considered in this paper are limited to one-hand gestures and therefore, only segments corresponding to the right part of the upper-body are mentioned. We considered the assumption that the unreferenced segments are labeled "None" and thus are not significant for the gestures. The **"Torso"** segment is mostly used as reference as it tends to be static or not significant during gestures; it corresponds to the upper torso, the IMU sensor is placed on the back of the neck of the subject. The **"Arm"** segment corresponds to the right upper-arm; the sensor is placed just above the elbow. The **"Forearm"** segment corresponds to the right forearm of the subject. The sensor is placed right before the wrist where the translation and rotation of the segment is maximal. Finally the **"Hand"** segment corresponds to the right hand of the subject; the sensor is placed in the palm of the hand to avoid providing visual clues to video sensors. Note that the arm, forearm and hand IMUs are placed such that they all bear the same orientation with z-axis upward when the subject performs a T-pose.

To classify the quantity of motion in several meaningful classes, a specific color code for the visual representation has been defined along with terms to describe each class. A **"Grey"** color represents the absence of signification of the segment for the considered gesture. A **"Black"** color represents a static component, the **"Red"** color represents a small quantity of motion, the **"Orange"** color represents a medium quantity of motion and the **"Green"** color represents a large quantity of motion.

Finally the visualization tool provides a mean to rapidly and intuitively visualize the final characterization of a particular gesture. A synthetic representation of the human upper right body part has been chosen, on top of which are displayed the processed information from the algorithm.

---

[2] http://msis.jsc.nasa.gov/

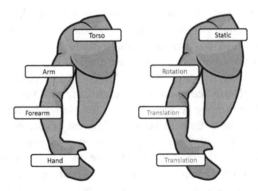

Fig. 1. An illustration of the visual representation tool. On the left, the template illustrating the considered body segments. On the right, a fictive example with a « dynamic pointing» gesture using the arm and hand only.

4   Method

4.1   Data Acquisition

The data has been acquired using the FEOGARM framework [4]. The FEOGARM software allows recording multiple sensors synchronously using distributed computers. To record the gesture database, subjects were carefully equipped with the accelerometers and asked to sit in front of a computer screen. The subject then had to read indication on what was going to happen. Once ready, the recording session started. The information was displayed to the user as in a slideshow. For each gesture, the name of the gesture was displayed on the screen along with a pre-recorded video showing the user what gesture he will have to perform; then after a short delay, the same video was replayed and the user mimics the movement simultaneously to the video. Such a method allows for automatic segmentation and annotation of the data across all sensors.

The dataset contains 10 commonly used gestures in the HCI literature recorded by 10 different subjects. Each gesture has been recorded twice per subject with 3 different resting postures and with two different lightning conditions. It contains a total of 1200 annotated gesture occurrences. The dataset contains all the raw data as acquired from the 4 Xsens MTw IMUs[3] and from one Microsoft Kinect for Windows[4].

4.2   Model Generation

In this work, in order to characterize a gesture, we processed the data acquired from the 4 IMUs. The algorithm developed automatically extracts the information corresponding to each gesture from the whole dataset using the provided annotations. Once

---

[3] http://www.xsens.com/en/mtw
[4] http://www.microsoft.com/en-us/kinectforwindows

extracted, we obtain, for each of the 10 gestures the 120 occurrences stored in a list. Each occurrence contains the data frames of the gesture and each data frame contains the measured linear acceleration (LinAcc), angular velocity (AngVel), Euler orientation and orientation quaternion. The linear acceleration is processed to remove the gravity component computed using the orientation quaternion.

The average quantities of motion for both the translation (1) and the rotation (2) are retrieved by summing the absolute values for each frame for a particular gesture and then averaging over all occurrences of a gesture.

$$AvgTr = \frac{\sum_0^f |LinAcc - gravity|}{f} \qquad (1)$$

$$AvgRot = \frac{\sum_0^f |AngVel|}{f} \qquad (2)$$

Then the principle motion components are defined according to the computed average quantities of motion. A simple comparison between the two motion values using specific threshold allows inferring the most significant component as described by the pseudo-code below:

```
if((AvgTr > TrTh) && (AvgRot > RotTh)){Complex-Trans&Rot}
if((AvgTr > TrTh) && (AvgRot < RotTh)){Translation}
if((AvgTr < TrTh) && (AvgRot > RotTh)){Rotation}
if((AvgTr < TrTh) && (AvgRot < RotTh)){Static}
```

Note that the "None" component could currently not be implemented due to the strong homogeneity of the dataset. The translation thresholds "TrTh" and the rotation thresholds "RotTh" have been inferred empirically by performing various gestures and recording their average motion quantities; in the pseudo-code, they correspond to the smallest values of the range "small motion quantity". Depending on the value of each component, the quantity of motion is characterized using the terms defined in section 3. The distinction between the classes is assessed using the following ranges; for the translation: static [0.0, 0.2], small]0.2, 0.5], medium]0.5,3] and large]3,infinite]; for the rotation: static [0.0, 0.3], small]0.3, 0.5], medium]0.5,1], large]1,infinite].

5   Results

The algorithms developed generated the values illustrated on Table 1, an intermediary phase before the automatic creation of the final visualization. Note that this table already presents the results; for example, the "WaveHello" gesture has the following translation values; large for the hand (4.94), medium for the fore-arm (1.34) and a small translation for the arm (0.30). This clearly indicates larger translation of the hand and that a sensor sensitive to translation should focus on that particular segment. However this representation in a table is not intuitive to read and complex to understand, therefore the algorithm converts it into a more human-friendly representation as shown on Fig. 2 and Fig. 3.

Table 1. The motion quantities obtained for the gesture and their motion components with respect to each segment (Hand, Fore-arm, Arm and Torso)

GestureName	Translation (H,F,A,T)	Rotation(H,F,A,T)
TakeFromScreen	(0.65,0.63,0.17, 0.07)	(0.31,0.26,0.27, 0.02)
PushToScreen	(0.61,0.62,0.26,0.09)	(0.26,0.23,0.20, 0.09)
CirclePalmRotation	(0.63,0.59,0.38, 0.13)	(0.65,0.58,0.29, 0.08)
CirclePalmDown	(0.54,0.46,0.25, 0.09)	(0.31,0.26,0.25, 0.08)
WaveHello	(4.96,1.34,0.30, 0.05)	(1.43,1.25,0.42, 0.013)
ShakeHand	(5,63,0.9,0.27, 0.07)	(2.24,1.45,0.86, 0.16)
SwipeRight	(0.55,0.51,0.10, 0.11)	(0.55,0.37,0.29,0.04)
SwipeLeft	(0.56,0.52,0.08, 0.12)	(0.57, 0.42,0.29, 0.04)
PalmUpRotation	(0.13,0.19,0.02, 0.08)	(0.76,0.65,0.13, 0.03)
PalmDownRotation	(0.05,0.03,0.08, 0.04)	(0.87,0.72,0.18, 0.03)

As previously stated, the final results are the pictures automatically generated for each gesture where the information is rapidly readable, even on a static display such as a sheet of paper.

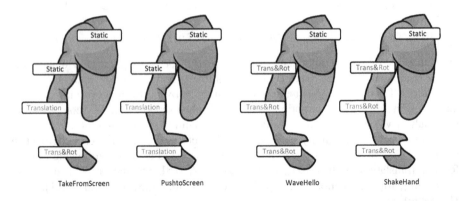

Fig. 2. The resulting characterization figures for the gestures "TakeFromScreen", "PushTo-Screen", "WaveHello" and "ShakeHand"

Using these representations, the developer can rapidly identify segments where the motion occurs and which motion component in particular is present and may pose problems or should be focused to optimize recognition. In Fig. 2 and Fig. 3, the gestures have been grouped by pairs of similar gestures. For example, "TakeFrom-Screen" and "PushToScreen" are very similar gestures where the user moved his arm towards and from the screen, the main difference being a small rotation of the hand for the first gesture while the other remains on the same posture. On the contrary, looking at representation of the gestures "PushToScreen" and "WaveHello", there is an obvious difference between the two gestures; for the latter the quantity of motion is larger and rotation occurs; a developer might infer that a gesture containing only small or medium rotation might be more difficult to recognize using video sensors that a gesture containing large translation.

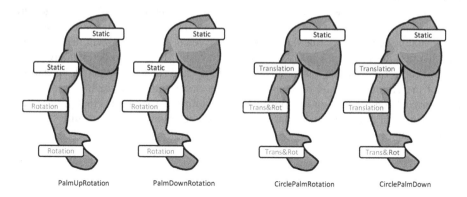

Fig. 3. The resulting characterization figures for the gestures "PalmUpRotation", "PalmDownRotation", "CirclePalmRotation" and "CirclePalmDown"

Such a representation also provides information on which segment should be monitored for optimal results. When designing gestures that should be portable across multiple types of sensors, such a representation should help to rapidly identify a common set of gestures, notably for sensors with specific body placement such as a smartwatch.

6   Conclusion and Future Work

In this paper we presented a simple and intuitive representation to characterize air gestures in the context of close human-computer interaction. The strength of the representation consists in providing developers a tool to identify the main motion components in a gesture; using this information, specific features might be added or removed in order to optimize gesture recognition with a particular type of sensor. Therefore it provides an interesting tool when designing gestures to be ported across multiple types of sensors by identifying, depending on the sensors capabilities, which segments and features to focus on. However some critics can be made about the current model; the tool does not process the data deep enough to clearly identify motion components on each plan and axes; this should be enhanced to provide more precise data and thus improve the visualization tool.

Simple and intuitive visualization and characterization tools for air gesture should become more spread as the algorithms are becoming standards. As the number of sensors on the market grows, the research should tend to focus on global gestures designed for all the heterogeneous sensors technologies used for human-computer interaction.

The algorithm should be applied to a larger dataset and more heterogeneous dataset to assess its reproducibility on other gestures and develop the "None" component class. The representation should also be enhanced to be more precise and provide more information to the developer. In order to improve the precision, the exact position and orientation of each body-segment should be computed for the whole gesture using a direct kinematic algorithm using the data from IMUs or using the skeleton data from the Kinect to define the exact space covered by each segment during a gesture. To provide more information, the obtained result should also be compared with state-of-the-art algorithms; once enough algorithms and sensors compared, an estimation of the global recognition complexity of a gesture with a particular sensor could be inferred from the motion components and quantity of a gesture. Finally, the practical utility of the visualization tool should be assessed by gesture designer/developers.

References

1. Varga, R., Prekopcsák, Z.: Creating a Database for Objective Comparison of Gesture Recognition Systems. In: Proceedings of the 15th International Student Conference on Electrical Engineering, pp. 1–6 (2011)
2. Morganti, E., et al.: A Smart Watch with Embedded Sensors to Recognize Objects, Grasps and Forearm Gestures. Procedia Engineering 4, 1169–1175 (2012)
3. Wachs, J.P., Kölsch, M., Stern, H., Edan, Y.: Vision-based hand-gesture applications. Communications of the ACM 54(2), 60 (2011)
4. Ruffieux, S., Mugellini, E., Lalanne, D., Khaled, O.A.: FEOGARMA Framework to Evaluate and Optimize Gesture Acquisition and Recognition Methods. In: Workshop on Robust Machine Learning Techniques for Human Activity Recognition; Systems, Man And Cybernetics, Anchorage (2011)
5. McNeill, D.: Language and Gesture. Cambridge University Press (2000)
6. Siegman, A.W., Pope, B.: Studies in dyadic communication. Pergamon general psychology series. Pergamon (1972)
7. Eisenstein, J., Davis, R.: Visual and linguistic information in gesture classification. In: Proceedings of the 6th International Conference on Multimodal Interfaces, ICMI 2004, p. 113. ACM Press, New York (2004)
8. Karam, M.: A framework for research and design of gesture-based human computer interactions. University of Southampton (2006)
9. Pavlovic, V.I., Sharma, R., Huang, T.S.: Visual interpretation of hand gestures for human-computer interaction: A review. IEEE Transactions on Pattern Analysis and Machine Intelligence 19(7), 677–695 (2002)
10. Sato, E., Yamaguchi, T., Harashima, F.: Natural Interface Using Pointing Behavior for Human–Robot Gestural Interaction. IEEE Transactions on Industrial Electronics 54(2), 1105–1112 (2007)
11. Aigner, R., Wigdor, D., Benko, H., Haller, M.: Understanding Mid-Air Hand Gestures: A Study of Human Preferences in Usage of Gesture Types for HCI. Microsoft Research Technical Report MSR-TR-2012-111 (2012)
12. Camurri, A., Mazzarino, B., Ricchetti, M., Timmers, R., Volpe, G.: Multimodal Analysis of Expressive Gesture in Music and Dance Performances. In: Camurri, A., Volpe, G. (eds.) GW 2003. LNCS (LNAI), vol. 2915, pp. 20–39. Springer, Heidelberg (2004)

13. Marshall, M., Peters, N.: On the development of a system for gesture control of spatialization. In: Proceedings of the International Computer Music Conference (2006)

14. Zhao, L., Badler, N.I.: Synthesis and acquisition of laban movement analysis qualitative parameters for communicative gestures. University of Pennsylvania, Philadelphia (2001)

15. Poppe, R.: Vision-based human motion analysis: An overview. Computer Vision and Image Understanding 108(1-2), 4–18 (2007)

16. Marechal, L., et al.: Measurement System for Gesture Characterization During Chest Physiotherapy Act on Newborn Babies Suffering from Bronchiolitis. In: 29th Annual International Conference of the IEEE Engineering in Medicine and Biology Society, EMBS 2007, pp. 5770–5773 (2007)

17. Huang, J.: Kinerehab: A kinect-based system for physical rehabilitation: a pilot study for young adults with motor disabilities. In: The Proceedings of the 13th International ACM SIGACCESS Conference on Computers and Accessibility, ASSETS 2011, pp. 319–320. ACM Press, New York (2011)

18. Dobrian, C.: A Method for Computer Characterization of "Gesture" in Musical Improvisation. In: International Computer Music Conference, pp. 494–497 (2012)

Annotate. Train. Evaluate. A Unified Tool for the Analysis and Visualization of Workflows in Machine Learning Applied to Object Detection

Michael Storz[1], Marc Ritter[1], Robert Manthey[1], Holger Lietz[2], and Maximilian Eibl[1]

Technische Universität Chemnitz
[1] Chair Media Informatics, Chemnitz, Germany
{michael.storz,marc.ritter,robert.manthey,
maximilian.eibl}@informatik.tu-chemnitz.de
[2] Professorship on Communications Engineering, Chemnitz, Germany
holger.lietz@etit.tu-chemnitz.de

Abstract. The development of classifiers for object detection in images is a complex task that comprises the creation of representative and potentially large datasets from a target object by repetitive and time-consuming intellectual annotations, followed by a sequence of methods to train, evaluate and optimize the generated classifier. This is conventionally achieved by the usage and combination of many different tools. Here, we present a holistic approach to this scenario by providing a unified tool that covers the single development stages in one solution to facilitate the development process. We prove this concept by the example of creating a face detection classifier.

Keywords: Model-driven Annotation, Image Processing, Machine Learning, Object Detection, Workflow Analysis.

1 Introduction

Object recognition has numerous application areas and can be applied in a variety of different fields like image retrieval, driver assistance systems or surveillance technology. In the context of Human Computer Interaction (HCI), object recognition can improve the interaction process leading to more proactive devices, for instance, by enabling technical devices to analyze and recognize properties of potential users like number, position, age or gender of persons. Devices could adapt their interface or their displayed content based on this context prior to the actual usage, thereby potentially leading to a higher usability.

To harness this increased interaction potential, developers need to either use existing technology or create their own object recognition classifiers. If existing technology is either ruled out because of their proprietary nature or because of the limited range of covered object categories, the latter option needs to be explored. State of the art results of the *PASCAL VOC 2012 Challenge* [1] show that object recognition techniques yield reasonable results with an average precision between 57.8% and 97.3 % on 20 different object classes. The chosen object classes are exemplary, so that object

M. Kurosu (Ed.): Human-Computer Interaction, Part V, HCII 2013, LNCS 8008, pp. 196–205, 2013.

recognition techniques can be applied to a far higher amount of object categories, most likely with comparable results.

The open source image database *ImageNet* [2] offers more than 14 million images of different object categories. It is organized according to the *WordNet* hierarchy [3]. Even very specific categories may contain quite large numbers of images. For example, *ImageNet* contains 2,022 images for the object category 'King Penguin' alone. The creation of classifiers for object recognition is a challenging task. One hand the amount of possible object categories is very large and available annotations lack detail, e.g. concerning the alignment and positioning of objects. Two, the training and evaluation processes are very time-consuming and repetitive tasks that require domain specific knowledge. This is mainly caused by large datasets and the high dimensionality of the classification problems. Due to the absence of tools for automated comparison of annotated data with actual classifier results evaluation it is also often a cumbersome process because many evaluations are done by counting results manually. In order to simplify this process, we have built a tool that captures the classifier development from intellectual annotation to training and evaluation in a single holistic environment.

In section 2 we state related work that covers parts of pattern recognition design process. Section 3 describes the developed system. In section 4 we apply our system to the specific task of face detection to proof the systems applicability in to this and related tasks. Finally section 5 outlines future research areas.

2   Related Work

The design of a pattern recognition system usually involves the processes: data collection and pre-processing, data representation, training, and decision making [4]. Our developed tool incorporates these processes and divides them into three stages called annotation, training and evaluation. Throughout this paper, data collection and pre-processing are represented in the annotation stage, which is concerned mostly by annotation of objects in images and by incorporating existing training and test datasets. Furthermore, we represent the algorithm specific parts of the design process, data representation and model creation inside the training stage. The decision making is situated in the evaluation stage and applies the trained model on test data to evaluate classifier performance.

2.1   Datasets

To create classifiers we need to represent the object that we want to classify. This is done by creating datasets made of positive and negative examples. The task of creating a dataset consists of gathering imagery and annotation about the desired object and marking the corresponding image region. The datasets created need to incorporate as much of the possible variation in appearance of an object or concept. Covering all possible variations of an object class, however, is not always possible and as such, can lead to extremely large datasets. The appearance of an object for example, a face, can vary due to intrinsic factors like age and identity and extrinsic factors like orientation and lighting conditions, which can obviously vary quite markedly   Covering all possible variations of an object class can be a near impossible task and oftentimes leads to extensive datasets. Thousands or ten thousands of object instances are not unusual, For example *Schneiderman* [5] collected 2,000 frontal and 2,000 profile faces for developing a face detector, and used a total of 2,000 images of cars for a car detector. Even more images were used in the training of a rotation invariant face detector, which was based on a total of 75,000 face samples [6]. For the prominent case of face detection and face recognition many datasets are available and well annotated. In combination they offer a large number of different faces like *FERET* [7] or *Caltech Webfaces* [8] as well as a large diversity in appearance e.g. *Yale* [9], *PIE* [10] and *Multi PIE* [11]. Prominent examples for datasets containing more than one object categories are *PASCAL VOC* [1], *Caltech 101* [12] and *Caltech 256* [13], *Image-Net* [2] and *LabelMe* [14].

2.2 Annotation Tools

We focus on annotating information that refers to the semantic content of the image to create classifiers that in the future can perform this task automatically. The understanding of the quality of annotation can vary from a simple text label to a comprehensive semantic description of image contents. This range of quality is also what differentiates available annotation tools. The *ESP Game* [15] can be understood as an annotation tool for simple text labels describing image content. In the game two players need to find the same keyword for a specific image in order to get a reward. If a keyword for one image is mentioned more than a certain threshold it is accepted as an annotation. The online accessible annotation tool *LabelMe* [14] allows the annotation of objects in uploaded images by marking the shape and by assigning a text label. The *LHI* annotation tool [16] allows a comparably sophisticated annotation. It includes among others segmentation, sketch, hierarchical scene decomposition and semantic annotation. The annotation tool can create object templates based on annotated data, which can simplify further annotations of the same object class. Unfortunately this tool is not publicly available.

The automatic labeling environment *ALE* [17] uses segmentation techniques (*Graph Cut* [18]) to simplify the annotation of regions. It computes the labels for each image pixel, based on statistics and co-occurrence of sets of labels. The so far presented annotation tools facilitate a very fixed form of annotation, like *text label* [15] or *shape and text label* [14]. The video annotation tool *ViperGT* [19] allows users to create templates that contain all necessary properties of an object. This customization approach allows the adaption to a specific object annotation case.

2.3 Training and Evaluation

Pattern recognition systems in the field of Machine Learning consist of two primary stages [4]. Within the training phase, a model or classifier is trained from given data samples to mathematically group or divide regions in the connected feature space. During the classification stage, which is also referred to as testing and evaluation, an unseen pattern that has not been used in the training stage is mapped by the classifier into the divided feature space where a final output value is usually assigned, e.g. positive or negative example. Naturally, this appears as a mere complex task that often requires expert knowledge from the application domain and involves critical decision making in the design of specific parts of the algorithms, in data representation as well as in the creation of the model inside the training component.

For the evaluation of the detection performance of a classifier, we use Receiver Operating Characteristic (ROC) curves. The ROC curve is usually defined by the relationship of specificity (false positive rate minus 1) and sensitivity (true positive rate). The analysis of those curves yields comparable results for two or more classifiers working on under different operating points (thresholds). The area under the ROC curve (AUC) provides a normalized overall-performance measure on the quality of the employed classification models. It is a measure to describe how well the classifier separates the two object classes. The higher the AUC value, the better the classifier. If the AUC value is equal to 1.0, the accuracy of the classifier applied on the test set is 100%. [20]

ROC analysis appeared in the context of signal detection theory in the field of psychophysics to describe how well a signal could be distinguished by a receiver from noise [21]. Since then it has grown to become the gold standard in medical data analysis [22] and weather forecasting [23] and is now used as a popular tool for analyzing and visualization of the performance of machine learning algorithms [24].

2.4   Are There Other Unified Approaches Yet?

To the best of our knowledge, there are only a few approaches to modeling the complete workflow from dataset creation in combination with annotation over the application of training and evaluation components from the area of machine learning in the field of image processing. *Schreiner et al.* [25] create a semi-automatic annotation tools in the context of driver assistance systems to annotate driving maneuvers. *Meudt et al.* [26] propose a system for multi-modal annotation that can be used to annotate video, audio streams as well as biophysical data supported by active learning environment to reduce annotation costs.

The processes of training and evaluation are directly related to the field of data mining. For instance, the WEKA [27] open source framework includes numerous state-of-the-art methods to analyze and process numerical comma separated data. The *Pattern Recognition Engineering* (PaREn) system [28] provides a holistic approach to evaluation and optimization. A major effort is the shift of the inherent complexity of machine learning methods from the potential academic and scientific end-user into the program itself in order to address the applicability to other programmers, what is achieved by automated application of different algorithms to a data mining problem yielding to adapted parameter sets.

During the last decade, sustainable software development has gained a lot of interest. *PIL-EYE* [29] is a system for visual surveillance that enables the user to build arbitrary processing chains on platform independent image processing modules. Despite the seemingly rich choice of available tools, most of the mentioned tools lack at least a dedicated component for the annotation or creation of datasets, and as such, considerable time and efforts needs to be invested to adapt the settings of a given application for the specific annotation task at hand. Therefore, we follow a more generic approach that enables us to handle more common object patterns that can be easily defined by an open modeling architecture.

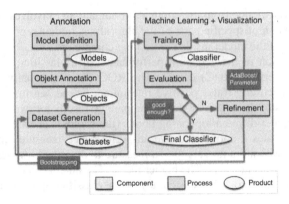

Fig. 1. Tool structure with its components, processes, intermediate products and links between processes that determine the possible sequences of the workflow

3   System Description

The proposed tool implements the design of a pattern recognition algorithm into two connected components: annotation and workflow. The annotation component includes customized annotation modeling, image annotation and dataset creation. The training and evaluation stages are integrated into the workflow component. The integration of processes like *Bootstrapping* [30] and *AdaBoost* [31] can lead to a non linear workflow. *Fig. 1* illustrates the system architecture.

3.1   Annotation Component

Following the example of *ViperGT* [19] our tools allows for the creation of annotation schema called models. They constitute a template for a certain object that needs to be annotated and can consist of an amount of the following elements: Point - coordinate (e.g. center of eye); Bounding Box - a rectangle; Ellipse - an ellipsoid; Polygon - a number of coordinates enclosing an image region; Text - a textual annotation; and Choice - a number of predefined text options. Since our annotation model is custo-mizable, common annotation models e.g. consisting of a polygon and a text label can also be created. Customization is especially advantageous if more than one locatable or several textual labels per object need to be annotated (e.g. body parts).

Additionally annotation guidelines can be added to the annotation model and its contained elements to help unify the annotation process.

Images are annotated by first importing them into the tool and then associating them with a specific model. Thereafter, the windows *Annotation Scheduler* and *Annotation Visualizer* (see *Fig. 2*) guide the annotation process. The scheduler allows the navigation through available annotation tasks - so called 'jobs'. A job is a single instance of a prede-fined model and refers to the annotation of a single object. The values of the model ele-ments are displayed in the scheduler and can be modified. The visualizer displays the locatable annotations (point, bounding box, and so on) of the current selected job and

Fig. 2. Annotation of several faces with a simple model using the *Scheduler* (right) and the *Visualizer* (top left)

others sharing the same model. In order to accelerate annotation the list of elements can be traversed automatically if an annotation is finished. Additionally the annotator can choose to annotate either all jobs subsequently or focus on the annotation of a single element in all jobs. To generate training and testing datasets, annotations can be stored in XML-format including or excluding images.

3.2   Machine Learning Workflow and Visualization

A workflow is made up by a number of separate processes that are sequentially executed and form a processing chain. In our workflow and visualization component we visualize all processes of the processing chain separately. This enables us to manipulate process parameters, visualize and store intermediate results and allow manual execution of processes. Making computations across a large number of images, as is common in the field of pattern recognition usually takes a lot of time. The visualization of intermediate results can help verify if parameter settings and also if underlying code is correct and thereby presumably avoid rerunning time-consuming computations. We allow the storage of the processing chain with its associated parameters and already computed intermediate results. This infrastructure makes it easy to pickup past computations without retraining and reconfiguring. The decomposition of a training algorithm into separate processes can lead to a high re-usability. Common algorithms encapsulated in such processes can be integrated into different processing chains.

Our tool visualizes the processing chain (see *Figure 3*) in form of a stack of square boxes that signify all involved processes. The boxes are labeled with the name of the process and can contain input fields for its parameters and a set of buttons to plot or preview intermediate results or to start associated computations. For convenience, the processing chain can either be composed using the GUI or by manipulating an XML configuration file. Evaluation results in like ROC curves can be visualized with the plot function.

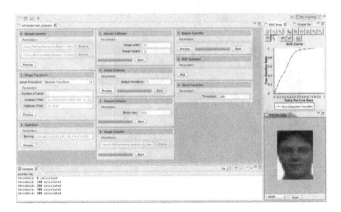

Fig. 3. Visualization of a processing chain: The GUI enables the composition of the processing chain (middle), the display of intermediate results (bottom right) and plotting of ROC curves (top right)

4 Proof of Concept at the Example of Face Detection

As a proof of concept we trained a face detection classifier using a simplified version of *Schneidermans* well known face detection method [32]. Annotation, creation of datasets, training and evaluation was done using the proposed tool.

The training process requires aligned faces. We created a simple face model containing coordinates for eyes, nose and mouth. Later on the alignment can be performed on these facial features. To create a training dataset we took faces from the *FERET* dataset [7] and created synthetic variations to increase the number of training samples. The first training dataset consisted of 50,000 positive and 50,000 negative examples. The *CMU test dataset* [33] was used for the evaluation; the images were re-annotated for that purpose.

The strategy of *Schneidermans* method [32] is to find face regions that have a high covariance and therefore are highly dependent on each other. To reduce dimensionality and to cover different resolutions and orientations a wavelet transform and a quantization are performed. Afterwards co-occurrence statistics for all wavelet coefficient pairs in all training data are gathered. The statistics are used to create subsets of coefficients that resemble highly dependent face regions. The values of wavelet coefficients in subsets form a feature value. Again these subsets are applied to the training data and all occurring feature values are gathered. To apply feature values in classification the feature values of an image region (e.g. from test dataset) need to be computed. The retrieved values can be compared with the occurrence inside the test dataset. If a specific feature value occurred more often in face samples than in non face samples the region is more likely to be a face. The classification score is computed by evaluating several feature values equal to the number of subsets.

For the training and evaluation process, we implemented the described method by constituting a processing chain of nine separate processes (see *Fig. 3*). For the optimization of the classifier, we used the bootstrapping strategy which adds false positives to the training dataset of the next iteration. *Fig. 4* shows the ROC curves for all three training iterations. Bootstrapping led to a significant increase in performance.

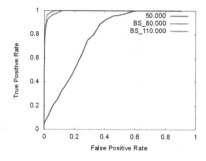

Fig. 4. Comparison of classifier performance of first iteration (50,000 training samples per class - red) with two bootstrapping iterations (additional 30,000 training samples per iteration - green and blue). The bootstrapped classifier with 80,000 has the best AUC value.

5   Summary and Future Work

We presented a generic approach to an all-in-one solution that covers the different development stages from dataset creation and model-driven annotation over training to evaluation and optimization in order to create more reliable classifiers for object detection. The proof of concept was demonstrated at the example of creating a face detection classifier. Future work will focus on predefined models of pedestrians, cars and other frequently emerging objects to provide some adaptability to customized data sets. Two final avenues we will explore in future are: an application to video annotation and analysis, and the incorporation of audiovisual models and features for speech and noise analysis.

Acknowledgments. This work was partially accomplished within the project ValidAX – Validation of the AMOPA and XTRIEVAL framework (Project VIP0044), funded by the Federal Ministry of Education and Research (Bundesministerium für Wissenschaft und Forschung), Germany and the Research Training Group CrossWorlds – Connecting Virtual and Real Social Worlds (Project GRK1780), funded by the DFG (Deutsche Forschungsgesellschaft), Germany.

References

1. Everingham, M., Van Gool, L., Williams, C.K.I., Winn, J., Zisserman, A.: The PASCAL Visual Object Classes (VOC) Challenge. International Journal of Computer Vision 88(2), 303–338 (2010)
2. Deng, J., Dong, W., Socher, R., Li, L., Li, K., Fei-Fei, L.: ImageNet: A large-scale hierarchical image database. In: IEEE International Conference on Computer Vision and Pattern Recognition, pp. 248–255 (2009)
3. Miller, G.A.: WordNet: A Lexical Database for English. Communications of the ACM 38(11), 39–41 (1995)
4. Jain, A.K., Duin, R.P.W., Gregory, R.L. (eds.): The Oxford Companion to the Mind, 2nd edn., pp. 698–703. Oxford University Press, Oxford (2004)

5. Schneiderman, H.A.: Statistical method for 3D object detection applied to faces and cars. PhD Thesis, Carnegie Mellon University (2000)
6. Huang, C., Ai, H., Li, Y., Lao, S.: High-Performance Rotation Invariant Multiview Face Detection. IEEE Transactions on Pattern Analysis and Machine Intelligence 29(4), 671–686 (2007)
7. Phillips, P., Moon, H., Rizvi, S., Rauss, P.: The FERET evaluation methodology for face-recognition algorithms. IEEE Transactions on Pattern Analysis and Machine Intelligence 22(10), 1090–1104 (2000)
8. Angelova, A., Abu-Mostafam, Y., Perona, P.: Pruning training sets for learning of object categories. In: IEEE International Conference on Computer Vision and Pattern Recognition, San Diego, CA, USA, pp. 494–501 (2005)
9. Georghiades, A., Belhumeur, P., Kriegman, D.: From few to many: illumination cone models for face recognition under variable lighting and pose. IEEE Transactions on Pattern Analysis and Machine Intelligence 23(6), 643–660 (2001)
10. Sim, T., Baker, S., Bsat, M.: The CMU Pose, Illumination, and Expression (PIE) database. In: Fifth IEEE International Conference on Automatic Face and Gesture Recognition, pp. 46–51 (2002)
11. Gross, R., Matthews, I., Cohn, J., Kanade, T., Baker, S.: Multi-PIE. Journal Image and Vision Computing 28(5), 807–813 (2010)
12. Fei-Fei, L., Fergus, R., Perona, P.: Learning generative visual models from few training examples: an incremental Bayesian approach tested on 101 object categories. Journal Computer Vision and Image Understanding 106(1), 59–70 (2007)
13. Griffin, G., Holub, A., Perona, P.: Caltech-256 Object Category Dataset. California Institute of Technology. Technical Report 7694 (2007)
14. Russell, B., Torralba, A., Murphy, K., Freeman, W.: LabelMe: a database and web-based tool for image annotation. International Journal of Computer Vision 77(1), 157–173 (2008)
15. Ahn, L., von, D.L.: Labeling images with a computer game. In: Proceedings of the 2004 Conference on Human Factors in Computing Systems, pp. 319–326 (2004)
16. Yao, B., Yang, X., Zhu, S.-C.: Introduction to a large-scale general purpose ground truth database: Methodology, annotation tool and benchmarks. In: Yuille, A.L., Zhu, S.-C., Cremers, D., Wang, Y. (eds.) EMMCVPR 2007. LNCS, vol. 4679, pp. 169–183. Springer, Heidelberg (2007)
17. Ladicky, L., Russell, C., Kohli, P., Torr, P.H.S.: Graph cut based inference with co-occurrence statistics. In: Daniilidis, K., Maragos, P., Paragios, N. (eds.) ECCV 2010, Part V. LNCS, vol. 6315, pp. 239–253. Springer, Heidelberg (2010)
18. Boykov, Y., Veksler, O., Zabih, R.: Fast approximate energy minimization via graph cuts. IEEE Transactions on Pattern Analysis and Machine Intelligence 23(11), 1222–1239 (2001)
19. Doermann, D., Mihalcik, D.: Tools and Techniques for Video Performance Evaluation. In: Proc. 15th International Conference on Pattern Recognition, vol. 4, pp. 167–170 (2000)
20. Lachiche, N., Flach, P.A.: Improving Accuracy and Cost of Two-class and Multi-class Probabilistic Classifiers Using ROC Curves. In: 20th International Conference on Machine Learning, pp. 416–423 (2003)
21. Tanner, W.P.J.R., Swets, J.A., Welch, H.W.: A New Theory of Visual Detection. Defense Technical Information Center, Electronic Defense Group, University of Michigan. Technical Reports, p. 42 (1953)
22. Metz, C.E.: Receiver operating characteristic analysis: A tool for the quantitative evaluation of observer performance and imaging systems. Journal of the American College Radiology 3(6), 413–422 (2006)

23. World Meteorological Organization (Eds.): Manual on the Global Data Processing System, part II, Attachments II.7 and II.8. 2010, Updated in 2012. Switzerland, p. 193 (2012)
24. Provost, F.J., Fawcett, T.: Robust Classification for Imprecise Environments. Machine Learning 42(3), 203–231 (2001)
25. Schreiner, C., Zhang, H., Guerrero, C., Torkkola, K., Zhang, K.: A Semi-Automatic Data Annotation Tool for Driving Simulator Data Reduction. In: Driving Simulation Conference, North America, p. 9 (2007)
26. Meudt, S., Bigalke, L., Schwenker, F.: Atlas Annotation tool using partially supervised learning and multi-view co-learning in human-computer-interaction scenarios. In: 11th International Conference on Information Science, Signal Processing and their Applications, pp. 1309–1312 (2012)
27. Hall, M., Frank, E., Holmes, G., Pfahringer, B., Reutemann, P., Witten, I.H.: The WEKA Data Mining Software: An Update. SIGKDD Explorations 11(1), 10–18 (2009)
28. Shafait, F., Reif, M., Kofler, C., Breuel, T.: Pattern Recognition Engineering. In: Rapid-Miner Community Meeting and Conference, Dortmund, Germany (2010)
29. Chang, H.J., Yi, K.M., Yin, S., Kim, S.W., Baek, Y.M., Ahn, H.S., Choi, J.Y.: PIL-EYE: Integrated System for Sustainable Development of Intelligent Visual Surveillance Algorithms. In: IEEE Digital Image Computing: Techniques and Applications, pp. 231–236 (2011)
30. Sung, K.K., Poggio, T.: Example-based learning for view-based human face detection. IEEE Transactions on Pattern Analysis and Machine Intelligence 20(1), 39–51 (1998)
31. Schapire, R., Freund, Y.: A decision theoretic generalization of on-line learning and an application to boosting. Journal of Computer and System Sciences 55(1), 119–139 (1997)
32. Schneiderman, H.: Learning statistical structure for object detection. In: Petkov, N., Westenberg, M.A. (eds.) CAIP 2003. LNCS, vol. 2756, pp. 434–441. Springer, Heidelberg (2003)
33. Rowley, H., Baluja, S., Kanade, T.: Neural network-based face detection. In: International Conference on Computer Vision and Pattern Recognition, pp. 203–208 (1996)

A New Real-Time Visual SLAM Algorithm Based on the Improved FAST Features

Liang Wang[1], Rong Liu[2], Chao Liang[1], and Fuqing Duan[3]

[1] College of Electronic Information and Control Engineering,
Beijing University of Technology, Beijing 100124, China
[2] Base Department, Beijing Institute of Fashion Technology, Beijing 100029, China
[3] College of Information Science and Technology,
Beijing Normal University, Beijing 100875, China
wangliang@bjut.edu.cn

Abstract. The visual SLAM is less dependent on hardware, so it attracts growing interests. However, the visual SLAM, especially the Extend Kalman Filter-based monocular SLAM is computational expensive, and is hard to fulfill real-time process. In this paper, we propose an algorithm, which uses the binary robust independent elementary Features descriptor to describe the features from accelerated segment test feature aiming at improving feature points extraction and matching, and combines with the 1-point random sample consensus strategy to speedup the EKF-based visual SLAM. The proposed algorithm can improve the robustness of the EKF-based visual SLAM and make it operate in real-time. Experimental results validate the proposed algorithm.

1 Introduction

The simultaneous localization and mapping (SLAM) is a problem that if it is possible, for a robot or autonomous vehicle to be placed at an unknown environment, to build a map of the environment and at the same time use this map to determine its location [1–3]. According the type of sensors, the SLAM techniques in literatures can be roughly categorized into laser-based, sonar-based, Global Positioning System (GPS)-based and vision-based techniques. In comparison with other sensors, the visual sensors are passive, which have high resolution, long range and low dependence on hardware, and can provide both depth and appearance information of the environment with one (image) measurement. So there is a growing interest in vision-based (or visual) SLAM.

Generally, there are two types of vision-based SLAM systems: the monocular SLAM and the binocular SLAM. In comparison with the latter, the former has less dependence on the hardware and more flexibility. Then the monocular SLAM is the focus of the research community. For simplicity, objects in the scene are usually assumed to be static. The monocular SLAM is performed by obtaining the depth information as follows. The camera captured images of features in the scene during its moving. Then with the knowledge of computer vision, the depth information of each feature observed in both current and last frame can be

M. Kurosu (Ed.): Human-Computer Interaction, Part V, HCII 2013, LNCS 8008, pp. 206–215, 2013.

computed. With the reconstructed 3D information of features, the environment map consisting of sparse features is building, and the localization information of the camera is also obtained.

Currently, most of the monocular SLAM algorithms are based on image point features. One representative work is the Parallel Tracking and Mapping (PTAM) algorithm proposed by G.Klein and D.Murray [4]. The PTAM algorithm splits the SLAM into two separate tasks: tracking and mapping. These tasks are processed in parallel threads: one thread deals with the task of robustly tracking camera motion, while the other produces a 3D map of feature points from previously observed images. With robust and sparse feature points, the tracking can robustly retrieve the camera trajectory and the asynchronously mapping can provide a dense map. This algorithm has a high precision. However its computational cost dramatically increases with the increase of the number of feature points. So it is limited to the indoor scene and applications with little movement in a small region, and not suitable for general applications. The other representative work is due to Davison et al. [1], which performs a recursive updating of scene structure and camera motion estimates using an Extended Kalman Filter (EKF). It has better environment adaptability and can fulfill task in large areas such as several street blocks, one city and so on [5]. However, the high computational cost is the bottleneck of the EKF-based SLAM. The main reason for its high computational cost comes from that most image point feature extraction methods used in the EKF-based SLAM cannot trade off the speed of extraction and the reliability of matching.

In this paper, we propose to use the Binary Robust Independent Elementary Features(BRIEF) descriptor to describe the Features from Accelerated Segment Test (FAST) aiming at improving feature points extraction and matching, and combine it with the 1-point RANdom Sample Consensus (RANSAC) strategy to improve the EKF-based visual SLAM. The BRIEF descriptor uses a binary sequence to describe the extracted feature point in the extraction process. Then in the matching process, the Hamming distance is calculated to perform features matching, which is just the XOR calculation of two sequences in bitwise. It can better trade off the speed of feature extraction and the reliability of feature matching. The 1-point RANSAC can reduce the number of the correction in the updating step of the EKF, i.e. reduce the dimension of the state, to improve the calculation speed. By incorporating the BRIEF and 1-point RANSAC strategy into the EKF-SLAM, we improve the robustness of the EKF-based SLAM, which is also satisfying real-time demands. Experimental results validate the proposed algorithms.

The paper is organized as follows: In Section 2 the computational framework of the EKF-based visual SLAM is briefly discussed. Then we show how to incorporate the improved FAST features and 1-point RANSAC strategy into the EKF-SLAM in Section 3 and 4 respectively. In Section 5 some experimental results are presented. Finally the paper is concluded in Section 6.

2   The EKF-Based Visual SLAM

The EKF is the extension of Kalman filter. It can be used to handle the non-linearity in transfer matrix and measurement function, and the nonlinear noise in the linear system. It is suitable to describe most of physical systems. The mathematical model of the EKF-based SLAM can be summarized as follows [2].

2.1   The State Description of the Visual SLAM System

In EKF-based monocular SLAM, the state vector of a camera, x_v, has 13 elements, which consists of a metric 3D position vector p^W, orientation quaternion q^{CW}, velocity vector v^W, and angular velocity vector ω^C relative to a fixed world frame W and camera frame C.

$$x_v = \begin{bmatrix} p^W \\ q^{CW} \\ v^W \\ \omega^C \end{bmatrix} \tag{1}$$

The state of the visual SLAM system can be described by a state vector $\hat{x}$ and its covariance matrix P. State vector $\hat{x}$ comprises the stacked state estimates of the camera $\hat{x}_v$ and feature points $\hat{y}_i$. Covariance matrix P is a square matrix which can be partitioned into submatrix elements as follows.

$$\hat{x} = \begin{bmatrix} \hat{x}_v \\ \hat{y}_1 \\ \hat{y}_2 \\ \vdots \end{bmatrix} \quad P = \begin{bmatrix} P_{xx} & P_{xy1} & P_{xy1} & \cdots \\ P_{y1x} & P_{y1y1} & P_{y1y2} & \cdots \\ P_{y2x} & P_{y2y1} & P_{y2y2} & \cdots \\ \vdots & \vdots & \vdots & \end{bmatrix} \tag{2}$$

2.2   The Motion Model of Camera

The "constant velocity, constant angular velocity model" is used to describe the motion of camera [2]. It is assumed that in each time step acceleration a^W and angular acceleration α^W observe the zero mean and Gaussian distribution. Then the velocity and angular velocity follow

$$n = \begin{bmatrix} V^W \\ \Omega^C \end{bmatrix} = \begin{bmatrix} a^W \Delta t \\ \Omega^C \Delta t \end{bmatrix} \tag{3}$$

Assuming that the covariance matrix of the vector n is diagonal, the state update procedure is

$$f_v = \begin{bmatrix} p^W + (V^W + v^W)\Delta t \\ q^{WC} \times q((V^W + v^W)\Delta t) \\ v^W + V^W \\ \omega^W + \Omega^W \end{bmatrix} \tag{4}$$

where $q((V^W + v^W)\Delta t)$ the quaternion corresponding to the angle-axis rotation vector $(V^W + v^W)\Delta t$.

In the EKF, the state uncertainty (covariance matrix) Q_v of the new state estimate $\mathbf{f}_v(\mathbf{x}_v, \mathbf{u})$ for the camera can be computed via the Jacobian matrix:

$$\mathbf{Q}_v = \frac{\partial \mathbf{f}_v}{\partial n} \mathbf{P}_n \frac{\partial \mathbf{f}_v}{\partial n}^T \tag{5}$$

where $\mathbf{p}_n$ is the covariance matrix of vector $\mathbf{n}$.

2.3   The Measurement Model of Camera

The measurement model uses the observed feature points in the scene and position of camera to predict the image position of each feature point before deciding which is to be measured. Then actively measure selected feature points to update the EKF. It firstly transforms feature point from the world coordinate system to the camera coordinate system

$$\mathbf{h}_i^C = \begin{bmatrix} \mathbf{x}_i^C \\ \mathbf{y}_i^C \\ \mathbf{z}_i^C \end{bmatrix} = \mathbf{R}^{CW}(\mathbf{y}_i^W - \mathbf{p}^W) \tag{6}$$

Then the image coordinates can be found using the standard pinhole camera model [6]

$$\mathbf{h}_i = \begin{bmatrix} u \\ v \end{bmatrix} = \begin{bmatrix} u_0 - f_u h_x/h_z \\ v_0 - f_v h_x/h_z \end{bmatrix} \tag{7}$$

where f_u, f_v is the focal length along u and v axis respectively, (u_0, v_0) is the principal point.

Finally, the predicted image coordinates of the feature can be determined with the distortion parameter K_1

$$\begin{cases} u_d = u_0 + \dfrac{u - u_0}{\sqrt{1 + 2K_1[(u-u_0)^2 + (v-v_0)^2]}} \\ v_d = v_0 + \dfrac{v - v_0}{\sqrt{1 + 2K_1[(u-u_0)^2 + (v-v_0)^2]}} \end{cases}. \tag{8}$$

2.4   The Iteration Process of the EKF

The iteration process of the EKF can be divided into two steps: prediction and updating [2]. In the prediction step, system can predict the state of the next step. Then in the updating step, the predicted parameters can be rectified with current measurements. The processes can be express as follows

$$\hat{\mathbf{x}}_{new} = \hat{\mathbf{x}}_{old} + W(\mathbf{z}_i - \mathbf{h}_i) \tag{9}$$

$$\hat{\mathbf{p}}_{new} = \hat{\mathbf{p}}_{old} - WSW^T \tag{10}$$

where $\mathbf{z}_i$ and $\mathbf{h}_i$ denotes the measured and predicted value of the feature point respectively, W denotes the Kalman gain,

$$W = p\frac{\partial \mathbf{h}_i}{\partial \mathbf{x}}^T S^{-1} = S^{-1} \begin{bmatrix} p_{xx} \\ p_{y1x} \\ p_{y2x} \\ \vdots \end{bmatrix} \frac{\partial \mathbf{h}_i}{\partial \mathbf{x}_v}^T + S^{-1} \begin{bmatrix} p_{xyi} \\ p_{y1yi} \\ p_{y2yi} \\ \vdots \end{bmatrix} \frac{\partial \mathbf{h}_i}{\partial \mathbf{x}_v}^T \tag{11}$$

where S denotes the uncertainty of features, the uncertainty of each feature can be expressed as

$$S_i = \frac{\partial u_{di}}{\partial \mathbf{x}_v} p_{xx} \frac{\partial u_{di}}{\partial \mathbf{x}_v}^T + \frac{\partial u_{di}}{\partial \mathbf{x}_v} p_{xyi} \frac{\partial u_{di}}{\partial \mathbf{x}_{yi}}^T + \frac{\partial u_{di}}{\partial \mathbf{x}_{yi}} p_{yix} \frac{\partial u_{di}}{\partial \mathbf{x}_v}^T + \frac{\partial u_{di}}{\partial \mathbf{x}_{yi}} p_{yiyi} \frac{\partial u_{di}}{\partial \mathbf{x}_{yi}}^T \quad (12)$$

where R is the noise matrix.

3   The Extraction and Matching of Point Feature

Most of the visual SLAM is based on point features, which takes image points corresponding to features in the scene as the landmark to construct 3D map and perform localization. So the quality of extraction and matching of feature points in images has great impact on the visual SLAM. Currently, the most popular feature point is the Scale-Invariant Feature Transform (SIFT) point [2]. The SIFT feature uses a vector with 128 elements to describe a feature point [8]. It is invariant to location, scale and rotation, and partially invariant to illumination changes. However the computational cost is too large to satisfy the real-time demands. The FAST [9] with BRIEF description [10] are used to perform feature extraction and matching.

3.1   Feature Extraction

The FAST point [9] is extracted from the scene to fasten the feature extraction. In real applications, each image is partitioned into several blocks and feature extraction is performed in each block. Firstly, randomly select one image coordinates. Take a 20×30 rectangular region whose center is the selected image coordinates. Then determine whether a feature point has been selected from this region. If there is no FAST point, perform FAST feature extraction and select the point with the highest precision as the FAST point of this region. Otherwise, re-select the image coordinates and repeat the former step until it reaches the threshold of the feature point number or the maximum iteration times. This method can reduce the computational cost and avoid selecting too dense features in some regions.

Once FAST points are extracted, the BRIEF descriptor [10] is used describe these feature points, which is a binary sequence. BRIEF is based on comparisons. Take a patch with a FAST point locating at the center of the patch. Then choose one point other than the center in this patch. Compare intensities of the selected point and the center. If the intensity of the selected point is larger than that of the center, assign the value '1', else '0'. Do that for each point in the patch except the center and end up with a string of boolean values. To improve the robustness, the image patch can be filtered by the Gaussian filter before comparison.

3.2   Feature Matching

Feature matching is performed by calculating the Hamming distance between BRIEF descriptors of two extracted FAST points. These two feature points come form the current image and the next image respectively. The Hamming distance is determined by bit-wise XOR operation. Only the two feature points whose computed Hamming distance is less than the threshold, are considered as the corresponding point. Due to the existence of regions with similarity texture, false matches inevitably occur. Since the motion of camera is continuous, positions of the corresponding feature point in current and next image should not be far away from each other. A 20×20 square region is used to narrow the match region. If two matching feature points are covered by a 20×20 square, the match is considered as a correct one, otherwise it is regarded as a false match and deleted.

3.3 Feature Fusion

After feature matching, the EKF can be updated by exploiting coordinates of the predicted feature points and coordinates of matched feature points. Do not delete those feature points without matches in the next image immediately. This is because in the real application, due to the influence of light changes or motion blur, some feature points would not be extracted in some images, however would appear again in later images. So in this case, these feature points are first marked, and cannot be used to update the EKF. The predicted values instead of coordinates of matched feature points are used to update the EKF. Once one feature does not find correct match in continuous five frames, the feature point is considered as lost and deleted.

4 Incorporating the 1-Point RANSAC into the EKF-Based SLAM

The computational framework of the EKF is introduced in Section 2. It is obvious that the computational cost is very high, especially in the updating step whose computational complexity is $O(M * N^2)$, where N is the dimension of the EKF state, M is the number of feature points. This is one important reason that the existing algorithms do not satisfy the real-time demands. In this section, we incorporating the 1-Point RANSAC into the EKF-based SLAM to reduce the computational cost.

The 1-Point RANSAC algorithm is due to Civera et al. [7]. The main contribution is that it greatly reduce computational cost of the updating step of the EKF. This is obtained by introducing the RANSAC scheme. Then the iteration number of updating step and the dimension number of the state are greatly reduced. It has following two steps.

4.1 Inlier Processing

The input of the 1-Point RANSAC algorithm is measured feature points z. Firstly the probability of measured feature point $\eta(h_K(\hat{\mathbf{x}}_{K|K-1}), S_K)$ is computed. Those feature points whose probability are equal to or greater than 99% are considered as inliers. Then make random sample from obtained inliers by the RANSAC scheme. In this process, to reduce the computational cost, only the state vector $\hat{\mathbf{x}}$, rather than the covariance matrix P, is updated. Finally, the number of iteration in updating step of the EKF is computed with

$$n_{hyp} = \frac{log(1-p)}{log(1-(1-\epsilon)^m)} \tag{13}$$

where p denotes the desired precision, ϵ is the inlier rate, $m = 1$.

4.2 Outlier Processing

The extracted feature points also contain the matched feature points with low precision and mismatched points besides inliers. The matched feature points with low precision can be divided into two categories: one class are feature points extracted in the first time whose low precision is due to the first extraction and limited number of iteration, the other are feature points near the camera. To improve the precision, we update each matched feature point with low precision and discard the mismatched feature points. First, use the state vector and covariance matrix obtained in inlier processing step to perform updating. Then validate all outlier. Only those feature points whose correlation rate is greater than 99% are regarded as inliers and the others are discarded.

(a) one frame of captured video

(b) feature points in one frame

(c) one frame with discarded feature point (d) one frame with unmatched feature point

Fig. 1. Sample frames of video captured by the camera and feature points used to perform the proposed algorithm

5   Experiments

To validate the proposed algorithm, some experiments have been performed. One of them is reported here. The data are from the University of Zaragoza [7]. The proposed algorithm is developed with Visual C++ and some functions of the OpenCV. It runs on a notebook computer with a P4 2.4GHZ CPU.

Figure 1(a) shows the first frame of the video captured by the camera performed the monocular SLAM. Figure 1(b) shows one frame of the video, in which feature points are extracted and matched and the red ellipse is used to show that the corresponding feature point is a inlier with high innovation. The size of the ellipse presents the predicted feature point within this ellipse with a likelihood of 99%. With the running of the algorithm, the precision of prediction increases. In figure 1(c), we can see that the size of red ellipse are greatly reduced. In figure 1(d), there is a magenta ellipse besides red ellipses, which is used to express the feature point discarded by the 1-Point RANSAC due to large error. In figure 1(c), there is a blue ellipse which is used to express the mismatched feature point. In experiments, we do not remove the feature points corresponding to the magenta and blue ellipses immediately. This is because some feature points would not be extracted in some images due to light changes or motion blur, however they would appear again in later frames. Figure 2 shows the final trajectory of the camera. Figure 3 shows the process time of each frame, where the largest one is about 47.8ms, the least one is about 5.6ms and the average is about 23.53ms. For real applications with a camera of 30 frames per second, the proposed algorithm can satisfy the real-time demands.

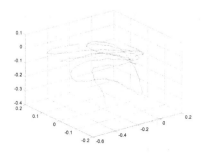

Fig. 2. Trajectory of the camera obtained by the proposed algorithm

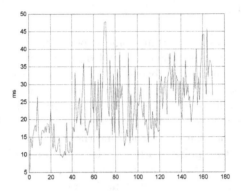

Fig. 3. Time consuming of the proposed algorithm

6   Conclusions

In this paper, we investigate the real-time monocular SLAM. A new algorithm is proposed to improve the EKF-based monocular SLAM. It incorporates the FAST feature detector with the BRIEF descriptor to extract and match image feature points and the 1-Point RANSAC strategy to improve the EKF-based SLAM. The proposed algorithm can improve the robustness of the EKF-based SLAM, which is also satisfying real-time demands. Experimental results validate the proposed algorithm.

Acknowledgment. This work was partially supported by the National Natural Science Foundation of China (Grant No.61101207).

References

1. Davison, A.J.: Real-time simultaneous localisation and mapping with a single camera. In: Proceedings of the International Conference on Computer Vision, Nice (2003)
2. Davison, A.J., Molton, N.D., Reid, I.D., Stasse, O.: MonoSLAM: Real-time single camera SLAM. IEEE Transactions on Pattern Analysis and Machine Intelligence 29(6), 1052–1067 (2007)
3. Civera, J., Davison, A.J., Montiel, J.M.: Inverse depth parametrization for monocular SLAM. IEEE Transactions on Robotics 24(5), 932–945 (2008)
4. Klein, G., Murray, D.W.: Parallel tracking and mapping for small AR workspaces. In: Proceedings of the International Symposium on Mixed and Augmented Reality, pp. 1–10 (2007)
5. Strasdat, H., Montiel, J.M., Davison, A.J.: Visual SLAM: Why Filter? Image and Vision Computing 30(2), 65–77 (2012)
6. Hartley, R., Zisserman, A.: Multiple view geometry in computer vision, 2nd edn. Cambridge University Press, Cambridge (2004)

7. Civera, J., Grasa, O.G., Davison, A.J., Montiel, J.M.: 1-Point RANSAC for EKF filtering: application to real-time structure from motion and visual odometry. Journal of Field Robotics 27(5), 609–631 (2010)
8. Lowe, D.G.: Object recognition from local scale-invariant features. In: Proceedings of the 7th International Conference on Computer Vision, Kerkyra, Herbert Bay (1999)
9. Rosten, E., Drummond, T.: Machine learning for high-speed corner detection. In: Leonardis, A., Bischof, H., Pinz, A. (eds.) ECCV 2006, Part I. LNCS, vol. 3951, pp. 430–443. Springer, Heidelberg (2006)
10. Calonder, M., Lepetit, V., Strecha, C., Fua, P.: BRIEF: Binary Robust Independent Elementary Features. In: Daniilidis, K., Maragos, P., Paragios, N. (eds.) ECCV 2010, Part IV. LNCS, vol. 6314, pp. 778–792. Springer, Heidelberg (2010)

A Coastline Detection Method Based on Level Set

Qian Wang[1], Ke Lu[1], Fuqing Duan[2], Ning He[3], and Lei Yang[4]

[1] College of Engineering and Information Technology,
Graduate University of Chinese Academy of Sciences, Beijing, 100049, China
[2] College of Information Science and Technology,
Beijing Normal University, Beijing, 100875, China
[3] School of Information, Beijing Union University, Beijing 100101, China
[4] National Satellite Meteorological Center, Beijing 100081, China
luk@gucas.ac.cn

Abstract. This paper proposes a level set based coastline detection method by using the template initialization and local energy minimization. It can complete the sea-land boundary detection in infrared channel image. This method is an improvement on the traditional level set algorithm by using the information of GSHHS to optimize the initialization procedure, which can reduce the number of iterations and numerical errors. Moreover, this method optimizes regional energy functional, and can achieve the rapid coastline detection. Experiments on the IR image of FY-2 satellite show that the method has fast speed and high accuracy.

Keywords: Edge detection, level set method, IR image processing.

1   Introduction

With the increase of the remote sensing satellite data, the preprocessing of satellite remote sensing data becomes more and more time consuming. Image navigation is the first step of preprocessing, which allows researchers to get the geographical latitude and longitude of each satellite image pixels. Automated image navigation makes satellite data processing efficiency improved significantly [1].The procedure of the automatic navigation of remote sensing images is as follows: firstly the real satellite image is matched with the template image to obtain the relative offset, and then adjust this satellite's attitude according to the offset . In the process, researchers usually realize the image matching using the satellite image coastline as features. Therefore, to achieve automatic navigation of remote sensing images, the first problem is: the automatic coastline detection of the remote sensing image.

The theory and technology of edge detection has made significant progress in recent years. Many new edge detection techniques have been developed, and they are based on neural networks, genetic algorithms, wavelet transform, morphology and partial differential equations (PDE) etc. Compared to other emerging methods, the method based on partial differential equations, has stronger local adaptability and higher flexibility [2]. PDE itself is based on the continuous image model, which makes the change at the current time of a pixel value in the image only depend on a

M. Kurosu (Ed.): Human-Computer Interaction, Part V, HCII 2013, LNCS 8008, pp. 216–226, 2013.

near infinite small neighborhood of the pixel point. In this sense, PDE method has nearly infinite local adaptive capacity. Considering the characteristics of satellite remote sensing images and the advantages of the PDE method, we use a method based on level set.

2   Related Works

Currently, in many mature edge detection method based on PDE, level set method performs well when processing complicated images and multi-target images, and it is easy to expand, Level set method can be categorized into two major classes: edge-based geometric active contour models [3-5] and region-based geometric active contour models [6-11].

The earliest level set model is the geometric active contour model proposed in 1993 by Caselles[3]. Since the model depends on the edge gradient information, the edge detection result is not ideal due to the tiny gradient. Moreover, level set model is sensitive to noise. Therefore, the contours detected by the model are discontinuous. In order to solve the problems of edge-based models, Chan [6] proposed a region-based model based on a simplified Mumford-Shah energy functional, referred as CV model. The model combines edge information and regional information, and no longer depends on the gradient information on the edges. The model is not sensitive to noise and applies to fuzzy edges or discontinuous edges. Subsequently, a number of researchers have made some improvements on the basis of this model. Li [7] propose a level set method without re-initialization. Lee[8] propose a level set method with stationary global minimum. Both of two models are based on the fact that image intensities are statistically homogeneous (roughly a constant) in each region. Therefore, they are known as the piecewise constant (PC) models. In fact, the real images often show intensity inhomogeneity, and PC model is not available for these images. To address this issue, Tsai[9] and Vese[10] independently propose two improved scheme to overcome the shortcomings of the PC model by piecewise smooth function, and make the edge extraction range expandable to multi-phase images. These models are called piecewise smooth (PS) model. However, the PS model also has the disadvantage of the large amount of calculation.

To efficiently perform the edge detection of images with intensity inhomogeneity, a new class of models has been proposed which not only utilize region-based techniques but also incorporate the benefits of local information. Brox[12] propose the idea of incorporating localized statistics into a variational framework, which shows that segmentation with local means is a first order approximation of the piecewise smooth simplification of the Mumford–Shah functional. Sum et al [13] minimize the sum of a global region-based energy and a local energy based on image contrast, thus construct a new hybrid-based level set model for the edge detection of vascular images with intensity inhomogeneity. Lankton et al[14] propose a new curve evolution scheme based on the combination of the geodesic active contour model and region-based method, which allows the region-based energy to be localized in a fully variaional way so that edges of objects with intensity inhomogeneity can be successfully extracted with the localized energies. Li[15] propose a variational level set method by introducing a local binary fitting (LBF) energy with a kernel function. By drawing upon spatially varying local region information, the LBF model is able to deal with intensity inhomogeneity. Recently, Li et al[16] improve the above formulation, and propose a distance regularized level set evolution(DRLSE) method. The method has an intrinsic capability of maintaining regularity of the level set function, particularly the desirable signed distance property in a vicinity of the zero level set, thereby avoids reinitialization procedures and its induced numerical errors, greatly reduces the number of iterations.

In order to adapt to the characteristics and the processing requirements of the infrared image, on the basis of the LBF method, we propose a new level set formulation by combining the template initialization and local energy minimization. The method optimizes initialization procedure by optimization of the information of GSHHS, reduces the number of iterations and numerical errors, improves the stability of the algorithm; optimizes regional energy functional, and improves the efficiency of the algorithm.

3   Coastline Detection Method Based on Level Set

In order to adapt the intensity inhomogeneous in IR images, this paper proposes a novel level set initialization method by combining the template initialization and local energy minimization. As shown in Fig.1, the system is composed of two sections: object region selection and edge detection. In the region selection section, we select a plurality of regions with the relative position and the same size respectively in the IR image and the template image by certain rules. In the edge extraction section, firstly, we initialize the object regions by the coastline contours in the template, and then we complete edge detection for object region through many times of iteration calculation. The detection results are the coastline contours in IR images.

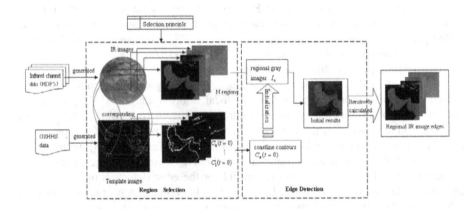

Fig. 1. Level set system based on template initialization

Object Region Selection. When selecting regions, we should pay attention to the exclusion of adverse factors, i.e. choosing the cloudless regions. At the same time, we should try to select the regions with better features too, such as the boundaries of selected regions with corners, continuous, not far away relative to the earth center.

Initialization Based on Template. Traditional level set algorithm is sensitive to the initialization conditions. Different initialization may lead to different evolution results. This paper generates the global coastline template images according to the Global Self-consistent Hierarchical High-resolution Shorelines (GSHHS) data, and then initializes level set by template images. In practice, we set the initial state to the coastline $C_n(t=0)$ in the evolution image I_n'. This means that the initial contour of the level set algorithm is consistent with template coastline. While matching, both the x-axis and y-axis offset of I_n relative to I_n' are no more than seven pixels. So we can find out, initial value of the level set algorithm is closed to the final evolution result, which can greatly reduce the number of iterations, thereby reduce the amount of computation. To some extent, this approach improves the accuracy and stability of the algorithm.

Region-Scalable Energy Model. In this section, we construct a region-scalable energy model [15] as the energy functional in level set method. Minimizing the energy functional can make the contour curves gradually approach the target boundary in the image, and thus achieve the purpose of the edge detection.

Consider a given vector valued image $I : \Omega \to \Re^d$, where $\Omega \subset \Re^n$ is the image domain, and is the dimension of the vector $I(x)$. Let C be a closed contour in the image domain Ω, which separates Ω into two regions: $\Omega_1 = outside(C)$ and $\Omega_2 = inside(C)$. For a given point $x \in \Omega$, we define the following local energy:

$$
\begin{aligned}
&e_x(C, f_1(x), f_2(x)) \\
&= \sum_{i=1}^{2} \lambda_i \int_{\Omega_i} K_\sigma(x-y) |I(y) - f_i(x)|^2 \, dy
\end{aligned}
\tag{1}
$$

where $K_\sigma(u) = \dfrac{1}{(2\pi)^{n/2} \sigma^n} e^{-|u|^2/2\sigma^2}$ is a Gaussian kernel, with a scale parameter $\sigma > 0$, λ_i is positive constant, and $f_i(x)$ is the approximation of image intensities in Ω_i.

In (1), the energy e_x is described by a weighted mean square error, with the weight $K_\sigma(x-y)$. Due to the localization property of the kernel function, the contribution of the intensity $I(y)$ to the energy e_x decreases increasingly as the point y goes away from the center point x. Therefore, only the points y in the neighborhood $\{y : | x - y | < 3\sigma\}$ are, $I(y)$ is dominant in the energy e_x.The values $f_1(x)$ and $f_2(x)$ approximate the image intensities in a region centered at the point x, whose size can be controlled by the scale parameter σ. The energy e_x with a small σ only involves the intensities within a small neighborhood of the point x, while the energy e_x with a large σ involves the image intensities in a large region centered at x. Therefore, the energy e_x is region-scalable.

Given a center point x, the energy e_x can be minimized when the contour C is exactly on the object boundary and the estimated values f_1 and f_2 optimally approximate the local image intensities on the two sides of C. To obtain the entire object boundary, we must find a contour C that minimizes the energy e_x for all x in the image domain Ω. This can be achieved by minimizing the integral of e_x over all the center points x in the image domain Ω, namely, $\int e_x(C, f_1(x), f_2(x))dx$. In addition, it is necessary to smooth the contour C by penalizing its length. Therefore, we define the following energy functional for a contour C :

$$
\begin{aligned}
& E(C, f_1(x), f_2(x)) \\
& = \int e_x(C, f_1(x), f_2(x))dx + \upsilon | C |
\end{aligned}
\tag{2}
$$

To handle topological changes, we will convert it to a level set formulation. Let the level set function ϕ take positive value outside the contour C and negative value inside C. Let H be the Heaviside function, then the energy functional $e_x(C, f_1(x), f_2(x))$ can be expressed as

$$
e_x(C, f_1(x), f_2(x)) = \sum_{i=1}^{2} \lambda_i \int_\Omega K_\sigma(x - y) | I(y) - f_i(x) |^2 M_i(\phi(y))dy
\tag{3}
$$

Let $M_1(\phi) = H(\phi)$ and $M_2(\phi) = 1 - H(\phi)$ in (3). Thus, the energy E in (2) can be written as

$$
\begin{aligned}
& E(C, f_1, f_2) \\
& = \sum_{i=1}^{2} \lambda_i \int(\int K_\sigma(x - y) | I(y) - f_i(x) |^2 M_i(\phi(y))dy)dx + \upsilon \int | \nabla H(\phi(x)) | dx
\end{aligned}
\tag{4}
$$

Where the last term $\int | \nabla H(\phi(x)) | dx$ computes the length of the zero level contour of ϕ, can be equivalently expressed as the integral $\int \delta(\phi(x)) | \nabla \phi(x) | dx$.

In practice, the Heaviside function H is approximated by a smooth function H_ε defined by

$$H_\varepsilon(x) = \frac{1}{2}[1 + \arctan(\frac{x}{\varepsilon})] \tag{5}$$

The derivative of H_ε is

$$\delta_\varepsilon(x) = H_\varepsilon^{'}(x) = \frac{1}{\pi}\frac{\varepsilon}{\varepsilon^2 + x^2} \tag{6}$$

The energy functional $E_\varepsilon(C, f_1, f_2)$ can be approximately written as

$$
\begin{aligned}
&E_\varepsilon(C, f_1, f_2) \\
&= \sum_{i=1}^{2} \lambda_i \int (\int K_\sigma(x-y)|I(y) - f_i(x)|^2 \, M_i^\varepsilon(\phi(y))dy)dx + \upsilon \int |\nabla H_\varepsilon(\phi(x))| \, dx
\end{aligned}
\tag{7}
$$

Where $M_1^\varepsilon(\phi) = H_\varepsilon(\phi)$ and $M_2^\varepsilon(\phi) = 1 - H_\varepsilon(\phi)$. To preserve the regularity of the level set function ϕ, we add the last term for the energy functional, defined as

$$P(\phi) = \int \frac{1}{2}(|\nabla\phi(x)| - 1)^2 \, dx \tag{8}$$

Therefore, we construct the final energy functional as

$$F(\phi, f_1, f_2) = E_\varepsilon(\phi, f_1, f_2) + \mu P(\phi) \tag{9}$$

where μ is a positive constant.

We use the steepest descent method to minimize the energy functional (9). For a fixed level set function ϕ, the functions $f_1(x)$ and $f_2(x)$ that minimize $F(\phi, f_1, f_2)$ satisfy the following Euler–Lagrange equations

$$\int K_\sigma(x-y)M_i^\varepsilon(\phi(y))(I(y) - f_i(x))dy = 0, \quad i = 1, 2 \tag{10}$$

From above (10), we obtain

$$f_i(x) = \frac{K_\sigma(x) * [M_i^\varepsilon(\phi(x))I(x)]}{K_\sigma(x) * M_i^\varepsilon(\phi(x))}, i = 1, 2 \tag{11}$$

Keeping f_1 and f_2 fixed, we minimize the energy functional $F(\phi, f_1, f_2)$ with respect to ϕ using the steepest descent method as follows:

$$\frac{\partial \phi}{\partial t} = -\delta_\varepsilon(\phi)(\lambda_1 e_1 - \lambda_2 e_2) + \upsilon \delta_\varepsilon(\phi) div(\frac{\nabla \phi}{|\nabla \phi|})$$

$$+\mu(\nabla^2 \phi - div(\frac{\nabla \phi}{|\nabla \phi|})) \tag{12}$$

Where $e_i(x) = \int K_\sigma(y-x) |I(x) - f_i(y)|^2 \, dy, i = 1, 2$, the above (12) is the level set evolution equation to be solved in the proposed method.

Implementation. In PDE (12), all of the partial derivative can be discretized into a finite difference, and then the left side of the equal sign can obtain a forward time difference of ϕ, so we can achieve the desired iterations according to the PDE (12). During initialization, according to the information provided by template, the value of land areas is initialized to the -2.0, the value of marine areas is initialized to 2.

4   Experimental Results

In this section, we present the experimental results of our model, and compare with the experimental results of other commonly used methods, such as Sobel operator, Canny operator, and traditional level set method. The proposed model was implemented on a computer with Intel Core i7-2920XM 2.5GHz CPU, 2G RAM, and Windows XP operating system. For all the experiments referred later in this section, we use infrared channel images of FY-2 satellite as processing objects; the sizes of the images are 2288×2288. We used the same parameters of $\sigma = 3.0$, $\varepsilon = 1.0$, $\lambda_1 = 1.0$, $\lambda_2 = 2.0$, time-step $\Delta t = 0.1$, $\mu = 1$ and $\upsilon = 0.004 * 255 * 255$ for all the experiments in this section.

Comparison with Common Edge Extraction Algorithm. In different regions of the infrared image, the degree of cloud coverage is different. In order to test the effect and accuracy of the algorithm from the cloudy and cloudless regions respectively, we randomly select plurality of regions to experiment. The regional boundaries in figure.2 and figure.3 are less affected by cloud, have a higher definition. The detected edge is discontinuity by using Sobel operator, not conducive to the subsequent processing. Canny operator can extract edge accurately, but due to the influence of the selected threshold, double-edge phenomenon occurs in some place. The image edge after processing by our algorithm is continuous, smooth, and with a higher accuracy.

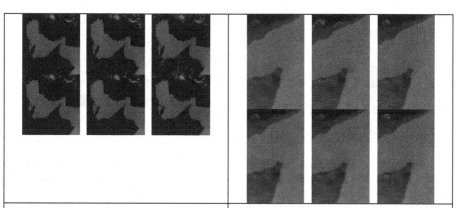

Fig. 2. Results of our method and common edge extraction methods for a regional IR image (cloudless-1). Column 1: up: Original image, and down: the result of Sobel operator; Column 2: up: the result of canny operator with the thresholds (43-129), and down: with the thresholds (87-261). Column 3: up: initial contours of our method, and down: final contours of our method.

Fig. 3. Results of our method and common edge extraction methods for a regional IR image (cloudless-2). Column 1: up: Original image, and down: the result of Sobel operator; Column 2: up: the result of canny operator with the thresholds (32-96), and down: with the thresholds (63-189). Column 3: up: initial contours of our method, and down: final contours of our method.

The regional boundaries in Figure.4 and Figure.5 are affected seriously by cloud, have a lower definition. The detected edge is not continuity by using Sobel operator, even only individual edge points. The detected edge by using canny operator is also discontinuous and there exist more useless edges. Using our algorithm can maximally avoid the interference of cloud, and obtain accurate, continuous and smooth edges.

Comparison with Traditional Level Set Method. In the traditional level set algorithm, we use standard coastline as the initial state of contour evolution. From Figure.6 and Figure.7, we can see the detected sea-land boundary is still missing after 2000 iteration steps. By comparison, our algorithm can detect the sea-land boundary more accurately only after 20 iteration steps, thereby has a great advantage in the computation and accuracy.

Finally, in order to compare our method with the traditional level set method quantitatively, we randomly select 25 regional images, respectively by using the two methods for edge extraction experiments. The computational time and iteration number for these images are listed in Table 1.

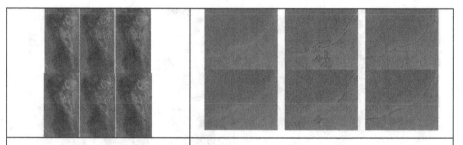

Fig. 4. Results of our method and common edge extraction methods for a regional IR image (cloudy-1). Column 1: up: Original image, and down: the result of Sobel operator; Column 2: up: the result of canny operator with the thresholds (16-48), and down: with the thresholds (22-66). Column 3: up: initial contours of our method, and down: final contours of our method.

Fig. 5. Results of our method and common edge extraction methods for a regional IR image (cloudy-2). Column 1: up: Original image, and down: the result of Sobel operator; Column 2: up: the result of canny operator with the thresholds (11-33), and down: with the thresholds (12-36). Column 3: up: initial contours of our method, and down: final contours of our method.

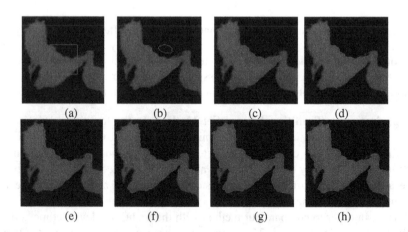

(a)   (b)   (c)   (d)

(e)   (f)   (g)   (h)

Fig. 6. Results of the traditional level set for a regional IR image. (a): Initial contours. (b): 20 iterations. (c): 100 iterations. (d): 200 iterations. (e): 400 iterations. (f): 800 iterations. (g): 1600 iterations. (h): Final contours, 2000 iterations.

Fig. 7. Results of our method for a regional IR image. Column 1: Initial contours. Column 2: Final contours, 20 iterations.

Table 1. The performance comparisons of our model and traditional level set model

	Our model	Traditional level set model
Average iteration number	21	2485
Average computational time (s)	2.16	273.35

5   Conclusions

With the increase of the remote sensing satellite data, the preprocessing of satellite remote sensing data becomes more and more time consuming. According to the characteristics of remote sensing images, we propose a new coastline detection method based on prior knowledge and level set method. This method uses the information of GSHHS to optimize the initialization procedure, and use an effective energy model, which is applicable to the infrared remote sensing image for sea-land boundary detection. The experiment results show that the proposed method can obtain satisfactory results, and can greatly reduce the computational complexity.

Acknowledgements. This work was supported by the Importation and Development of High-Caliber Talents Project of Beijing Municipal Institutions, China Special Fund for Meteorological-scientific Research in the Public Interest (GYHY201106044), NSFC (Grant nos. 61103130, 61070120, 61271435); National Program on Key basic research Project (973 Programs) (Grant nos. 2010CB731804-1, 2011CB706901-4). The Importation and Development of High-Caliber Talents Project of Beijing Municipal Institutions (no. IDHT20130225).

References

1. Yang, L., Yang, Z.: The Automated Landmark Navigation of the Polar Meteorological Satellite. J. of Applied Meteorological Science 20(3), 329–336 (2009) (in Chinese)
2. Wang, D., Hou, Y., Peng, J.: The partial differential equations method in image processing. Science Press, Beijing (2010)
3. Caselles, V., Carte, F., Coil, T., et al.: A geometric model for active contours. Numerische Mathematik 66, 1–31 (1993)
4. Kimmel, R.: Fast edge integration. In: Osher, S., Paragios, N. (eds.) Geometric Level Set Methods in Imaging, Vision, and Graphics, pp. 59–78. Springer (2003)

5. Wang, L., Li, C., Sun, Q., et al.: Active contours driven by local and global intensity fitting energy with application to brain MR image segmentation. Computerized Medical Imaging and Graphics 33, 520–531 (2009)
6. Chan, T., Vese, L.: Active contours without edges. IEEE Trans. on Image Processing 10(2), 266–277 (2001)
7. Li, C., Xu, C., Gui, C., et al.: Level set Evolution without re-initialization: A new variational formulation. In: Proc. IEEE International Conference on Computer Vision and Pattern Recognition (CVPR), vol. 1, pp. 430–436 (2005)
8. Lee, S.H., Seo, J.: Level set-based bimodal segmentation with stationary global minimum. IEEE Transactions on Image Processing 15(9), 2843–2852 (2006)
9. Tsai, A., Yezzi, A., Wiilsky, A.S.: Curve evolution implementation of the Mumford-Shah functional for image segmentation, denoising, interpolation, and magnification. IEEE Transactions on Image Processing 10(8), 169–1186 (2001)
10. Vese, L., Chan, T.: A multiphase level set framework for image segmentation using the Mumford and Shah Model. Int. Journal of Computer Vision 50(3), 271–293 (2002)
11. Chen, S., Sochen, N.A., Zeevi, Y.: Integrated active contours for texture segmentation. IEEE Trans. on Image Processing 15(6), 1633–1646 (2006)
12. Brox, T., Cremers, D.: On the Statistical Interpretation of the Piecewise Smooth Mumford-Shah Functional. In: Sgallari, F., Murli, A., Paragios, N. (eds.) SSVM 2007. LNCS, vol. 4485, pp. 203–213. Springer, Heidelberg (2007)
13. Sum, K.W., Cheung, P.Y.S.: Vessel extraction under non-uniform illumination: a level set approach. IEEE Transactions on Biomedical Engineering 55(1), 358–360 (2008)
14. Lankton, S., Tannenbaum, A.: Localizing region-based active contours. IEEE Transactions on Image Processing 17(11), 2029–2039 (2008)
15. Li, C., Kao, C.-Y., Gore, J.C., Ding, Z.: Minimization of Region-Scalable Fitting Energy for Image Segmentation. IEEE Transactions on Image Processing 17(10), 1940–1949 (2008)
16. Li, C., Xu, C., Gui, C., et al.: Distance Regularized Level Set Evolution and Its Application to Image Segmentation. IEEE Transactions on Image Processing 19(12), 3243–3254 (2010)

Tracking End-Effectors for Marker-Less 3D Human Motion Estimation in Multi-view Image Sequences*

Wenzhong Wang[1], Zhaoqi Wang[2], Xiaoming Deng[3], and Bin Luo[1]

[1] Computer Science Department, Anhui University, China
[2] Institute of Computing Technology, Chinese Academy of Sciences
[3] Institute of Software, Chinese Academy of Sciences
{wenzhong,luobin}@ahu.edu.cn, zqwang@ict.ac.cn,
xiaoming@iscas.ac.cn

Abstract. We propose to track the end-effectors of human body, and use them as kinematic constraints for reliable marker-less 3D human motion tracking. In the presented approach, we track the end-effectors using particle filtering. The tracked results are then combined with image features for 3D full pose tracking. Experimental results verified that the inclusion of end-effectors' constraints improves the tracking performances.

Keywords: end-effectors, motion tracking, particle filtering.

1 Introduction

Estimation of 3D human motion from multi-view image sequences has been a very active research topic in the late decades [1]. The most successful approaches are those based on marker tracking, such as the VICON system. These marker-based methods have proven to be very accurate in estimating 3D body poses. However, these methods face several practical difficulties and inconveniences: markers attached to the subjects hinder their motion; erroneous marker reconstructions need to be manually fixed; et al. Highly accurate motion can hardly be obtained without skilled operators and intensive labors. The rather cumbersome processes of marker-based motion capture systems limit their applications. These drawbacks have led to the vast research on marker-less motion capture. Marker-less motion capture aims to estimate 3D human motion directly from image sequences with no attachments on the subjects. It proves to be very hard due to the image noises, motion variations, and high degrees of freedom (D.O.F.) in human motions.

Most of the published research on this problem can be categorized into two groups, namely, the bottom-up approaches and the top-down approaches [1].

The former ones estimate human motion directly from image features. These methods either assemble the local image features (such as joint locations, limb edges) to 3D poses or learn a map from image features to 3D poses. In the first case, it is hard to detect discriminating and unambiguous local features from the noisy images, and

---

* This work is partially supported by the NSFC (project No. 61005039).

M. Kurosu (Ed.): Human-Computer Interaction, Part V, HCII 2013, LNCS 8008, pp. 227–235, 2013.

the estimated poses are rather coarse. In the later case, the mappings from image features to 3D poses are multi-valued and it is unlikely to learn such mappings without plentiful training data. The reliance on training data makes these approaches only applicable on specific motions.

The generative methods, on the other hand, do not presume any pre-captured training data. These approaches utilize a 3D human body model, and minimize an error function which measures how well this model fits the images. Many different error functions have been devised; most of them are based on the residuals between the 2D projections of 3D body model and the observed image features (such as silhouettes and edges). Some other error functions represent the alignment error of the body model to the voxel reconstructions of multi-view body silhouettes. The optimal 3D pose is found by minimizing these functions. Due to image noise and ambiguities, these functions are very peaky, rendering a difficult optimization problem. Still further, the high D.O.Fs of human pose makes the optimization even harder. There are mainly two categories of methods to this optimization problem, the one based on local optimization and the other one based on stochastic optimization. Local optimization methods start from an initial guess and iteratively find a descent direction of the error function. These methods guarantee convergence to local minimums. Their performance relies heavily on the initial values. They can hardly recover from errors during tracking. The stochastic methods approximate the posterior of poses by a set of weighted samples and can thus represent multi-mode distributions. This representation power makes the recovery from tracking error probable. The downside of these methods is that the effective number of samples increases exponentially with the D.O.Fs. Another difficulty with these methods is that the high dimensional prior distribution of poses cannot be effectively modeled, and this may results in many wrong samples which deteriorate the optimization performance.

In this paper, we try to explore high level information to facilitate top-down motion tracking. Specifically, we propose to track human limbs (a.k.a. end-effectors) firstly. This limb information is then used in a stochastic optimization process yielding optimal poses.

The motivation of our approach is based on these facts: the end effectors can significantly narrow down the solution space by the rule of inverse kinematics (IK); and this will compensate the weakness of the widely used low-level image features for pose optimization.

2   Related Works

We briefly discuss some works on human motion tracking using end-effectors' constraints.

Ganapathi [2] et al. detects the head, hands and feet in depth image, and generates poses from these detections using inverse kinematics.

Pons-Moll [3] et al. attaches inertial sensors on the arms and tibias, and use orientation cues derived from these devices to sample particles from the manifold of valid poses.

In [4], Hauberg et al. propose to track human motion in the end-effectors space. In their approach, the full 3D poses are modeled as normal distributed around the mean obtained from the end-effectors' goals using inverse kinematics. Samples are then drawn from this distribution and evaluated using stereo image data. The mean of these weighted samples is used as the pose estimation.

Andreas Baak [5] et al. estimates five feature points from depth images; these feature points are assumed to be the head, hands and feet. These points are then used to retrieve similar 3D poses from a pre-recorded motion database.

These works use depth image data or other devices to facilitate the human pose estimation. Our work differs from the above ones in that we do not rely on any supplementary devices to obtain the end-effectors positions. We'd rather estimate this information directly from multi-view image data.

3   Proposed Method

Our method consists of two interleaved processes: one for end-effectors tracking and the other for 3D pose tracking. Both of these processes are implemented using particle filters.

Let g_t, x_t be the end-effectors goals and pose at time instance t, respectively. We denote the observed image at time t as z_t. We formulate the pose tracking problem in a Bayesian filtering framework [6] as depicted in Fig1.

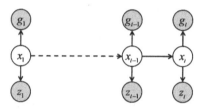

Fig. 1. Probabilistic graphical model for our pose estimation problem

In Bayesian filtering, we are interested in the posteriori of x_t, $P(x_t|G_t, Z_t)$:

$$P(x_t|G_t, Z_t) = k_t P(g_t, z_t|x_t) \int P(x_t|x_{t-1}) P(x_{t-1}|G_{t-1}, Z_{t-1}) d\, x_{t-1} \qquad (1)$$

where we have defined $G_t = \{g_1, \cdots g_t\}$, $Z_t = \{z_1, \cdots z_t\}$ and k_t is a normalization constant independent of x_t.

3.1   Particle-Based Estimation of Human Pose

The posteriori in (1) is approximated as a set of weighted particles and estimated using particle filtering.

Since the end-effectors' goal g_t and image z_t are conditionally independent given the pose x_t, (1) can be written as:

$$P(x_t|G_t, Z_t) = k_t P(g_t|x_t) P(z_t|x_t) \int P(x_t|x_{t-1}) P(x_{t-1}|G_{t-1}, Z_{t-1}) d\, x_{t-1} \qquad (2)$$

We now define the image observation model $P(z_t|x_t)$, forward kinematics model $P(g_t|x_t)$ and motion dynamical process $P(x_t|x_{t-1})$.

The Image Observation Model $P(z|x)$. We use a simple human body model shown in Fig2. The whole body is decomposed into 10 rigid parts, and each part is modeled as a circular truncated cone. This is a very coarse simplification of real human body. The benefit of such a model is that it is very easy to calculate its projection and check its limb occlusions.

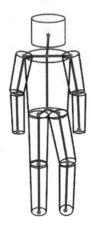

Fig. 2. 3d human body model

Having specified the 3D body model, we now define the image observation model as:

$$P(z|x) \propto \exp \left(-\frac{\epsilon^2(x)}{\sigma_z^2}\right) \tag{3}$$

where $\epsilon(x)$ is the discrepancy between the projection of body model and the observed image z.

We use edges as image features and define this error as:

$$\epsilon(x) = D^{M(x)} \cdot E^I + D^I \cdot E^{M(x)} \tag{4}$$

where $E^{M(x)}$ and $D^{M(x)}$ are , respectively, edge map and chamfer distance map of the projected image of of body model in pose x, and E^I is the edge map of observed image z and D^I its distance map. $\cdot$ denotes pixel-wise product operation.

The Forward Kinematics Model $P(g|x)$. The end-effectors' positions are deterministic functions of full pose x: $g = f(x)$. However, in order to account for the errors in our estimations of g, we model the kinematics constraint in a probabilistic way:

$$P(g|x) = N(f(x), \Sigma_g) \tag{5}$$

The Dynamical Process $P(x_t|x_{t-1})$. Each body pose is composed of 36 degrees of freedom: $x = (r, o, \theta)$, where $r \in R^3$ is the root position, $o = (o_x, o_y, o_z)$ is the root orientation, and $\theta = (\theta_1, \cdots \theta_{30})$ is 30 joint angles.

Since r, o, θ are independent of each other. We factorize $P(x_t|x_{t-1})$ into three parts:

$$P(x_t|x_{t-1}) = P(x_t|x_{t-1})P(o_t|o_{t-1})P(\theta_t|\theta_{t-1}) \tag{6}$$

The parameters are learned from a prerecorded human motion database.

In our implementation, we firstly retarget the training joint angles θ_j to the body model of the subject to be tracked, and then extracted all the joint positions $y_j \in R^D$. Other than learn a probabilistic model in the angular space (which is not a vector space), we learn the dynamical process in the joint position space.

The original high dimensional position states y_j's are projected into a lower dimensional space using PCA:

$$y_j = \bar{y} + W\tilde{y}_j$$

where $\bar{y}$ is the mean state of training data $\{y_j\}_{j=1}^n$, $W \in R^{D \times K}$ is the projection matrix consisting of the first K principle components. In our experiments, $K = 6$. So the high dimensional state θ_j is replaced with a low-dim vector $\tilde{y}_j \in R^6$.

We will track the poses in the joint space of r, o and $\tilde{y}$. The dynamical process in (6) now becomes

$$P(x_t|x_{t-1}) = P(r_t|r_{t-1})P(o_t|o_{t-1})P(\tilde{y}_t|\tilde{y}_{t-1}) \tag{7}$$

The three components in (7) are modeled as first order Gaussian auto-regression processes:

$$\begin{cases} P(r_t|r_{t-1}) = N(r_{t-1}, \Sigma_r) \\ P(o_t|o_{t-1}) = N(o_{t-1}, \Sigma_o) \\ P(\tilde{y}_t|\tilde{y}_{t-1}) = N(\tilde{y}_{t-1}, \Sigma_y) \end{cases} \tag{8}$$

The covariance matrices $\Sigma_r, \Sigma_o, \Sigma_y$ are estimated from training data.

3.2 Tracking the States of End-Effectors

The above statements presume the end-effectors' positions are known. In this subsection, we present an approach for tracking these positions.

The end-effectors we used are the four lower limbs (two tibias and two forearms) of the human body. The state of each end-effector is described by its position and orientation in the world coordinate system.

The state of the four end-effectors is given by $g = \{p_i, \alpha_i, \beta_i\}_{i=1}^4$. Where $p_i \in R^3$ is the position of the proximal end of limb i, and (α_i, β_i) is pointing direction of limb i (in a cylindrical parameterization).

The state g is estimated using particle filtering. We now define the dynamical process of g and the observation process of image z given g as follows:

The dynamical process of g, $P(g_t|g_{t-1})$, is factorized into three processes, $P(g_t|g_{t-1}) = P(p_t|p_{t-1})P(\alpha_t|\alpha_{t-1})P(\beta_t|\beta_{t-1})$ and defined as below:

$$\begin{cases} P(p_t|p_{t-1}) = N(p_{t-1}, \Sigma_p) \\ P(\alpha_t|\alpha_{t-1}) = N(\alpha_{t-1}, \sigma_\alpha) \\ P(\beta_t|\beta_{t-1}) = N(\beta_{t-1}, \sigma_g) \end{cases} \qquad (9)$$

The observation process is defined as:

$$P(z|g) \propto exp\big(-\alpha_a d_a(g,z) - (1 - \alpha_a)d_c(g,z)\big) \qquad (10)$$

where $d_a(g,z)$ and $d_c(g,z)$ are two energy terms representing the mismatch of the limb projections with the image z.

The appearances of lower limbs are represented using eigen-templates [7]. We manually labeled about 30 image patches that corresponds to each limb, and project the scale normalized patches onto a low dimensional space using PCA[7], yielding eigen-template representations of each limb. Since the left and right limbs are very similar in appearance, they share the same eigen-templates model. Fig3 shows the learned eigen-templates of forearm. The first panel shows the mean patch and the rests are the first 15 principle components.

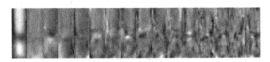

Fig. 3. Eigen-templates of forearm

The appearance energy $d_a(g,z)$ is calculated in the low dimensional space. The image area covered by the limb projection is extracted, rotated and rescaled, and then is projected onto the low dimensional space spanned by the eigen-templates. $d_a(g,z)$ is then calculated as done in [7].

The edge energy $d_c(g,z)$ is calculated as the edge mismatch between image edges and projected limb contours, defined as equation (4).

Occlusion Handling. One must be careful when evaluating the particles because the limbs may be occluded by the body. So we introduce the visibility parameter for each limb in each camera view.

In order to predict the limb visibility, we project the estimated pose in previous frame onto each view, and calculate the visible area of each limb. We then project a single limb onto each view, and calculate the area it covers. Wu use the ratio between these two areas as the visibility measurement.

The visibility of limb i in camera view k, $v_{i,k}$, is defined as

$$v_{i,k} = \frac{A_{i,k}}{A_{i,0}}$$

where $A_{i,k}$ is the visible projection of limb i in camera view k in previous pose, and $A_{i,0}$ is the area covered by limb i in camera view c. Fig4 illustrates the projections of left forearm, the white area in the left and right images are $A_{i,k}$ and $A_{i,0}$, respectively. Note that in the left image, the left forearm is partially occluded by the rest of the body (red area).

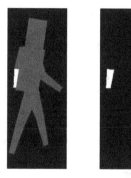

Fig. 4. Calculating limb visibility

Having determined the visibility of each limb, we can accumulate the energy terms in different camera views:

$$\begin{cases} d_a(g_i, z) = \sum_{k=1}^{C} v_{i,k} d_a\big(g_i^k, z^k\big) \\ d_c(g_i, z) = \sum_{k=1}^{C} v_{i,k} d_c\big(g_i^k, z^k\big) \end{cases} \tag{11}$$

where g_i^k is the projection of g_i onto image k, and z^k is the image in camera view k.

4   Experiments and Discussions

We've tested the above method on the HumanEvaI [8] dataset. In order to illustrate the improvement made by the end-effectors' constraints, we also run the tests without using these constraints (i.e. traditional particle filtering). All other parameters are the same for these tests.

We perform experiments using walking videos of subject S1 in cameras C1, C2, and C3. The parameters of dynamical processes (eqn. 8, eqn. 9) are learned from the other 3D motion data of S1. We've tried to estimate these parameters from motion data of other subjects, but found that the tracking results significantly deteriorated. This indicates that the motion model is crucial for robust tracking.

The per-joint position errors are shown in Fig5. We've achieved better results than state-of-the-art method (particle filtering). We attribute the performance improvement to the adoption of the end-effectors' constraints. The end-effectors' information can significantly reduce the ambiguities in images, and bias the search of optimal pose towards those satisfying end-effectors' constraints.

The tracking results for end-effectors are shown in Fig6. We've found that tracking performance highly relies on the motion model (eqn.9) and the limbs' visibility. The end-effectors can hardly be tracked reliably with less than three cameras. In our experiments, each limb can be clearly observed from at least two cameras. Due to the ambiguities in between the left and right limbs, the tracker may be confused by the similar image data. This would happen when the two tibias cross over each other.

We show motion tracking results in Fig7. These results are visually appealing. It proves very difficult for reliable tracking when there are severe occlusions. In these cases, the image data are ambiguous and the tracker will fail without precise motion models.

In conclusion, we've verified that end-effectors constraints could be helpful for reliable tracking.

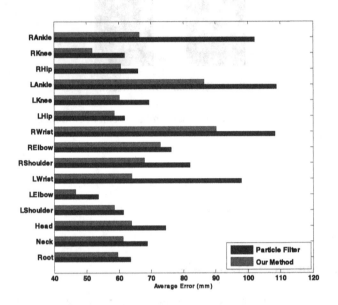

Fig. 5. Averaged joint error

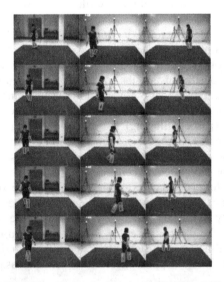

Fig. 6. Tracked end-effectors in C1,C2 and C3

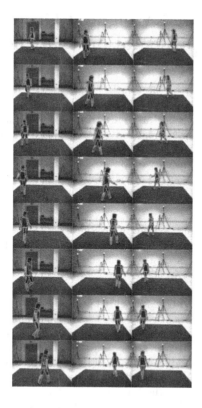

Fig. 7. Human pose tracking results in C1,C2 and C3

References

1. Moeslund, T.B., et al. (eds.): Visual analysis of humans: looking at people. Springer (2011)
2. Ganapathi, V., et al.: Real time motion capture using a single time-of-flight camera. In: IEEE Conference on Computer Vision and Pattern Recognition (CVPR). IEEE (2010)
3. Pons-Moll, G., et al.: Outdoor human motion capture using inverse kinematics and von mis-es-fisher sampling. In: IEEE International Conference on Computer Vision, ICCV. IEEE (2011)
4. Hauberg, S., Pedersen, K.S.: Predicting articulated human motion from spatial processes. International Journal of Computer Vision 94(3), 317–334 (2011)
5. Baak, A., et al.: A data-driven approach for real-time full body pose reconstruction from a depth camera. In: IEEE International Conference on Computer Vision (ICCV). IEEE (2011)
6. Isard, M., Blake, A.: Condensation—conditional density propagation for visual tracking. International Journal of Computer Vision 29(1), 5–28 (1998)
7. Moghaddam, B., Pentland, A.: Probabilistic visual learning for object representation. IEEE Transactions on Pattern Analysis and Machine Intelligence 19(7), 696–710 (1997)
8. Sigal, L., Balan, A.O., Black, M.J.: Humaneva: Synchronized video and motion capture dataset and baseline algorithm for evaluation of articulated human motion. International Journal of Computer Vision 87(1), 4–27 (2010)

Kernel Based Weighted Group Sparse Representation Classifier

Bingxin Xu[1], Ping Guo[1,*], and C.L. Philip Chen[2]

[1] Image Processing and Pattern Recognition Laboratory
Beijing Normal University, Beijing, China
[2] Faculty of Science and Technology
University of Macau, Macau, China
xbing@mail.bnu.edu.cn, pguo@ieee.org, philipchen@umac.mo

Abstract. Sparse representation classification (SRC) is a new framework for classification and has been successfully applied to face recognition. However, SRC can not well classify the data when they are in the overlap feature space. In addition, SRC treats different samples equally and ignores the cooperation among samples belong to the same class. In this paper, a kernel based weighted group sparse classifier (KWGSC) is proposed. Kernel trick is not only used for mapping the original feature space into a high dimensional feature space, but also as a measure to select members of each group. The weight reflects the importance degree of training samples in different group. Substantial experiments on benchmark databases have been conducted to investigate the performance of proposed method in image classification. The experimental results demonstrate that the proposed KWGSC approach has a higher classification accuracy than that of SRC and other modified sparse representation classification.

Keywords: Group sparse representation, kernel method, image classification.

1 Introduction

Sparse representation or sparse coding has been successfully applied to many computer vision tasks, including face recognition [1], image super-resolution [2] and image classification [3]. Sparse representation classification (SRC) is a new framework which assumes that the test sample can be represented by linear combination of training samples which belong to the same class. However, SRC only uses L1-norm as regularization item to control the sparsity of linear coefficients. L1-norm treats each training sample equally and doesn't consider the cooperation of training samples from the same class. Zou [4] proves that L1-norm only selects a single sample from a group of correlated training samples. Therefore, when the test sample has a similar training sample which label is different with it, this training sample will possibly lead to wrong classification result.

---

* Corresponding author.

M. Kurosu (Ed.): Human-Computer Interaction, Part V, HCII 2013, LNCS 8008, pp. 236–245, 2013.

In order to solve the problem mentioned above, the group sparse classifier (GSC) is proposed by Majumdar [5]. GSC employs a L1-norm mixed L2-norm as regularization item which make coefficients belong to the same group are dense but among groups are sparse. Even though GSC considers the cooperation of training samples in the same group, it is not perfect either. The reason is that the inner L2-norm of mixed norm selects all the training samples from a particular class [6]. Actually, representing a test sample does not need all the training samples even thought they are in the same class. Due to the diversity of training samples, only some of them are more similar with test sample and these samples play an important role in linear representation.

A kernel based weighted group sparse classifier (KWGSC) is proposed in this paper to solve the shortcomings of SRC and GSC. The kernel method is successfully used in many algorithms for pattern analysis and the famous one is support vector machine (SVM). The effect of kernel method is mapping the original data into a high dimensional feature space by non-linear transformation. In order to improve the representation ability of SRC, kernel based sparse representation classification (KSRC) algorithms have been proposed in [7] [8]. However, these algorithms only replace the original space to the kernel space to compute distance. Actually, the value of kernel function which computing the inner product of a pair of data indicates the similarity of these data in the kernel space. Therefore, kernel function is directly applied to feature extraction in this paper. For the weight of each group, we also consider the influence of the similarity between test sample and members of each group.

The rest of this paper is organized as follows: In section 2, the background of this work is discussed. The proposed kernel based weighted group sparse classifier is described in section 3. Experimental design and results are presented in section 4 and the conclusion is presented in section 5.

2   Related Work

In this section, we briefly introduce the SRC method proposed in [1] and GSC method proposed in [5]. Finally, the kernel trick is reviewed.

2.1   Sparse Representation Classifier

Sparse representation classifier assumes that the training samples from a single class do lie on a subspace. Ideally, a test sample from one class can be represented by a linear combination of training samples from the same class. Specifically, given n_i training samples from the i-th class, the samples' feature vectors are stacked as columns of a matrix $\mathbf{F}_i = [\mathbf{f}_{i,1}, \mathbf{f}_{i,2}, \dots, \mathbf{f}_{i,n_i}] \in R^{m \times n_i}$. Any new test sample $\mathbf{y} \in R^m$ from the same class can be represented as:

$$\mathbf{y} = x_{i,1}\mathbf{f}_{i,1} + x_{i,2}\mathbf{f}_{i,2} + \dots + x_{i,n_i}\mathbf{f}_{i,n_i} = \mathbf{x}_i\mathbf{F}_i, \tag{1}$$

where $\mathbf{x}_i$ is the coefficient vector of linear representation. Since the class label of the test sample is unknown, a new matrix $\mathbf{F}$ is defined by concatenating all the classes:

$$\mathbf{F} = [\mathbf{F}_1, \mathbf{F}_2, \ldots, \mathbf{F}_i, \ldots, \mathbf{F}_c], \tag{2}$$

where $i = 1, 2, \ldots, c$. Then the linear representation of $\mathbf{y}$ can be rewritten in terms of all the training samples as:

$$\mathbf{y} = \mathbf{F}\mathbf{x} \in R^m, \tag{3}$$

where $\mathbf{x}$ is the vector of coefficients. If $\mathbf{y}$ belongs to i-th class, the entries of $\mathbf{x}$ are expected to be zero except some of those associated with this class. Namely $\mathbf{x} = [0, 0, \ldots, \mathbf{x}_i, \ldots, 0]^T \in R^N$ and N is the total number of training samples. In SRC, the objective function is formulated as a convex programming problem:

$$\min \|\mathbf{y} - \mathbf{F}\mathbf{x}\|_2 + \lambda \|\mathbf{x}\|_1, \tag{4}$$

where $\|.\|_1$ denotes the L1-norm. After computing the coefficient vector, the test sample is classified according to the reconstruction error between $\mathbf{y}$ and its approximations. The i-th class approximation is computed by using only the coefficients associated with class i and assigning zeros to other entries.

2.2 Group Sparse Classifier

The group sparse classifier has the same assumption as SRC. The difference is that GSC intends to ensure all the coefficients for the correct class are selected. It means that the coefficients in the same group or class should be zeros or non-zeros simultaneously. L1-norm minimization cannot satisfy this condition, so the objective function of GSC is formulated as:

$$\min \|\mathbf{y} - \mathbf{F}\mathbf{x}\|_2 + \lambda \sum_{j=1}^{c} \|\mathbf{x}_{G_j}\|_2, \tag{5}$$

where $\mathbf{x}_{G_j}$ is the coefficients associated with group $\mathbf{G}_j$. Although GSC can ensure the group structure of coefficients, this may not be always desired. In practice, we expect most of the correct samples can be selected which are more similar to the test sample and exclude the samples which are different to the test sample even thought they are in the same class. The proposed method can solve this problem and will be detailed in section 3.

2.3 Kernel Trick

The kernel trick is a very well-know technique in machine learning [8]. It has been widely applied to pattern recognition and function approximation, such as support vector machine [9] [10], kernel-based clustering methods [11] [12] and kernel principal component analysis (KPCA) [13]. The kernel trick attracts much interest because it can easily generalize a linear algorithm to a non-linear algorithm. Actually, kernel method is a feature projection method which can map the original feature space into a higher feature space by a non-linear algorithm.

In the kernel space, the distribution of samples will be changed. The different class's samples are more separate and the same class's samples are more gathering. Usually, a Mercer kernel k can be expressed as:

$$k(x, y) = \phi(x)^T \phi(y), \tag{6}$$

where x and y are any two points, and ϕ is the implicit nonlinear mapping function associated with the kernel function k. The trick of kernel function is that we don' t need to know the expression of ϕ and just use kernel function to instead of its inner product. The commonly used kernel function are polynomial kernel: $k(x, y) = (1 + \langle x, y \rangle)^p$ and radial basis function (RBF) kernel: $k(x, y) = exp(-\|x - y\|_2^2)/\sigma^2$. p and σ are parameters need to be predefined before used.

3   Kernel-Based Weighted Group Sparse Classifier

3.1   Kernel-Induced Feature

The proposed method is different to other kernel-based methods for transforming the objective function into an inner product form. The kernel trick is directly used for feature extraction. In the kernel space, equation (6) is a distance measure between $\mathbf{x}$ and $\mathbf{y}$. This kernel-induced distance is a non-Euclidean distance measure in original data space and has been used in kernel clustering [14]. Therefore, the different value of kernel function can describe the similarity between points. Specifically, the training samples construct the dictionary $\mathbf{X} = [\mathbf{x}_1, \mathbf{x}_2, \ldots, \mathbf{x}_N]$, N is the total number of training samples. For data $\mathbf{x}_i$, the feature transformation can be formulated as:

$$\mathbf{f}_i = k(\mathbf{X}, \mathbf{x}_i) = [k(\mathbf{x}_1, \mathbf{x}_i), k(\mathbf{x}_2, \mathbf{x}_i), \ldots, k(\mathbf{x}_N, \mathbf{x}_i)]^T, \tag{7}$$

where k is predefined kernel function and different kernel function transforms the original data into a different feature space. For the feature vector $\mathbf{f}_i$, if the point $\mathbf{x}_j \in \mathbf{X}$ is closer to $\mathbf{x}_i$, the value of $k(\mathbf{x}_j, \mathbf{x}_i)$ will be larger than others. Because the dictionary $\mathbf{X}$ contains all the class samples, the kernel-induced feature vectors are discriminative in different classes.

3.2   Group Construction

In the proposed method, we construct a new group members based on the original training samples and the numbers of each group are adaptive. For each class's training samples, a kernel matrix is computed and using the minimal value of the matrix as a threshold to select the members of this group. Specifically, $\mathbf{X}_i = [\mathbf{x}_{i1}, \mathbf{x}_{i2}, \ldots, \mathbf{x}_{in_i}]$ is the i-th class samples and n_i is the total number of samples in this class. The kernel matrix of i-th class is computed as follows:

$$K_i = \begin{bmatrix} k(\mathbf{x}_{i1}, \mathbf{x}_{i1}) & k(\mathbf{x}_{i1}, \mathbf{x}_{i2}) & \ldots & k(\mathbf{x}_{i1}, \mathbf{x}_{in_i}) \\ k(\mathbf{x}_{i2}, \mathbf{x}_{i1}) & k(\mathbf{x}_{i2}, \mathbf{x}_{i2}) & \ldots & k(\mathbf{x}_{i2}, \mathbf{x}_{in_i}) \\ \ldots & \ldots & \ldots & \ldots \\ k(\mathbf{x}_{in_i}, \mathbf{x}_{i1}) & k(\mathbf{x}_{in_i}, \mathbf{x}_{i2}) & \ldots & k(\mathbf{x}_{in_i}, \mathbf{x}_{in_i}) \end{bmatrix}. \tag{8}$$

For this class, a threshold is defined as:

$$m_i = \min(K_i(:)), \tag{9}$$

which used for selecting the members of this group. For a test sample $\mathbf{y}$, only the training samples which are similar to it will be selected to construct the group. Others which are different from it are considered that has little effect in representing the test sample. For example, in the i-th class, the kernel-induced feature of $\mathbf{y}$ on $\mathbf{X}_i$ is computed as $f_{i,y} = k(\mathbf{X}_i, \mathbf{y})$ by equation (7). If $k(\mathbf{x}_{i,j}, \mathbf{y})$ is bigger than the threshold m_i, it represents the sample $\mathbf{x}_{i,j}$ is similar to $\mathbf{y}$ and should be selected in group i. Therefore, the number of members in each group for test sample $\mathbf{y}$ is dependent on the data. The proposed method doesn't use all the training samples as dictionary mechanically, but according to the relationship between the test sample and the training samples to select the member of each group.

3.3   Weighted Group Sparse Classifier

Assume that the new dictionary $\mathbf{G}$ is partitioned into c disjoint groups $\mathbf{G}_1, \mathbf{G}_2, \ldots,$ $\mathbf{G}_c$, c is the number of groups and $\mathbf{G}_i \cap \mathbf{G}_j = \varnothing$ when $i \neq j$. For different test sample, the dictionary $\mathbf{G}$ is not fixed and computed by the method in section 3.2. In each group $\mathbf{G}_i$, there are N_i training samples to represent the test data. The objective function of proposed method is defined as:

$$\min \|\mathbf{k}(\mathbf{G}, \mathbf{y}) - \mathbf{k}(\mathbf{G}, \mathbf{G})\mathbf{x}\|_2 + \sum_{i=1}^{c} w_i \|\mathbf{x}_{G_i}\|_2, \tag{10}$$

where $k(\mathbf{G}, \mathbf{y})$ is the kernel-induced feature vector of $\mathbf{y}$ and the column of $k(\mathbf{G}, \mathbf{G})$ is the kernel-induced feature vector of each member of $\mathbf{G}$. $\mathbf{x}_{G_i}$ is the coefficients associated with group i and w_i is the weight of this group which represents the importance of this group for reconstruction the test sample. w_i is defined as:

$$w_i = \frac{N_i}{N} + sum(k(\mathbf{G_i}, \mathbf{y})), \tag{11}$$

where N is the total number of members in dictionary $\mathbf{G}$. The first item N_i/N describes the ratio between the numbers of group i and the total number of all the groups. If the number of samples in group i is more, w_i should be larger to indicate i-th group samples are more important to represent the test sample. The second item of equation (11) has the same effect to control the weights. If the samples in group i are similar to the test data $\mathbf{y}$, the value of this item will be larger and lead to the weight becomes larger. In order to avoid noise or outlier make the value of second item big, the first item can balance the weight to an appropriate value. After computing the coefficients, the test sample can be classified to the class that minimizes the residual:

$$\min R_i(\mathbf{y}) = \|k(\mathbf{G}, \mathbf{y}) - k(\mathbf{G}, \mathbf{G})\delta_i(x)\|_2. \tag{12}$$

$\delta_i(x)$ defines the coefficient vector which only retains the coefficients related to group i and set other entries to zero. The complete algorithm is described as follows:

1. Input: a matrix of training samples $\mathbf{F} = [\mathbf{F}_1, \mathbf{F}_2, \ldots, \mathbf{F}_c] \in \mathbf{R}^{m \times N}$ for c classes and a test sample $\mathbf{y}$.
2. Compute the kernel matrix $\mathbf{K}_i$ and threshold m_i for each class by equation (8) and (9).
3. Compute the test sample's kernel-induced feature on each class, $\mathbf{f}_{i,y} = k(\mathbf{F}_i, \mathbf{y})$.
4. Construct the group $\mathbf{G}_i$ for each class and compute the weight w_i for group $\mathbf{G}_i$ by equation (11).
5. Solve the minimization problem defined by equation (10) and compute the residual $R_i(\mathbf{y})$ of each class.
6. Output: identity$(\mathbf{y}) = \arg \min(R_i(\mathbf{y}))$.

4   Experiments

In this section, the effectiveness of the proposed kernel-based weighted group sparse classifier (KWGSC) algorithm is carried out on three facial image dataset, namely, ORL dataset [15], Extended Yale B dataset [16] [17] and AR dataset [18]. Downsampled is used for reducing the dimension of original image. In our experiments, polynomial kernel function is used for all the dataset and the parameter p is set to 2. We compare the classification ability of KWGSC algorithm with SRC [1], GSC [5] and KSRC [8]. The optimization method used to solve the group sparse problem is alternating direction method which proposed in [19]. The experiments are carried out ten times and the average classification rate is the final result.

4.1   ORL Database

ORL database has 10 different images for each class and contains 40 distinct individuals. The images of each individual are variations in pose and facial expression. The size of each face image is 112×92 with 256 gray levels per pixel. The downsampling ratio is $1/8$ and the resulting standardized input space dimension is 168. Fig.1 shows sample images of one person. In the experiment, five images are random selected from each individual as training samples and the rest as testing samples. Therefore, the number of images for both training and test are 200. The experimental results are presented in table 1. In order to only compare the performance of different classifier, we set the same to original feature dimension and kernel function. From table 1, it can be discovered that GSC has worse result than SRC. This is because in ORL database, the size of each group is small with only five members and the benefit of group sparsity is not significant. In according with the conclusion proposed in [20], the result is that group sparsity favors large sized groups. However, the proposed KWGSC

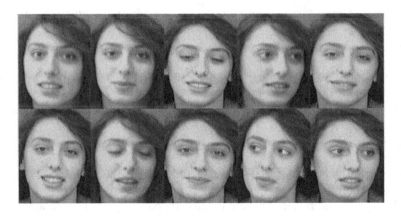

Fig. 1. Sample images of one person with different expression in ORL database

Table 1. Comparison results of various methods on the ORL database

Method	SRC	KSRC	GSC	KWGSC
Classification rate	91.5%	92.5%	87.5%	94.1%
Feature Dimension	168	168	168	168

has better result than SRC and GSC. This indicates that an appropriate weight of group can overcome the shortcoming of small sized group and improve the performance of group sparsity.

4.2   Extended Yale B Database

The Extended Yale B database consists of 2414 frontal face images of 38 individuals (about 64 images per category) captured under various laboratory controlled lighting conditions. For each category, we randomly select 32 images for training with the remaining images for testing. Therefore, the total number images for training and test is 1216 and 1198. The size of each face image is 192×168 with 256 gray levels per pixel. The downsampling ratio is $1/16$ and the downsampled image is 12×11. Therefore the original standardized input space dimension is 132. Fig.2 shows sample images of one person with different lighting condition. The experimental results are presented in Table 2. The proposed method KWGSC is superior to others. Although the images in this database have much different lighting condition, KWGSC can select more important training samples automatically to represent the test image. It means that if the test image is very dark, then the training samples in the same lighting condition will be more effective than others even in the well lighting condition.

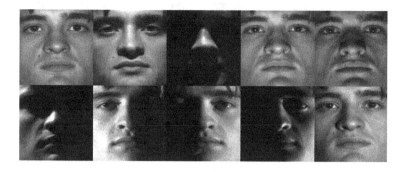

Fig. 2. Sample images of one person with different lighting condition in Extended Yale B database

Table 2. Comparison results of various methods on the Extended Yale B database

Method	SRC	KSRC	GSC	KWGSC
Classification rate	93.4%	94.8%	94.5%	95.6%
Feature Dimension	132	132	132	132

4.3   AR Database

The AR database consists of over 4,000 frontal images for 126 individuals. For each individual, 26 pictures were taken in two separate sessions [16]. These images include more facial variations including illumination change and expressions than the Extended Yale B database. The same with other methods, only a subset of the data are chosen to consist the database which contains 50 male individuals and 50 female individuals. For each individual, only 14 images with illumination change and expressions are selected. The image are cropped with dimension 165×120 and converted to gray scale [1]. The downsampling ratio is $1/12$ and the downsampled image is 14×10. Therefore the original standardized input space dimension is 140. Fig.3 shows sample images of one person which are in different lighting condition and expression. We also partition each class averagely into training set and testing set. Table 3 lists the classification rates of each method. From Table 3, it can be concluded that KWGSC also has better result than others. The difficult of AR database is that the total number of individuals is large and the number of training samples in each class is small. However, KWGSC can avoid these problems. Although the total number of individuals is large, KWGSC excludes some training samples which have little relationship to the test sample in the group construction process.

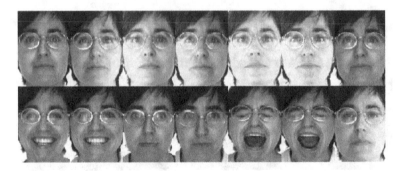

Fig. 3. Sample images of one person with different expression and lighting condition in AR database

Table 3. Comparison results of various methods on the AR database

Method	SRC	KSRC	GSC	KWGSC
Classification rate	89.4%	90.7%	91%	91.5%
Feature Dimension	140	140	140	140

5 Conclusion

In this paper, a kernel-based weighted group sparse classifier (KWGSC) algorithm is proposed. For KWGSC, samples are mapped from original feature space into a kernel-induced feature space, and then construct a new dictionary which depends on the test sample. The new dictionary contains many groups which have much relationship to test sample and excludes others based on kernel-induced distance measure. We compare the proposed method with SRC, GSC and KSRC on different facial image database. The experimental results indicate the effectiveness of KWGSC.

Acknowledgment. The research work described in this paper was fully supported by the grants from the National Natural Science Foundation of China (Project No. 90820010, 60911130513).

References

1. Wright, J., Yang, A.Y., Granesh, A.: Robust Face Recognition via Sparse Representation. IEEE Transactions on Pattern Analysis and Machine Intelligence 31(2), 210–227 (2009)
2. Yang, J.C., Wright, J., Huang, T., Ma, Y.: Image super-resolution as sparse representation and raw patches. In: Proceedings of IEEE International Conference on Computer Vision and Pattern Recognition, pp. 1–8 (2008)
3. Xu, B.X., Hu, R.K., Guo, P.: Combining affinity propagation with supervised dictionary learning for image classification. Neural Computing and Applications (in press), doi:10.1007/s00521-012-0957-7

4. Zou, H., Hastie, T.: Regularization and variable selection via the Elastic Net. Journal of the Royal Statistical Society, Series B 67(2), 301–320 (2005)
5. Majumdar, A., Ward, R.K.: Classification via group sparsity promoting regularization. In: Proceedings of IEEE International Conference on Acoustics, Speech and Signal Processing, pp. 861–864 (2009)
6. Majumdar, A., Ward, R.K.: Improved group sparse classifier. Pattern Recognition Letters 31(13), 1959–1964 (2010)
7. Yin, J., Liu, Z.H., Jin, Z., Yang, W.K.: Kernel sparse representation based classification. Neurocomputing 77(1), 120–128 (2012)
8. Zhang, L., Zhou, W.D., Chang, P.C., Liu, J., Yan, Z., Wang, T., Li, F.Z.: Kernel sparse representation-based classifier. IEEE Transactions on Signal Processing 60(4), 1684–1695 (2012)
9. Vapnik, V.N.: Statistical learning theory. Wiley-Interscience, New York (1998)
10. Burges, C.J.C.: A tutorial on support vector machines for pattern recognition. Data Mining Knowledge Discovery 2(2), 121–167 (1998)
11. Kin, D.W., Lee, K.Y., Lee, L.K.H.: Evaluation of the performance of clustering algorithms in kernel-iinduced feature space. Patern Recoginition 38(4), 607–611 (2005)
12. Graves, D., Pedrycz, W.: Kernel-based fuzzy clustering and fuzzy clustering: A comparative experimental study. Fuzzy Sets and Systems 161(4), 522–543 (2010)
13. Scholkopf, B., Smola, A.J., Muller, K.R.: Nonlinear component analysis as a kernel eigenvalue problem. Neural Computation 10(5), 1299–1319 (1998)
14. Chen, S.C., Zhang, D.Q.: Robust image segmentation using FCM with spatial constraints based on new kernel-induced distance measure. IEEE Transactions on Systems, Man, and Cybernetics, part B: Cybernetics 34(4), 1907–1916 (2004)
15. Samaria, F.S., Harter, A.C.: Parameterisation of a stochastic model for human face identification. In: Sarasota, F.S. (ed.) 2nd IEEE Workshop on Applications of Computer Vision, pp. 138–142 (1994)
16. Georghiades, A., Belhumeur, P., Kriegman, D.: From few to many: Illumination cone models for face recognition under variable lighting and pose. IEEE Transactions on Pattern Analysis and Machine Intelligence 23(6), 643–660 (2001)
17. Lee, K., Ho, J., Kriegman, D.: Acquiring linear subspaces for face recognition under variable lighting. IEEE Transactions on Pattern Analysis and Machine Intelligence 27(5), 684–698 (2005)
18. Martinez, A., Benavente, R.: The AR face database. CVC Tech. Report. 24 (1998)
19. Deng, W., Yin, W.T., Zhang, Y.: Group sparse optimization by alternating direction method. Technical Report TR11-06, Department of Computational and Applied Mathematics, Rice University (2011)
20. Huang, J.Z., Zhang, T.: The benefit of group sparsity. Annals of Statistics 38(4), 1978–2004 (2010)

Kernel Fuzzy Similarity Measure-Based Spectral Clustering for Image Segmentation

Yifang Yang[1], Yuping Wang[1], and Yiu-ming Cheung[2]

[1] School of Computer Science and Technology, Xidian University, Xi'an 710071, China
yangyifang@xsyu.edu.cn, ywang@xidian.edu.cn
[2] Department of Computer Science, Hong Kong Baptist University, Hong Kong
ymc@comp.hkbu.edu.hk

Abstract. Spectral clustering has been successfully used in the field of pattern recognition and image processing. The efficiency of spectral clustering, however, depends heavily on the similarity measure adopted. A widely used similarity measure is the Gaussian kernel function where Euclidean distance is used. Unfortunately, the Gaussian kernel function is parameter sensitive and the Euclidean distance is usually not suitable to the complex distribution data. In this paper, a novel similarity measure called kernel fuzzy similarity measure is proposed first, Then this novel measure is integrated into spectral clustering to get a new clustering method: kernel fuzzy similarity based spectral clustering (KFSC). To alleviate the computational complexity of KFSC on image segmentation, Nyström method is used in KFSC. At last, the experiments on three synthetic texture images are made, and the results demonstrate the effectiveness of the proposed algorithm.

Keywords: spectral clustering, kernel fuzzy-clustering, image segmentation, Nyström method.

1 Introduction

Image segmentation is just to segment an image into different sub-images with different characters and get some interested objects. It is an important process of image analysis and image understanding [1], and plays a fundamental role in computer vision as a requisite step in such tasks as object detection, classification, and tracking [2]. As one of the key methods in image segmentation, clustering algorithms have been widely used in image segmentation [3,4,5,6,7]. In the past few decades, spectral clustering algorithms [4,8,9,10,11] have shown great promise in image segmentation and attracted more and more interest due to its high performance. It realizes dimension reduction by transforming the original dataset into a new one in a lower-dimensional eigenspace, then performs clustering by utilizing the eigenvectors of the normalized similarity matrix derived from the original dataset in lower-dimensional eigenspace.

However, there are still some open problems in spectral clustering algorithms. First of all, the similarity measure is usually constructed by the Gaussian kernel function based on Euclidean distance, while the Gaussian kernel function is parameter sensitive and the Euclidean distance is often not suitable to the complex distribution data [12], therefore, it is crucial to design suitable similarity measures for spectral clustering algorithms. In order to overcome the influence of the scale parameter, Zelnik-Manor and

M. Kurosu (Ed.): Human-Computer Interaction, Part V, HCII 2013, LNCS 8008, pp. 246–253, 2013.
© Springer-Verlag Berlin Heidelberg 2013

Perona [13] proposed a self-tuning spectral clustering algorithm (STSC) that utilized a local scale for each data point to replace the single scale parameter, however, the effect of the local scale parameter is similar to that in Gaussian kernel function and the distance is still Euclidean distance. Thus the algorithm cannot overcome the drawbacks of the original spectral clustering [20]. Fischer, B. et al.[21] proposed a path-based similarity. This similarity considers two points in one cluster if they are connected by a set of successive points in dense regions. This is intuitively reasonable. However, it is not robust enough against noise and outliers[22]. Feng Zhao et al.[23] proposed a fuzzy similarity measure by utilizing the prototypes and partition matrix obtained by fuzzy c-means clustering algorithm, however, This similarity is susceptible to the result of fuzzy c-means clustering algorithm. In addition, when the scale n of the data set is relatively large, the overall time complexity and space complexity of standard spectral clustering can reach $O(n^3)$ and $O(n^2)$, respectively [9], which make it difficult to store and decompose a large similarity matrix. Fowlkes et al. [4] presented the Nyström approximation technique to alleviate the computational burden of spectral clustering algorithms.

Several papers reported that the kernel fuzzy-clustering method (KFCM) had better performance than the standard fuzzy clustering method (FCM). Authors in [24] reported the higher classification rate of kernel fuzzy-clustering algorithm on a 2-dimensional non-linearly separable synthetic data set. Also, the kernel-based clustering algorithm performs better on data with non-spherical clusters such as the ring clusters [25].

In this paper, a novel kernel fuzzy similarity measure is proposed and a new spectral clustering algorithm based on this measure is used to the image segmentation. To alleviate the computational complexity, time and space complexity of the algorithm, the Nyström method is applied to the algorithm.

The rest of this paper is organized as follows. In Section 2, we present a short overview about techniques of KFCM, and construct the kernel fuzzy similarity measure. Nyström approximation technique is briefly introduced and the proposed KFSC method is described in Section 3. Section 4 contains parameter setting, computational experiments results, the analysis and discussion. Finally, conclusions are made in Section 5.

2   Kernel Fuzzy Similarity Measure and KFSC Method

2.1   Kernel Fuzzy Similarity Measure

Given a data set $X = \{x_1, x_2, \cdots, x_c\}$ in the p-dimensional space R^p. The KFCM algorithm partitions X into c fuzzy subsets by minimizing the following objective function:

$$J_m(U, V) = 2 \sum_{i=1}^{c} \sum_{k=1}^{n} (u_{ik})^m (1 - K(x_k, v_i)) \tag{1}$$

where c is the number of clusters , n is the number of data points, $U = (u_{ik}|i = 1, 2 \cdots, c, k = 1, 2 \cdots, n)$ is partition matrix with $u_{ik}(1 \leq i \leq c, 1 \leq k \leq n)$ being the fuzzy membership of kth pixel belonging to the ith cluster and satisfying $\sum_{i=1}^{c} u_{ik} = 1$,

$V = \{v_1, v_2, \cdots, v_c\} \subset R^p$ is cluster centers with $v_i (1 \leq i \leq c)$ being the centroid of ith cluster. $K = e^{-\| x-y \|^2 / t}$ with t being a positive number called Kernel width.

By minimizing $J_m(U, V)$, one can get the partition matrix U and clustering center V [18]

$$\mu_{ik} = \frac{\{1/(1 - K(x_k, v_i))\}^{1/(m-1)}}{\sum\limits_{j=1}^{c} \{1/(1 - K(x_k, v_j))\}^{1/(m-1)}} \tag{2}$$

$$v_i = \frac{\sum\limits_{k=1}^{n} (\mu_{ik})^m K(x_k, v_i) x_k}{\sum\limits_{k=1}^{n} (\mu_{ik})^m K(x_k, v_i)} \tag{3}$$

If the element u_{ij} of matrix U is seen as the probability that the i-th data point belonging to the j-th cluster, we can reasonably assume that two data points belonging to the same cluster have higher similarity, while two data points belonging to different cluster have lower similarity. Let $\mathbf{u}_i$ be the ith column vector of matrix U. It is the probability of data point x_i belonging to c clusters. We can also assume that the greater the inner product of $\mathbf{u}_i$ and $\mathbf{u}_j$, the higher the similarity of data points x_i and x_j. Based on this idea, a new kernel fuzzy similarity measure is proposed

Algorithm 1: A New Kernel Fuzzy Similarity Measure
Input: .data set X to be clustered.
Output: Obtain the similarity matrix S of the dataset.

Step 1. Cluster the data set X into c clusters via KFCM, and obtain the partition matrix U.
Step 2. Let $U = \{\mathbf{u}_1, \cdots, \mathbf{u}_i \cdots, \mathbf{u}_n\}$, here, $\mathbf{u}_i$ is the ith column vector of matrix U, it consist of the membership value that data point x_i belonging to c clusters.
Step 3. For each x_i and x_j,
    If x_i and x_j belonging to the same cluster,
    $s_{ij} = 1$
    Else s_{ij} is equal to the inner product of $\mathbf{u}_i$ and $\mathbf{u}_j$.
Step 4. Finally, the similarity matrix S is obtained.

2.2 KFSC Method for Texture Image Segmentation

In order to apply our method to texture image segmentation, NSCT (Non-subsampled Contourlet Transform) texture feature extracted from the image are used as a suitable image representation. We perform three-level discrete wavelet decomposition on texture images to extract a ten-dimension energy feature using a 16×16 window, which can be written as

$$E = \frac{1}{MN} \sum_{i=1}^{M} \sum_{j=1}^{N} |coef(i, j)| \tag{4}$$

where $M \times N$ is the subband size and $|coef(i, j)|$ is the coefficient in the ith row and jth column of the subband.

When the size of the dataset is large, spectral clustering encounters an expensive computation and large matrix storage in pairwise similarity construction. Moreover, the algorithm requires much time and memory to compute the eigenvectors of the similarity matrix. Suppose the number of the dataset is N. We select $n(n \ll N)$ sample points randomly, which can define an n-by-n affinity matrix A. B denotes the similarity matrix between sample points and the remaining points with $B \in R^{n \times m}$ and $m = N - n$. The original affinity matrix W can be rewritten as

$$[A \quad B; B^T \quad C]$$

To alleviate the computation and storage burden, Nyström method calculates it by

$$W = \begin{bmatrix} A & B \\ B^T & B^T A^{-1} B \end{bmatrix}$$

Note that $B^T A^{-1} B$ is easy to obtain. Thus the method reduces the computational complexity of spectral clustering. Based on all these, an outline of KFSC for texture image Segmentation is as follows:

Algorithms 2: KFSC
Input: An $M \times M$ image I to be segmented; the number k of the image segments; kernel parameter σ in NJW (Nyström) and number l of samples in NJW (Nyström); $l > k$.

Step 1. Extract Feature by using NSCT, compute the texture features of each pixel in the image by Eq.(4), and obtain the texture feature dataset X with M^2 instances and 10 attributes.

Step 2. Construct the similarity matrix S using proposed kernel fuzzy similarity measure.

Step 3. Compute the dominant eigenvectors matrix $V \in R^{n \times n} (n = M^2)$ by adopting Nyström approximation technique.

Step 4. Let $1 = \lambda_1 \geq \lambda_2 \geq \cdots \lambda_k$ be the k largest eigenvalues of L and $p^1, p^2 \cdots, p^k$ be the corresponding eigenvectors. Form the matrix $P = [p^1, p^2 \cdots, p^k] \in R^{n \times k}$ and here p^i is the column vector.

Step 5. Form the matrix Y from P by renormalizing each of s rows to have unit length (i.e., $Y_{ij} = \{P_{ij}/\sum_j (P_{ij})^2\}$).

Step 6. Treat each row of Y as a point in R^k , and cluster them into k clusters via K-means algorithm to obtain the final segmentation result of image I.

3   Experimental Results and Analysis

3.1   Experimental Setup and Performance Measure

To study the performance of KFSC, it is tested on three synthetic texture images in this section, and compared with three other algorithms: FCM [], KMEANS [], and NJW []. The experiments are implemented in MATLAB 7.10 (R2010a) on a computer with Intel (R) Xeon (R) CPU, 2.53 GHz and Windows XP Professional.

In the experiments, kernel parameter t is set using fast bandwidth selection rule in [19]. For dataset $X = \{x_1, x_2, \cdots, x_N\}$, the data center is denoted by $\bar{x}$. The mean distance is denoted by $\bar{d}$. The bandwidth is defined by $t = \frac{1}{N-1} \sum_{i=1}^{N} (d_i - \bar{d})^2$. In FCM: the maximum number of iterations was 100 and the fuzzy exponent was 2.0. In NJW: the scale parameter varied in the interval [0.15 1] with step length 0.02. In KFSC: the centroids obtained by KMEANS are taken as the initial centroids of KFSC to reduce the instability in initialization. For all the methods, we performed 10 independent runs. The minimum average CE rates were listed in Table 1.

Clustering Error (CE) is used to evaluate clustering performance and defined by [26]

$$CE = 1 - \frac{\sum_{i=1}^{n} \delta(y_i, map(c_i))}{n},$$ where n is the number of samples and y_i and c_i denote the true label and the clustering label of the algorithm, respectively, $\delta(y, c)$ equals one if $y = c$, and zero, if else. $map(\cdot)$ maps each cluster label to a category label. The smaller the Clustering Error, the better the performance.

3.2   Analysis of Experimental Results

In this section, the three synthetic texture images with two, three, and four categories from the Brodatz album of the University of Southern California are used in the experiments. The statistical results (average CE rates) are shown in Table 1.

Table 1. Comparison of the Minimum average CE rates (%) obtained by compared algorithms

	$Texture2$	$Texture3$	$Texture4$
FCM	10.38	8.25	5.35
KMEANS	8.93	8.10	5.12
NJW (Nyström)	5.69	7.66	4.92
KFSC	5.07	7.08	4.84

It can be seen from Table 1 that KFSC outperforms the other three algorithms. Therefore, the proposed KFSC is effective and non sensitive to the scale parameter on spectral clustering.

Fig. 1 shows the segmentation results of the synthetic texture image with two categories of the compared algorithms. It can be seen that there are some misclassified spots in black regions in Fig. 1(c) and (d); whereas, the segmented result in Fig. 1(e) is the better, and the result in Fig.1(f) is the best. This coincides with the results in Table 1. From the segmentation results for two categories texture images, KFSC method obtains the best performance.

Fig. 2 shows the segmentation results of the synthetic texture image with three categories. The similar conclusion can be obtained.

Fig. 3 shows the segmentation results of the synthetic texture image with four categories. In Figs. 3(c) and 3(d), there are also some misclassified points on the upper left of the figures. Thus, FCM and KMEANS perform the worst, NJW performs better than FCM and KMEANS, and KFSC obtain the best performance.

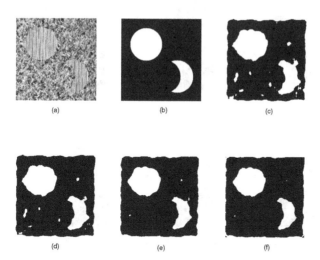

Fig. 1. The segmentation results of the synthesized texture image with two categories. (a) The original image (256 256 pixels); (b) The ideal segmentation, (c) FCM , (d) KMEANS, (e) NJW (Nyström) , (f) KFSC at t=0.23.

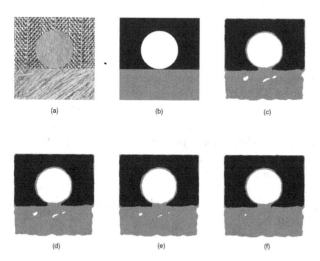

Fig. 2. The segmentation results of the synthesized texture image with three categories. (a) The original image (256 256 pixels); (b) The ideal segmentation, (c) FCM , (d) KMEANS, (e) NJW (Nyström) , (f) KFSC at t=0.25.

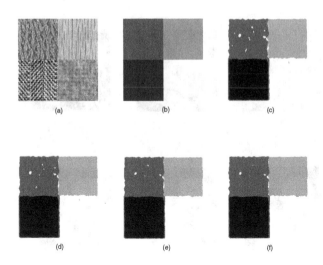

Fig. 3. The segmentation results of the synthesized texture image with four categories. (a) The original image (256 256 pixels); (b) The ideal segmentation, (c) FCM , (d) KMEANS, (e) NJW (Nyström) , (f) KFSC at t=0.28.

As seen from Fig. 1-3 and Table 1, KFSC have obtained more satisfying segmentation results than other three algorithms. Thus, we can conclude that KFSC is efficient and robust.

4   Conclusion

In order to solve the drawbacks of the similarity measure based on Euclidean distance, a novel image segmentation method (KFSC) is proposed based on a novel kernel fuzzy similarity measure. The proposed KFSC can not only avoid the influence of the scale parameter on spectral clustering effectively, but also overcome the drawback of Euclidean distance measure. In addition, the Nyström method is applied to alleviate the computational complexity. Experiments on three synthetic texture images indicate the proposed algorithm is efficient and robust.

Acknowledgement. This work was supported by the National Natural Science Foundation of China under Grant No. 61272119.

References

1. Shi, J., Malik, J.: Normalized cuts and image segmentation. In: IEEE Conf. Computer Vision and Pattern Recognition, vol. 3(2), pp. 731–737. IEEE Computer Society (1997)
2. Yilmaz, A., Javed, O., Shah, M.: Object tracking: A survey. ACM Computing Surveys 38(4), 1–45 (2006)

3. Chen, S.C., Zhang, D.Q.: Robust image segmentation using FCM with spatial constraints based on new kernel-induced distance measure. IEEE Transactions on Systems Man and Cybernetics Part B: Cybernetics 34(4), 1907–1916 (2004)
4. Fowlkes, C., Belongie, S., Chung, F., Malik, J.: Spectral grouping using the Nyström method. IEEE Transactions on Pattern Analysis and Machine Intelligence 26(2), 214–225 (2004)
5. Shi, J., Malik, J.: Normalized cuts and image segmentation. IEEE Transactions on Pattern Analysis and Machine Intelligence 22(8), 888–905 (2000)
6. Zhao, F., Jiao, L.C., Liu, H., Gao, X.B.: A novel fuzzy clustering algorithm with nonlocal adaptive spatial constraint for image segmentation. Signal Processing 91(4), 988–999 (2011)
7. Liu, H., Zhao, F., Jiao, L.: Fuzzy spectral clustering with robust spatial information for image segmentation. Applied Soft Computing 12, 3636–3647 (2012)
8. Liu, H.Q., Jiao, L.C., Zhao, F.: Non-local spatial spectral clustering for image segmentation. Neurocomputing 74(1-3), 461–471 (2011)
9. Gou, S.P., Zhuang, X., Jiao, L.C.: Quantum Immune Fast Spectral Clustering for SAR Image Segmentation. IEEE Geoscience and Remote Sensing Letters 9(1) (January 2012)
10. Chen, W., Feng, G.: Spectral clustering with discriminant cuts. Knowledge-Based Systems 28, 27–37 (2012)
11. Rebagliati, N., Verri, A.: Spectral clustering with more than K eigenvectors. Neurocomputing 74, 1391–1401 (2011)
12. Su, M.C., Chou, C.H.: A modied version of the K-means algorithm with a distance based on cluster symmetry. IEEE Trans. Pattern Anal. Mach. Intell. 23, 674–680 (2001)
13. Zelnik-Manor, L., Perona, P.: Self-tuning spectral clustering. In: Advances in Eighteenth Neural Information Processing Systems (NIPS), Vancouver, Canada, pp. 1601–1608 (2004)
14. Kim, D.W., Lee, K.Y., Lee, D., Lee, K.H.: Evaluation of the performance of clustering algorithms in kernel-induced feature space. Pattern Recognit. 38(4), 607–611 (2005)
15. Graves, D., Pedrycz, W.: Performance of kernel-based fuzzy clustering. Electron. Lett. 43(25), 1445–1446 (2007)
16. Graves, D., Pedrycz, W.: Kernel-based fuzzy clustering and fuzzy clustering: A comparative experimental study. Fuzzy Sets Syst. 161(4), 522–543 (2010)
17. Chen, L., Chen, C.L.P., Lu, M.: A Multiple-Kernel Fuzzy C-Means Algorithm for Image Segmentation. IEEE Transactions on Systems, Man, and Cybernetics Part B: Cybernetics 41(5), 1263–1274 (2011)
18. Zhang, D.Q., Chen, S.C.: A novel kernelized fuzzy C-means algorithm with application in medical image segmentation. Artif. Intell. Med. 32(1), 37–50 (2004)
19. Tsai, D.-M., Lin, C.-C.: Fuzzy C-means based clustering for linearly and nonlinearly separable data. Pattern Recognition 44, 1750–1760 (2011)
20. Zhang, X., Li, J., Yu, H.: Local density adaptive similarity measurement for spectral clustering. Pattern Recognition Letters 32, 352–358 (2011)
21. Fischer, B., Buhmann, J.M.: Path-based clustering for grouping of smooth curves and texture segmentation. IEEE Trans. Pattern Anal. Machine Intell. 25(4), 513–518 (2003)
22. Chang, H., Yeung, D.-Y.: Robust path-based spectral clustering. Pattern Recognit. 41(1), 191–203 (2008)
23. Zhao, F., Liu, H., Jiao, L.: Spectral clustering with fuzzy similarity measure. Digital Signal Processing 21, 701–709 (2011)
24. Zeyu, L., Shiwei, T., Jing, X., Jun, J.: Modified FCM clustering based on kernel mapping. In: Proc. of the Internat. Society for Optical Engineering, vol. 4554, pp. 241–245 (2001)
25. Graves, D., Pedrycz, W.: Kernel-based fuzzy clustering and fuzzy clustering: A comparative experimental study. Fuzzy Sets and Systems 161, 522–543 (2010)
26. Bach, F., Jordan, M.: Learning spectral clustering. In: Proceedings of NIPS 2003, pp. 305–312 (2003)

Depth Camera Based Real-Time Fingertip Detection Using Multi-view Projection

Weixin Yang, Zhengyang Zhong, Xin Zhang[*], Lianwen Jin,
Chenlin Xiong, and Pengwei Wang

School of Electronic and Information Engineering,
South China University of Technology, Guangzhou, P.R. China
yang.wx@mail.scut.edu.cn, timothy7784@gmail.com,
eexinzhang@scut.edu.cn, eelwjin@scut.edu.cn, xcl_722@163.com,
eepwwang@163.com

Abstract. We propose a real-time fingertip detection algorithm based on depth information. It can robustly detect single fingertip regardless of the position and direction of the hand. With the depth information of front view, depth map of top view and side view is generated. Due to the difference between finger thickness and fist thickness, we use thickness histogram to segment the finger from the fist. Among finger points, the farthest point from palm center is the detected fingertip. We collected over 3,000 frames writing-in-the-air sequences to test our algorithm. From our experiments, the proposed algorithm can detect the fingertip with robustness and accuracy.

Keywords: Kinect, depth image, finger detection, fingertip detection, multi-view projection.

1 Introduction

Natural Human Computer Interface (N-HCI) has long been a heated research topic. Traditional finger detection methods are usually vision-based. When the finger is pointing out, finger area is relatively small and traditional vision-based fingertip detection methods face several challenges. The local maximum curvature [2] and the fingertip template matching [3, 4] are typical successful vision-based methods. These methods heavily relied on good segmentation results and are sensitive to noise and occlusion. Therefore, prior information is introduced, like the arm-shoulder structure [9] and hand skeletal model [10]. Such approaches can provide the finger joint location and estimate the hand pose, but it costs extra computational load. When the hand is close to the skin or skin-like color objects, they could be regarded as hand, which may affect the result. Furthermore, it is difficult to detect fingertip precisely with traditional vision-based methods, due to the variety of position and angle of fingers. The launch of Microsoft Kinect [1] camera provides new possibilities of natural human computer interaction. Several recent researches have made promising progress in HCI using depth information

---

[*] Corresponding author.
[1] http://www.microsoft.com/en-us/kinectforwindows/

M. Kurosu (Ed.): Human-Computer Interaction, Part V, HCII 2013, LNCS 8008, pp. 254–261, 2013.
© Springer-Verlag Berlin Heidelberg 2013

provided by Kinect [5-8]. In order to improve effectiveness and usability of device-free HCI systems, robust methods of fingertip detection become important.

In this work, we propose a real-time fingertip detection method via multi-view projection using depth information. Our method has the following main steps: extracting the user from the background, segmenting the hand from user's torso using depth threshold, employing multi-view projection to detect candidate pixels of the finger, finding the fingertip according to the distance between candidate fingertip pixels and the palm center. Our algorithm can robustly and effectively locate the fingertip of a pointing-out finger, which overcomes traditional vision-based problems like hand pose variation, skin-color similarity and hand self-occlusion. Moreover, the accurate fingertip detection framework enables us to further design a fingertip-based device-free writing system.

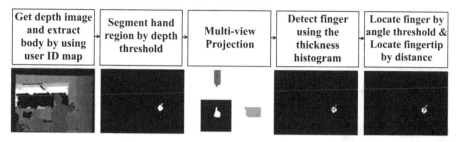

Get depth image and extract body by using user ID map	Segment hand region by depth threshold	Multi-view Projection	Detect finger using the thickness histogram	Locate finger by angle threshold & Locate fingertip by distance

Fig. 1. The flowchart of proposed algorithm

2   Proposed Framework

When the finger is pointing out, finger area is relatively small and traditional vision-based fingertip detection methods face several challenges. Our multi-view projection method converts frontal view depth image into top and side view depth images. By combining different views, we can effectively distinguish the fist and pointing-out finger using histograms of hand thickness from different views. Further, we employ a simple physical model to eliminate points that are wrongly classified as finger. Finally, the farthest finger point from the palm center is marked as the fingertip. The flowchart is shown in Fig.1.

2.1   Hand Segmentation

We get depth image captured by the Kinect camera and extract the user from the background using user ID map provided by OpenNI[2]. And then depth threshold is employed to segment hand from user's torso on the basis of the assumption that the writing hand is always in front of torso. By increasing the threshold of hand segmentation, we can get extra forearm pixels and mark the spatial center of them as forearm point for further process.

2.2   Finger Identification

For the proposed multi-view projection, we firstly convert pixels of depth image into 3D points [1] and project each point onto top and side view (Fig.2 (a)-(c)). When a finger is

---

[2] http://www.openni.org/Downloads/OpenNIModules.aspx/

pointing out, it is reasonable to assume that the finger is thinner than palm and wrist (Fig.2 (d)). Thickness histograms can roughly segment fist and finger. Thickness histograms are generated by calculating the number of pixels of each thickness value. According to our observation, there are two different histogram patterns. As shown in Fig. 3, without the pointing-out finger, thickness histograms from different views have only one peak; while in the other case, the histogram of one view might show only one peak but histogram of the other view would have two peaks, indicating the fist and finger. We select finger pixels accordingly for further process.

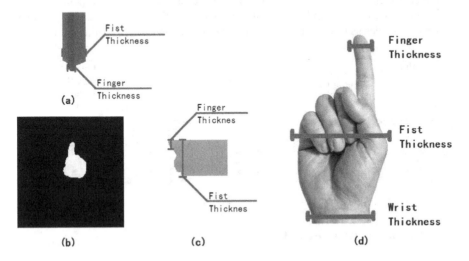

Fig. 2. Three views demonstration and hand thickness illustration: (a) top view (b) front view (c) side view (d) thicknesses of finger, fist and wrist

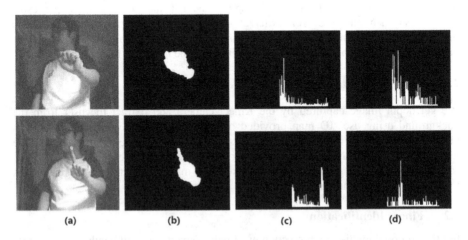

Fig. 3. Thickness histograms of two different situations: (a) RGB image, (b) Extracted hand, (c) Thickness histogram of side view, (d) Thickness histogram of top view

2.3   Fingertip Identification

According to our experiments, some fist points would be wrongly classified as finger points. This might lead to false fingertip detection. We take the following steps to remove these interference points. Considering the writing gesture of hand, the angle between finger and forearm is greater than 90° and this becomes a simple physical constraint for finger point classification, as shown in Fig.4. The palm center is defined as the average of palm pixels, marked as blue triangle in Fig. 4. By enlarging the threshold of hand segmentation, we can get extra forearm pixels and calculate the average position of these points, and then mark it as forearm point, as shown in fig.4. By connecting forearm point and candidate finger point with palm center, we have two vectors and define the angle between them as θ. Candidate points which formed angles less than 90° are regarded as interference points.

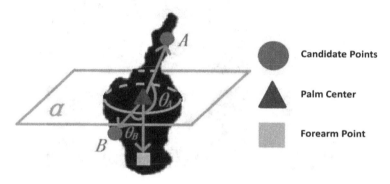

Fig. 4. Physical constraint for further finger point classification using angle-based threshold. (*A* is the finger point and *B* is wrongly classified fist point.)

P(x,y,z) is a candidate point, then finger point can be identified by the formula below:

$$P(x, y, z) = \begin{cases} 1 & (\theta \geq 90°) \\ 0 & (\theta < 90°) \end{cases}$$

As illustrated in Fig.5, point A and point B are both candidate points and formed two angles: θA and θB. Angle θA is greater than 90° while θB is less than 90°, so point A is finger point and point B is interference point. Plane α is formed by 90°angles, and is employed to distinguish space of finger points from the space of interference points.

Among finger points, the farthest point from palm center is the detected fingertip. Detailed illustration and final fingertip detection results are shown in Fig.5. Fig.5 (a) is the color image with detected fingertip (yellow point) corresponding to the recognition result in Fig.5 (c). Fig.5 (b) is the depth image with candidate finger points (red points) generated by multi-view projection and thickness histogram. Fig.5 (c) indicates the process of fingertip identification.

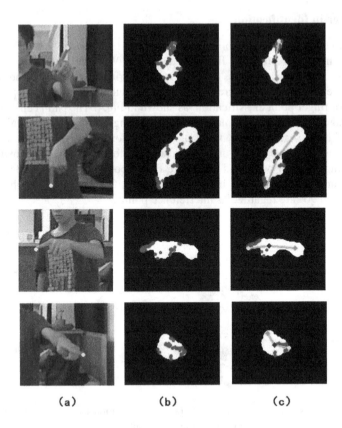

Fig. 5. Finger point classification and final fingertip detection result. (a) Color image (b) Candidate finger points (RED) (c) Results (RED for finger points, BLUE for palm center, GREEN for forearm point and YELLOW for fingertip).

3   Experimental Result and Discussion

Testing data was captured at 30 fps using OpenNI, with a resolution of 640×480. We focus on the hand and finger tracking, therefore the test subject was required to stand straight to the camera and write in the air with one finger out. We collected 3,010 frames (shown in Table 1) writing-in-the-air sequences, including writing in numbers, English and Chinese characters. We manually labeled these videos. The data includes location of fingertip and a bounding box of the hand in each frame.

Table 1. Overview of testing data set

	Uppercase	Lowercase	Character	Number	Total
Frame Number	507	485	1503	515	3,010

Detailed comparison is shown in Table 2. Curvature fitting [2] and template matching [3, 4] methods have been proposed for the vision-based fingertip detection. Both methods face common 2D vision challenges like noisy and clutter background, similar skin color and hand pose variation. Nearest method uses the depth image and aims at finding the pixel with the smallest depth value. It has largest pixel error, because fingertip is not always the closest point to the camera. The cluster method [1] is the most recent and relevant work using depth data. It has relatively few pixel errors but its performance is unstable. Pixel error greatly increases when dealing with complex inputs like Chinese character. The proposed method has the best fingertip detection accuracy rate of 93.69% (within 10 pixels) and an average deviation of 4.76 pixels with a stable performance for various inputs.

Table 2. Accuracy comparison within 10 pixels (<10) and 5 pixels (<5)

Method	Nearest	Curvature[2]	Template[3,4]	Cluster[1]	Proposed
<10	71.05%	81.04%	88.59%	90.41%	**93.69%**
<5	57.48%	52.24%	64.65%	**67.58%**	64.82%

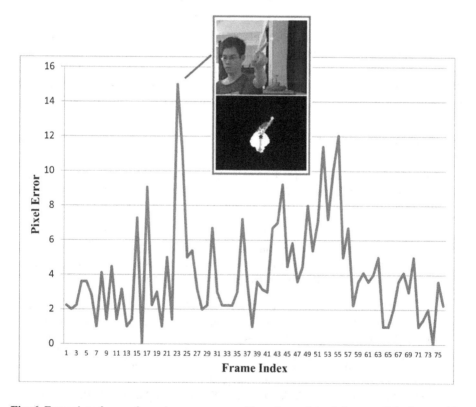

Fig. 6. Error plot of every frame in one sequence. The color and depth images of the frame with the highest pixel error are shown.

An example of distribution of pixel errors in a sequence is shown in Fig. 6. The errors are distributed on an average of 4.15 pixels. The color image and depth image of the frame with the most error are shown. Due to the time synchronization issue between color and depth images when the user is writing fast, the detected fingertip (yellow point) from the depth image is different from the fingertip in the color image. Skin color detection is an available method to solve this issue. The finger pixels that are not skin-colored can be removed from the fingertip candidate pixels.

4   Conclusion

This paper proposes a real-time fingertip detection method that overcomes traditional vision-based problems like hand pose variation, skin-color similarity and hand self-occlusion. We employed depth threshold to segment hand from user's torso, and then use the depth information of front view to generate depth map of top view and side view. Due to the difference between finger thickness and fist thickness, thickness histogram is employed to segment the finger from the fist, and the farthest finger point from the palm center is marked as the detected fingertip. According to the result of our experiments, the proposed algorithm can detect the fingertip with robustness and accuracy.

Acknowledgement. This work is supported by National Undergraduate Innovative and Entrepreneurial Training Program (No.111056115), National Science and Technology Support Plan (2013BAH65F01, 2013BAH65F04), GDSTP (No. 2012A010701001), Natural Science Foundation of China (No. 61202292) and Fundamental Research Funds for Central Universities of China (No. 2012ZM0022).

References

1. Feng, Z., Xu, S., Zhang, X., Jin, L., Ye, Z., Yang, W.: Real-time fingertip tracking and detection using Kinect depth sensor for a new writing-in-the air system. In: International Conference on Internet Multimedia Computing and Service (ICIMCS 2012), pp. 70–74 (2012)
2. Malik, S., Laszlo, J.: Visual touchpad: a two-handed gestural input device. In: Proceedings of international Conference on Multimodal Interfaces, pp. 289–296. ACM (2004)
3. Jin, L., Yang, D., Zhen, L., Huang, J.: A novel vision based finger-writing character recognition system. Journal of Circuits, Systems, and Computers (JCSC) 16(3), 421–436 (2007)
4. Crowley, J., Berard, F., Coutaz, J.: Finger tracking as an input device for augmented reality. In: International Workshop on Gesture and Face Recognition, pp. 195–200 (1995)
5. Minnen, D., Zafrulla, Z.: Towards robust cross-user hand tracking and shape recognition. In: IEEE International Conference on Computer Vision Workshops, pp. 1235–1241 (2011)
6. Pugeault, N., Bowden, R.: Spelling it out: Real-time asl fingerspelling recognition. In: IEEE International Conference on Computer Vision Workshops, pp. 1114–1119 (2011)
7. Shotton, J., Fitzgibbon, A., Cook, M., Sharp, T., Finocchio, M., Moore, R., Kipman, A., Blake, A.: Real-time human pose recognition in parts from single depth images. In: IEEE Conference on Computer Vision and Pattern Recognition (CVPR), vol. 2, p. 7 (2011)

8. Tang, Y., Sun, Z., Tan, T.: Real-time head pose estimation using random regression forests. In: Sun, Z., Lai, J., Chen, X., Tan, T. (eds.) CCBR 2011. LNCS, vol. 7098, pp. 66–73. Springer, Heidelberg (2011)

9. Wu, A., Shah, M., Da Vitoria Lobo, N.: A virtual 3D blackboard: 3D finger tracking using a single camera. In: International Conference on Automatic Face and Gesture Recognition, pp. 536–543 (2000)

10. Kang, S., Nam, M., Rhee, P.: Color based hand and finger detection technology for user interaction. In: International Conference on Convergence and Hybrid Information Technology, pp. 229–236 (2008)

Evaluation of Hip Impingement Kinematics
on Range of Motion

Mahshid Yazdifar, Mohammadreza Yazdifar, Pooyan Rahmanivahid,
Saba Eshraghi, Ibrahim Esat, and Mahmoud Chizari

School of Engineering and Design
Brunel University
United Kingdom
Mahshid.YazdiFar@brunel.ac.uk

Abstract. Femoroacetabulare impingement (FAI) is a mechanical mismatch between femur and acetabulum. It would bring abnormal contact stress and potential joint damage. This problem is more common on people with high level of motion activity such as baler dancer and athletics. FAI causes pain in hip joints and consequently would lead to reduction in range of motion. This study investigates whether changing the kinematics parameters of hip joint with impingement can improve range of motion or not. Hip joint model is created in finite element environment, and then the range of motion was detected. The original boundary conditions are applied in the initial hip impingement model. Then gradually the gap between femur and acetabulum in the model was changed to evaluate the changing kinematics factors on range of motion.

Mimics (Materialise NV) software was used to generate the surface mesh of three-dimensional (3D) models of the hip joint from computerised tomography (CT) images of the subject patients diagnosed with FAI. The surface mesh models created in Mimics were then exported to Abaqus (Simulia Dassault Systems) to create a finite element (FE) models that will be suitable for mechanical analysis. The surface mesh was converted into a volumetric mesh using Abaqus meshing modules. Material properties of the bones and soft tissues were defined in the FE model. The kinematic values of the joint during a normal sitting stance, which were obtained from motion capture analysis in the gait lab, were used as boundary conditions in the FE model to simulate the motion of the hip joint during a normal sitting stance and find possible contact at the location of the FAI. The centre of rotation for a female hip model with impingement was changed and range of motion was measured in Abaqus. The results were compared to investigate the effect of centre of rotation on range of motion for hip with femoroacetabular impingement. There was a significant change on range of motion with changing the gap between femur and acetabulum. Decreasing the distance between femur and acetabulum decreases the range of motion. When the distance between femur and acetabulum changes the location of impingement shifted. Increasing the distance between femur and acetabulum, there is no noticeable change in the location of impingement. This study concludes that changing the kinematics of hip with impingement changes the range of motion.

Keywords: hip joint, femoroacetabular impingement, finite element, kinematics.

M. Kurosu (Ed.): Human-Computer Interaction, Part V, HCII 2013, LNCS 8008, pp. 262–269, 2013.
© Springer-Verlag Berlin Heidelberg 2013

1   Introduction

Femoroacetabular impingement (FAI) is deformability of femur or acetabulum which causes pain and decreasing range of motion [6]. In fact, the aetiology of osteoarthritis of the hip has long been considered secondary to congenital or developmental deformities or primary whilst presuming some underlying abnormality of the articular cartilage [4]. However, recent information supports the hypothesis that the so-called primary osteoarthritis is also secondary to subtle developmental abnormalities.

Two types of impingement have been distinguished based on the origin and the mechanism of impingement commonly referred as pincer and cam [5], [9]. Cam impingement results in deterioration of the hip joint [11]. This is triggered by chondral avulsion, damage of the acetabular cartilage and labural detachment and degeneration [11]. The effects of pincer impingement are similar to those caused by cam impingement. The difference in its effect lies in the characteristic of the damage and the failure mechanism. In this study, we focused on cam impingement. During cam impingement, the labrum remains uninvolved over a rather long period [4].

Various articles reported about the location and angle of impingement [3, 7, 10]. In this study, we investigated about the effect of the distance/gap between the femur and acetabulum on the hip impingement of a 35 years old female patient with a height of 180 cm and weight of 86.04 kg at the time of surgery.

It was hypothesised that: the impingement would occur at the anterosuperior rim of the acetabulum, the risk of impingement would be affected by varying the gap between the femur and acetabulum with a subsequent change in the kinematics. The femur was moved 3 mm, 2mm and 1 mm both inward and outward the acetabulum. The effect of the kinematics on the FAI contact zone of the six different cases were compared with that of the patient's hip at the time of surgery.

2   Methodology

Mimics software was used to read the Digital Imaging and Communications in Medicine (DICOM) files of the patient. Segmentation and editing tools allow the manipulation of data to select the bone and soft tissue. After generating the 3D model it was surface meshed using triangular elements. The surface mesh was then exported to Abaqus and converted to a volume mesh (tetrahedral elements) prior to post processing.

2.1   Methods for the 3D Modelling

The anthropometrical data of a 35 years old female patient with a height of 180 cm and weight of 86.04 kg at the time of surgery with bone abnormalities indicative of cam-type deformity at the anterosuperior head-neck junction of her right hip was acquired using GE Medical systems/Light speed VCT computed tomography scan. The patient was scanned from the pelvic girdle to the distal end of the femur in the supine position. The scans were made up of 1027 cross-sectional slices with a slice distance of 0.625 mm and a field of view (FOV) of 380 mm. The DICOM images generated during the CT scan were processed using Mimics software to obtain the primary 3D model using density segmentation techniques. 1027 image slices in the DICOM format were selected and automatically imported to Mimics.

The image slices were then stacked and converted to be displayed in the coronal, sagittal and axial views. However, the orientation of each view was defined before continuing. Once the 3D model of the hip joint was obtained, the mesh discretisation process was carried out in Mimics and subsequently material constitutive, loads and boundary conditions, and kinematics constraints were defined in Abaqus software.

Thresholding. Thresholding based on the Hounsfield scale was used to separate both the cortical bone of the femur and acetabulum from the surrounding muscles and tissues. CT images are a pixel map of the linear X-ray attenuation coefficient of tissue [11, 8]. Mimics software allows changing the predefined threshold to the required threshold by choosing compact bone of an adult from a drop down menu. Therefore, a lower limit of 662 Hounsfield units (HU) and upper limit of 1652 HU were defined in order to select only the cortical bones from the surrounding tissues (Figure 1).

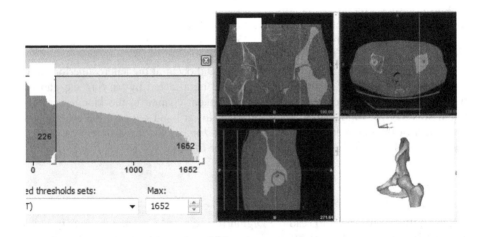

Fig. 1. (a) Hounsfield threshold value for bone. (b) corresponding histogram which shows the separation of bone from soft tissue.

Region Growing. The femur and the pelvic girdle were then selected separately and assigned different masks through region growing operation.

Segmentation Mask. The region growing has split the segmentation into separate entities. This enables to generate separate geometrical files and subsequently the 3D model. Noisy pixels and artefacts were eliminated manually. Cavity filling operation was performed in order to produce density masks of both the femur and the pelvic girdle. A 3D computation was then carried out on both the density masks. Both the 3D models were wrapped and conservatively smoothened using a smooth factor of 0.4 in order to have a good surface topography without affecting the anthropometrical data of the patient.

2.2   Generating 3D Model

The 3D models were separately divided into triangular elements and nodes (surface meshing) in Mimics remesher module. A quality threshold of 0.3 was defined for the triangular elements. The height/base parameter was used to measure the ratio between the height and the base of a triangle.

Operations for the Quality Improvement of the Triangular Elements. The aim was to improve the quality of the triangular elements through different operations prior to autoremeshing of the 3D models. The amount of detail of the femur was reduced by conservatively smoothening it without compromising the faithfulness of the model.

Surface Mesh to Volumetric Mesh. The head of the femur and the osseous bump at the head-neck junction was manually refined with the triangles smaller than 2.5 mm in order to get an accurate result at the contact interface. The head of the femur was acting as the slave surface mesh whereas the acetabulum was behaving as the master surface mesh in order to respect the master and slave formulation in Abaqus. Moreover, the head of the femur was smoothened and refined relative to the pelvic girdle in order to prevent any sharp corners in the model that might cause convergence problems and to prevent any penetrations of master nodes into the slave surface from going undetected [1].

When the surface meshes of the femur and pelvic girdle satisfactorily passed the mesh quality test, they were saved as 'inp' files format and exported to Abaqus separately. There, the triangular elements were converted to tetrahedral elements. The femur and the pelvic girdle consisted of 125 376 and 169 483 tetrahedral elements, respectively.

2.3   Load and Boundary Conditions

The femur was then instanced with the pelvic girdle using part/model instance option in Abaqus. In this study, a sphere was wrapped around and fitted on the femur to approximately locate the hip joint centre which was defined as a reference point.

Material property of the finite element (FE) model: the bone was assumed to behave as nonlinear elastic and an isotropic Poisson's ratio of 0.3 was used which is sufficient for the study of stress and strain [4]. An equivalent bone material property was defined for the finite element (FE) model with an average value of elastic modulus of 750 MPa and an average density of 1.281 g/cm3.

Contact Algorithm and Boundary Conditions. In this study, the general automatic contact algorithm was defined in Abaqus. The contact between the acetabulum and the femur was assumed to be ideally frictionless with no bonding. As the acetabulum and femur comes into contact, the penetrations are detected and the contact constraints are applied according to the penalty constraint enforcement method [1].

The medial pelvic wall was pinned in the x, y and z directions. The pelvic bone support of the acetabulum was non-rigid [12, 2].

The centre of rotation of the femur was coupled with the whole section of the femur.

3   Results and Discussion

Seven different cases were defined to examine the effect of kinematics on the FAI as shown in Table 1. Case 4 was fixed as a control which centre of rotation was same as centre of femur. Cases 1, 2 and 3 represented the condition when the centre of rotation was moved toward acetabulum whereas cases 5,6 and 7 centre of rotation was moved outward the acetabulum as shown in Figure 2.

Table 1. Seven different location of centre of rotation were introduced between femur and acetabulum to examine the effect of kinematic on the FAI

Cases	Centre of rotation according to the centre of femur
1	-1 mm
2	-2 mm
3	-3 mm
4	0 mm
5	+1 mm
6	+2 mm
7	+3 mm

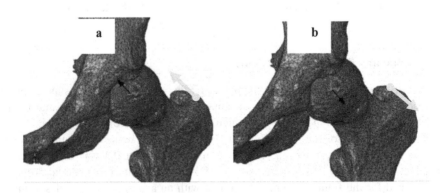

Fig. 2. Schematic of centre of rotation displacement a) toward (-) and b) outward (+) the acetabulum

The impingement was applied to measure the impingement angle for all of the models. Figure 3 illustrates the flexion, adduction and internal rotation for all cases at the contact point.

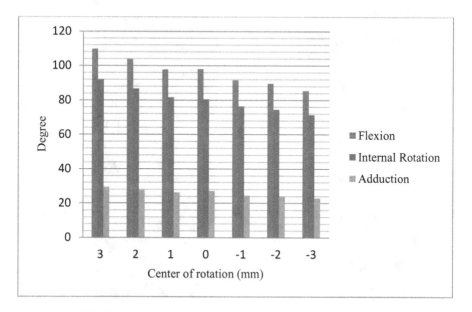

Fig. 3. Flexion, adduction and internal rotation for different cases

When the centre of rotation was moved outward the acetabulum, flexion, adduction and internal rotation increased correspondingly resulting to lower risk of impingement. While varying the centre of rotation between the femur and the acetabulum changed the angle at which impingement happened.

Figure 4 shows the contact area of the hip impingement for all of the cases.

When the centre of rotation was moved inward the acetabulum (cases 1, 2 and 3), the location of impingement shifted. The impingement occurred earlier on the rim of the acetabulum as shown in Figure 4c and Figure 4d as compared to Figure 4a. However, when the centre of rotation was moved outward the acetabulum, there was no noticeable change in the location of impingement as shown in Figure 4e to 4g. From a clinical point of view, this can be beneficial to patients as the impingement would occur at a higher range of motion without the need of removing any osseous bumps provided that the congruency or conformity of the joint is not compromised. Changing the kinematic can be done by pins or microstructure layer between femur and acetabular.

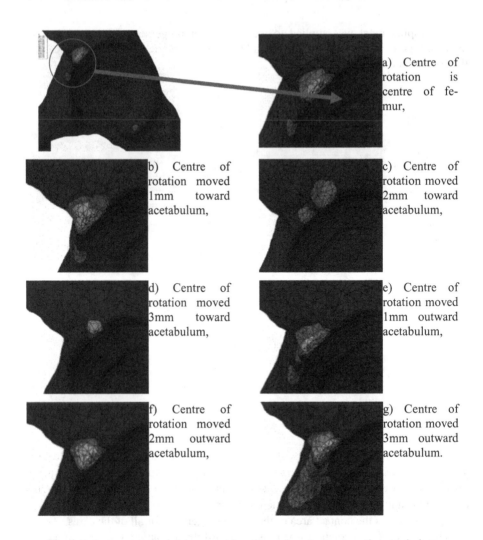

a) Centre of rotation is centre of femur,

b) Centre of rotation moved 1mm toward acetabulum,

c) Centre of rotation moved 2mm toward acetabulum,

d) Centre of rotation moved 3mm toward acetabulum,

e) Centre of rotation moved 1mm outward acetabulum,

f) Centre of rotation moved 2mm outward acetabulum,

g) Centre of rotation moved 3mm outward acetabulum.

Fig. 4. Femoroacetabular impingement location and contact area on the acetabulum

4 Conclusion

Our investigation demonstrates that the changing the centre of rotation affects the kinematics of a hip joint with cam-type FAI. It is observed that by moving the centre of rotation outward the acetabulum causes a significant reduction in the risk of impingement with an increase in the range of motion.

References

1. Abaqus (Simulia Dassault Systems), User's manual, version 6.10
2. Beau Paul, E., Zaragoza, E., Motamedi, K., Copelan, N., Dorey, F.J.: Three-dimensional computed tomography of the hip in the assessment of femoroacetabular impingement. Journal of Orthopaedic Research 23, 1286–1292 (2005)
3. Clohisy John, C., Knaus Evan, R., Hunt Devyani, M., Lesher John, M., Harris, H.M., Heidi, P.: Clinical Presentation of Patients with Symptomatic Anterior Hip Impingement. Clin. Orthop. Relat. Res. 467(3), 638–644 (2009)
4. Ganz, R., Leunig, M., Leunig-Ganz, K., Harris, W.H.: The etiology of osteoarthritis of the hip: an integrated mechanical concept. Clin. Orthop. Relat. Res. 466(2), 264–272 (2008)
5. Ganz, R., Parvizi, J., Beck, M., Leunig, M., Notzli, H., Siebenrock, K.A.: Femoroacetabular impingement: a cause for osteoarthritis of the hip. Clin. Orthop. Relat. Res. 417, 112–120 (2003)
6. Ingvarsson, T.: Prevalence and inheritance of hip osteoarthritis in Iceland. Acta Orthop. Scand. Suppl. 298, 1–46 (2000)
7. Langer Kubiak, M., Moritz, T., Murphy, S.B., Siebenrock, K.A., Langlotz, F.: Range of Motion in Anterior Femoroacetabular Impingement. Clinical Orthopaedics and Related Research 458, 117–124 (2007)
8. Mimics Materialise, User's manual, version 13.1
9. Horisberger, M., Brunner, A., Herzog, R.F.: Arthroscopic Treatment of Femoroacetabular Impingement of the Hip. Clinical Research 468(1), 182–190 (2010)
10. Philippon Marc, J., Brian Maxwell, R., Johnston, T.L., Schenker, M., Briggs, K.K.: Clinical presentation of femoroacetabular impingement. Knee Surg. Sports Traumatol Arthrosc. 15, 1041–1047 (2007)
11. Banerjee, P., Mclean, C.R.: Femoroacetabular impingement: a review of diagnosis and management. Curr. Rev. Musculoskelet Med. 4(1), 23–32 (2011)
12. Russell Mary, E., Shivanna Kiran, H., Grosland Nicole, M., Pedersen, D.R.: Cartilage contact pressure elevations in dysplastic hips: a chronic overload model. Journal of Orthopaedic Surgery and Research 1, 6 (2006)

Tracking People with Active Cameras

Alparslan Yildiz[1], Noriko Takemura[1], Yoshio Iwai[2], and Kosuke Sato[1]

[1] Osaka University, 1-1 Yamadaoka Suita, 565-0871, Japan
[2] Tottori University, 4-101 Koyamacho-minami, 680-8550, Japan

Abstract. In this paper, we introduce a novel method on tracking multiple people using multiple active cameras. The aim is to capture as many targets as possible at any time using a limited number of active cameras.

In our context, an active camera is a statically located PTZ (pan-tilt-zoom) camera. Using active cameras for tracking is not researched thoroughly, since it is relatively easier to use increased number of fully static cameras. However, we believe this is costly and a deeper research on the employment of active cameras is necessary.

Our contributions include the removal of necessity for the detection of each person individually in an efficient way and estimating the future states of the system using a simplified fluid simulation.

Keywords: multiple view, tracking, active cameras.

1 Introduction

There is a large amount of research on tracking multiple targets with multiple cameras. Most research in the literature focus on tracking with static cameras, which are fixed in terms of position and viewing angle throughout the entire life of the application they are being used for [1] [2]. However, there are situations where fixing camera position and viewing angle would introduce disadvantages such as not being able to surveille all the targets, or in most cases most of the targets. One would expect to surveille more efficiently using active cameras rather than fully static cameras. An active camera in our definition is a statically located PTZ (pan-tilt-zoom) camera. It is arguable that, increasing number of static cameras would solve the problem mentioned above. However, increasing number of cameras can become quite costly, and it also introduces new problems, such as increased computation power demand and more complex connections. For such reasons, we propose to employ active cameras to automatically track unknown number of people in a closed environment. In the current literature, research on tracking with active cameras is still not thoroughly investigated. Xie et.al [3] presented a tracking method with a PTZ camera for a given target. They use patch based feature learning to track the target, which lets them ignore the background changes on the PTZ camera. Bimbo and Pernici [4] presented a tracking method with the prediction on the tracked target. Huang and Fu's recent work [5] is closest to our research. They presented a multi-target tracking with multiple PTZ cameras. However, our research differs in the following ways;

M. Kurosu (Ed.): Human-Computer Interaction, Part V, HCII 2013, LNCS 8008, pp. 270–279, 2013.

we do not need to detect the targets individually which makes our method independent of the amount of targets in complexity and formulation, and the camera movement decision in our research is more centralized.

In our method, we make the common assumption that people move on a fixed ground plane, and we track people by means of tracking the *occupancy* of the people on this ground plane. The occupancy of the ground plane is computed using the method described in [6]. There are some advantages in tracking the occupancy of the people rather than the people themselves. Firstly, the people in the scene are not required to be detected individually in each camera view. Hence, the system is independent of the number of people in the scene. The formulation becomes simpler and the system can run faster. Secondly, the decision on the active camera movements is not affected by the number of cameras or the people under surveillance. This is possible by tracking the occupancy of the people and with a centralized decision for the active cameras. Thirdly, computing the occupancy of the ground plane merges all the information from multiple cameras into a single occupancy map. This makes the tracking easier, more unified and more suitable for high level functionalities.

On the other hand, using active cameras requires some attention on the design of the system. One problem is the response time of the cameras. As the active cameras are controlled programmatically from a computer, when the computer sends a command to a camera, such as pan or tilt action, it will take some time for the camera to complete the given command. This should be taken into account and the system should account for the latency of the cameras. For this purpose, the estimation of the future states of the system becomes necessary. In this paper, we propose a novel method for estimating the future states of the ground occupancy map. We treat the whole occupancy map as an evolving 2D field of particles. We then run a simplified fluid simulation on this 2D field to estimate the future state of the occupancy map. Details of this method is presented in the following section.

The rest of this paper is organized as follows; in Section 2 we present the details of our system, in Section 3 we present the experimental results and in Section 4 we give the conclusions and discuss the future work.

2   The Method

The main advantage of our tracking method is that, we do not need to detect and track each and every person in the scene. We achieve this by computing the occupancy map of the ground plane and track the occupancy map itself. The occupancy map is computed with the virtual planes method presented in a related work [6]. Once we have the occupancy map, our main goal is to compute camera directions in terms of pan and tilt values in order to surveille as many people as possible. This is achieved by maximizing the total amount of occupancy under the viewing cones of the cameras. An illustration is given in Fig. 1. Working with camera viewing cones simplifies the problem of keeping the subjects under surveillance and leads to a unified formulation for multiple cameras.

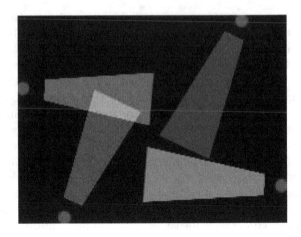

Fig. 1. Sample viewing cones of the cameras (red dots)

In practice, active cameras such as [8] have noticable response times when a pan/tilt command is issued. Thus, we need to compute the camera movement decisions in advance and we need to estimate the future states of the occupancy map.

We divide the problem of deciding the camera movements into two parts; estimation of the future states of the occupancy map and making the camera movement decisions using the camera viewing cones on this estimated occupancy map. We address these two sub-problems in the following two subsections.

2.1   Estimating Future Time Steps

Using active cameras requires some extra care because an issued pan/tilt command cannot be executed immediately. Thus, we need to estimate the future states of the system in order to give camera commands in advance. Depending on the camera response times, which can be calibrated easily, we need to estimate the system's state a few time steps into the future to issue a pan/tilt command.

Tracking each person in the scene individually would let us predict the future positions of the people using particle filtering or similar methods. In our study, we propose a novel method to estimate the future states of the ground occupancy map itself, without the necessity of performing detection. First, we treat the whole occupancy map as a 2D field composed of particles (occupancies). Each discrete location on the occupancy map corresponds to a particle, and each particle carries a single property, the occupancy value. Next, we let the 2D field evolve in time using a simplified fluid simulation [7], in which we do not have any viscosity or external forces. The fluid simulation requires the instantaneous velocities of the particles. However, it is relatively easy to achieve since we have the occupancy maps for previous time steps, and we simply compute a dense optical flow between the current

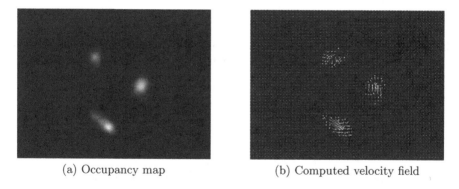

(a) Occupancy map   (b) Computed velocity field

Fig. 2. Instatneous velocity computation with optical flow

and the previous occupancy maps. Fig. 2 shows a sample of velocity computation for the occupancy map. Finally, we run the fluid simulation for the 2D field and obtain the estimation of the occupancy maps for future time steps. A sample output of this estimation is given in Fig. 3.

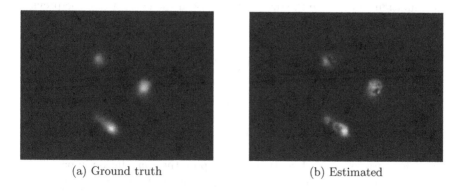

(a) Ground truth   (b) Estimated

Fig. 3. Estimating future time steps

We use fluid simulation to estimate futures states of the occupancy map because it is relatively well defined for 2D fields and a throughly studied topic. There are also GPU implementations [7] which run extremely fast and is adequate for our purposes. In contrast to simply extrapolating the velocities of the particles into future frames, using a fluid simulation lets us capture non linear motions on the occupancy map. However, using this method to estimate future states of the occupancy map is not very stable for longer intervals into the future. Luckily, we need to estimate the future occupancy maps only to catch the response time of the active cameras. A typical active camera (Sony EVI-D100 [8]) has a response time less than a second for a pan-tilt action of 90 degrees. In practice, the cameras are expected to move a maximum of 10 to 20 degrees in 1 second.

2.2   Camera Movement

The main decision about the camera movement is done once the ground occupancy for the future time steps are estimated. Given the camera locations, we only need to compute the pan and tilt values for the estimated ground occupancy map. This is modeled as

$$\max_{\theta} \sum_{i}^{N_c} \sum_{\Omega} F * M_i(\theta) \,, \tag{1}$$

where θ encapsulates all the pan-tilt parameters of all the cameras, N_c is the number of cameras, Ω is the domain of the occupancy map, F is the ground occupancy map and M_i is the mask that represents the viewing cone of the i-th camera. With this formulation, we try to maximize the total amount of occupancy under the viewing cones of the cameras.

Eq. 1 may lead to solutions with high diversity in successive configurations for a camera. As the active cameras cannot move instantly, this situation should be avoided and the successive configurations for a camera should be smooth to achieve quick response. For this purpose, we introduce a simple penalty term into the formulation to penalize the diversities between the current and the previous camera configurations.

$$\max_{\theta} \sum_{i}^{N_c} \sum_{\Omega} F * M_i(\theta) - \lambda D(\theta_{prev}, \theta) \,. \tag{2}$$

Here $D(\theta_{prev}, \theta)$ simply evaluates the total difference between the two configurations and λ is a scalar to adjust the importance of the penalizing.

The mask $M(\theta)$ in the above equations is a 2D mask where only the points in the camera viewing cone are nonzero. We pre-compute the masks for each camera configuration, and at runtime we compute the convolution in parallel. In Fig. 4 precomputed masks for different camera configurations are presented. The length of the viewing masks on the occupancy map domain are truncated by the average height of the targets on the corresponding configuration of a camera. If a target's location is further than a predefined distance, where a standing target's height is half of the average target height on that view, the targeted object is considered out of view and the viewing mask is truncated.

Maximizing Eq. 2 is done by discretization of the parameter space by 10 degrees intervals and evaluating the summation for each configuration. However, this direct approach is not suitable for real time purposes as the parameter space must cover all possible combinations of each camera configuration. Thus, we maximize Eq. 2 for each camera separately and we process the cameras in a fixed random order. Different orders result only in little difference in practice. Moreover, to remove redundancy, once a camera is processed, we reduce the occupancy that has been observed by that camera on the occupancy map. This lets us to efficiently guide the cameras to divide the ground occupancy among them.

Fig. 4. Camera viewing masks for discrete panning configurations

Discretizing the pan-tilt parameter space also addresses two subtle problems. First one is the tracking of the background image for each camera view. The computation of the occupancy map requires a simple background subtraction performed on each camera view. However, as the cameras change their view with pan-tilt commands, the background images also change. We propose a simple yet effective method to address this problem. As we discretize the pan-tilt parameter space of the cameras, we have the set of all possible configurations for each camera. This lets us to collect the background images for all the possible configurations as an initialization phase. For an active camera with a 640×480 image resolution, and 18 pan and 3 tilt configurations, it takes around 16 MB of memory to store all the background images.

The second subtle problem is the calibration of the cameras. Computation of the occupancy map also requires the homography matrices to map the camera views to the rectified ground, i.e the occupancy map domain. The initialization phase to collect the background images also lets us collect image features to compute the necessary homography matrices. Computing all the rectifying homography matrices during the initialization phase addresses the problem of geometric calibration, and storing all the homography matrices would only take a few kilobytes of memory. Computing the camera viewing masks is also achieved by warping the full camera view back onto the ground plane using the inverses of the rectifying homographies. Thus, we only require the ground rectifying homographies, which map the ground plane viewed on a camera image to the occupancy map domain, as the geometric calibration of the system.

3   Experiments

We performed simulation experiments to evaluate the performance of our method. To measure the performance, we define *coverage* as the total occupancy under the camera viewing cones over all the occupancy on the ground occupancy map. We compute the average *coverage* over the length of the experiment. For simulation experiments, we also computed the maximum achievable *coverage* using dynamic programming, and we report the accuracy of the experiments as the ratio of the achieved *coverage* to the maximum achievable *coverage*.

We made simulations with a rectangular environment without any obstacles and let the subjects move freely and randomly. The subjects can leave the room and come back in. We performed several trials with random trajectories and present the average results. We aimed to make simulations realistic by implementing the camera response times and acquisition speeds based on the active cameras we have [8].

To show the effectiveness of the active cameras, we run the same experiments with static cameras which have fixed viewing angles. In the reported results, *Best empty* represents the static camera case where the cameras are set to maximize the *coverage* in an empty ground occupancy map, *Best prior* represents the case where the camears are set to maximize the *coverage* over the average locations of the objects on the ground occupancy map. The average locations are computed in during the generation of the simulation data.

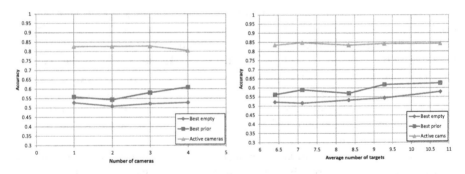

Fig. 5. System performance (*coverage* ratio)

In Fig. 5 we present the graphical results to show the performance of our method. We report the accuracy against the number of cameras and against the average number of targets. In the latter case, we report the mean accuracy over all number of cameras. Note that during the execution of the system, we do not compute the number of targets at any time. Corresponding numerical results are presented in Table 1 and Table 2.

In Fig. 5 it is clear that our method, when the coverage ratio against the maximum achievable coverage is considered, is not affected by the number of cameras or the number of targets in the environment.

To show the applicability of our method, we also performed real experiments in a closed environment. For real experiments, we used 3 Sony EVI-D100 [8] cameras. We used 9 evenly spaced pan positions of the cameras as configurations. We did not use tilt positions as the cameras could cover all the experiment

Table 1. System performance against number of cameras (*coverage* ratio)

# of cameras	Best empty	Best prior	Active cameras
1	0.527	0.558	0.827
2	0.509	0.543	0.827
3	0.522	0.581	0.828
4	0.528	0.610	0.804

Table 2. System performance against number of targets (*coverage* ratio)

Average # of targets	Best empty	Best prior	Active cameras
6.403	0.522	0.562	0.834
7.118	0.516	0.589	0.848
8.338	0.532	0.569	0.834
9.283	0.543	0.618	0.841
10.772	0.578	0.626	0.843

space with panning only. The experimental environment is shown in Fig. 6 as a panaroma created from the view of one of the cameras. The other two cameras are marked with red circles in the panaroma. The panaroma image is created from all the images of 9 configurations.

Fig. 6. Experiment environment

For real experiments we had a limited space, for which the cameras can easily cover all the space with little movement. Thus, to make the experiments more challenging, we used ROI masks on the camera images to crop off half of the image area closer to image borders. This way, the cameras had smaller viewports and were forced to pan to cover more space. We also restricted the camera panning to neighbor configurations to speed up the computation. With this restriction the regulazation in Eq. 2 is implemented as checking at most 3 configurations and picking the best.

Table 3. System performance for real experiments with 2 and 3 cameras

# of objects	3 cameras	2 cameras
1	0.92	0.90
2	0.88	0.81
3	0.77	0.71

In Table 3, we present the computed *coverage* of the real experiments. Since we do not have the ground truth locations of the targets, the *coverage* in real experiments simply corresponds to the value computed using Eq. 1 and is likely to be higher than the actual *coverage*. Due to limited space, we could work with upto 3 subjects at most. To compare the accuracy with the simulations, we shut off one of the cameras and repeated the experiments with 2 cameras only. Table 3 shows that there is not big difference between 2 and 3 camera cases for real experiments. However, the decrease in the accuracy as the number of objects increase is considerably high. This is due to many practical difficulties including; inaccuracies in the background subtraction, the homography computation for ground plane recitfication and the occupancy computation, and the mechanical inaccuracy of the camera movement. These practical difficulties lead to inaccuracy in the computed occupancy map.

4   Conclusions and Future Work

We have presented a novel method for tracking people using multiple active cameras, and our method can also be used to surveille any kind of object as long as the ground occupancy map can be computed. In our method, we removed the necessity of detection of subjects by tracking the occupancy of the subjects, instead of the subjects themselves. We performed synthetic experiments to show the potential and real experiments to show the real-time capabilities and the applicability of our system. We also presented a fast and reliable future time step estimation for the occupancy map using simplified fluid simulation.

As working with occupancy maps instead of detecting individuals makes the process simpler, it also removes some control on the system. One disadvantage is that, now it is not trivial to assign importance on individuals. This option may be desirable in some applications where an operator may need to select subjects of interest. A workaround would use color tracking based methods on a selected camera view and increase the occupancy values for the corresponding ground positions on the occupancy map.

The presented work here includes novelties and some different approaches. We believe, however, that more detailed research is necessary for the optimization of camera control decisions. Our future work includes such detailed research, as well as detailed experiments in bigger and more complex environments. Another future work would be the involvement of the zoom property of the active cameras. In broader environments with limited number of cameras, zooming is necessary to achieve better surveillance quality.

References

1. Kim, K., Davis, L.S.: Multi-camera Tracking and Segmentation of Occluded People on Ground Plane Using Search-Guided Particle Filtering. In: Leonardis, A., Bischof, H., Pinz, A. (eds.) ECCV 2006. LNCS, vol. 3953, pp. 98–109. Springer, Heidelberg (2006)
2. Khan, S.M., Shah, M.: A multiview approach to tracking people in crowded scenes using a planar homography constraint. In: Leonardis, A., Bischof, H., Pinz, A. (eds.) ECCV 2006. LNCS, vol. 3954, pp. 133–146. Springer, Heidelberg (2006)
3. Xie, Y., Lin, L., Jia, Y.: Tracking Objects with Adaptive Feature Patches for PTZ Camera Visual Surveillance. In: International Conference on Pattern Recognition (2010)
4. Bimbo, A.D., Pernici, F.: Uncalibrated 3D Human Tracking With a PTZ-Camera Viewing a Plane. In: 3DTV Conference: The True Vision - Capture, Transmission and Display of 3D Video (2008)
5. Huang, C., Fu, L.: Multitarget Visual Tracking Based Effective Surveillance With Cooperation of Multiple Active Cameras. IEEE Transactions on Systems, Man, and Cybernetics Part B: Cybernetics 41(1) (2011)
6. Yildiz, A., Akgul, Y.: A Fast Method for Tracking People with Multiple Cameras. In: Third Workshop on HUMAN MOTION Understanding, Modeling, Capture and Animation, Greece (2010)
7. Nvidia GPU Gems, `http://http.developer.nvidia.com/GPUGems/gpugems_ch38.html`
8. Sony EVI-D100 PTZ Camera, `http://pro.sony.com/bbsc/ssr/cat-industrialcameras/cat-robotic/product-EVID100/`

Classification Based on LBP and SVM
for Human Embryo Microscope Images

Yabo Yin[1], Yun Tian[1], Weizhong Wang[2], Fuqing Duan[1],
Zhongke Wu[1], and Mingquan Zhou[1]

[1] College of Information Science and Technology, Beijing Normal University, Beijing, China
[2] Reproductive Medicine Center, Navy General Hospital, PLA, Beijing, China
yinyabo0612@163.com

Abstract. Embryo transfer is an extremely important step in the process of in-vitro fertilization and embryo transfer (IVF-ET). The identification of the embryo with the greatest potential for producing a child is a very big challenge faced by embryologists. Most current scoring systems of assessing embryo viability are based on doctors' subjective visual analysis of the embryos' morphological features. So it provides only a very rough guide to potential. A classifier as a computer-aided method which is based on Pattern Recognition can help to automatically and accurately select embryos. This paper presents a classifier based on the support vector machine (SVM) algorithm. Key characteristics are formulated by using the local binary pattern (LBP) algorithm, which can eliminate the inter-observer variation, thus adding objectivity to the selection process. The experiment is done with 185 embryo images, including 47 "good" and 138 "bad" embryo images. The result shows our proposed method is robust and accurate, and the accurate rate of classification can reach about 80.42%.

Keywords: embryo microscope images, feature extraction, automatic classifier, local vector pattern, support vector machine.

1   Introduction

Nowadays, the efficiency of single embryo transfer remains relatively poor. In order to improve the odds of a successful pregnancy, it is common to transfer more than one to the uterus per cycle, but this often results in multiple pregnancies (twins, triplets, etc), which are associated with significantly elevated risks of serious complications [1]. How to recognize viable embryos remains a big challenge. Most current methods of embryo viability assessment are based on doctors' subjective visual analysis of the embryos' morphological features [2], and tend to be subjective and imprecise. More-over, the ability of human's eye is so limited that some important but unintuitive information cannot be identified. Aiming at this problem, some researchers [3, 4] attempted to identify viable embryos with the help of the pattern recognition methods, and they identified embryo images into two classes: those suitable for procreation and those not suitable, by designing a decision support system. So the problem will be

M. Kurosu (Ed.): Human-Computer Interaction, Part V, HCII 2013, LNCS 8008, pp. 280–288, 2013.
© Springer-Verlag Berlin Heidelberg 2013

simplified into two sub-problems. One is to get the feature vectors that are significantly different between two classes. The other is to design a good classifier with a high accuracy rate. Patrizi et al.[5] presented the TRACE algorithm, which recognized embryos as belonging to one of the two classes. The pattern vector of embryo features is comprised of 10 moments calculated based on the embryo image histogram. The classification algorithm used training sets to establish centers of classes and classification was then performed based upon a measure of distance from the class centers. In tests on 165 images, the average accuracy was claimed up to 0.85. Morales et al. [6] developed a decision support system based on a Bayesian classifier. The feature vector that describes each embryo image consists of variables based on morphology and on the clinical data of the patients. In tests on 63 cases, using different types of Bayesian classifiers, the accuracy of the systems was claimed from 63.49% to 71.43%. These classifiers greatly depend on the experiment datasets and may be improper to other data, although good for theirs.

The feature extraction of embryo images has a very big influence on the classification, so finding a valid description of the images is a key step for classifying the embryos. Some commonly used feature extraction methods in pattern recognition, such as principal component analysis (PCA) [7] or linear discriminant analysis (LDA) [8], mostly depict an image from the overall point of view, which can well extract the global properties of the image but are sensitive to light and position. Moreover, the feature extraction method based on central moments [5, 9] which has been used by some previous researches can quantify the key characteristics of embryo images in some extent, but it fails to consider the difference in the light intensity of embryo images. The LBP algorithm is not sensitive to light and can well extract the local texture of an image. It was first put forward by Ojala et al. [10], and it is widely used in texture analysis, such as face recognition. However, in this paper, we first employ the LBP algorithm to analyze human embryo images.

Designing a well-performed classifier is the other key step to ensure the recognition rate. Since SVM shows a superior performance in learning from a small number of samples, the classifier designed in this paper is based on this algorithm. In addition, we also use another classical pattern recognition method—K-NN [11] to design a classifier as a contrast. These classifiers will make use of our first-hand embryo images, which are all from fresh transfer cycles. Most of the images (about 90%) are first used to train the classifier, and then the rest to make a test on the basis of the rule learnt in training.

2   Methodology

2.1   Feature Extraction Based on LBP

The LBP algorithm is defined as a gray-scale invariant texture measure, derived from a general definition of texture in a local neighborhood. This algorithm is a non-parametric kernel which summarizes the local spatial structure of an image.

Moreover, it is invariant to monotonic gray-scale transformations. Hence the LBP representation may be less sensitive to changes in illumination. This is a very interesting property in embryo microscope image recognition. Almost all the state-of-the-art embryo image recognition algorithms are based on statistical classifier and local image features, which are noise sensitive and hardly to deliver perfect recognition performance. We have developed a novel feature extraction method, using the histogram of local binary pattern for global embryo image texture representation.

Fig.1 describes the basic LBP algorithm. Fig.1 (a) is a 3×3 rectangular area. First take its center pixel grayscale value as the closed value, and then compare its neighborhood eight pixels. If the center pixel value is less than its adjacent pixel value, we will set the value of this adjacent pixel as 1, otherwise set it as 0, so to produce the binary code of the region. As shown in Fig.1 (b), this model can be described by eight binary codes. Convert the binary code to a decimal number, and we obtain the LBP code of the center pixel, as shown in Fig.1 (c). The corresponding LBP code of each pixel reflects its neighborhood gray distribution. The LBP code histogram of certain region of an image can be used to describe the regional texture structure.

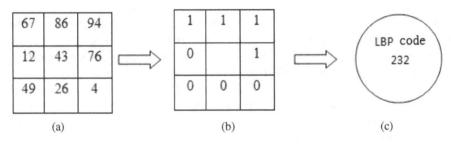

Fig. 1. The transformation process of LBP. (a) 3×3 pixels with different grayscale value ;(b) eight binary codes; (c) LBP code of the center pixel.

Fig.2 shows the results of applying LBP algorithm to extract features of different images. After calculating every pixel's corresponding LBP code, we do a frequency count for all the LBP code to produce a new gray histogram, and finally extract a feature vector by 1 * 256.

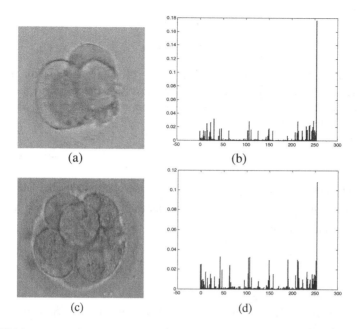

Fig. 2. LBP histograms of different images. (a) a "bad" embryo image;(b) LBP histogram corresponding to (a); (c) a "good" embryo image;(d) LBP histogram corresponding to (c).

2.2   Classifier Design

The SVM algorithm [12, 13] is based on the statistical learning theory and the principle of minimum structural risk. It seeks an optimal compromise between the complexity of the model (i.e. the accuracy of studying pending training samples) and the learning ability (i.e. no wrong recognition ability of arbitrary sample) according to the limited sample information, so as to get the best promotion ability. The SVM aims at finite samples and the sample size is small in this paper, so it is proper to be employed to classify the microscope embryo images.

Considering a model classifier, the training set is given, x_i is a vector of the input space, and y_i is the classification identification, $y_i \in \{-1,1\}$. If the input vector set is linearly separable, then they can be set apart by a hyperplane $w * x_i - b = 0$ (w is the normal vector of the classification hyperplane; b denotes the offset).If this vector set was correctly set apart by the hyperplane and the distance between the nearest vector from the hyperplane and the hyperplane is the longest, then the vector set is divided by the optimal hyperplane. In order to describe the classification hyperplane, we use the following form:

$$\begin{cases} w * x_i - b \geq 1 & if \quad y_i = 1 \\ w * x_i - b \leq -1 & if \quad y_i = -1 \end{cases} \tag{1}$$

According to the minimum structural risk (SRM) principle, looking for the biggest interval hyperplane is just looking for the smallest VC dimension to minimize the confidence interval. So by solving the following optimization problem, SVM obtains the optimal classification hyperplane:

$$\min \frac{1}{2} \| w \|^2 \quad s.t \; y_i [w * x - b] \geq 1 (i = 1, 2, ..., l) \tag{2}$$

In order to construct the optimal classification hyperplane when the data is linearly inseparable, we introduce the nonnegative variables $\xi_i > 0$ to construct a weak and indirect optimal hyperplane:

$$\min \frac{1}{2} \| w \|^2 + C (\sum_{i=1}^{l} \xi_i) \quad s.t \; y_i [w * x - b] \geq 1 - \xi_i (i = 1, 2, ..., l, \xi_i \geq 0) \tag{3}$$

where C is a regular parameter.

For the linearly inseparable problem, the SVM method maps the input space to a high-dimension feature space through nonlinear mapping, and then constructs the optimal hyperplane in the high-dimension feature space. SVM converts the inner product operation into the calculation in the input space by introducing the kernel function, namely $K(x_i, x_j) = [\varphi(x_i), \varphi(x_j)]$, where $K(x_i, x_j)$ is the kernel function.

So solving SVM is solving the optimization of the following problem:

$$\max W (a) = \sum_{i=1}^{l} a_i - \frac{1}{2} \sum_{i,j=1}^{l} a_i \, l \, a_j y_i y_j K (x_i, x_j); (0 \leq a_i; i = 1, ...l) \tag{4}$$

$$s.t \quad \sum_{i=1}^{l} a_i y_i = 0$$

The corresponding decision-making function for classification is:

$$f (x) = \text{sgn}(\sum_{i=sv} y_i a_i^0 K (x_i, x) - b_0) \tag{5}$$

Here, we use gaussian kernel to be the kernel function. We randomly select about 90% of the embryo images in each class to train the classifier, and use cross validation to get the optimal classifier parameters. Finally, the classifier is tested with the rest images.

3 Experiments and Analysis

This paper uses 385 fresh embryo images, 185 transfer cycles, from assisted reproductive medical center at navy general hospital, as original datasets. They were photographed 3 days after fertilization. In each transfer cycle, two or three embryos were transferred into the uterus. Clinical outcome showed that among these 185 embryo transfer cycles, 138 failed, which means, no embryo gave birth in the end. And the other 47 cycles succeeded, that is to say, at least one embryo became a baby in the end.

Considering that the labels of those images from the good class (at least one embryo gave birth) are imprecise, for it is difficult to know which one on earth became a baby in the end. In order to simplify the problem, this paper assumes that all images used in the experiment belong to a certain class for sure. Thus, there are some limitations when picking the images. In this paper, we randomly select one embryo from each transfer cycle. Consequently, we obtain 185 images, 138 of which are from the bad class, and the other 47 are from the good class.

According to our direct visual observation, there is no significant difference in morphological features between the two kinds of images. One possibility is that all these embryos have already been picked out according to their morphological features. Thus it is important to have the aid of pattern recognition methods to further select the viable embryos. Figure 3 shows some of the typical and atypical images from the two classes.

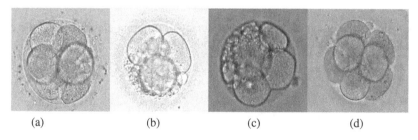

(a)   (b)   (c)   (d)

Fig. 3. Some of the "good" and "bad" images. (a), (b), (c) and (d),typical "good", typical "bad", atypical "good" and atypical "bad" embryo image, respectively.

Table 1 shows the classification results of the K-NN and SVM methods with different feature vectors. We do three groups of experiments for each classification method. They are named as group1, group 2 and group 3. In each group, the classification algorithm is carried out for 10 times with randomly chosen training and testing samples. And we estimate the performance of the classifier according to the classification accuracy of the test. Finally, we calculate the average accuracy of the three groups for each method.

Table 1. Classification accuracy of different classification methods

Feature extraction—Classification	Group No.	Mean	Max	Min
Central moments—K-NN	1	0.5278	0.7222	0.2778
	2	0.6000	0.7778	0.4444
	3	0.6833	0.8889	0.5556
	The average accuracy of three group:0.6037			
LBP—K-NN	1	0.6333	0.7778	0.4444
	2	0.7000	0.8889	0.5000
	3	0.6556	0.7778	0.5000
	The average accuracy of three group:0.6630			
LBP—SVM	1	0.7964	0.8571	0.7143
	2	0.8143	0.8571	0.7143
	3	0.7964	0.8214	0.6786
	The average accuracy of three group:0.8024			

Fig.4 shows the graph of the average accuracy of different classifiers and the comparison chart of two feature extraction methods with K-NN.

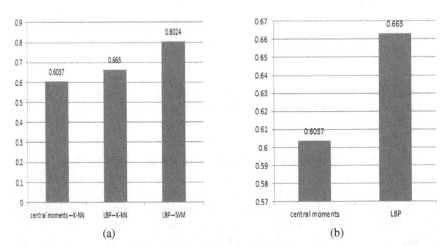

(a)   (b)

Fig. 4. Graph of the average accuracy of different classifiers. (a) average classification accuracy of the three classification methods; (b) average classification accuracy of the K-NN classifier with   two different feature extraction methods.

From the Fig.4, we can see that LBP algorithm as the feature extraction method is better than that based on central moments. And obviously, SVM algorithm has the absolute advantage in classifying our datasets. Not only that the average accuracy of classification by using SVM is as high as 0.8024, but also that the accuracy is over a small range. In addition, since most classifiers are sensitive to the number of the training samples, especially when the difference between different classes is not significant, too few samples will greatly decrease the performance of the classifier.

However, the performance of these classifiers is not very ideal in the whole compared with other applications. As for the reasons, the first one is that whether an embryo is good or not is not the single factor to decide the possibility of giving a birth. Some other factors, such as the gestational age, the number of pregnant, et al, will also greatly influence the result. Secondly, only one embryo image actually cannot reflect all information of the embryo, thus we cannot make an objective assessment of the embryo's quality just by one image.

4   Discussion

Challenge is enormous. First of all, related researches about the classification of the embryo images are relatively rare. Moreover, the few researches greatly depend on the data they use and a classification method with a relatively high accuracy for some data may not be proper for other data. In this paper, a new classification solution is proposed based on the LBP and SVM algorithms. The method has an average accuracy of 0.8024, and is superior to that based on central moments and K-NN. However, the proposed method still has some aspects to be improved, and if some other morphological features, such as the thickness of zona pellucida or the number of blastomeres et al, are taken into account, the classification result may be more excellent.

In addition, from the aspect of designing the classifier itself, there are some problems to be solved. First, as mentioned above, it is a semi-supervised classification problem, which is difficult to deal with. Secondly, there is no significant morphological difference between the two classes, which makes the feature extraction difficult and negatively influences the performance of the classifier. Thirdly, the number of embryo images available in this experiment is too small and even a little unbalanced between the two classes, which may lead to the classifier not stable.

Acknowledgements. This work is partly supported by the National Natural Science Foundation of China (No.61003134, 61170170) and the Key Program of Beijing Natural Science of Beijing (No. 4081002). The authors would like to thank the anonymous reviewers who gave valuable suggestions that have helped to improve the quality of the manuscript.

References

1. Santos Filho, E., Noble, J.A., Wells, D.: A Review on Automatic Analysis of Human Embryo Microscope Images. Open Biomed. Eng. J. 4, 170–177 (2010)
2. Santos Filho, E., Noble, J.A., Wells, D.: Toward a Method for Automatic Grading of Microscope Human Embryo Images. In: IEEE International Symposium on Biomedical Imaging, pp. 1289–1292 (2010)

3. Ning, F., Delhomme, D., LeCun, Y., Piano, F., Bottou, L., Barbano, P.E.: Toward Automatic Phenotyping of Developing Embryos fromVideos. IEEE Trans. Image Process 14(9), 1360–1371 (2005)
4. Siristatidis, C., Pouliakis, A., Chrelias, C., Kassanos, D.: Artificial Intelligence in IVF: A Need. Syst. Biol. Reprod. Med. 57(4), 179–185 (2011)
5. Patrizi, G., Manna, C., Moscatelli, C., Nieddu, L.: Pattern Recognition Methods in Human-assisted Reproduction. Int. Trans. Oper. Res. 11(4), 365–379 (2004)
6. Morales, D.A., Bengoetxea, E., Larranaga, P., et al.: Bayesian Classification for the Selection of in Vitro Human Embryos Using Morphological and Clinical Data. Comput. Meth. Prog. Bio. 90(2), 104–116 (2008)
7. Behdad, M., French, T., Barone, L., Bennamoun, M.: PCA for Improving the Performance of XCSR in Classification of High-dimensional Problems. In: Proceedings of the 13th Annual Conference Companion on Genetic and Evolutionary Computation, pp. 361–368 (2011)
8. Kobayashi, H., Zhao, Q.F.: Face Detection Based on LDA and NN. In: Proceedings of the 2007 Japan-China Joint Workshop on Frontier of Computer Science and Technology, pp. 146–154 (2007)
9. Morales, D.A., Bengoetxea, E., Larrabaga, P.: Selection of Human Embryos for Transfer by Bayesian Classifiers. Comput. Biol. Med. 38(11), 1177–1186 (2008)
10. Ojala, T., Pietikainen, M., Maenpaa, T.: Multiresolution Gray-Scale and Rotation Invariant Texture Classification with Local Binary Patterns. IEEE Trans. PAMI 24(7), 971–987 (2002)
11. Kou, H.Z., Gardarin, G.: Study of Category Score Algorithms for K-NN Classifier. In: Proceedings of the 25th Annual International ACM SIGIR Conference on Research and Development in Information Retrieval, pp. 393–394 (2002)
12. Christopher, J., Burges, C.: A Tutorial on Support Vector Machines for Pattern Recognition. Data Min. Knowl. Disc. 2(2), 121–167 (1998)
13. Wan, H.L., Chowdhury, M.: Image Semantic Classification by Using SVM. JSW 14(11), 1891–1899 (2003)

Semantic Annotation Method of Clothing Image

Lu Zhaolao, Mingquan Zhou, Wang Xuesong[*], Fu Yan, and Tan Xiaohui

College of Information Science and Technology, Beijing Normal University, Beijing, China
{lzl,mqzhou,wangxs,fuyan}@bnu.edu.cn, xiaohuitan@163.com

Abstract. Semantic annotation is an essential issue for image retrieval. In this paper, we take the online clothing product images as sample. In order to annotate images. we first segment the image into regions, then remove the background and noise information. The illumination and light interference is considered too. Cloth position and region are determined by rules. Images are translated into some features. Visual words are prepared by human and calculate methods. Finally, Image features are mapped to different visual words. Preprocessing and post-processing steps which uses face recognition method and background rule analysis are applied. Finally, some segmentation and annotation results are given to discuss the method.

Keywords: Semantic annotation, Image segmentation, Graph cut.

1   Introduction

Nowadays, with the growing of Internet and digital camera equipment, the number of digital images steps into a geometric progression growth. It is estimated, images stored in the Web has reached ten billion orders of magnitude. On the one hand, the vast amounts of web images have result the disaster of information retrieval. It is difficult to obtain the desired image from the Internet. On the other hand, it is also a powerful way to spread the e-commerce rapidly. In order to meet people's need, semantic annotation is focused.

1.1   Image Segmentation for Semantic Annotation

The semantic annotation is the basis of the image semantic retrieval, and adds keywords to reflect the image semantics by a specific algorithm to the image, and use the marked images or other information automatically learn the semantic and visual feature space, and established relationships marked image. The main image annotation methods include two types: global features of image annotation (Global-Based Image Annotation) and by region-based image annotation methods (Region-Based Image Annotation). The global features of images based on global statistical characteristics marked the [1-5], does not require segmentation and object-oriented analysis, but provide only coarse-grained semantic description, cannot provide more image detail semantics; marked by region-based image by image segmentation, semantic

---

[*] Corresponding author.

M. Kurosu (Ed.): Human-Computer Interaction, Part V, HCII 2013, LNCS 8008, pp. 289–298, 2013.

understanding can provide a low-level visual features of the regional level, more and more researchers focus, forming a series of effective methods [6-10].

In this paper, we take the online clothing product images as sample. The method establishes the global and region features of the images too. Pre-processing and post-processing steps which uses face recognition method and background rule are applied.

1.2   Related Works

Image segmentation is the basic work and an important part of [11-13] image annotation and analysis It is one of the hot research fields in computer image processing and machine vision. Automatic segmentation has still many difficulties and problems, but has been made part of the research results. Segmentation method using graph theory can change complex image processing problems into computable problem.

Image segmentation method, but there is no common, universal segmentation method. Mainstream image segmentation method include: regional competition and merger-based segmentation method [3], segmentation method based on graph theory [14-16], atlas-based segmentation method [17], based on gray level threshold segmentation method [19,20], edge detection-based segmentation method, based on target feature model segmentation method [18], based on level set segmentation of the Level Set method [21,22] based on Active Contour (Snake) [23, 24] segmentation method.

1.3   Graph-Cut Based Image Segmentation

Based on graph theory, image segmentation technology has made great progress in recent years. It has become hot spots of image segmentation [25]. Fixed thresholds and local features of the earlier graph cut algorithm to partition Zahn [26] using the minimal spanning tree method of image segmentation, the weight of edges in the graph using the gray value of the pixel distance calculation, divided the criterion of a maximum weight of minimum spanning tree (MST) segmentation method.

Wu and Leahy [27] proposed a map segmentation method, using the minimal cut for image segmentation by finding the minimal cut minimizes the similarity between the segmented pixels as segmentation criteria. Shi and Malik [28] proposed an improved Normalized Cut segmentation method. The method is also a minimum cut, but is calculated for all the right side of the connection value in the calculation of connection in the entire map edge centralization value of the share component.Grady and Schwartz [29] calculated for Ncut method for complex image size.Felzenszwalb [30,31] consider the overall visual effect, the use of global information and efficient function of the weights and the segmentation criteria, map segmentation by region merging.

2   Image Representation and Regional Definition

2.1   Image HSI Model Represents

Segmentation depends on the expression of the color image, color models, including RGB, HSI, due to the complexity of natural images, simple gray-scale information, likely to cause over-segmentation or under segmentation. HSI model based on the characteristics of human vision is used for it is closest to the human visual point of view to describe the image.

The statistical characteristics of the clothing image are relatively stable and very suitable for image segmentation as a graph theory basis for the weights. Red, green and blue components of the image pixels, respectively, for the R, G, B, and H, S, I can define the image RGB to the HSI conversion functions.

2.2   Image with a Weighted Undirected Graph

The image data is a two-dimensional matrix of pixels, in the context of graph theory, the figure is the number of nodes and connections between different nodes, the edge of the geometry of. The graph of the graph theory used to represent relationships between objects, the object itself as a graph node v_i, the connection between the object node to the edge e_i. Graph theory algorithms used in image processing pixel map mapping [34, 35].

To image P, Each pixel as a vertex $v_i \in V$, Set V is a set of all vertices; any two pixels constitute an edge $e(v_i, v_j) \in E$, marked e_{ij}, Set E is the collection of all edges ; the whole image corresponding to the undirected graph $G = (V, E)$。

If each side of the figure corresponds to a relative weight $w(v_i, v_j) \in W$, marked as w_{ij}, Set W is the collection of all edge weights set ; ; ; ; ; whole image corresponding to a weighted graph is expressed as $G = (V, E, W)$.

2.3   Image Map on the Partition Defined

Graph theory, image segmentation method, in fact, the threshold to complete the fuzzy transformed into optimization to solve problems, form the point of clustering results under certain conditions.

In a variety of graph theory of image segmentation method, the pixel as a node, as a side to the correlation between the pixels, the image is converted to weight undirected graph processing. Our image segmentation using graph theory to obtain the best segmentation of images.[32]

The graph theory of image segmentation, in fact transformed into a weighted diagram $G = (V, E, W)$, calculation for the minimal cut sets. Vertices of the graph corresponds to each pixel in the image P, the image P segmentation is equivalent to the set of vertices V divided into disjoint subsets. The definition of Graph Cut can be expressed as[36]:

$$V = \bigcup_{i=1}^{k} V_i, V_i \cap V_j = \Phi, i \neq j, i, j = 1, 2, ..., k, \mathrm{S}(V_1, V_2, ..., V_k)$$ is a segmentation of V, k is the number of blocks.

The goal of the vertex set V partitioning is to split a subset of nodes within the same class difference between pixels, the pixels of different classification between node dissimilarity. If all nodes of the graph is divided into m subsets $V_1, V_2, ..., V_m$,, segmentation criteria [36]: (1) Partition of the set V_i of internal vertices relevance and similarity; (2) The relevance and similarity between V_i and V_j different subsets is low.

Image segmentation, the graph G 'belongs to G, G' is divided into A and B, two-part cost function:

$$S\ (A,B) = \sum_{u \in A, v \in B} w(u, v)$$

The smallest division of the cost function S corresponding graph G is optimal binary divide, the split graph G, corresponding to the best split of the original image.

2.4   Image Feature Representation

Image segmentation to extract the appropriate low-level visual features, said to facilitate follow-up semantic annotation and similarity calculation. Contour and texture of the clothing commodity characteristics with the figure standing, layouts, block, etc., are very different, therefore the characteristics of the image area is the most stable color characteristics, combined with the macro issues the analysis is complete.

The entire image overall features, you can target prospects after removing the background area, quantified HSI space, computing 128-dimensional histogram of the overall feature of the dress pictures. $L = \{l_1, l_2, ..., l_{128}\}$, 其中 $i = 1, 2, ..., 128$

3   Image Segmentation Method

3.1   Image Preprocessing

Picture of online apparel goods into two categories, one category is the single items of clothing of the online store, usually the better picture quality, the background is relatively simple; pictures of people with complex background and dress models to deal with relatively complex.

Targeted to deal with the clothing image, the first category of images need to remove the background, the second category of images need to remove the interference of the face, hair, etc. For clothing commodities image, we set two types of rules:

- **Rule 1** simple background rules

Assume that the commodity picture for a simple background image, then the edge of the area of the larger connected region as the background area, the central region of the prospects for apparel region;

- **Rule 2** figure background rules

Assumed that the image has a character background, character clothing worn should be consistent with the rules of human proportion, costumes position and the position of the face with a specific topology.

3.2   Face Detection and Location and Background Removal

Apparel goods pictures often include models, how to extract the pure clothing area, there will be conducive to the subsequent segmentation and labeling. The most effective method to detect figures is to detect face in the image. It can help to determine the human faces' location, size, and numbers in the given images.

The face detection is using a template matching algorithm with default template. In order to improve the reliability and validity, we limit the minimum pixel scale and the largest proportion of the face images.

3.3   Weight Calculation of Graph Cuts

After the split of a subset, the degree of similarity need to quantify the calculation and description, usually expressed as a weight function w. The similarity degree function between the different pixel nodes, usually calculated by the distance of node.

Weight function and similarity function can take many forms and methods of calculation. The theory is mapping actually pixel to a feature space to calculate the quantitative distance in the space. The common form is based on the pixel gray value of the weight function. The function using local feature analysis of similarity:

Analysis of the commodity image pixel distance and color information is very important. It needs a variety of information in order to achieve effective segmentation. Each pixel mapping feature point (x, y, and h, s, i), where (x, y) pixel location information, (h, s, i) is the hue brightness of the pixel message.

HUE functions between pixel values and adjacent from the constructor the right functions, consider the global and local features [36], constructing an image pixel chrominance distance function:

$$w_{ij} = \begin{cases} \exp(-\dfrac{\|F_i - F_j\|_2^2}{\sigma_i^2}) \times \exp(-\dfrac{\|X_i - X_j\|_2^2}{\sigma_x^2}) & if \left\|X_i - X_j\right\|_2^2 < r \\ 0 & others \end{cases} \tag{1}$$

Where F is the quantization of the chrominance hue brightness function used to represent the color properties of pixels; X represents the pixel location, r is the associated pixel radius of the specified pixel, the two values σ is the color and distance of the Gaussian smoothing parameter. The function of the weights between the two pixels are color values the closer the greater the similarity between two pixels, the closer the distance between two pixels.

3.4   Graph Segmentation Method

Usually graph theory, image segmentation using split larger regional, and gradually obtained the spin-off results, the discriminated function to determine the minimal cut sets to make splitting the cost minimum. The method due to uncertainty in the segmentation process, it is difficult to optimize the partition.

We adopt a bottom-up merging the idea [31], to initialize a smaller area, gradually merged to form suitable region segmentation. An image corresponding to a weighted graph $G = (V, E, W)$, suppose there are n nodes and m connections, to get the corresponding image segmentation results.

4   Image Region Analysis Rule

4.1   Product Image Segmentation Implementation

The core algorithm of graph theory have been listed in the front, the original image processing and analysis need to pretreatment, post-region coloring, format conversion and a series of auxiliary work. The overall process of the algorithm is summarized as follows: (1) Read the given image, convert the data format, structure weighted directed graph G.(2)The right value for a given function, the weights for calculation of weight W values;(3)according to the weight W value, the consolidation of the computational domain to obtain the final image region segmentation results;(4)Segmentation second merger absorption of discrete points of the regional;(5)Reasonable detection of the image area, remove the background and foreground region;(6)Get the image segmentation results of the analysis and feature extraction, the output feature vector.

4.2   Graph Theory Segmentation of Short-Range Absorption Algorithm

For apparel merchandise picture image segmentation, focus on the analysis of the target area for example .we determine the location and area of the clothing. As part of the dress patterns and designs, easy to region embedded within the texture segmentation. To eliminate too small parts, we use large area contains a small region to merge.

The region merging post-processing criteria of the need to maintain the original segmentation results, while meeting the basic criteria: (1) to guarantee the value of the similarity between the different degrees and classes in the class, meet the threshold of the threshold L requirements on a smaller area to absorb; (2) the size of the region of smaller area to less than 1/10 of the larger size of the area.

4.3   Rules Limit Divided by Graph Theory

Image segmentation forms a series of block regional. According to the rules described in the chapter 2.1 template methods for authentication. Images contain faces, according to four times the face width, 12 times of face height to determine the clothing active area for the region in the color segmentation block. if its center is located outside the region, discarded the block as a complex background.

For color segmentation of the image area, if the image area that contains the edge pixels, and the central location of the regional block located on the edge of the 1/9 of the region, background removal happened for the face image rely on edge detection.

Image feature calculation includes the overall characteristics of the image and the regional characteristics. Feature description of the main block is a 128-dimensional feature for each sub-block together with the center block and the number of pixels, as well as the main color values can be further described.

5   Image Annotation

5.1   Product Image Main Color Annotation

To each image, we use to clear the background after the first hue and the second hue representative of the color properties, due to the presence of Person class apparel pictures in color feature extraction, on the picture, in accordance with the upper and lower portions to be divided. The the calculated color color label said, can form a multi-level representation. Clothing color accuracy requirements are not very high, you can select a a less standard color expression.

5.2   Product Image Classify and Attribute Annotation

Different goods belonging to a specific category, such as:T-shirt, jacket, men windbreaker, etc. Through extracting and validating the tag words and further image content verification, we get the vocabulary to annotate online commodity pictures. The result should be a relatively accurate one.

In addition to the generic commodities, garments, due to the design style and color, style difference, the formation of a range of potential properties, such as: leisure, sports, etc. described in words. These words are easy online analytical process, to provide more personalized search and shopping guide service.

6   Image Segmentation and Annotation Results Analysis

6.1   Character Image Segmentation Results

The experimental results are the Gaussian smoothing parameter 0.5.[30] In our algorithm contains the threshold of the independence of the small area in order to reduce over-segmentation problem. For product images with human, the original image in Fig.1(a) Picture of experimental results in Fig.1(b).

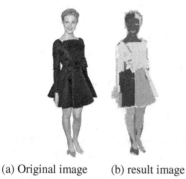

(a) Original image   (b) result image

Fig. 1. Image with human Results

(a) original image   (b) result image

Fig. 2. Pure clothing image Results

Pure Clothing Image segmentation Results are shown in Fig.2. Segmentation results in Fig.2.(a) of the original image, Fig.2.(d) the results of this article HSI distance function graph cut.

6.2   Image Annotation Results

We download images from the web, the color annotation results. We got color vocabulary {red} for Fig.1 image and {Pink, Light brown} for Fig.2 image. Each image, we got a classify annotation results. In our example, the classify vocabulary is respectively named {women dress} and {Tight skirt}.

After we got the classify words, attribute annotation results are reached by image content and tag word analysis. We got some attribute for above samples, {beautiful, charming, and young} for Fig.1 and {professional, fresh, sleeveless} for Fig.2.

7   Conclusion

Graph theory for image segmentation can achieve good results. Experimental results show that the HSI model can be very effective with such image analysis needs.Using the machine learning methods, we can get the statistic modal characteristics of the image HIS features. Segmentation of the image results based on statistical knowledge to judge divided region need further research. Image semantic annotation should be treated in more detail.

Acknowledgements. This paper is supported by "the Fundamental Research Funds for the Central Universities". Some example data is provided by BlueBerrytech inc., China.

References

1. Mori, Y., Takahashi, H., Oka, R.: Image-to- word transformation based on dividing and vector quantizing images with words. In: Proc. of Intl. Workshop on Multimedia Intelligent Storage and Retrieval Management (MISRM 1999), Orlando (October 1999)
2. Oliva, A., Torralba, A.: Modeling the shape of the scene: A holistic representation of the spatial envelope. Int. J. Compute Vision 42(3), 145–175 (2001)
3. Hare, J.S., Lewis, P.H.: Saliency-based models of image content and their application to auto-annotation by semantic propagation. In: Proceedings of the Second European Semantic Web Conference (ESWC 2005), Heraklion, Crete (May 2005)
4. Yavlinsky, A., Schofield, E., Rüger, S.M.: Automated Image Annotation Using Global Features and Robust Nonparametric Density Estimation. In: Leow, W.-K., Lew, M., Chua, T.-S., Ma, W.-Y., Chaisorn, L., Bakker, E.M. (eds.) CIVR 2005. LNCS, vol. 3568, pp. 507–517. Springer, Heidelberg (2005)
5. Lavrenko, V., Manmatha, R., Jeon, J.: A model for learning the semantics of pictures. In: Proc. of Advances in Neural Information Processing Systems (NIPS 2003) (2003)

6. Yang, C., Dong, M., Hua, J.: Region-based image annotation using asymmetrical support vector machine-based multiple-instance learning. In: Proc. of IEEE Int. Conf. on Computer Vision and Pattern Recognition (CVPR 2006), New York, USA, pp. 2057–2063 (June 2006)

7. Feng, S.-H., Xu, D.: Transductive Multi Instance Multi Lab el Learning Algorithm with Application to Automatic Image Annotation. Expert Systems with Applications 37(1), 661–670 (2010)

8. Lu, H., Zheng, Y.-B., Xue, X.-Y., et al.: Content and context-based multi label image annotation. In: IEEE Computer Society Conference on Computer Vision and Pattern Recognition Works hops. Miami, FL, USA, pp. 61–68 (2009)

9. Liu, J., Li, M., Liu, Q., et al.: Image annotation via graph learning. Pattern Recognition 42(2), 218–228 (2009)

10. Wang, Y., Mei, T., Gong, S.G., et al.: Combining global, regional and con textual features for automat ic image annotation. Pattern Recognition 42(2), 259–266 (2009)

11. Noble, J.A., Boukerroui, D.: Ultrasound Image Segmentation: A Survey. IEEE Transactions on Medical Imaging 25(8), 987–1010 (2006)

12. Luccheseyz, L., Mitray, S.K.: Color Image Segmentation: A State-of-the-Art Survey. Image Processing, Vision, and Pattern Recognition 67(2), 207–221 (2001)

13. Comaniciu, D., Meer, P.: Robust analysis of feature spaces: color image segmentation. In: Proceedings of IEEE Conference on Computer Vision and Pattern Recognition, San Juan, Puerto Rico, pp. 750–755 (1997)

14. Urquhart, R.: Graph theoretical clustering based on limited neighborhood sets. Pattern Recognition 15(3), 173–187 (1982)

15. Shi, J., Malik, J.: Normalized cuts and image segmentation. In: Proceedings of the IEEE Conference on Computer Vision and Pattern Recognition, San Juan, Puerto Rico, pp. 731–737 (1997)

16. Wu, Z., Leahy, R.: An optimal graph theoretic approach to data clustering:Theory and its application to image segmentation. IEEE Transactions on Pattern Analysis and Machine Intelligence 15(11), 1101–1113 (1993)

17. Weiss, Y.: Segmentation using Eigenvectors: A Unifying View. In: Proceedings of the IEEE International Conference on Computer Vision, Kerkyra, Greece, vol. 2, pp. 975–982 (1999)

18. Comaniciu, D., Meer, P.: Mean shift analysis and applications. In: Proceedings of IEEE Conference on Computer Vision and Pattern Recognition, New York, USA, pp. 1197–1203 (1999)

19. Ng, H.-F.: Automatic thresholding for defect detection. Pattern Recognition Letters 27(14), 1644–1649 (2006)

20. Zhang, Y.J., Gerbrands, J.J.: Transition region determination based thresholding. Pattern Recognition Letters 12(1), 13–23 (1991)

21. Osher, S., Sethian, J.A.: Fronts Propagating with Curvature-Dependent Speed: Algorithms Based on Hamilton–Jacobi Formulations. Journal of Computational Physics 79(1), 12–49 (1988)

22. Adalsteinsson, B., Sethian, J.A.: The fast construction of extension velocities in level set methods. Journal of Computational Physics 148(1), 2–22 (1999)

23. Kass, M., Witkin, A., Terzopoulos, D.: Snakes: Active Contour Models. International Journal of Computer Vision 1(4), 321–331 (1988)

24. Li, P.-H., Zhang, T.-W.: Review on active contour model (Snakemodel). Journal of Software 11(6), 751–757 (2000)

25. Boykov, Y., Jolly, M.: Interactive graph cuts for optimal boundary & region segmentation of objects in n-d images. In: Proc. IEEE Conf. Computer Vision, Vancouver, Canada, vol. 1, pp. 105–112 (2001)

26. Zahn, C.T.: Graph-theoretic methods for detecting and describing gestalt clusters. IEEE Transactions on Computing 20(1), 68–86 (1971)

27. Wu, Z., Leahy, R.: An optimal graph theoretic approach to data clustering: Theory and its application to image segmentation. IEEE Transactions on Pattern Analysis and Machine Intelligence 15(11), 1101–1113 (1993)

28. Shi, J., Malik, J.: Normalizaed Cuts and Image Segmentation. IEEE Transactions on Pattern Analysis and Machine Intelligence 22(8), 888–905 (2000)

29. Grady, L., Schwartz, E.L.: Isoperimetric Graph Partitioning for Image Segmentation. IEEE Transactions on Pattern Analysis and Machine Intelligence 28(3), 469–475 (2006)

30. Felzenszwalb, P., Huttenlocher, D.: Image segmentation using local variation. In: Proceedings of IEEE Conference on Computer Vision and Pattern Recognition, Santa Barbara, California, USA, pp. 98–104 (1998)

31. Felzenszwalb, P.: Effcient Graph-Based Image Segmentation. International Journal of Computer Vision 59(2), 167–181 (2004)

32. Boykov, Y., Funka-Lea, G.: Graph Cuts and Efficient N-D Image Segmentation. International Journal of Computer Vision 70(2), 109–131 (2006)

33. Cooper, M.C.: The tractability of segmentation and scene analysis. International Journal of Computer Vision 30(1), 27–42 (1998)

34. Sarkar, S., Soundararajan, P.: Supervised learning of large perceptual organization:graph spectral and learning automata. IEEE Transactions on Pattern Analysis and Machine Intelligence 22(5), 504–525 (2000)

35. Ding, C.H.Q., He, X., Zha, H., et al.: A min-max cut algorithm for graph partitioning and data clustering. In: ICDM 2001, pp. 107–114. IEEE Computer Society, Los Alamitos (2001)

36. Li, X.-B., Tian, Z., Liu, M.-G.: Weighted Cut Based Image Segmentation. Acta Electronica Sinica 36(1), 76–80 (2008)

Part III

Emotions in HCI

Audio-Based Pre-classification
for Semi-automatic Facial Expression Coding

Ronald Böck[1], Kerstin Limbrecht-Ecklundt[2], Ingo Siegert[1],
Steffen Walter[2], and Andreas Wendemuth[1]

[1] Cognitive Systems Group, Otto von Guericke University Magdeburg, Universitätsplatz 2,
39106 Magdeburg, Germany
[2] Medical Psychology, Ulm University, Frauensteige 6, 89075 Ulm, Germany
ronald.boeck@ovgu.de
http://www.cognitive-systems-magdeburg.de

Abstract. The automatic classification of the users' internal affective and emotional states is nowadays to be considered for many applications, ranging from organisational tasks to health care. Developing suitable automatic technical systems, training material is necessary for an appropriate adaptation towards users. In this paper, we present a framework which reduces the manual effort in annotation of emotional states. Mainly it pre-selects video material containing facial expressions for a detailed coding according to the Facial Action Coding System based on audio features, namely prosodic and mel-frequency features. Further, we present results of first experiments which were conducted to give a proof-of-concept and to define the parameters for the classifier that is based on Hidden Markov Models. The experiments were done on the *EmoRec I* dataset.

1 Introduction

Dispositions and emotions are substantial elements of daily life as they influence the communication and way of interaction as well as they can induce the willingness to act in a specific way. Humans are able to analyse specific cues (facial expression, gesture, speech, etc.) in order to interpret emotional states in themselves and others, i.e. mainly generating hypotheses on how another person might feel or react. Modern technical systems increasingly occupy a wide range of daily activities like organisational tasks, calendar synchronisation, entertainment, etc. Therefore, researchers intend to optimise the usability of such cognitive, technical systems [22] in such a way that they will provide people not only with helpful information, but also to support them during their decision making processes. Hence, technical systems have to identify emotional cues properly during Human-Computer Interaction (HCI) [22].

Emotions are usually expressed by multiple modalities like verbal and paralinguistic speech expression, speech content, facial expressions, and gestures. Thus, a cognitive technical system has to consider a large amount of data [9]. In recent years, various emotion recognition methods and an enormous number of multimodal datasets were generated (cf. [8, 14, 24]) and therefore, a strong need for efficient labelling strategies arises [9, 16, 17]. In this paper, we are presenting first steps towards a semi-automatic annotation framework for multimodal datasets. The main idea is to use audio analyses to

M. Kurosu (Ed.): Human-Computer Interaction, Part V, HCII 2013, LNCS 8008, pp. 301–309, 2013.
© Springer-Verlag Berlin Heidelberg 2013

identify relevant affective sequences which can be aligned with the corresponding video material and hence, provide a pre-classification for the Facial Action Coding System (FACS) (cf. Fig. 1). From the schematic workflow of the framework we can see that the recodings have to be multimodal, for the framework itself, but as well such modalities vary from user to user and from situation to situation. In fact, this is the case for almost all corpora which are currently recorded and we assume that this will be true for those which will be generated. Further, the utilised audio features should be relatively general, so that a wide range of domains and audio conditions are covered. This was investigated in parts in [1, 2]. The relevant video sequences can be afterwards determined by marking the time stamps of the video material; an idea influenced by forced aligment in speech recognition. The identification of the sequence further provides the opportunaty to pre-classify the FACS. Thus, human annotators are just asked to label debatable sequences which reduces the manual effort of annotation. A detailed visualisation of the pre-selection process is given in Fig. 2. The general idea was influenced by [5] where mimicry cues, e.g. poses, were related with verbal utterances. The audio features used here were previously identified in [1, 2]. So far, we do not intent to and are not able to specify a set of pre-selected Action Units (AUs) from audio analysis.

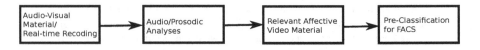

Fig. 1. Workflow to establish a pre-classification of video material as it is proposed

Certified human FACS coders achieve a hit rate of at least 76% when annotating facial expressions manually [10]. But unfortunately this is a very time consuming task. Therefore, researchers are very interested in developing a computer based approach for facial expression analysis. From literature, we know that there are systems which already deal with an automatic analysis of video material to support FACS coding, for instance [4, 12, 18]. In contrast to those systems our framework overcomes several disadvantages. Most systems fail if additional characteristics are in the face, for instance, glasses, fringe, or sensors. Further, video based facial analysis can just work if the face is oriented towards the camera. In natural HCI this is not the case all the time. Therefore, we apply a different modality, namely audio, to get a selection of relevant sequences, independent from the video features.

In this paper, we present the results for the automatic analyis of audio sequences to identify relevant parts related to FACS in the audio modality. As this paper presents a kind of proof-of-concept, we used material which is already manually annotated and thus, are able to evaluate our framework as well as the applied audio classifiers. Further, to show the generalisation of the framework we already applied the methods in a Leave-One-Speaker-Out (LOSO) validation (cf. Sect. 3.2 and 4). In future non-annotated material has to be processed; not considering any kind of evaluation.

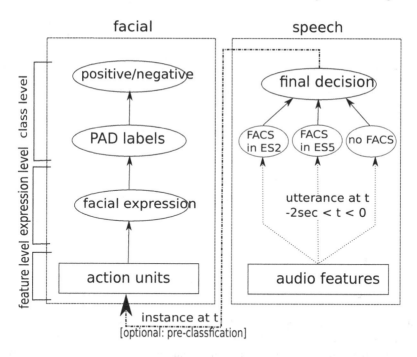

Fig. 2. Flow of the features to get a classification. The speech part presents the steps getting a classification based on audio features. Pre-selected instances are fed into the facial annotation part. Using Action Units and facial expressions a classification can be derived.

2   Dataset

To evaluate the framework, we applied it on the EmoRec I experiment, a realistic speech driven HCI introduced in [21]. The material was collected in a Wizard-of-Oz experiment which represents an interaction with a computer-based memory trainer relying on the design of the game "Concentration". In this experimental setup the user was passed through specific pleasure-arousal-dominance (PAD) space [11] octants, representing the user's emotions, in a controlled fashion. Although the emotional state was induced in the experiment by design, a prediction of the specific user reaction, no matter whether by way of interaction with the system or by emotional expressiveness, was not possible. Hence, we assume that the dataset represents a natural, realistic HCI. The experiment's workflow is given in Fig. 3, where each Experimental Sequence (ES) represents certain octants in the PAD space. To see the differences in the user reactions, we investigated so far only ES-2 that is assumed to be positive and ES-5 which is negative that are interpretations of the PAD values. For a detailed description of the experiment and its parameters see [21].

In general, 125 subjects participated in the experiment. For this case study, we analysed the video material of the *EmoRec I* dataset - 20 subjects (ten female, ten male) - which is so far manually FACS coded. As not everybody showed facial expression in

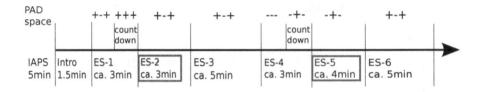

Fig. 3. Workflow of *EmoRec I* experiment

the ESs we reduced the number of subjects to 13 individuals who provided material for training and testing of classifiers. Despite performance loss, this enabled us to evaluate the framework's performance.

3   Methods

3.1   Facial Action Coding System

Ekman & Friesen [6] developed the FACS since facial expressions can be defined as a sequential set of facial movements caused by the underlying muscle activities. Facial expressions are defined by Action Units. The stimuli used in this *EmoRec I* experiment were selected based on EmFACS [7]. The manual labelling process for the 20 subjects took 70-80 hours in total. In particular, for the two ESs it took 13-19 hours for a video time of roughly $20 \cdot 7min = 140min$ (cf. Figure 3) which indicates the necessity of a semi-automatic annotation.

As it was intended to classify and identify positive and negative emotions, facial expressions occurring during the interaction with the technical system were checked for AUs indicating expressions of happiness and anger (cf. Table 1). The analysis revealed that especially negative emotions were shown most frequently in HCI. Therefore, we decided to identify negative reactions first. Moreover, this is feasible as for positive expressions almost no material was available, neither as facial expression nor as speech samples. Furthermore, only for a subset of AUs enough material occurred in the dataset. Hence, we opt for a combination of AUs and a selection of those which are mainly exhibited in the sequences. All facial expressions were coded by a certified FACS coder not involved in the experiment. It can be assumed that facial activity is less expressive in HCI as no additional value for the interaction partner (in this case a technical system) is expected [21].

3.2   Audio Features

Based on previous work [1], we selected prosodic features which are quite expressive in emotion analyses: the first to third formant and their corresponding bandwidth, pitch, jitter, and intensity [13, 14, 19]. Further, those features represent or are related to negative and high aroused emotions [5, 14, 15]. Moreover, we combined this feature set of prosodic features with Mel Frequency Cepstral Coefficients (MFCCs) to enrich the

Table 1. Basic emotions combined by Action Unit where a number represents the muscle activity and a letter the strength of it (A ... low; E ... high) (cf. [1])

Emotion	Action Unit Combination
Happiness	[6 and/or 7] with 12CDE
Sadness	1 + 4 + [6 and/or 7] + 15ABC + 64
Disgust	9 + [10 and/or 16] + 19 + 26
Anger	4CDE + 5CDE + 7CDE + 17 + 23 + 24
Fear	1 + 2 + 4 + 5ABCDE + 7 + 20ABCDE + 26
Surprise	1CDE + 2CDE + 5AB + 26

set as a positive effect was figured out in preliminary tests. To handle time dependencies derivatives were added, too; namely Δ and $\Delta\Delta$. Those features are meaningful and widely used in the speech emotion recognition community (cf. e.g. [2, 15, 24]).

To extract the features we applied PRAAT [3] and the Hidden Markov Toolkit (HTK) [23] on frame-level to speech samples which are a combination of letters and numbers, for instance, "C 2", "C 4" representing the commands of the "Concentration" game. For classification purposes we used Gaussian Mixture Models (GMMs) as they are commonly used in the community. In total, we had three classes (cf. Fig. 2): *FACS in ES-2*, *FACS in ES-5*, and *no-FACS* (cf. Fig. 2). For each class a GMM is trained on audio features on utterances which are 2 seconds before the facial expression (cf. Fig. 4). Again, so far we concentrated on the negative user's reactions in each ES, only. In testing we classified audio samples according to defined classes and compared these to the FACS coded facial expressions (cf. [1]). For training and testing the HTK [23] was used. The evaluation strategy was LOSO that means one speaker's material was left out from training and used only in testing. With this method, the generalisation purpose of the framework can be shown.

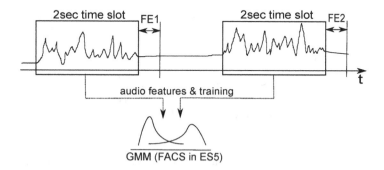

Fig. 4. A 2 seconds slot before the facial expression (FE) starts is used to define an utterance. Its extracted features are used to train GMMs. In the figure it is visualised exemplarily with *FACS in ES5*.

4   Results

In these experiments we applied a LOSO strategy on the three class issue introduced in Sect. 3.2 and in Fig. 2 as so far this is the only possibility for us to assess our framework. To evaluate the classification results, we rely on the Weighted Average accuracy (WA) which reflects the class-wise accuracy calculated for each class separately and afterwards averaged over all classes. This can be compared to the Unweighted Average accuracy (UA) calculated based on all samples and therefore, a representation of the sample's distribution is not considered. Thus, the WA is the more meaningful measure as we have just a low amount of samples for relevant sequences but a lot for no-FACS.

At first, we examined the parameter setting for the classifier. Based on previous experiments reported in [20] we opt for GMMs. Further, the training is executed according to specifications in [2]; that is, having five iterations. The number of mixtures is varied in the range of [6, 15] which was driven by [20] and the search for the number of mixtures was stopped when the performance decreased whereas the performance was evaluated in a LOSO manner. The corresponding accuracy values are given in Tab. 2. From this, we will operate the recognisier in future applications with the following parameters: 9 Gaussian Mixtures, 5 training iterations. Nevertheless, a kind of tuning towards specific conditions like noise, echo, etc. might be necessary.

Table 2. Classification results according to the number of mixtures used in a GMM in percent

Number of mixtures	Weighted Average accuracy	Unweighted Average accuracy
6	72.6	69.9
9	**73.9**	**75.6**
12	73.4	84.2
15	70.4	84.2

Table 3. Classification results in percent as Weighted Average accuracy over all LOSO experiments combining ES-2 and ES-5. The bold value represents the false acceptance rate whereas the italic on is the false rejection rate.

Classifier \ true	FACS in ES	no-FACS
FACS in ES	61.9	**18.8**
no-FACS	*38.1*	81.2

To evalute the system, we are interested in the performance as a notifier; this means, we are observing the false acceptance and false rejection rates (cf. Tab. 3). False acceptance indicate how many samples are identified as a *FACS in ES* though it was *no-FACS*; false rejection is defined the otherway around. Despite we use three GMMs we combined the two classes of ES as we do not give any indication to the coder to keep him unaffected in the annotation process. Therefore, we discuss the acceptance rates based on two classes. From Tab. 3 it can be seen that the distinct classes can be distinguished.

Further, the false acceptance rate is 18.8%. So far, the false rejection rate is relatively high with 38.1%. This has tow reasons: 1) the number of samples in ES is low in comparison to *no-FACS* which has influence of the performance and 2) in some LOSO experiments the distinguishing of the classes is confused.

For this, we inspected the confusion matrices of each run of LOSO manually as we were interested in the distribution of recognition results. From this we found that the class *FACS in ES-5* was recognised reliably. On the other hand, no-FACS and FACS in ES-2 are often confused. In particular, for a few number of subjects FACS in ES-2 was not recognised at all which also results in a low performance of the system. To evaluate this issue we looked into the recordings and found that the particular subjects showed only slight facial expressions and, from a human's point of view, almost no reactions in speech. As even for humans the recognition of *FACS in ES-2* is quite hard, we will concentrate our research on finding suitable features and methods which can deal with such slight, but natural emotional reactions, and further handle positive reactions properly as well.

5 Conclusion

In this paper, we presented and discussed a framework towards audio-based semi-automatic selection of video sequences for labelling of emotional facial expressions. Based on classified emotional utterances a pre-selection of sequences is given that afterwards have to be annotated according to the regulations of FACS. Having this selection process the manual effort of annotation is reduced. For instance, watching 30-40 minutes per subject for each subexperiment in *EmoRec I* is reduced to a few minutes (varies according to the certain subject) in total which have to be viewed and annotated. Furthermore, automatic recogniser can also handle slight and natural reactions of subjects which are even hard to realise by humans. Results from an evaluation were presented and discussed.

In future, we apply the framework to the *EmoRec II* material which is not labelled, yet, but has quite similar characteristics as *EmoRec I*. We expect to reduce the manual labelling effort drastically due to the low amount of emotional facial expressions in natural HCI.

Acknowledgement. We acknowledge continued support by the Transregional Collaborative Research Centre SFB/TRR 62 "Companion-Technology for Cognitive Technical Systems" funded by the German Research Foundation (DFG).

References

[1] Böck, R., Limbrecht, K., Siegert, I., Glüge, S., Walter, S., Wendemuth, A.: Combining mimic and prosodic analyses for user disposition classification. In: Wolff, M. (ed.) Proceedings of the 23rd Konferenz Elektronische Sprachsignalverarbeitung, Cottbus, Germany, pp. 220–228 (2012)

[2] Böck, R., Hübner, D., Wendemuth, A.: Determining optimal signal features and parameters for hmm-based emotion classification. In: Proceedings of the 15th IEEE Mediterranean Electrotechnical Conference, pp. 1586–1590. IEEE, Valletta (2010)

[3] Boersma, P.: Praat, a system for doing phonetics by computer. Glot International 5(9/10), 341–345 (2001)

[4] Cohn, J.F., Zlochower, A.J., Lien, J., Kanade, T., Analysis, A.F.: Automated face analysis by feature point tracking has high concurrent validity with manual facs coding. Psychophysiology 36(1), 35–43 (1999)

[5] De Looze, C., Oertel, C., Rauzy, S., Campbell, N.: Measuring dynamics of mimicry by means of prosodic cues in conversational speech. In: 17th International Congress of Phonetic Sciences, Hong Kong, China (2011)

[6] Ekman, P., Friesen, W.: Facial Action Coding System: Investigators Guide, vol. 381. Consulting Psychologists Press, Palo Alto (1978)

[7] Ekman, P., Friesen, W.: Emfacs facial coding manual. Human Interaction Laboratory, San Francisco (1983)

[8] Gunes, H., Pantic, M.: Automatic, dimensional and continuous emotion recognition. International Journal of Synthetic Emotions 1(1), 68–99 (2010)

[9] Koelstra, S., Muhl, C., Patras, I.: Eeg analysis for implicit tagging of video data. In: 3rd International Conference on Affective Computing and Intelligent Interaction and Workshops, pp. 1–6. IEEE, Amsterdam (2009)

[10] Limbrecht-Ecklundt, K., Rukavina, S., Walter, S., Scheck, A., Hrabal, D., Tan, J.W., Traue, H.: The importance of subtle facial expressions for emotion classification in human-computer interaction. Emotional Expression: The Brain and The Face 5(1) (in press, 2013)

[11] Mehrabian, A.: Pleasure-arousal-dominance: A general framework for describing and measuring individual differences in Temperament. Current Psychology 14(4), 261–292 (1996)

[12] Pantic, M.: Automatic facial expression analysis and synthesis. In: Symposium on Automatic Facial Expression Analysis and Synthesis, Proceedings Int'l Conf. Measuring Behaviour (MB 2005), pp. 1–2. Wageningen, The Netherlands (2005)

[13] Scherer, K.R.: Appraisal considered as a process of multilevel sequential checking. In: Appraisal Processes in Emotion: Theory, Methods, Research, pp. 92–120 (2001)

[14] Schuller, B., Vlasenko, B., Eyben, F., Rigoll, G., Wendemuth, A.: Acoustic emotion recognition: A benchmark comparison of performances. In: Proceedings of the IEEE Automatic Speech Recognition and Understanding Workshop, ASRU 2009, Merano, Italy, pp. 552–557 (2009)

[15] Schuller, B., Vlasenko, B., Eyben, F., Wollmer, M., Stuhlsatz, A., Wendemuth, A., Rigoll, G.: Cross-corpus acoustic emotion recognition: Variances and strategies. IEEE Transactions on Affective Computing I, 119–131 (2010)

[16] Siegert, I., Böck, R., Philippou-Hübner, D., Vlasenko, B., Wendemuth, A.: Appropriate Emotional Labeling of Non-acted Speech Using Basic Emotions, Geneva Emotion Wheel and Self Assessment Manikins. In: Proceedings of the IEEE International Conference on Multimedia and Expo, ICME 2011, Barcelona, Spain (2011)

[17] Siegert, I., Böck, R., Wendemuth, A.: The influence of context knowledge for multimodal annotation on natural material. In: Böck, R., Bonin, F., Campbell, N., Edlund, J., de Kok, I., Poppe, R., Traum, D. (eds.) Joint Proc. of the IVA 2012 Workshops, Otto von Guericke University Magdeburg, Santa Cruz, USA, pp. 25–32 (2012)

[18] Soleymani, M., Lichtenauer, J., Pun, T., Pantic, M.: A multimodal database for affect recognition and implicit tagging. IEEE Transactions on Affective Computing 3(1), 42–55 (2012)

[19] Vlasenko, B., Philippou-Hübner, D., Prylipko, D., Böck, R., Siegert, I., Wendemuth, A.: Vowels formants analysis allows straightforward detection of high arousal emotions. In: 2011 IEEE International Conference on Multimedia and Expo (ICME), Barcelona, Spain (2011)

[20] Vlasenko, B., Prylipko, D., Böck, R., Wendemuth, A.: Modeling phonetic pattern variability in favor of the creation of robust emotion classifiers for real-life applications. Computer Speech & Language (2012) (in press)

[21] Walter, S., Scherer, S., Schels, M., Glodek, M., Hrabal, D., Schmidt, M., Böck, R., Limbrecht, K., Traue, H.C., Schwenker, F.: Multimodal emotion classification in naturalistic user behavior. In: Jacko, J.A. (ed.) Human-Computer Interaction, Part III, HCII 2011. LNCS, vol. 6763, pp. 603–611. Springer, Heidelberg (2011)

[22] Wendemuth, A., Biundo, S.: A companion technology for cognitive technical systems. In: Esposito, A., Esposito, A.M., Vinciarelli, A., Hoffmann, R., Müller, V.C. (eds.) COST 2102. LNCS, vol. 7403, pp. 89–103. Springer, Heidelberg (2012)

[23] Young, S., Evermann, G., Gales, M., Hain, T., Kershaw, D., Liu, X., Moore, G., Odell, J., Ollason, D., Povey, D., Valtchev, V., Woodland, P.: The HTK Book, version 3.4. Cambridge University Engineering Department (2009)

[24] Zeng, Z., Pantic, M., Roisman, G.I., Huang, T.S.: A survey of affect recognition methods: Audio, visual, and spontaneous expressions. IEEE Transactions on Pattern Analysis and Machine Intelligence 31(1), 39–58 (2009)

Sentimental Eyes!

Amitava Das[1] and Björn Gambäck[2]

[1] SAIT Lab., Samsung Research India, Bangalore, India
[2] Norwegian University of Science and Technology, Trondheim, Norway
amitava.santu@gmail.com, gamback@idi.ntnu.no

Abstract. A closer look at how users perform search is needed in order to best design a more efficient next generation sentiment search engine and understand fundamental behaviours involved in online review/opinion search processes. The paper proposes utilizing personalized search, eye tracking and sentiment analysis for better understanding of end-user behavioural characteristics while making a judgement in a Sentiment Search Engine.

Keywords: Sentiment Analysis, Sentiment Search, Eye Tracking.

1   Introduction

Broad access to an abundance of information is one of the defining characteristics of today's web search environment. Internet search engines act as intermediaries between users' information needs and the massive number of potentially relevant pages on the Web. Still, users are largely unsuccessful in finding their desired information, with failure rates often approaching 50% [1][2]. Clearly, this presents a significant dilemma for online searches – why are users only modestly successful in formulating their search queries, and what can be done to improve the situation?

Several attempts have been made to better understand user behaviour during the search process, for example through *personalized information retrieval*. Personalized web search is crucial in today's world of information overflow, to provide only relevant information – depending on the context – such that users get the correct information when they need it. According to Schwartz (2006), "choice is the critical sign that we have freedom and autonomy" [3]. Most of the time, choice is good and more choice is better. With the accessibility of more information, we have more choice, and presumably more freedom, autonomy, and self-determination, than ever before. It would seem that increased choice increases well-being; however, studies have shown that this is not the case: there is a need for good (personalized) information retrieval systems that help the user to take good decisions without decreasing her well-being. Generally, personalization methodologies can be divided into two complementary processes: user information collection, used to describe the user interests [4], and inference of the gathered data to predict the closest content to the user expectation [5]. Hence, future generation web interfaces by necessity need to be more intelligent in order to understand the end-users' sentimental needs and preferences.

There has been a rapid development in sentiment analysis techniques during the last two decades and sentiment search is one of the most promising futuristic

M. Kurosu (Ed.): Human-Computer Interaction, Part V, HCII 2013, LNCS 8008, pp. 310–318, 2013.

technologies with immense commercial value. The main driving necessity behind sentiment search systems is that whenever we need to make a decision, we may want opinions from others. Some sentiment/review/opinion search fall into the informational genre by definition, with queries for this kind of task being classified into two basic genres: direct search and comparative search, which could be instantiated by queries such as "iPhone 5" and "iPhone5 vs. Samsung Galaxy", respectively. Unfortunately, there is no publicly available system which gives satisfactory output for this type of querying, and posing opinionated queries to a general purpose search engine leads to navigational surfing for users because desired information is distributed over several pages.

Today's search engines keep logs of user browsing data and effectively use that data to produce satisfactory results for particular users. In addition, we suggest to keep logs of user eye-tracking data, in order to understand and track the user's sentiment while working with the Web-search interfaces. Most laptops, smart phones and tablets have good quality cameras built in, while eye-and-gaze tracking technology has reached a quality level where such single-camera tracking without external light or infrared sources is feasible (for example, an already available system such as YouEye[1] is capable of tracking eye-gaze and facial emotions just using a standard web camera). Allowing for such non-intrusive eye-tracking is central to the possibility of utilizing the technology in a range of tasks, including the tracing of user search behaviour. Hence, eye tracking and sentiment analysis could have a great significance on the next generation of Human Computer Interfaces.

In this paper, we report some initial experiments on using eye tracking information as a knowledge source for sentiment analysis. It is an on-going task and this paper should overall be seen as a position paper. The rest of the text is laid out as follows: the next section discusses some relevant previous research efforts. Section 3 in turn introduces the basis for the current text, the long-term research questions that need to be addressed. The experimental setup is described in Section 4, while Section 5 presents some initial results. Finally, Section 6 sums up the discussing and points to areas of future interest.

2   Related Work

The application of eye tracking to online search has recently received a considerable amount of attention from research scientists, search engine companies, marketing firms, and usability professionals, even though no previous work has focused directly on using eye-tracking for Internet search sentiment analysis. The first use of eye tracking to investigate Internet search behaviour comes from Granka *et al.* (2004) who analysed users' basic eye movements and sequence patterns throughout ten different search tasks performed on Google [6]. Lorigo *et al.* (2006) [7] examined user eye tracking patterns through fixation on classified areas of interest (AOI) such as title, abstract, and metadata.

It has long been known in neuropsychology that the retinal image is transmitted to the brain during fixations, but not during saccades (rapid eye movements); hence, it is

---

[1] https://www.youeye.com/

the fixations that represent the acquisition and processing of information [8]. During normal reading, a reader does not fixate upon each word in sequence, but rather makes a rapid series of fixations followed by saccades, which may skip over some words entirely. Saccades commonly occur 3-4 times per second. In addition, approximately 15% of all the saccades occur backwards, to earlier text – a phenomenon known as a regression.

Cognitive psychologists have studied how viewers examine printed advertisements, and in particular how different aspects of the ads and the users' goals interact to influence viewing behaviour. In recent years eye-tracking technology has been utilized for these studies, in order to automatically determine how much time readers devote to specific areas of interest in an ad. Rayner *et al.* (2008) asked readers to rate how much they liked an ad and then examined the correlation between these ratings and how much time the readers spent on the ad, as well as how the viewing time was divided between textual objects and images, concluding that the user needs and the actual users' profiles matters most for how much attention is devoted to an ad [9]. A related problem which has been studied is how eye-tracking can be utilized to trace the processes underlying user decision making. Glaholt & Reingold (2011) has proposed that the dwell duration is central there, i.e., that users tend to look longer at items they prefer, while dwell frequency – how many times they look at an item – is less important [10].

3   Empiricism on Eye-Tracking and Sentiment Analysis

The research motivation of the present work is to reach better understanding of user behaviour during the sentiment search process. This can be formulated into basic level objective questions based on what has been suggested in previous studies [6]:

- How long does it take searchers to select a document?
- How many abstracts do searchers look at before making a selection?
- Do searchers look at abstracts ranked lower than the selected document?
- Do searchers view abstracts linearly?
- Which parts of the abstract are most likely to be viewed?

However, there are contrastive differences between general search and sentiment search engines. In order to adapt eye-tracking methods to sentiment search, those research questions have to be extended in the following two directions, which are the key contributions of this paper.

1. Overall Searching Behaviour

  — How long does it take users to select a document based on a direct query vs. a comparative query?
  — How long does it take users to select a document from a general query vs. from a sentiment query?
  — How much time do we spend viewing each abstract?

2. Overall Viewing Behaviour

- — How many times does a user look at the query word(s)?
- — How many times does a user look at the sentiment word(s)?
- — How many times does a user look at the domain specific query word(s)?
- — How many times does a user look at his/her preferred sentiment word(s)?

The answers to these empirical questions will help to improve search interfaces in the future and will also help in personalizing them according to an end-user's preferences. It could be argued that our preferences of sentimental word choices or websites preferences do not differ in practice. However, a separate study on social network personality by Kosinski et al. (2012) has already reported that user website preferences change with user personality [11]. To support this argument, our experimental results are presented in the next section.

4   Experiments

A set of very initial experiments was carried out at the Department of International Business Communication, Copenhagen Business School, using an EyeLink 1000[2] eye tracker from SR Research Ltd. The experimental setup was developed by using the Experiment Builder[3] software that comes with the eye tracker toolkit.

Three test participants were instructed to formulate four types of queries each, one from each category listed below. The participants were asked to formulate queries that very restricted to the specific domain: the direct sentiment queries should be restricted to be within the movie domain, while the comparative sentiment queries should be in the electronic domain.

- • General domain, non-sentiment queries
  - — Informational (Ex: *USA president*)
  - — Navigational (Ex: *tourist info Lyon France*)
- • Sentiment queries
  - — Direct Sentiment (Ex: *Skyfall review*)
  - — Comparative (Ex: *iPhone 4 vs. Galaxy*)

The actual document retrieval was carried out by the Google search engine. A typical search page consists of title, URL, text snippets, images, video links, and metadata. For the task fixation, the title, abstract, URL and metadata were classified as areas of interest (AOI).

5   Analysis

The end-user search patterns significantly change depending on the query type. It could be observed that most of the time people do not look at the result snippets

---

[2] http://www.sr-research.com/EL_1000.html
[3] http://www.sr-research.com/accessories_EL1000_eb.html

linearly except for during general domain informational queries. Furthermore, for that type of queries people generally do not look beyond the top-5 results. Even only for 40% cases people have click on any URL. We discussed this with the participants and deduced that their informational needs generally were fulfilled by the text snippets obtained in the search results.

5.1   Sentiment Query vs Non-sentiment Query

Interestingly, we observed that almost everybody randomly moved their eyes over the search results for sentimental queries. For example, for movie reviews people had preferences for the sites IMDB[4] or Rottentomatoes[5]. It did not matter much how those sites were ranked in the results returned by the search engine: people generally jumped to those search results and fixed their eyes.

The direct sentiment queries often needed to be re-formulated and the participants in general added their preferred aspects of choices like acting, direction, academy award, etc. The observed characteristic differences between the general domain queries and sentiment queries are reported in Table 1.

Table 1. Differences between general domain queries and sentiment queries

Query type	Average time to complete	Snippets linearly visited	Reformulation of query
Informational	10-20 sec	Mostly	No
Navigational	60-90 sec	Varies	Mostly and after each 20-30 sec
Direct Sentiment	30-40 sec	No	For only 40% of the cases
Comparative	60-90 sec	No	Mostly and after each 30 sec

5.2   Term Preferences

In addition to the fundamental behavioural differences between general search queries and sentiment queries, we analysed people eye fixations on the *query word(s), domain specific word(s)* and *sentiment word(s)*. The observations are reported in Table 2.

Two domain dictionaries were created for the experiments. The first dictionary is in movie domain and consists of 100 domain ontologies like acting, direction, academy award, cinematography, etc. The second dictionary is on computer and electronic products and other associated terms like apps, display, battery, software, etc. It was created semi-automatically by automatically merging two online dictionaries[6,7] and manually validating the result. The dictionary has 5K terms altogether; the words carrying sentiment were restrieved from SentiWordNet 3.0[8].

---

[4] http://www.imdb.com/
[5] http://www.rottentomatoes.com/
[6] http://www.alphadictionary.com/directory/
   Specialty_Dictionaries/Electronics/
[7] http://www.interfacebus.com/Glossary-of-Terms.html
[8] http://sentiwordnet.isti.cnr.it/

Table 2. Eye fixation on categorical terms

Query type	Query Words	Domain Specific Words	Sentiment Words
Direct Sentiment	30%	18%	43%
Comparative	36%	26%	32%

Google highlights the query terms automatically, so we were quite surprised to find that the participants did not look at the query words very often. Rather, the fixation statistics show that a user stop at either domain specific or sentiment words. Post-experiment discussions with the participants revealed that they generally had a prior expectation on exactly what they were looking for. For example, the "skyfall review" was given to the three participants and they revealed that they had had the following expectations before initiating the search:

1st Participant: Interested in Acting: whether the new Bond is better than Pierce Brosnan / Sean Connery. Also interested in Academy Awards.

2nd Participant: Interested in Bond gadgets specifically!

3rd Participant: Interested in action sequences only!

The same is true for the comparative sentiment queries. In the context of iPhone vs. Galaxy, people generally want to look at feature-based sentiment comparisons for electronic products like aps, display, and battery life. For that reason people are more concerned with the domain terminologies during comparative sentiment queries than during direct sentiment queries.

5.3 Structural Preferences

Search results generally have a typical structure. In order to investigate the users structural preferences, we considered the following items: title, URL, metadata, image/video thumbnail. The text snippets were excluded, as they are content features and had been analysed separately in the term preference part described above.

Table 3 shows how a typical user's eye-fixation stops at each structural aspect. The percentage calculation is based on the time spent on structural aspects divided by the overall time spent on the page before clicking on any link.

Table 3. Structural Preferences

	Title	URL	Metadata
Direct Sentiment	1-2 %	32%	66%
Comparative	1-2 %	10 %	23 %

Heat maps were generated from each user's browsing data for further analysis. Two very relevant example heat maps can be seen in Figures 1 and 2. The first heat map is on "skyfall review" and it is clear that users stop his/her eyes on the star-rating by Rottentomatoes. The second heat map is from a comparative sentiment query and it is clear that the user is more interested in the image/video link.

We discussed these issues with participants after the experiments and understood that they looked at particular structural aspect based on the topic(s) of their search. For example, if it is a general knowledge topic they tend to look at Wiki URLs; for movies they look either at Rottentomatoes or IMDB; and for products they have their personal choices for reading reviews.

Fig. 1. Heat map for a direct sentiment query: *"skyfall review"*

Fig. 2. Heat map for a comparative sentiment query: *"iPhone 4 vs. Galaxy"*

6   Conclusions and Future Aspects

In conclusion, the paper has reported some incipient work on understanding user eye movements and fixations based on their sentimental preferences during online search. There is a huge potential for this research when moving towards the next generation of Human Computer Interfaces, since eye-and-gaze tracking technology has reached a quality level where it now is feasible to utilize remote, non-intrusive single-camera eye-tracking, using a standard web camera without external light or infrared sources.

This paper only reports the initial study to understand the relations between the eye movements and user sentiment search patterns. At the next level, we are working towards personalized sentiment search by creating user profiles with the technique, and with the intension to add facial emotions to the eye-tracking. No similar work has been attempted so far, but a US patent application has outlined an architecture using brain-computer interface (BCI) technology for sentiment tracking [12]. It suggests that the BCI system could be complemented by measuring eye and face movement activation signals. Thus in a quite intrusive manner – and using an as-of-yet fairly unreliable input method. The present paper in contrast proposes to induce the user sentiment in a totally non-intrusive manner and by utilizing quite mature and cheap off-the-shelf eye-tracking techniques.

Acknowledgements. The experiments were carried out at Copenhagen Business School, Denmark. We want to thank Prof. Michael Carl for providing the laboratory facilities equipped with the eye-tracker.

References

1. Sherman, C.: Why search engines fail. Search Engine Watch (August 29, 2002)
2. http://searchenginewatch.com/article/2068006/
   Why-Search-Engines-Fail
3. Nordlie, R.: User revealment – a comparison of initial queries and ensuing question development in online searching and human reference interaction. In: Proceedings of SIGIR, New York, USA, pp. 11–18 (1999)
4. Schwartz, B.: The paradox of choice – why more is less. Google Tech Talks (2006), http://www.youtube.com/watch?v=6ELAkV2fC-I
5. Winter, S., Tomko, M.: Translating the web semantics of georeferences. In: Web Semantics and Ontology, pp. 297–333. Idea Publishing (2006)
6. Koutrika, G., Ioannidis, Y.: A unified user-profile framework for query disambiguation and personalization. In: Proceedings of the Workshop on New Technologies for Personalized Information Access, Edinburgh, UK, pp. 44–53 (2005)
7. Granka, L., Joachims, T., Gay, G.: Eye-tracking analysis of user behavior in WWW search. In: Proceedings of the SIGIR Conference on Research and Development in Information Retrieval, New York, USA, pp. 478–479 (2004)
8. Lorigo, L., Pan, B., Hembrooke, H., Joachims, T., Granka, L., Gay, G.: The influence of task and gender on search and evaluation behavior using Google. Information Processing and Management 42(4), 1123–1131 (2006)

9. Glaholt, M.G., Wu, M.-C., Reingold, E.M.: Predicting preference from fixations. Psych-Nology Journal 7(2), 141–158 (2009)
10. Rayner, K., Miller, B., Rotello, C.M.: Eye movements when looking at print advertisements: the goal of the viewer matters. Appplied Cognitive Psychology 22, 697–707 (2008)
11. Glaholt, M.G., Reingold, E.M.: Eye movement monitoring as a process tracing methodology in decision making research. Journal of Neuroscience, Psychology, and Economics 4(2), 125–146 (2011)
12. Kosinski, M., Stillwell, D., Kohli, P., Bachrach, Y., Graepel, T.: Personality and Website Choice. In: Proceedings of WebSci 2012, Evanston, Illinois, USA (2012)
13. Pradeep, A., Knight, R.T., Gurumoorthy, R.: Neurological sentiment tracking system. United States Patent Application Publication, US 2011/0270620 A1 (November 3, 2011)

Developing Sophisticated Robot Reactions by Long-Term Human Interaction

Hiromi Nagano, Miho Harata[1], and Masataka Tokumaru[2]

[1] Graduate School of Kansai University
3-3-35 Yamate-cho, Suita-shi, Osaka 564-8680, Japan
[2] Kansai University, 3-3-35 Yamate-cho, Suita-shi, Osaka 564-8680, Japan
{k757820,k149657,toku}@kansai-u.ac.jp

Abstract. In this study, we proposed an emotion generation model for robots that considers mutual effects of desires and emotions. Many researchers are developing partner robots for communicating with people and entertaining them, rather than for performing practical functions. However, people quickly grow tired of these robots owing to their simplistic emotional responses. To solve this issue, we attempted to implement the mutual effects of desires and emotions using internal-states, such as physiological factors. Herein, the simulation results verified that the proposed model expresses complex emotions similar to humans. The results confirmed that the emotions expressed by the proposed model are more complex and realistic than those expressed by a reference model.

1 Introduction

Many researchers are attempting to develop partner robots. Communication is a quality required for robots to coexist with humans. These partner robots are need the ability to communicate which is essential for the robots to coexist with humans. These robots are problematic in that people quickly grow tired of them because of their simplistic emotion generation algorithms. To solve this issue, many models have attempted to generate more complex expressions of emotion. However, none of these studies have focused on "the growth of the robot." This feature can result in the robot growing in a manner similar to humans. Previously, we proposed a growth model for emotions that involved changing the structure of a self-organizing map (SOM)[2]. Also, the study attempted to make the expression of emotions more sophisticated using growth functions in a multilayer perceptron neural network (NN). Although our earlier model closely simulated the development of emotions in genetic psychology, it is still imperfect, because robots always express the same emotions when they receive the same input from a user, because their output emotions are only influenced by these inputs. A possible solution is to provide robots with some internal factors that are related to the natural and realistic expressions of emotion.

Herein, we applied the mutual effects of desires and emotions to the model. These effects are defined as a "consecutive cycle" in which experiences influence

M. Kurosu (Ed.): Human-Computer Interaction, Part V, HCII 2013, LNCS 8008, pp. 319–328, 2013.
© Springer-Verlag Berlin Heidelberg 2013

future actions. Also, the proposed model distinguishes emotions from feelings according to their characteristics. We expressed the mutual effects of desires and emotions using the internal-states of the robot. For example, along with saving a hunger state and an exhaustion state, the robot also saves the states of desires and feelings as numerical data. Desires and emotions can be mutually influenced by using the internal-states to generate different desires and emotions. With this model, a partner robot can express different emotions even though it receives the same input from users for solving the predictability problem observed in conventional partner robots. Herein, using numerical simulations, we verified that the model expressed various natural emotions similar to humans. We also used four types of input sets to examine the variations in the emotions expressed by the experimental model.

2   Proposed Model

This study aims to construct a robot that communicates better with humans via the functioning of emotions. We propose an emotion generation model for robots that considers the mutual effects of desires and emotions to create more complex and life-like emotions. Hence, it was important for the simulated emotions and desires that are used in the proposed model to be supported by proper psychological reasoning. Also, we needed to combine a person's desires, which are very closely related to his/her body, with the emotions that express these mutual effects. We distinguish emotions from feelings using a neuroscience perspective and consider the relationship among emotions, feelings, and desires in a human's body.

Moreover, we present the development of emotions and their relationship to robots. To introduce the capability of emotions and desires into robots, we use a self-organizing map (SOM) to represent and generate those emotions and desires. We also construct a function for the development of emotions in the proposed model using examples from our previous study, because our previous study demonstrated the effectiveness of an emotion-growing model. The emotional expression of the proposed model is based on M. Lewis's study on the differentiation and development of emotions as an emotional genetic model in psychology [1]. The generation of desires is based on Maslow's hierarchy of needs and is also represented using SOMs [3]. Also, we provide the robots with internal-states to enable them to have the equivalent of human body functions.

In the next section, we explain in detail the proposed emotion generation model and the methods used in this model.

2.1   Structure of the Proposed Model

Figure 1 shows the structure of the proposed model. It contains an internal-state with a section for emotion generation and a section for desire generation. Also, there is an external environment.

An input is generated by external stimulations and is received by the robot's internal-state. The desire and emotion generation networks then receive their inputs from the internal-state and generate new desires and emotions. Emotions generated by the emotion generation network are then expressed by the robot. Also, the internal-state is updated by the generated desires and emotions. This means that the internal-state now includes the influence of both the desire generation network and the emotion generation network. The mutual effects of desires and emotions are expressed by connecting the two networks in this way. Thus, the robot can generate emotions that are more similar to those expressed by humans, because the model is influenced by simulated physiological factors and the robot's current state of desire.

2.2 Self-Organizing Map(SOM)

We used SOMs to generate desires and emotions. Figure 2 shows a standard SOM. This type of an NN tends to treat data as vectors that have a characteristic classification of multidimensional data. Learning is unsupervised and uses the commonly used Euclidean distance. The Euclidean distance is the geometric distance between two points in a straight line and is defined by the Pythagorean theorem. An SOM searches for neuron i that assumes a Euclidean distance with x being a minimum when an input vector x is given. We assumed that the neuron had a reference vector (m_c), which assumes a Euclidean distance with x being a minimum toward the winner unit. The winner unit and the units around the outskirts learn the input vector using Eq. (1).

$$m_i(t+1) = m_i(t) + h_{ci}(t)[x(t) - m_i(t)] \qquad (1)$$

Here, h represents the neighborhood function, and t is an input step number. The neighborhood function can be expressed using Eq. (2) [4].

$$h_{ci}(t) = a(t) \cdot \exp\left(-\frac{\|r_c - r_i\|^2}{2\sigma^2(t)}\right) \qquad (2)$$

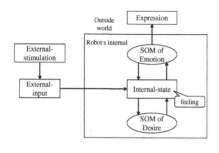

Fig. 1. Structure of the proposed model   **Fig. 2.** Self-organizing maps(SOMs)

Here $a(t)$ is the learning rate coefficient and parameter $\sigma(t)$ is the neighborhood radius. The functions $a(t)$ and $\sigma(t)$ are monotone decreasing functions of time. Using this neighborhood function, the neighborhood radius is large for the first learning process, and it gradually reduces as learning converges. By repetitive learning for all vectors, similar units come together for each input vector.

2.3   Network Evolution

From a psychological perspective, it is believed that various human emotions increase with age. Prior research has proposed a method by which the evolution of an NN can replicate this process [2]. Network evolution increases the amount of information that a model can control by increasing the number of combinations of neurons in the NN. In the early stages of evolution, the output layer neurons are not connected with all input layer neurons (shown by dashed lines in Figure 3(a). The output layer neurons only become combined with unconnected input layer neurons during evolution (development). After the evolution of the NN, the output layer neurons are combined with all input layer neurons. Figure 3(b) shows the increase in information at the inputs, and w_{11} - w_{34} represents the combined load between each neuron. Each neuron interval is a non-combination, that is, the combination load between each neuron expresses a 0 state.

Herein, we applied SOMs to control the evolution of the model, and we used an SOM as a desire and emotion generation network to represent the change associated with the development in humans. Hence, we developed an emotion model based on the evolution of the network as follows. First, the cosine degree of resemblance of the input data x_n and data inputs(x_{n-1}, x_{n-2}, ..., x_0) are found. Next, the network is evolved when data is judged to resemble the input data based on the cosine degree of resemblance, with x_n being greater than a predetermined number of times. Here we express the cosine degree of resemblance as two vectors in terms of the numerical values from Eq. (3).

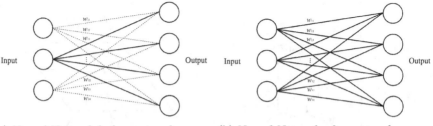

(a) Neural Network before network evolution

(b) Neural Network after network evolution

Fig. 3. Network evolution

$$sim = \frac{x_0 \cdot y_0 + x_1 \cdot y_1 + \cdots + x_n \cdot y_n}{\sqrt{(x_0^2 + x_1^2 + \cdots + x_n^2)(y_0^2 + y_1^2 + \cdots + y_n^2)}}$$

$$= \frac{x \cdot y}{|x| * |y|} \tag{3}$$

2.4   Emotion Development Model

The proposed model is required to consider the connections between the body and the emotions to adopt a desire. Here the body means a function that causes desires such as physiological needs. Thus, we distinguish feelings from emotions as two states of mind, each having strictly different properties. For example, A. E. Damasio, a brain scientist and philosopher, defines emotions as being external and public and feelings as being internal and private [5]. Based on these definitions, emotions are generated in the NN, while feelings are controlled by the internal-state of the robot. Also, Lewis's study of the differentiation and development of emotions has attracted attention in recent years [1]. Figure 4 shows results from Lewis's study. Lewis explains that infants experience contentment, interest, and distress by nature. Furthermore, he explained that from eight months of age, infants can express joy, surprise, sadness, disgust, anger, and fear, which eventually leads to nine basic emotions. To differentiate between emotions and developmental functions, we use Lewis's emotional development model (Figure 5).

2.5   Desire Model

In this study, we model the concept of desire using NNs. A. H. Maslow's hierarchy of needs, commonly known as the Maslow hierarchy of needs, is famous as the theory of hierarchical human desire [3]. Maslow's hierarchy of needs is shown in Figure 6.

We propose a model for desire to realize an increasing hierarchy of needs (Figure 7). This model includes seven desires from the first to the fourth stage;

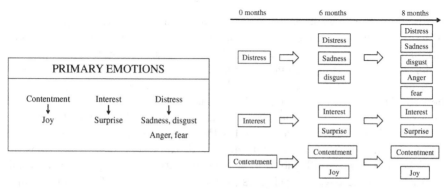

Fig. 4. Lewis's study   **Fig. 5.** Emotion development model

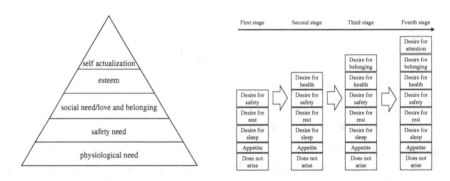

Fig. 6. Maslow's hierarchy of needs   **Fig. 7.** Desire model

namely, appetite and desires for sleep, rest, safety, health, belonging, and attention.

2.6   Internal-States and External-Stimulation

In the proposed model, robots have internal-states, which is similar to a memory holder in which the robot saves its current state. This internal-state is similar to the main states of being for a human. For example, the robot saves hunger and exhaustion, which are physiological factors, as numerical values in its internal-state. The factors of internal-states are shown in Table 1.

Also, we consider external-stimulations. An external-stimulation is defined as one from a robot's environment that influences it. Because there are many such causes of external-stimulation in the real world, we cannot consider all possible causes. Therefore, in this study, we consider only common external-stimulations. An example of the type of external stimulation that is used for this study is shown in Table 2.

3   Simulation

3.1   Simulation of Emotion Generation

In this simulation, we compared the emotional expression of the proposed model to that of a reference model to determine the usefulness of the proposed model. Although the reference model does not have the characteristics of the proposed model, such as the mutual effects of desires and emotions, its emotion generation network is identical to that of the proposed model. Each parameter for the two SOMs used for the simulation is shown in Table 3. Also, we used four types of input sets to examine variations in the expressed emotions for our proposed model. These sets were created with the assumption that the robot and human would have a long-term interaction. Of the four types, express positive (Table 4) and negative (Table 5) images in each other. The other two are a mix of positive and negative image inputs (Tables 6 and 7). These four types of input

Table 1. Factors of internal-states

Factor of internal-states	Category	
Hunger		
Sleep		
Exhaustion		
Safety		
Growth	States influenced by desires	
Comfort		
Love		
Injury		
Distress		
Interest		
Contentment		
Sadness		
Disgust	Feelings	
Surprise		
Joy		
Anger		
Fear		
Euphoria		
Enjoyment	States changed by feelings	
Contempt		
Curiosity		
Expectation	Feelings of will generated by desires	
Volition		
Appetite		
Desire for sleep	Physiological need	
Desire for rest		
Desire for safety		Container for desires
Desire for health	Safety need	
Desire for belonging	Love and belonging	
Desire for attention	Esteem	

Table 2. Types of external-stimulation

Type of external-stimulation	Category
Praise	
Slap	
Scold	
Call	
Show	Stimulation by users
Prepare a meal	
Play	
Amuse	
Heal	
Cry	
Move	
Sleep	Stimulation by action
Motion with emotions (ME)	
Brightness	
Sound	Stimulation by environment
Temperature	

Table 3. SOM parameters

Common parameter	
Number of times of learning	2,000
Number of times of external-input	−1.0 - 1.0
Parameter of emotion generation SOM	
Combination load	−1.0 - 1.0
Inputlayer neuron	30
Output layer neuron	Initial value 1, maximum 9
Parameter of desire generation SOM	
Inputlayer neuron	30
Output layer neuron	Initial value 4, maximum 8

sets consist of stimulations of 20 steps. Each step consists of an interference stimulation, parent's emotion, action stimulation, and environment stimulation. Here a "step" is the time taken by a robot to recognize an important input from a number of real-world inputs. Therefore, the time between steps is not uniform.

3.2   Simulation Results and Discussion

Only some of the results of the emotion generation simulation can be discussed here owing to space limitations.

Figure 8 shows the differentiation state of this simulation data, from which we confirm that each emotion is differentiated into one or more sub-emotions. That is, each of these emotions belongs to a system of emotions. In this figure, we assign colors to units in each system so as to distinguish between systems. Tables 8 and 9 and Figure 9 show the results of the emotion generation simulation.

Table 8 shows the changes in the emotional expressions in the proposed model with respect to the positive and negative image inputs. The results show that the proposed model expresses biased emotions when given biased inputs. Also, we confirmed that emotional expression and all the input sets in the reference model have a one-to-one relation between this figure and Table 4. Table 9 shows the changes in the emotional expression in the proposed model for mixed inputs and demonstrates the variation in the expressed emotions when mixed inputs are received from both the positive and negative input sets. Figure 9 shows the changes in the internal-state when the proposed model received positive image inputs. This confirms that the robot's emotions are influenced by previously expressed emotions, the present state, and the present desires of the robot. Clearly,

Table 4. Positive image inputs

	Interference stimulation	Parent's emotion	Action stimulation	Environment stimulation
$t = 1$	None	None	Cry	Temperature
$t = 2$	Amuse	Contentment	Cry	Brightness
$t = 3$	Prepare a meal	Contentment	ME	Brightness
$t = 4$	None	None	Sleep	Temperature
$t = 5$	Amuse	Interest	ME	Sound
$t = 6$	Praise	Joy	ME	Temperature
$t = 7$	None	None	Cry	Sound
$t = 8$	Play	Joy	Move	Sound
$t = 9$	Heal	Interest	Move	Sound
$t = 10$	None	None	Sleep	Brightness
$t = 11$	None	None	Cry	Temperature
$t = 12$	Amuse	Contentment	Cry	Brightness
$t = 13$	Prepare a meal	Contentment	ME	Brightness
$t = 14$	None	None	Sleep	Temperature
$t = 15$	Amuse	Interest	ME	Sound
$t = 16$	Praise	Joy	ME	Temperature
$t = 17$	None	None	Cry	Sound
$t = 18$	Play	Joy	Move	Sound
$t = 19$	Heal	interest	Move	Sound
$t = 20$	None	None	Sleep	Brightness

Table 5. Negative image inputs

	Interference stimulation	Parent's emotion	Action stimulation	Environment stimulation
$t = 1$	None	None	Cry	Temperature
$t = 2$	None	None	Cry	Brightness
$t = 3$	Prepare a meal	Interest	ME	Brightness
$t = 4$	None	None	Cry	Temperature
$t = 5$	None	None	ME	Brightness
$t = 6$	None	None	ME	Brightness
$t = 7$	None	None	Cry	Sound
$t = 8$	Amuse	Contentment	Cry	Brightness
$t = 9$	None	None	Cry	Sound
$t = 10$	None	None	ME	Brightness
$t = 11$	None	None	Cry	Sound
$t = 12$	None	None	Cry	Temperature
$t = 13$	Prepare a meal	Joy	ME	Sound
$t = 14$	None	None	ME	Sound
$t = 15$	None	None	ME	Sound
$t = 16$	None	None	ME	Sound
$t = 17$	None	None	Cry	Sound
$t = 18$	Play	Joy	Cry	Sound
$t = 19$	None	Joy	Sleep	Sound
$t = 20$	None	Joy	ME	Sound

Table 6. Positive and Negative image inputs

	Interference stimulation	Parent's emotion	Action stimulation	Environment stimulation
$t = 1$	None	None	Cry	Temperature
$t = 2$	Amuse	Contentment	Cry	Brightness
$t = 3$	Prepare a meal	Contentment	ME	Brightness
$t = 4$	None	None	Sleep	Temperature
$t = 5$	Amuse	Interest	ME	Sound
$t = 6$	Praise	Joy	ME	Temperature
$t = 7$	None	None	Cry	Sound
$t = 8$	Play	Joy	Move	Sound
$t = 9$	Heal	Interest	Move	Sound
$t = 10$	None	None	Sleep	Brightness
$t = 11$	None	None	Cry	Sound
$t = 12$	None	None	Cry	Temperature
$t = 13$	Prepare a meal	Joy	ME	Sound
$t = 14$	None	None	ME	Sound
$t = 15$	None	None	ME	Sound
$t = 16$	None	None	ME	Sound
$t = 17$	None	None	Cry	Sound
$t = 18$	Play	Joy	Cry	Sound
$t = 19$	None	Joy	Sleep	Sound
$t = 20$	None	Joy	ME	Sound

Table 7. Negative and Positive image inputs

	Interference stimulation	Parent's emotion	Action stimulation	Environment stimulation
$t = 1$	None	None	Cry	Temperature
$t = 2$	None	None	Cry	Brightness
$t = 3$	Prepare a meal	Interest	ME	Brightness
$t = 4$	None	None	Cry	Temperature
$t = 5$	None	None	ME	Brightness
$t = 6$	None	None	ME	Brightness
$t = 7$	None	None	Cry	Sound
$t = 8$	Amuse	Contentment	Cry	Brightness
$t = 9$	None	None	Cry	Sound
$t = 10$	None	None	ME	Brightness
$t = 11$	None	None	Cry	Temperature
$t = 12$	Amuse	Contentment	Cry	Brightness
$t = 13$	Prepare a meal	Contentment	ME	Brightness
$t = 14$	None	None	Sleep	Temperature
$t = 15$	Amuse	Interest	ME	Sound
$t = 16$	Praise	Joy	ME	Temperature
$t = 17$	None	None	Cry	Sound
$t = 18$	Play	Joy	Move	Sound
$t = 19$	Heal	interest	Move	Sound
$t = 20$	None	None	Sleep	Brightness

our simulation results show that the emotions expressed by the proposed model are more complex and realistic than those expressed by the reference model. Also, the results show that the proposed model can generate appropriate and a range of various emotions when it receives biased image inputs.

4 Conclusion

Herein, we discussed the use of an emotion model to systematize a robot's emotions. We noted that human desires and emotions affect each other, and we proposed an emotional development model. Also, we observed that desire and emotion should develop in the same way as experienced in humans. Thus, we constructed an emotion generation model for robots that depended on physiological factors. Furthermore, by evolving its SOM network, the proposed model mimicked emotional development in humans and could generate more complex emotions, which further developed with growth. Finally, we confirmed that emotions expressed by the proposed model were more complex and realistic than those expressed by the reference model.

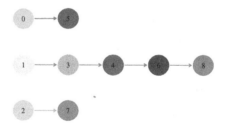

Fig. 8. Differentiation state of the simulation data

Table 8. Result of emotional expression: biased inputs (positive or negative)

	Positive image inputs : Proposed model			Negative image inputs : Proposed model			Positive image inputs : Reference model		
	3 months	6 months	8 months	3 months	6 months	8 months	3 months	6 months	8 months
Step1	0	0	0	0	0	0	0	4	4
Step2	1	1	1	0	5	5	1	1	1
Step3	1	1	1	1	5	5	2		
Step4	2	2	2	1	5	5	2	3	7
Step5	2	2	7	0	5	5	2		
Step6	2	2	7	0	5	5	2	2	2
Step7	2	2	7	0	5	5	0	4	8
Step8	0	5	5	0	5	5	0	4	4
Step9	0	5	5	0	5	5	0	0	0
Step10	2	2	7	0	5	5	2	3	7
Step11	1	2	2	0	5	5	0	4	4
Step12	2	3	7	0	5	5	1	1	1
Step13	2	4	7	0	5	5	2		
Step14	2		7	0	5	5	2	3	7
Step15	2		7	0	5	5	2		
Step16	2		2	0	5	5	2	2	2
Step17	2		7	0	5	5	0	4	8
Step18	2		7	0	5	5	0	4	4
Step19	2		7	0	5	5	0	0	0
Step20	2		7	0	5	5	2	3	7

Table 9. Result of emotional expression: mixed inputs (positive and negative)

	Positive image inputs : Proposed model			Positive and negative image inputs : Proposed model			Negative and positive image inputs : Proposed model		
	3 months	6 months	8 months	3 months	6 months	8 months	3 months	6 months	8 months
Step1	0	0	0	0	0	0	2	2	2
Step2	1	1	1	1	1	1	1	5	5
Step3	1	1	1	1	1	1	1	5	5
Step4	2	2	2	2	2	2	1	5	5
Step5	2	2	7	2	2	7	2	5	5
Step6	2	2	7	2	2	7	2	5	5
Step7	2	2	7	2	2	2	2	5	5
Step8	0	5	5	0	5	5	0	5	5
Step9	0	5	5	0	5	5	0	5	5
Step10	2	2	7	2	2	7	2	2	2
Step11	1	2	2	1	2	7	2	5	5
Step12	2	3	7	1	3	3	2	5	5
Step13	2	4	7	1	4	4	2	5	5
Step14	2		7	1			2	2	2
Step15	2		7	2		8	2	2	2
Step16	2		2	2		8	2	2	7
Step17	2		7	1		8	2	5	5
Step18	2		7	1		8	0	5	5
Step19	2		7	2		8	0	5	5
Step20	2		7	2	4	4	2	2	2

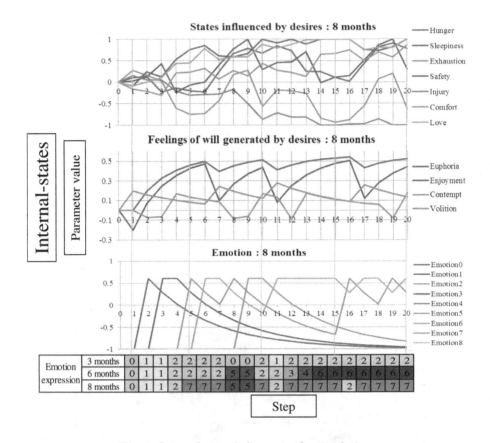

Fig. 9. Internal-states' change in the simulation

In future studies, we will apply the proposed method to an actual robot system.

References

1. Lewis, M., Jeannette M., Haviland-Jones: Handbook of Emotions, 2nd edn. Guilford Press (2000)
2. Harata, M., Sumitomo, H., Tokumaru, M.: Emotion generation model for a robot with a growth function. In: 12th International Symposium on Advanced Intelligent Systems, ISIS 2011, Suwon, Korea, pp. 319–322, 2011-10 (2011)
3. Goble, F.: The Third Force. The Psychology of Abraham Maslow, Maurice Bassett (1970)
4. Mera, K., Ichimura, T.: Personal Mental State Transition Network using Self Organizing Map. In: 27th Fuzzy System Symposium, Japan, pp. 1281–1286 (2011)
5. Damasio, A.R.: Looking for Spinoza: Joy, Sorrow, and Feeling Brain, Harcourt (2007)

An Awareness System for Supporting Remote Communication – Application to Long-Distance Relationships

Tomoya Ohiro, Tomoko Izumi, and Yoshio Nakatani

The Graduate School of Science and Engineering,
Ritsumeikan University, Kusatsu city, Shiga, 525-8577 Japan
is009086@ed.ritsumei.ac.jp,
{izumi-t,nakatani}@is.ritsumei.ac.jp
http://www.sc.ics.ritsumei.ac.jp/

Abstract. Recently, the methods of conducting long distance communication have dramatically changed due to improvements in communication technology including TV phones, e-mail, and SNS (Social Networking Services). However, people still have difficulty in enjoying sufficient long distance communication because subtle nuance and atmosphere are difficult to be felt in a distant place. For example, there are many romantic partners with feelings of anxiety about long-distance relationships. This is because an environment that allows the partners to understand each other has not been sufficiently supported. The purpose of this study is to help people separated by a long distance to understand each other by enabling the sensing of a partner's feelings from the partner's behavior. Our target is long-distance romantic partners. When people feel, sense, or are conscious of another person's existence or state, this ability or state is called "awareness.". Awareness is nonverbal communication. Awareness sharing among people is very important for managing relationships successfully, especially for people separated by a long distance. This is because a partner will develop feelings of unease if awareness sharing is not adequate. Our approach is as follows. First, examine what kind of action is useful for representing the feeling of love. Next, monitor these actions in partners. Third, summarize actions to quantitative indications. The prototype system was evaluated through evaluation experiments. Three pairs of partners used the system for two weeks. The result verified the effectiveness of this system as it promoted mutual communication.

Keywords: long distance communication, nonverbal communication, awareness.

1 Introduction

The development of information technology in recent years has come to allow communication between remote places to be conducted smoothly. Up until recently, people had been used communicating via letters and telephones. However, methods of

M. Kurosu (Ed.): Human-Computer Interaction, Part V, HCII 2013, LNCS 8008, pp. 329–338, 2013.

communication underwent significant changes with the development of e-mail and SNS (Social Networking Services). As a result, restrictions of time and a place have become lost. And through our improvement in information literacy, as long as an environment allowing connection to a network exists, anyone can perform remote communication easily. However, people still have difficulty in having sufficient long distance communication because subtle nuance and atmosphere are difficult to be felt in a distant place. This problem is generally found more in cases containing a strong emotional aspect than logical settings, such as business meetings. For example, there are many romantic partners who have feelings of anxiety about long-distance relationships. This is because an environment that allows partners to understand each other has not been sufficiently supported. It is generally said that long-distance relationships are not often successful. The opportunity for partners in a long-distance relationship to engage in communication is restricted. Therefore, they are forced to depend on language-based communication such as phone calls, e-mails, and letters. However, these media do not help individuals in understanding their partners' feelings easier. This is especially true for Asian people, whose cultures tend to rely more heavily on picking up on non-verbal communication clues. And they cannot actualize a feeling of staying close. The purpose of this study is to help people separated by long distances to understand each other by enabling a sensing of the feelings of their partners from the behavior of their partners. Our target is romantic partners involved in long-distance relationships.

2   Associated Research

Examples of previous studies on awareness shared systems do exist. Kajio's research is on a system for family members who lead separate lives [1]. This system uses a drawer as a device. The contents of this drawer synchronize with the partner's user. In this system, the contents of one partner's drawer are displayed on the other partner-user's drawer as a photograph. This function has actualized virtual living together. Through the operation of opening a drawer, the user has achieved an intuitive operation.

Furthermore, research exists on long-distance romantic relationships. Tsujita says that the couple in a long-distance relationship needs a strong bond [2]. Therefore, Tsujita proposed a method wherein, for daily items which two persons possess, when one partner uses an item, the other partner synchronizes by undergoing the same action. Some synchronized daily items are prepared, such as a trash box, a lamp, and a television. Each of these is an item that is used in usual living. Thereby, the mapping of a partner's action becomes simple. For example, if the trash box in one location is opened, the trash box in the remote location will be opened. This method provided the user with the matter-of-fact feeling of being together with the partner. These studies have features which give the user an intuitive operation. This has the merit of giving the feeling of being together virtually. Moreover, since the two systems use daily items as a device, anyone can use them. However, since the devices are always synchronized, there is the problem of infringing on the partner's privacy. Furthermore, if a system is used when the relationship with a partner is bad, it may become worse. Therefore, for supporting a long-distance relationship, the present situation with the partner must be considered. A long-distance relationship may easily develop into a situation where a small misunderstanding develops into a larger problem. Therefore, long-distance relationship support requires the function in which a partner's feeling can be guessed.

3   The System Proposal

In this chapter, an awareness sharing system for the couple involved in a long-distance relationship is proposed. Flexible correspondence which understands the condition with the partner and is appropriate in each individual instance is required for a long-distance relationship. Therefore, in long-distance relationship support, flexible support which considers the conditions of both parties at the time of each communication is required.

3.1   Support Using a Smartphone

The feeling of living together virtually has been achieved in previous studies. However, the methods therein did not take privacy into consideration. Furthermore, the partner's feeling is not transmitted. Moreover, as mentioned above, it is said that most couples who are involved in long-distance relationships do not engaged in awareness sharing well, and this is the cause which makes those relationships fail easily compared with couples who are not involved in long-distance relationships. Negative feelings, such as uneasiness and doubts, appear notably in long-distance relationships. So, this study proposes a method for supporting relationships using mobile terminals currently in general use. People may act unconsciously when thinking about someone. From such knowledge, unconscious action is also made applicable to evaluation in this study. Therefore, as a device used by this study, almost all persons thought that a mobile terminal that is always carried was optimal. In this research, a system which supports relationships by guessing the present condition of the self and the partner according to the usage situation of each function of a smartphone is proposed. A partner's uneasiness may be able to be canceled by showing the usage situation of the smartphone of the partner who is present in a remote place. Moreover, the act itself might become a cause which raises awareness. From the usage situation of the smartphone, this system evaluates and notifies nonchalantly about it. Thereby, the effect of promoting the sharing of awareness is also expected.

3.2   Selection of Love Action

"Love action" was selected based on opinions obtained by a questionnaire. Whether or not intimacy is felt regarding a certain action is different for different people. Therefore, in order to give subjectivity to the system, generality is given by taking and equalizing distribution based on the results obtained by the questionnaire.

First, some actions performed when thinking of the partner from the function of the smartphone were extracted. Next, those actions were surveyed in a questionnaire designed to discover whether or not the respondents feel intimacy with the partner based on the actions. The following table lists the functions of a smartphone mentioned as the example.

(a)Make a call

(b)Send e-mail

(c)A past e-mail is read.

(d)Music with meaning to both partners is listened to.

(e)A photograph taken with the partner is viewed.

(f)A schedule with the partner is adjusted.

This questionnaire survey was conducted with 36 college students. The questionnaire evaluated in five steps whether or not the above mentioned actions would inspire a feeling of intimacy for a partner. This survey obtained the standard deviation and distribution, and it confirmed whether or not the actions mentioned above could be considered common.

Shimizu and Obo describe that "intensity" and "diversity" are the important things for keeping a relationship positive [3]. In this study, "frequency" is used for evaluation in addition to "intensity" and "diversity." A couple's condition is quantitatively evaluated using these three elements.

Frequency: The sum total value of the frequency in which the couple performed the love action.

Intensity: The evaluation value of the love action computed by the questionnaire.

Diversity: The kind of love action performed within the measurement period.

The evaluation result was evaluated using these three variables. The formula is given in (1) below.

Intensity of love action x Frequency of love action ÷ Total of the frequency of the two persons' love action   ...(1)

3.3   The Method of Notifying Users

Users can be made to feel a sharing of awareness by receiving notifications in a casual manner. The partner's feelings are notified to the user by color changes. Specifically, the present state is indicated by changing the background color of the home screen of the system. This allows an awareness which cannot be obtained through conventional communication to be felt by the user.

4   Details of the System

This system is provided as an iPhone application. Therefore, Objective-C was used as the development language. Regarding the usage situation of the user's iPhone, if a user performs love action, the system will record an action history. The history acquired by the system is evaluated. The partner is notified of this result. The user guesses the partner's feelings and the present condition of the partner according to the evaluation value received from the partner-user. Fig. 1 is a composition figure of this system.

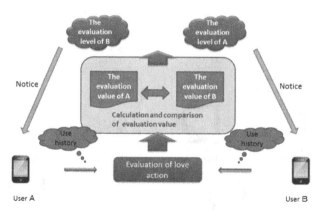

Fig. 1. System configuration diagram

4.1   Calculation of Evaluation Values

The system will perform an evaluation from the history after 1 day passes from commencement of use. The system calculates each user's sum total value. The value of the difference of the two sum total values is calculated. It finds the gap between the two persons' feeling by taking the difference of the two persons' sum total value. If the evaluation value of the two persons differs greatly, this means there is a gap in their feelings. Since the sum total value of this evaluation changes with couples, generality is lost. Therefore, the result of dividing the difference and average value of the sum total value of evaluation is the evaluation value. The evaluation level is calculated from this value and each user is notified of their partner's value. By being notified of the partner's evaluation level, the user can guess the partner's feelings. Users are notified of their partner's previous day evaluation level. The system changes the color of the background of the system with an evaluation level. This evaluation level is expressed in five steps.

For example, if the evaluation value of the system is low, it will display and carry out red. This is a report that the relation with the partner is not good. Conversely, a blue background will be displayed if the evaluation value is high. This is a report that the relation with the partner is good. Thus, it is surmised that the relationship of couples will improve because users check the transfer of dependence with their partner mutually daily. This is why the system has reported to the user the evaluation result of the previous day.

It is conjectured that notifying of changes in a partner's feelings that are difficult to sense in long-distance relationships raises a couple's level of awareness.

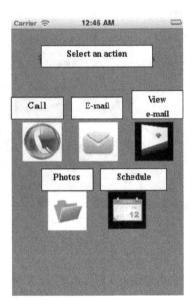

Fig. 2. Notification using the background color

5   Evaluation Result

Before performing the experiment, we hypothesized the effect upon the user that can be expected through the use of this system.

The system was expected to enhance awareness by receiving information that is usually not transmitted from partner to partner.

The user is expected to be able to resolve uneasy feelings of wondering what the partner is thinking and wondering if the partner is really thinking about the user.

This system is expected to be used as a new medium of communication.

In order to validate these hypotheses, we performed experiments and examined the hypotheses through the survey results.

We conducted the evaluation experiment for approximately one week among three couples who are involved in long-distance relationships.

The 1st set consists of the couple made up of Examinee A (a 21 year old male student who lives alone in Shiga) and Examinee A' (a 22 year old female student who lives with her parents in Okayama). This couple has been in a relationship for 5 years and 8 months. This is the 4th year that they have been in a long-distance relationship. The 2nd set consists of the couple made up of Examinee B (a 26 year old male who worker who lives alone in Kagawa) and Examinee B' (a 22 year old female student who lives with her parents in Okayama). They have been in a relationship for 2 years and 5 months. This is the 2nd year of their long-distance relationship. The 3rd set consists of the couple made up of Examinee C (a 22 year old male student who lives with his parents in Okayama) and Examinee C' (a 21 year old female student who lives alone in Fukuoka). They have been in a relationship for 6 months. They have been in a long-distance relationship for 6 months. The examination method employed the system for one week after notifying the examinees of the background and the purpose of this research. The examinees filled out a questionnaire after using the system. Moreover, the log data of the evaluation values of action under investigation and evaluation levels were constantly recorded, and we investigated whether the system was being used effectively.

5.1   Analysis of Log Data

Next we will discuss the log data that the server recorded during the experiment period. For the couple composed of Examinees A and A' and the couple composed of Examinees C and C', perhaps because the members of both couples were students, the sum total of love actions was high and the total value of actions were also high. In contrast to this, the tendency for the couple composed of Examinees B and B' was low for both members regarding the sum total of love actions and low for the total value of actions as well. The couple composed of Examinees A and A' showed a high evaluation level regarding the result for the first day of use, and the relationship was not in a good state at that point. However, the evaluation level showed a trend of converging on a low value from the second day onward, and it can be said that the members of the couple began to communicate more with a focus on sensing the feelings of each other based on the values generated by the system. Also, the couple composed of Examinees C and C' experienced a convergence towards low values for each other as well, and the same level was recorded for 6 days out of 1 week. As mentioned later, this was a result of the couple composed of Examinees C and C' using this system as a new medium of communication and attempting to synchronize their evaluation levels in the spirit of playing a game. Examinees B and B' do not correspond with each other very much in general, and the total value of love actions was low with an overall high evaluation level. Examinee B is very busy at his job, and there were many days when Examinee B' engaged in love actions on her own. Also, Examinee B did not pay much attention to this system, and as a result, not much of an effect was seen from this system in his case.

5.2   Analysis through Questionnaire Results

In order to validate the beneficial nature of the system, examinees were asked to fill out questionnaires after the survey period. The questionnaires consisted of 6 items and a free area to write opinions and feedback about the system.

In response to Question 1 (Were there any changes in the frequency of love actions before and after using this system?), we received many responses stating that the number of times of engage in love actions increased through the use of this system. We received feedback that reported that using the system led to engaging in actions that were not engaged in before using the system, such as looking at photos again or rereading e-mail messages, and that a feeling of newness that existed during the beginning of the relationship had been recaptured.

In response to Question 2 (Did you make phone calls or send e-mail messages as a result of this system?), we received many responses stating that communication had increased as a result of this system. Feedback was also received stating that this system became a topic of interest in and of itself, and that checking on the partner's evaluation level by phone or checking to see the type and number of actions the partner had engaged in during the day were conducted. Examinees C and C' used this system in the spirit of playing a game, each checking to see how many love actions the partner engaged in in order to synchronize their evaluation levels. This system can be considered to have acted as a new medium leading to increased communication.

In response to Question 3(How did you feel about this system?), we received favorable evaluations from most of the examinees. Regarding examinees who answered that they felt relieved or experienced feelings of happiness, a feeling of security can be thought to have arisen from the ability to assess what the partner is feeling about the examinee. The Examinee who answered that the system was irritating reported a feeling that the evaluations were all bad, and that was upsetting.

In response to Question 4 (Did this system enable you to feel close to your partner?), we received many responses stating that the examinees were able to feel close. From this we can suppose that awareness toward each other was enhanced through the system's displaying of awareness information that is not generally transmittable to partners in distant locations. Also, we received feedback stating that the act of speculating about a partner's feelings in and of itself provides a chance to feel close to the partner.

In response to Question 5 (Did communication increase through this system?), all Examinees stated that it increased or increased somewhat. This can be attributed to the system leading to an enhancement in communication because the frequency naturally increases due to the fact that the system uses calling and e-mail messaging as evaluation criteria. Also, the reasons obtained from Question 2 also play a large role in this.

Regarding Question 6 (Were the changes in emotion as indicated by color changes in the system communicated to you effectively?), the questionnaire feedback indicates that the changes were communicated effectively, and the method of notification can be considered useful.

5.3   Verification of the Hypotheses

Hypothesis 1: The system may enhance awareness through the communication of information that is generally not able to be communicated between partners. From responses to Question 5 in the previous section, this system can be considered to have enhanced awareness of the users through usage. The reason that this system was the cause for enhancing awareness is that users accepted the displaying by the system, through the format of smartphone usage conditions, of atmosphere information that cannot generally be assessed through long-distance communication. Also, as mentioned above, awareness can be considered to have been further enhanced by making it easy for the user to infer the feelings of the partner through not displaying evaluation values directly to users but instead notifying users through the method of evaluation levels. These considerations verify the usefulness of this system as an awareness sharing system.

Hypothesis 2: Usage of this system may be able to eliminate feelings of anxiety about what the partner is thinking or if the partner is really thinking about me. From the responses to Question 4 in the previous section, it can be assumed that users were put at ease by using this system. Examinee feedback stating that, "our feelings were more in sync than I thought, and I was relieved," and, "I was happy to receive an evaluation showing that my partner was thinking about me," indicates that using this system put users at ease. We also received feedback stating that, "the evaluation was never good and the system itself was irritating." The system puts users at ease when the evaluations are positive, but can be annoying and irritating when the evaluations are negative. However, another examinee reported in the free feedback area that, "When evaluations were negative, talking with each other about why they were negative eliminated anxiety." This indicates that the system is useful as a tool for eliminating feelings of anxiety.

Hypothesis 3: This system could be used as a new medium of communication. Responses to Questions 1, 2, 3, and 6 indicate that this tool was actively used as a communication tool. Among the responses to Question 1, a couple that did not normally engage in love actions, which are the criteria for evaluation in the system, stated that they began to engage in them frequently due to using the system and restored the feeling of newness that existed at the beginning of the relationship. And, as in the case of Examinees C and C', there were users who used this system with the goal of achieving a high evaluation through cooperating with each other. This indicates that the system successfully functioned as a new communication tool.

The observations mentioned above all indicate that hypotheses 1 through 3 can be considered verified and that the system can be considered useful as an awareness sharing system with the purpose of supporting long-distance relationships.

6   Issues and Future Prospects

As mentioned above, several areas that could be improved in the system were discovered through the survey experiment.

The current experiment surveyed couples who normally engage in communication and a couple with a low frequency of communication. For the couples with a high frequency of communication (the couple composed of Examinees A and A' and the couple composed of Examinees C and C'), the log data analysis indicated that the system is useful. However, for the couple with a low frequency of communication (the couple composed of Examinees B and B'), the total value of love actions was low at times, making the importance of each action high, and large gaps in total values emerged. Because of this, the actual evaluation values and the evaluation levels converged at a high value for most of the days. However, there were also many days on which the evaluation levels converged towards the same level, and it seems that the emotions of the partners were in sync with each other. The same criteria values were set for all couples in the experiment, but in reality, the methods and frequency of communication is not the same for all couples. Thus, a function that can adjust the criteria for evaluation levels for each couple after conducting a pre-experiment survey is considered necessary. In the future, it will be necessary to create a flexible system that can adjust to the characteristics of the users.

Also, we received feedback stating that, "It was inconvenient to have to select love actions through the system screen each time." The system was designed to be an application for the iPhone, and due to restrictions for developing iPhone applications, it was difficult to give this system the functionality of recording the history of love actions automatically by having it operate in the background of the iPhone operating system. Because of this, the system was designed with the functionality of recording action history by allowing the home screen for this system to act as a quasi iPhone home screen. That feature did not give users a very good impression, and as a result, it made it difficult for the system to provide a feeling of casualness. Though this problem is somewhat irresolvable due to the choice of making the system an iPhone application, imbedding the system as an application in other android terminals that allow easier interaction with applications would probably solve the problem. Also, the current experiment was limited to an iPhone application in terms of the device for measuring love actions, but in the future we are considering a method wherein tags are attached to furnishings and items on or around the body to measure evaluation values from daily life activities.

Regarding the prospect for the future, based on an awareness of points for improvement obtained through conducting these experiments, we want to improve the system so that it can provide flexible support that couples with various different types of communication styles can use and a more casual type of support that feels effortless to the user.

References

1. Kajio Jim Rowan, I., Mynatt, E.: Digital Décor: The interaction by the strengthened furniture. In: The Information Processing Society of Japan Symposium Series, Interaction 2003 Collected Papers, vol. 2003(7), pp. 41–42 (2003) (in Japannese).
2. Tshujita, H., Tsukada, K., Kajio, I.: Appliances to Arouse Mutual Awareness between Close People Separated by Distance "SyncDecor". Computer Software Academic Journal 26(1), 25–37 (2009) (in Japanese)
3. Shimizu, Y., Obo, I.: The structure of interaction in romantic relationships: hierarchical data analysis of inter-subjectivity between partners. The Japanese Journal of Psychology 78(6), 575–582 (2008) (in Japanese)

Emotion Sharing
with the Emotional Digital Picture Frame

Kyoung Shin Park[1], Yongjoo Cho[2], Minyoung Kim[3],
Ki-Young Seo[4], and Dongkeun Kim[2]

[1] Department of Multimedia Engineering, Dankook University, Korea
[2] Division of Digital Media, Sangmyung University, Korea
[3] Department of Computer Science, Sangmyung University, Korea
[4] Department of Computer Science, Dankook University, Korea
kpark@dankook.ac.kr, ycho@smu.ac.kr, pupleshine@gmail.com,
windzard@empas.com, dkim@smu.ac.kr

Abstract. This paper presents the design and implementation of emotional digital picture frame system, which is designed for a group of users to share their emotions via photographs with their own emotional expressions. This system detects user emotions using physiological sensor signals in real-time and changes audio-visual elements of photographs dynamically in response to the user's emotional state. This system allows user emotions to be shared with other users in remote locations. Also, it provides the emotional rule authoring tool to enable users to create their own expression for audio-visual element to fit their emotion. In particular, the rendering elements of a photograph can appear differently when another user's emotion is received.

Keywords: Emotional Digital Picture Frame, Emotional Intelligent Contents, Emotional Rule Authoring Tool.

1 Introduction

Emotion stimulates five senses of a human being and it plays a critical role while enjoying multimedia contents, such as movie, game and picture. Typically such contents are created by designers beforehand with one or more emotional themes (such as sadness, happiness, suspense and so on). They do not dynamically change its emotional theme in real-time according to user's emotional feeling. However, recent research has emphasized the role of emotion in information technology, and a number of emotion-related studies are underway.

As shown in Fig. 1, Emotionally Intelligent Content (EIC) is the digital content that goes beyond pre-defined emotional theme. For instance, the emotionally intelligent game detects if a user tends to be too immersed in playing a violent game, and it would change its visual and aural aspects to distract the user [1]. Thus, EIC recognizes user emotional states in real-time and responses to them to give certain emotion or make them more engaged into the content. Also, EIC allows each individual user to express or change its visual and aural elements based on user emotional states.

M. Kurosu (Ed.): Human-Computer Interaction, Part V, HCII 2013, LNCS 8008, pp. 339–345, 2013.

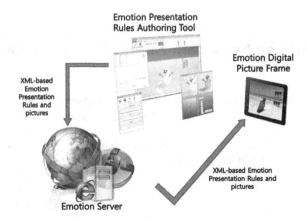

Fig. 1. The Conceptual Diagram of the Emotional Digital Picture Frame

Photographs often capture a user's special memories and moments. Photographs are also great medium for sharing emotion with others. When a family sees a photo of a child riding a bike in a garden, it may trigger a memory for their private thoughts or emotional feeling at the time, even though the details may not be shown in the picture. In this paper we focus on such emotional sharing via an emotional digital picture frame in which the system allows altering the photo's visual and aural presentation customized for each user's preference. In this system, user's own emotion rules specify how the visual and aural properties of the picture such as color, brightness as well as sound would be rendered to other users.

Fig. 1 shows a conceptual diagram of emotional digital picture frame system. As shown in Fig. 1, if a new picture with emotional rules is uploaded to the server, it is automatically shown on other remote users. This system uses PPG, GSR, SKT physiological signals to retrieve user emotional states. The audio-visual elements are then rendered in response to user emotions. Using the Emotion Rule Authoring Module, users can create XML-based emotional rules that define the audio-visual elements to fit their emotions. In this system, changing user emotions can be verified through the picture appearance, such as the saturation, luminosity and brightness. This mechanism allows the remote user to get the other user's emotion in real-time.

We will first look at related work and discuss the design and implementation of emotional digital picture frame system and its authoring tool. Then, we will describe our prototype and analyze the results of the test conducted on the emotion images. Finally we offer our conclusions and discuss steps for future research.

2   Related Works

There are many studies on sharing emotion using photo frames. Sondhi and Sloane developed a picture-sharing system for a ubiquitous smart home environment [2]. In this system, users are able to input descriptive emotional words onto the pictures that are shared.

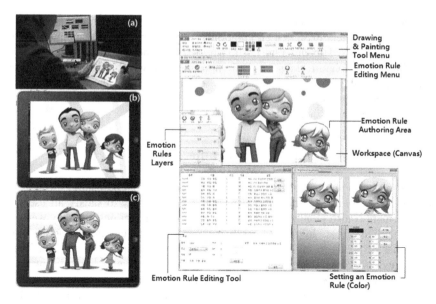

Fig. 2. A Prototype of Emotional Digital Picture Frame with Emotional Rule Authoring Tool

LumiTouch [3] is an emotional communication device which consists of a pair of interactive picture frames. It is used for enhancing the communication between loved ones. In LumiTouch, light-emitting elements and pressure sensors are attached to the photo frame. The sensors then detects when a user took hold of the photo frame with his/her hands, and then different colored lights are displayed on the other user's frame depending on how, how long and where a user holds the frame. Also, when a user is standing in front of the frame, the other LumiTouch device emits ambient light glows to indicate the presence of user. The features of LumiTouch allow users to display their emotions and develop an abstract emotion language for basic communication.

The EmoHeart project by University of Tokyo took Second Life one step further by displaying emotions (via inputted text) on the avatars' chests [4]. Second Life is an online virtual world where user avatars utilize clients to meet and interact. It is used to foster business and personal relationships and as a medium for the exchange of new ideas. The EmoHeart project requires the active analysis of inputted text to discover user emotions, which are then displayed on the avatar's face or as a heart on the avatar's chest.

The Emoti-Picture Frame added a tangible user interface in a frame to convey emotional features to remote users [5]. This frame is divided into two parts: the picture area (on the left) displays the remote user's emotion and the feeling area (on the right) consists of emotional buttons and heart-shaped emotion indicator, which allows the current user to express his/her own emotions. When a user pressed an emotional button, the relevant emotion is transmitted to the other user and the related picture is displayed on the frame along with the heart emotional indicator blinking.

All these systems allow users to share emotions, but they only work for two users and do not provide any infrastructure to make them scalable for more users. Also, the way emotions are understood in these cases have either been pre-determined or a new language must be established that both users have agreed to, which makes it difficult for users to adequately express their emotions. Another problem is that emotions must be actively expressed in order to be transmitted. On the other hand, our proposed emotional digital picture frame system not only allows for the sharing of user emotions, but utilizes physiological signals so that emotions can be transmitted and shared without any work on the part of the user. It also allows users to express and understand emotions in their own individualized way.

3   Emotional Digital Picture Frame

Each person has a unique personality and often shows different reactions to the same contents. As such, a uniform emotion response is not likely to satisfy all users [6]. Schneiderman[7] also pointed out that users preferred interactions that they were able to control themselves. In this research we introduce an emotional picture frame system that responds to individual user emotions by user-specified emotional rules. Unlike previous works, this system enables for users to be able to directly control how audio-visual elements on a picture will change depending on their emotions.

Fig. 2 shows a prototype of the emotional digital picture frame system built for sharing photographs among group members with individualized emotion. We conjured up a scenario of a family separated by distance. In particular, we envisioned this system allowing a father separated from his family to be able to share emotions. Once the father touches the frame, his emotional state is read by the sensors on the frame and his emotion is then displayed on the family's picture frame. Conversely, a daughter touches her picture frame to project her emotional states onto the father's frame.

In this current prototype, a user attaches the sensors to his/her finger and ear to view the emotional picture as shown in Fig. 2 (a). While he/she is looking at the picture, the emotions he/she feels are transmitted to the remote user via the server. The remote user is looking at the same picture that is slightly altered according to his/her emotion rules. Fig. 2 (b) and (c) show how the individualized emotion rules affect the screen differently. Fig. 2 (b) shows the picture of the daughter on the father's screen. Fig. 2 (c) shows the father's emotional state. As we can see, the color of the daughter's hair and clothes as well as the background image appears different from each other's view. Fig. 2 (d) shows the emotional rule authoring tool to customize individual's emotional expression.

4   Design and Implementation

Fig. 3 shows the system architecture of the emotional digital picture frame system. This system is built on top of Emotional Content Framework that consists of the Emotion Recognizer Module, the Emotion Server, the Emotional Rule Authoring Module and the Emotional Contents Player Module.

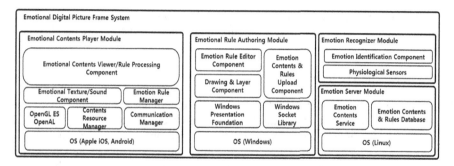

Fig. 3. The System Architecture of the Emotional Digital Picture Frame

4.1   Emotion Recognizer Module

The Emotion Recognizer Module receives user's physiological signals via PPG (pulse wave), SKT (skin temperature), and GSR (skin resistance) sensors attached to the BIOPAC MP100 system. It then applies the emotion deduction algorithm on the signals using our own LabView (Windows platform) program [CE] to classify into one of the nine emotional states: alertness, unpleasant, pleasant, relaxed, stressed, excited, sluggish, tired or neutral, in a two-dimensional space based on Russell's emotion model [8].

4.2   Emotion Server

The Emotion Server is a REST (Representational State Transfer)-based service running on a Linux-based Apache web server. It manages user's emotional states (retrieved from the Emotion Recognizer Module) and the emotional rules suited for each individual and the contents he/she created (using the Emotional Rule Authoring Module) stored on the MySQL database system. In order to do this, a unique identification number is given to each user along with the user's emotional states at given dates and times. The server also notifies the current user's emotional states to Emotional Digital Picture Frame clients.

4.3   Emotional Rule Authoring Tool

The Emotional Rule Authoring Tool is used for creating individualized emotional rules. This module is implemented using C# and Windows Presentation Foundation (WPF) library. Users login with their identification number to update their individualized emotion rules in personalized spaces in the server. When users upload new pictures, sharing (among Emotional Digital Picture Frame systems) commences. These shared pictures reflect the emotion rules that were saved for the user. If no emotion rules were specified, the server applies default emotion settings. Fig. 2 (d) shows specifying the emotional rules by selecting the field in the workspace where the rules will be applied. The emotional rule field is where a user can alter color, saturation and light luminosity and overall brightness on the image to reflect individualized emotion. Once the field is selected, a user selects the emotion response from the toolbox to specify the emotion rules. This authoring tool provides several primary drawing features for easy manipulation of rules on the picture and layers for defining more complex rules.

4.4   Interaction Scenario

The Emotional Contents Player Module displays emotional pictures on devices like the Apple iOS and Google Android. It receives new pictures and emotion rules on the server and then renders the customized visual and aural aspects on an image in response to user's emotional states. This module renders graphics using OpenGL ES 3D mobile-platform graphics library and sounds using the OpenAL library. The libcURL library was used for communication between the Emotional Contents Player Module and the Emotion Server to make use of the various Web service features. In addition, open source libXML was used to handle XML for emotion rules and contents expression.

5   Conclusions

This paper introduces a prototype of the Emotional Digital Picture Frame system constructed with Emotional Content Framework designed for creating emotional pictures by using individualized emotion rules. In this prototype, user's emotions are read by analyzing physiological sensors and these emotions are sent to remote users via the server. The picture changes its color and lighting dynamically depending on the specified emotion rules according to user emotions. In this prototype, users can utilize the Emotional Rule Authoring Tool to specify individualized emotion rules. In future research, we plan to conduct a systematic experiment to evaluate how much the users would be affected by sharing emotional pictures. Also, we will improve other elements to determine the effect of emotion contents and emotion sharing features.

Acknowledgements. This research was supported by Basic Science Research Program through the National Research Foundation of Korea (NRF) funded by the Ministry of Education, Science and Technology (Grant Number: 2012R1A1A1010815).

References

1. Kim, M., Park, K.S., Cho, Y.: Design and Implementation of the iOS-based Game Framework for Emotional Character Expression. Journal of Korean Society for Computer Game 24(1) (March 2011)
2. Sondhi, G., Sloane, A.: Digital Photo Sharing and Emotions in a Ubiquitous Smart Home. In: Venkatesh, A., Gonsalves, T., Monk, A., Buckner, K. (eds.) Home Informatics and Telematics: ICT for the Next Billion. IFIP, vol. 241, pp. 185–200. Springer, Boston (2007)
3. Chang, A., Resner, B., Koerner, B., Wang, X., Ishii, H.: LumiTouch: an emotional communication device. In: Extended Abstracts on ACM Human Factors in Computing Systems, CHI (2001)

4. Neviarouskaya, A., Prendinger, H., Ishizuka, M.: EmoHeart: Conveying Emotions in Second Life Based on Affect Sensing from Text. In: Advanced in Human-Computer Interaction (2010)
5. Neyem, A., Aracena, C., Collazos, C.A., Alarcón, R.: Designing Emotional Awareness Devices: What One Sees is What One Feels. Revista Chilena de Ingenieria 15(3), 227–235 (2007)
6. Richard, S.: Lazarus, Emotion and Adaptation, pp. 15–29. Oxford University Press (1991)
7. Schneiderman, B.: Designing the User Interfaces: Strategies for Effective Human-Computer Interaction, 3rd edn. Addison-Wesley, Reading (1998)
8. Posner, J., Russell, J.A., Peterson, B.S.: The circumplex model of affect: An integrative approach to affective neuro science, cognitive development and psychopathology, vol. 17, pp. 715–734. Cam bridge University Press (2005)

Vision Based Body Dither Measurement
for Estimating Human Emotion Parameters

Sangin Park, Deajune Ko, Mincheol Whang, and Eui Chul Lee[*]

Department of Computer Science & Department of Emotion Engineering,
Sangmyung University, Seoul, Republic of Korea
ini0630@naver.com, kodeajune@gmail.com, {whang,eclee}@smu.ac.kr

Abstract. In this paper, we propose a new body dither analyzing method in order to estimating various kinds of intention and emotion of human. In previous researches for quantitatively measuring human intention and emotion, many kinds of physiological sensors such as ECG, PPG, GSR, SKT, and EEG have been adopted. However, these sensor based methods may supply inconvenience caused by sensor attachment to user. Also, therefrom caused negative emotion can be a noise factor in terms of measuring particular emotion. To solve these problems, we focus on facial dither by analyzing successive image frames captured from conventional webcam. For that, face region is firstly detected from the captured upper body image. Then, the amount of facial movement is calculated by subtracting adjacency two image frames. Since the calculated successive values of facial movement has the form of 1D temporal signal, all of conventional temporal signal processing methods can be used to analysis that. Results of feasibility test by inducing positive and negative emotions showed that more facial movement when inducing positive emotion was occurred compared with the case of negative emotion.

Keywords: Body dither measurement, Emotion recognition, Image subtraction.

1 Introduction

In previous researches for quantitatively measuring human intention and emotion, many kinds of physiological sensors (ECG (electrocardiography), PPG (photoplethysmography), GSR (galvanic skin response), SKT (skin temperature), and EEG (electroencephalography)) based methods have been adopted [1]. Conventionally, in case of using ECG or PPG, heart rate can be measured by analyzing successive pulse to pulse intervals [2]. Also, amplitude levels are analyzed in order to respectively measure skin response and skin temperature in case of using GSR and SKT. Especially, EEG data can be interpreted by using various kinds of method at time domain or frequency domain. However, the above mentioned methods may supply inconvenience caused by sensor attachment to user. Also, therefrom caused negative emotion can be a noise factor in terms of measuring particular emotion.

---

[*] Corresponding author.

M. Kurosu (Ed.): Human-Computer Interaction, Part V, HCII 2013, LNCS 8008, pp. 346–352, 2013.
© Springer-Verlag Berlin Heidelberg 2013

Recently, some camera vision based physiological data acquisition methods were proposed which is instead of the above mentioned conventional physiological sensors. The Cardio-Cam was proposed for measuring human heart rate by ICA (Independent Component Analysis) based color channel analyses without any sensor attachment [3].♦For the same purpose, many smartphone applications have been released which can real-timely measure heart rate by using built-in backside camera and white illuminator. In these applications, they used a concept that the brightness levels of successive images were changed because the amount of illuminative reflection is continuously and regularly changed according as the blood flow. Even though the mentioned methods are meaningful in terms of no sensor attachment, they only can decode heart rate.

To solve these problems, we focus on various kinds of dither of human body. In our analyses, we define the hierarchical model of human body as shown in figure 1 which can clarify dependency (or independency) of each part's dither. For convenience in this paper, only facial dither is analyzed.

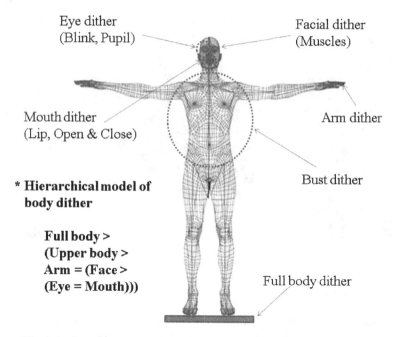

Fig. 1. Definition of various dithers of human body and hierarchical model

To confirm the feasibility of dither analyzing based emotion measurement, facial dither is analyzed in this paper. In captured upper body image, face region is firstly detected by using Adaboost (adaptive boosting) method. Then, the amount of facial movement is calculated by subtracting successive two images. At result of comparing two groups which respectively inducing positive and negative emotions during 10 minutes, more dither was occurred in case of positive emotion group than the case of negative emotion.

2   Proposed Method

Firstly, face region is detected in the 1st frame of upper body image by using OpenCV Adaboost face detector. Face detections of every frames do not needed because our propose method extracts the amount of facial motion by subtracting pixels of same position of two successive image frames. The Adaboost method uses a strong classifier generated by combining simple weak classifiers to detect face on an input image [4]. Although this algorithm takes much training time, it has advantages such as rapid time required for detection and good detecting performance. It took 29ms per an image in average to detect facial region. Examples of face detection in our experimental images are shown in red rectangles of Fig. 2.

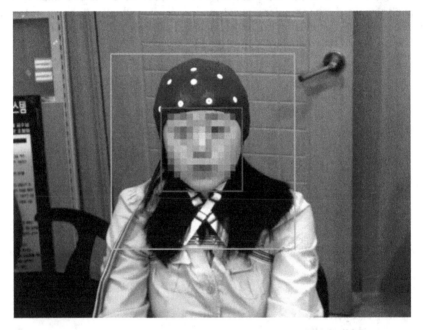

Fig. 2. Face detection results using Adaboost. (Red: Detected facial region, Green: Defined candidate region of facial dither).

After face detection shown in Fig. 2, the candidate region for subtracting image to calculate facial dither is defined by expanding 80 pixels directed to four direction of facial region rectangle as shown in green rectangles of Fig. 2.

To measure the amount of dither, the camera vision analysis program was implemented. In dither analyses, the captured color image is converted to gray level one because color component is not important in terms of estimating motion. The average amount of facial dither (M) can be calculated as following equation.

$$M = \frac{1}{WH} \sum_{j=y}^{H} \sum_{i=x}^{W} |I_n(i,j) - I_{n-1}(i,j)| \tag{1}$$

In equation (1), W and H are the horizontal and the vertical length of the facial dither candidate region respectively. And $I_n(i, j)$ means pixel value of ith column and jth row of nth image frame. Example of successive image frames and their dither extraction result is shown in Fig. 3.

After all, since the extracted continuous M values in equation (1) generate 1D temporal signal as shown in figure 4, it can be analyzed by using same way of conventional signal analyses methods. Although previous background subtraction methods for object detection have problem of continuously changed background or complex background modeling, our proposed is independent upon background changes because the proposed method uses only the latest two image frames.

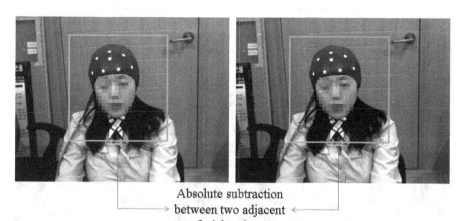

Absolute subtraction
between two adjacent
facial regions

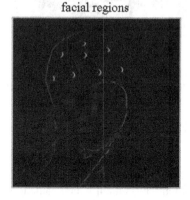

Fig. 3. Example of successive image frames (left and middle) and their extraction result (right)

The proposed method can be used to any part of human body mentioned in figure 1. If the previous implemented particular body part detection method is combined, their dither and muscle movement can be analyzed. For example, if Adaboost (Adaptive boosting) based face detection method is used, only face part dither can be analyzed. Fig. 5 shows many kinds of dither detecting results by using the program. In the figures, bright regions are regions in dither.

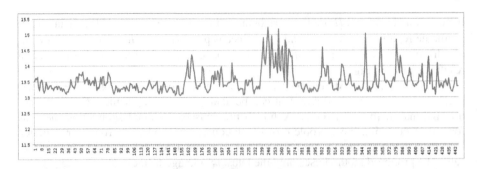

Fig. 4. Example of the successive value of facial dither

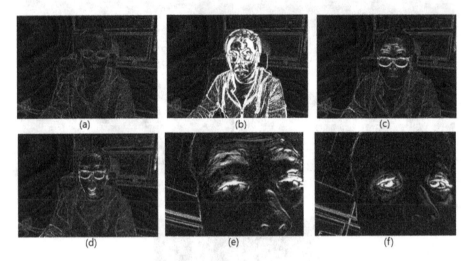

Fig. 5. Dither detection results. (a) No dither. (b) Bust dither. (c) Upper facial muscle movement. (d) Mouth movement. (e) Eye blink. (f) Changing gaze direction.

3   Experimental Result

To confirm the feasibility of our proposed method, a test for classification between positive and negative emotions was performed. For that, 10 persons were participated in which each subject heard announcement which causing positive and negative emotions during 10 minutes. During hearing the announcement, upper body image was captured by using a conventionally used webcam as resolution of 640 pixels by 480 pixels and 15 frames per second. Consequently, our method analyzed the amount of dither at 15Hz frequency band.

The average result of the dither amounts for two groups is shown in Fig. 6. According to this result, we found that the amount of dither for inducing positive emotion was greater than the case of negative emotion. Also, the difference between two cases for causing two contrary emotions was statistically significant in terms of t-test based average difference validation [5].

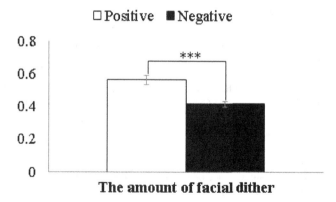

Fig. 6. The average amount of dither for two groups such as inducing positive and negative emotions (***: statistically significant at the confidence level of 99%)

4   Conclusion

In this paper, we proposed a new body dither analyzing method in order to estimating various kinds of intention and emotion of human. To solve problems of previously performed bio-signal based emotion measuring methods, we focused on facial dither by analyzing successive image frames captured from conventional webcam. After that, the amount of facial dither was measured by subtracting adjacency two image frames. Because the measured successive values of facial movement were the form of 1D temporal signal, all of conventional temporal signal processing methods might be used. Results of feasibility test by inducing positive and negative emotions showed that more facial movement when inducing positive emotion was occurred compared with the case of negative emotion.

In future works, we will experimentally validate connectivity between each part's dither of human body and various kinds of conventional physiological responses. For example, we will analyze correlation between pulse to pulse interval and the amount of bust dither after acquiring both ECG signal and bust dither for specific visual stimulus. Also, our dither analysis will be performed in terms of various frequency bands.

Acknowledgements. This work was supported by the by the Global Frontier R&D Program on <Human-centered Interaction for Coexistence> funded by the National Research Foundation of Korea grant funded by the Korean Government (MEST) (2012-055701).

References

1. Wu, N., Jiang, H., Yang, G.: Emotion recognition based on physiological signals. In: Zhang, H., Hussain, A., Liu, D., Wang, Z. (eds.) BICS 2012. LNCS, vol. 7366, pp. 311–320. Springer, Heidelberg (2012)
2. Chang, F.C., Chang, C.K., Chiu, C.C., Hsu, S.F., Lin, Y.D.: Variations of HRV analysis in different approaches. In: Proc. of Computers in Cardiology, pp. 17–20 (October 2007)
3. http://www.livescience.com/15469-cardiocam-mirror-mit-siggraph.html (accessed on March 17, 2013)
4. Han, P., Liao, J.: Face detection based on Adaboost. In: Proc. of Apperceiving Computing and Intelligence Analysis, pp. 337–340 (October 2009)
5. Johnson, D.H.: The insignificance of statistical significance testing. The Journal of Wildlife Management 63(3), 763–772 (1999)

Evaluating Emotional State during 3DTV Viewing Using Psychophysiological Measurements

Kiyomi Sakamoto[1], Seiji Sakashita[1], Kuniko Yamashita[2], and Akira Okada[2]

[1] R&D Division, Panasonic Corporation,
3-1-1 Yagumo-naka-machi, Moriguchi City, Osaka 570-8501, Japan
[2] Department of Human Life Science, Osaka City University,
3-3-138 Sugimoto, Sumiyoshi-ku, Osaka, 558-8585 Japan
{sakamoto.kiyomi,sakashita.seiji}@jp.panasonic.com,
{yamasita,okada}@life.osaka-cu.ac.jp

Abstract. Using a 50-inch 3DTV, we experimentally estimated the relationship between TV viewers' emotional states and selected physiological indices. Our experiments show complex emotional states to be significantly correlated with these physiological indices, which comprise near-infrared spectroscopy (NIRS), representing central nervous system activity, and the low frequency/high frequency ratio (LF/HF), representing sympathetic nervous system activity. These are useful indices for evaluating emotional states that include "feeling of involvement."

Keywords: emotional states, physiological and psychological measurements, NIRS, HR variability, 3DTV, TV viewing.

1 Introduction

To develop TVs that create a powerful sense of enjoyment and involvement, it is essential to make measurements of user's emotional states. Current methods of doing this depend chiefly on subjective reports by participants, but these evaluations often show considerable variation between individuals. Thus, to be able to improve the accuracy of these types of evaluations, it is necessary to measure emotional states objectively. Previous studies on objective estimations of users' psychological states have utilized physiological indices, including heart rate (HR), electroencephalograms (EEGs), and galvanic skin response (GSR)[1]-[3]. However, few previous studies have explored the physiological and psychological measurement of emotional states with any precision, using various types of video content during TV viewing or the relationship between physiological index and complex emotional states. Therefore, in a prior study [4], we used both physiological and psychological strategies to objectively estimate emotional states in response to 3DTV video content. Our results indicate that near-infrared spectroscopy (NIRS), representing nervous system activity, can provide useful indices for evaluating the emotional states of "stressed—relaxed" and "feeling of involvement." In our prior study [4], ten minutes of content was divided into three scenes. The scores obtained in the questionnaire correspond to the average

M. Kurosu (Ed.): Human-Computer Interaction, Part V, HCII 2013, LNCS 8008, pp. 353–361, 2013.
© Springer-Verlag Berlin Heidelberg 2013

of the physiological data value for each scene. However, the fact that participants' emotional states might show rapid changes within each scene prompted us to carry out further investigations to gain more precise insight into the relationship between emotional states and physiological indices. We therefore explored experimental and statistical analysis methods of estimating emotional state using psychophysiological measurements, and attempted to clarify the influence of stereoscopic vision on emotional state during 3DTV viewing. In our physiological evaluation, we used NIRS, brain waves, blink rate, heart rate, and a sympathetic nerve activity index; and in the psychological evaluation, we employed questionnaires and interviews.

2   Experiment

2.1   Methods

Participants: Twelve adults aged from their 20s to 30s participated in this experiment.

Measurements: The following items were investigated.

(1) Psychological state ("like—dislike," "stressed—relaxed," "feeling of involvement—bored," and "comfortable—uncomfortable"), reported on a scale of 3 to –3, through questionnaires and interviews. Four psychological items were additionally defined by the results of the pilot experimental interviews and those of our prior study [4].   In this study, the opposite emotional state to that of "feeling of involvement" was defined as "bored." The emotional state for "feeling of involvement" was defined as the feeling on the part of the user that they could watch TV with concentration and engagement [5], and that for "bored" was defined as the feeling of not being able to watch TV with interest or absorption.

(2) Near-infrared spectroscopy (NIRS): brain activity based on total hemoglobin or oxyhemoglobin, obtained by NIRS (NIRS detectors were placed on the left and right sides of the participant's forehead);

(3) Heart rate (HR) and heart rate variability (level of sympathetic nerve activity: LF/HF defined as the ratio of the low frequency band (LF: 0.04 - 0.15 Hz) to the high frequency band (HF: 0.15 - 0.5 Hz) [6][7], calculated employing FFT analysis using the R-R interval based on an electrocardiogram);

(4) Blinking rate obtained by electrooculogram (EOG);

(5) β/α (Electroencephalogram: EEG; Cz reference);

(6) Respiration rates were calculated by monitoring a respiratory sensor unit attached to the thorax. RR(Respiration rates) was defined as the number of respiration per minute obtained by the respiration curve.

Apparatus

(1) The display device was a 50-inch plasma TV (PDP) (Panasonic, TH-P50VT33; resolution: HD 1920 x 1080, aspect ratio: 16:9, width: 1106 mm, height: 622 mm).

(2) The viewing distance was set at 165 cm, the same as that used in previous studies [4].

(3) Test room conditions were maintained at a constant level: ambient temperature of 23 °C, relative humidity of 50%, and illumination of 150 lx. Humidity, which affects blinking rates, was strictly controlled. Illumination was set at 150 lx to simulate the average light level of a Japanese living room based on JIS standardization.

Procedure: Participants engaged in TV viewing of two kinds of video content (action horror and SF fantasy). Each content lasted 10 - 12 minutes. Two minutes' rest time was inserted before the 10 - 12 minute viewing test. Two viewing tests with different content were performed continuously. Each content set included two initial minutes of dummy video material, followed by five different scenes. To eliminate unstable physiological data, a dummy video was inserted between each scene. The measurement items adopted were brain activity obtained by NIRS, HR and LF/HF by ECG, blinking rates by EOG, and β/α by EEG and RR. Physiological indices were monitored while the participants underwent the viewing test. The participants gave a subjective assessment of their psychological state ("like—dislike," "stressed—relaxed," "feeling of involvement—bored," and "comfortable—uncomfortable," on a score of 3 to –3 after watching each scene. To eliminate the order effect, the order of viewing content was different for each participant.

2.2   Results and Discussion

A Pearson product-moment correlation coefficient analysis was performed to analyze the statistical relationship among the physiological indices (LF/HF, HR, NIRS, β/α, RR, blinking rates) and the participants' evaluation of their psychological state while viewing each type of content.

Correlation among NIRS, HR, LF/HF, blinking rate, β/α and the participants' evaluation of their psychological state:

Tables 1 (a) and (b) show the correlation between physiological and psychological states for content 1 (action horror). Tables 2 (a) and (b) show the correlation between those for content 2 (SF fantasy). S1 - S12 in Table 1 and 2 indicate each participant's identification number. P, N, and NULL indicate the following. P: significant positive correlation, N: significant negative correlation, NULL: no significant correlation. The level of significance was set at a coefficient of correlation of > 0.4 or $< –0.4$. For example, "N," a value on S1's "comfortable—uncomfortable" axis and "NIRS" column in Table 1 (a), shows a significant negative correlation between NIRS and this participant's scores for the "comfortable—uncomfortable" psychological state, reflecting a general pattern of NIRS being significantly lower in participants who gave higher scores for "comfortable." Tables 3 (a) and (b) show the sum of "like—dislike" score points for each scene.

The results show NIRS values to be lower in participants who gave higher scores for "comfortable," with a significant correlation shown in six of the twelve participants, shown in Table 1 (a), and to be higher in participants who gave higher scores for "feeling of involvement," with a significant correlation in nine of the twelve participants, as shown in Table 1 (a); and to be higher in participants who gave higher scores for "stressed," with a significant correlation in eight of the twelve participants in Table 1 (b). However, this was not the case for "like—dislike, NIRS" in Table 1 (b). Some participants (S1, S6, S11, S12) gave higher scores for "like" when viewing "action horror" and showed a higher NIRS value. Other participants (S2 - S5, S10) gave higher scores for "dislike" when viewing "action horror" and showed a higher NIRS value.

The results showed LF/HF values to be higher in participants who gave higher scores for "feeling of involvement," with a significant correlation in seven of the twelve participants, as shown in Table 1 (a), and also to be higher in participants who gave higher scores for "stressed," with a significant correlation in seven of the twelve participants, as shown in Table 1 (b). However, as with the NIRS results, this was not the case for "like—dislike, LF/HF" in Table 1 (b). Some participants (S1, S6, S9, S11, S12) gave higher scores for "like" when viewing "action horror" with a higher LF/HF. Other participants (S2 - S4, S8, S10) gave higher scores for "dislike" when viewing "action horror" and showed a higher LF/HF.

The results showed NIRS and LF/HF values to be significantly higher in participants who gave higher scores for "feeling of involvement" or "feeling stressed" when viewing action-horror content, irrespective of whether the content preference of the participants was clearly "like" or "dislike" (Table 3 (a)). However, the results showed NIRS, LF/HF and HR to be significantly higher in participants who gave higher scores for "comfortable" if the participant's content preference was "like," and NIRS and HR were significantly higher in participants who gave higher scores for "uncomfortable" when the participants' content preference was "dislike."

On the other hand, for content 2 (SF fantasy) where the content preference of the participants was not so clearly separated into "like" or "dislike" (Table 3 (b)), NIRS was higher in participants who gave higher scores for "comfortable," with a significant correlation seen in nine of the twelve participants, as shown in Table 2 (a). However, NIRS values were significantly higher in some participants who gave higher scores for "feeling of involvement—bored" (Table 2 (a)), and "stressed—relaxed" (Table 2 (b)); and NIRS values were significantly lower in other participants who gave higher scores for those emotions. In addition, as with NIRS, LF/HF was significantly higher in some participants who gave higher scores for those emotions, and LF/HF was significantly lower in other participants who gave higher scores for those emotions.

Figure 1 (a) shows the relationship between the two psychological axes, NIRS and LF/HF, when viewing a "liked" scene with any content, and Figure 1 (b) shows the relationship between the two psychological axes NIRS and LF/HF when viewing a "disliked" scene with any content. When viewing "liked" scenes with any content, the psychological evaluations of the experimental participants tended towards "comfortable" and "feeling of involvement." There was a positive significant relationship between complex emotional states and NIRS and the autonomic nervous activity level estimated by LF/HF. In other words, NIRS and LF/HF were significantly higher in participants who gave higher scores for "feeling of involvement" or "comfortable" when viewing "liked" scenes. On the other hand, when viewing "disliked" scenes, the psychological evaluations of participants tended towards "uncomfortable" and "feeling of involvement." However, there was a significantly negative correlation between "comfortable—uncomfortable" and NIRS and autonomic nervous activity level estimated by HR. NIRS and HR were significantly higher in participants who gave higher scores for "uncomfortable," irrespective of whether the content preference of the participants was clearly separated into "like" or "dislike." However, our results showed NIRS and LF/HF to be significantly higher in participants who gave higher scores for "comfortable" if the participant's content preference was "like," and was significantly higher in participants who gave higher scores for "uncomfortable" when the participants' content preference was "dislike."

Figure 1(a) and (b) show that participants' feelings were separated "like" or "dislike," or "comfortable" or "uncomfortable" for the two kinds of content used in this study, but most felt a "feeling of involvement" in the content. In the light of the results of the interviews, it appears that the effect of 3D video tended to increase "feeling of involvement." Further investigations will be needed to gain a more precise picture of the relationship between emotional states and physiological and objective psychological measurements.

Table 1. Correlation between physiological and psychological states for content 1 (action horror)

S1-S12 indicates each participant's identification number. The signs P, N, and NULL indicate the following. P: significant positive correlation, N: significant negative correlation, NULL: no significant correlation. The level of significance was set at a coefficient of correlation of > 0.4 or < -0.4.

	comfortable-uncomfortable			feeling of involvement-bored		
	NIRS	HR	LF/HF	NIRS	HR	LF/HF
S1	N	NULL	P	P	NULL	NULL
S2	N	NULL	N	P	P	P
S3	N	NULL	N	P	NULL	P
S4	N	N	N	P	P	P
S5	N	N	NULL	P	P	NULL
S6	P	P	NULL	P	P	P
S7	NULL	NULL	NULL	P	P	P
S8	NULL	NULL	N	P	P	P
S9	P	NULL	P	P	NULL	P
S10	N	N	N	N	N	N
S11	P	P	P	N	P	NULL
S12	P	P	P	NULL	N	NULL

(a)

Table 1. (*continued*)

	like –dislike			stressed- relaxed		
	NIRS	HR	LF/HF	NIRS	HR	LF/HF
S1	P	P	P	P	N	P
S2	N	NULL	N	P	NULL	P
S3	N	NULL	N	P	NULL	P
S4	N	NULL	N	P	NULL	P
S5	N	N	NULL	P	NULL	P
S6	P	P	P	NULL	NULL	P
S7	NULL	NULL	NULL	P	NULL	NULL
S8	NULL	NULL	N	NULL	NULL	NULL
S9	NULL	NULL	P	P	NULL	NULL
S10	N	N	N	P	P	P
S11	P	NULL	P	N	NULL	N
S12	P	P	P	N	N	N

(b)

Table 2. Correlation between physiological and psychological states for content 1 (SF fantasy)

	comfortable-uncomfortable			feeling of involvement-bored		
	NIRS	HR	LF/HF	NIRS	HR	LF/HF
S1	P	NULL	N	P	NULL	N
S2	P	P	P	P	P	P
S3	P	P	N	N	NULL	P
S4	P	NULL	P	N	N	N
S5	N	N	NULL	N	P	N
S6	P	NULL	N	P	NULL	P
S7	NULL	P	P	NULL	N	N
S8	P	NULL	P	P	NULL	P
S9	P	N	N	P	P	NULL
S10	P	NULL	N	P	NULL	P
S11	P	NULL	P	P	NULL	NULL
S12	NULL	NULL	NULL	NULL	NULL	NULL

(a)

Table 2. (*continued*)

	like –dislike			stressed- relaxed		
	NIRS	HR	LF/HF	NIRS	HR	LF/HF
S1	P	NULL	N	N	N	N
S2	P	P	P	P	P	P
S3	P	P	N	N	NULL	P
S4	P	P	P	N	NULL	NULL
S5	N	N	P	P	P	NULL
S6	P	P	P	P	NULL	NULL
S7	NULL	NULL	P	N	NULL	N
S8	NULL	NULL	P	P	NULL	P
S9	P	N	N	NULL	P	NULL
S10	P	NULL	P	P	NULL	P
S11	P	N	NULL	NULL	NULL	N
S12	P	P	N	NULL	NULL	P

(b)

Table 3. Sum of "like—dislike" score points for each scene

S1	S2	S3	S4	S5	S6
1	-3	-7	-5	-4	7

S7	S8	S9	S10	S11	S12
4	2	-5	1	7	-6

(a) for content 1 (action horror)

S1	S2	S3	S4	S5	S6
1	2	-2	-1	-1	5

S7	S8	S9	S10	S11	S12
3	7	6	6	12	3

(b) for content 2 (SF fantasy)

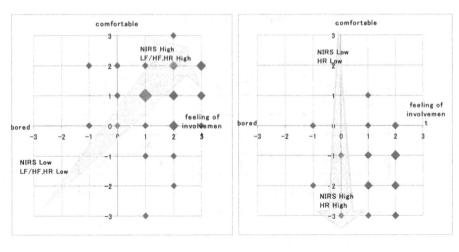

(a) The relationship between the two psychological axes, NIRS and LF/HF, in the case of viewing "like" scene.
X-axis (score for feeling of involvement—bored)
Y-axis (score for comfortable—uncomfortable)

(b) The relationship between the two psychological axes, NIRS and LF/HF, in the case of viewing "dislike" scene.
X-axis (score for feeling of involvement—bored)
Y-axis (score for comfortable—uncomfortable)

Fig. 1. (a) in Figure 1 show the relationship between the two psychological axes in the case of viewing "like" scene. And (b) in Figure 1 show the relationship between the two psychological axes in the case of viewing "dislike" scene. The X-axis for (a) and (b) in Figure 1 indicates the participants' scores for the "feeling of involvement—bored" psychological state. The Y-axis for (a) and (b) in Figure 1 indicates scores for ("comfortable—uncomfortable"). The font size of the plot markers (◆) on the graph grid indicates the number of points that were selected by participants as their psychological scores. The total number of points for (a) was 52 and that for (b) was 34. The scale of font sizes is 1 to 7 points.

3   Conclusion

Our results indicated that near-infrared spectroscopy (NIRS) and low frequency/high frequency ratio (LF/HF) to be potentially useful indices for evaluating emotional states that include "feeling of involvement," "stressed—relaxed" and "comfortable—uncomfortable." Moreover, they indicate the potential for improving accuracy of evaluation by combining complex emotional states into a physiological index, and that both physiological and psychological approaches can be used to obtain objective measurements of emotional states.

Further investigations will be needed to gain a more precise picture of the relationship between emotional state and physiological and psychological measurements when viewing various types of content and with analyzing for each separated group by participants' preference.

References

1. Wiederhold, B.K., Jang, D.P., Kim, S.I., Wiederhold, M.D.: Physiological Monitoring as an Objective Tool in Virtual Reality Therapy. CyberPsychology & Behavior 5(1), 77–82 (2002)
2. Shimono, F., Ohsuga, M., Terashita, Y.: Method for assessment of mental stress during high-tension and monotonous tasks using heart rate. respiration and blood pressure. The Japanese Journal of Ergonomics 34(3), 107–115 (1998) (in Japanese)
3. Yoshida, T.: Special issue: Significance of 1/f fluctuation from the viewpoint of brain wave levels. Japanese Journal of Medical Electronics and Biological Engineering 8(10), 29–35 (1994) (in Japanese)
4. Sakamoto, K., Asahara, S., Sakashita, S., Yamashita, K., Okada, A.: Influence of 3DTV video contents on physiological and psychological measurements of emotional state ISCE2012. In: Proceedings of The 16th IEEE International Symposium on Consumer Electronics, pp. 251–254 (June 2012)
5. Nojiri, Y.: Ultimate broadcasting system conveying a strong sensation of reality. NHK Technical Annual Report 2005, pp.6-6 (2005) (in Japanese)
6. Ishibashi, K., Kitamura, S., Kozaki, T., Yasukouchi, A.: Inhibition of Heart Rate Variability during Sleep in Humans By 6700 K Pre-sleep Light Exposure. Journal of Physiological Anthropology 26(1), 39–43 (2007)
7. Ishibashi, K., Ueda, S., Yasukouchi, A.: Effects of Mental Task on Heart Rate Variability during Graded Head-Up Tilt. Journal of Physiological Anthropology 18(6), 225–231 (1999)

Affect-Based Retrieval of Landscape Images Using Probabilistic Affective Model

Yunhee Shin[1], Eun Yi Kim[2], and Tae-Eung Sung[3]

[1] Department of Patent Valuation & Licensing,
Korea Invention Promotion Association, Seoul, Korea
[2] Visual Information Processing Lab., Konkuk University, Seoul, Korea
[3] Department of Technology Commercialization Information,
Korea Institute of Science and Technology Information, Seoul, Korea
yhshin827@gmail.com,
eykim@konkuk.ac.kr, ts322@kisti.re.kr

Abstract. We consider the problem of ranking the web image search using human affects. For this, a Probabilistic Affective Model (PAM) is presented for predicting the affects from color compositions (CCs) of images, then the retrieval system is developed using them. The PAM first segments an image into seed regions, then extracts CCs among seed regions and their neighbors, finally infer the numerical ratings of certain affects by comparing the extracted CCs with pre-defined human-devised color triplets. The performance of the proposed system has been studied at an online demonstration site where 52 users search 16,276 landscape images using affects, then the results demonstrated its effectiveness in affect-based image annotation and retrieval.

Keywords: Affect-based image retrieval, probabilistic affective model, mean-shift clustering, color image scale.

1   Introduction

With increasing the importance of affective computing, it becomes necessary to retrieve and process images according to human affects or preference, as even same categorized image can be differently interpreted depending on mood and affects. In particular, among several image domains such as photographic, medical, and artistic, it is very important to use affective meanings in landscape images.

However, judging such affective qualities of images is not easy task as affective meanings are always hidden, i.e. there is no direct mapping from the image to the meanings. Thus, to discover the hidden affective meanings from observable visual image features is a key step towards an affect-based image retrieval.

Generally, images provide color, texture, shape, and pattern information. Thus, various studies have been proceeded to investigate the relationships between these visual features and human affects and to identify certain visual features that predict human affects [1-4], and some retrieval systems have been developed [1,3,4].

M. Kurosu (Ed.): Human-Computer Interaction, Part V, HCII 2013, LNCS 8008, pp. 362–371, 2013.
© Springer-Verlag Berlin Heidelberg 2013

Datta et al. investigated the relationships between several visual features and the aesthetic quality in photographic images [1], where color, texture and color compositions were extracted. Based on those relationships, they have developed a classification model to assign aesthetic quality of a photographic image, thereafter have applied this model to Photo.net, where is photo sharing community site on Web. In [3], to predict 8 affective classes such as anger, despair, interest, joy, pleasure, pride and sadness from landscape images, they used the dominant colors of an image, and developed multiple linear regressions to establish the mapping between such colors and 8 affective classes. From them, the affective classification system was implemented on Web. In [4], they investigated relationship between visual features, such as color and pattern, and human affects. With these features, they automatically predict human affects associated with a textile image using machine learning algorithms. And, they have implemented textile retrieval system on Web.

Among various visual features above-mentioned, the most widely used feature is color including dominant colors and color compositions [1-4]. Then, it is hard to represent various affects, based solely on color, so this study focuses on using color compositions that constitute an image.

In this paper, a novel affect-based retrieval of landscape images using a Probability Affective Model (PAM) is presented. The PAM is developed for automatically predicting the affects from color compositions of images, and it is used for annotating Web images in our retrieval system. Our system was tested with 16,276 online landscape images the affective judgments of 52 users on an online demonstration site. Then, it produced the performance of 85.22% in annotation and that of 62.5% in retrieval. Consequently, these results proved the potential of the proposed system in affect-based annotation and retrieval of web images.

2   Image Annotation

For affect-based image retrieval, it should be first performed to annotate a given image using human affects. In this section, we define affective classes and then introduce PAM to automatically predict human affects from visual image features.

2.1   Affective Classes

For affective classes to annotate photographic images, this study defines 15 affective classes based Kobayashi's vocabularies, where they selected 180 affective adjectives that widely used in applications such as visual media, design and art [2].

To find representative words among such 180 adjectives, a survey on Yahoo! search results was performed. First, we collected 8,300,000 landscape images from Yahoo!, and investigated affective adjectives to be used in annotating images. Then, the affective adjectives were ranked according to the frequencies of occurrence in image tags. Based on these results, we re-defined 15 representative affects: {pretty, colorful, dynamic, gorgeous, wild, romantic, natural, graceful, quiet, classic, dandy, majestic, pure, cool and modern}.

2.2   Probabilistic Affective Model

A Probabilistic Affective Model (PAM) is used to annotate a given image using affective features. Let denote $W = \{w_1, w_2,...,w_L\}$ is affective classes, where L is 15. As
such, the affective features belonging to an image i are denoted by a 15-D vector $e(i)$
$= (e_{i,1},..., e_{i,L})$, where $e_{i,m}$ means the probability that the image i belongs to the corresponding m_{th} affective class.

Given an image i, each probability of affective feature $e_{i,m}$ is computed as follows:

$$e(i) = \left(e_{i,1}, e_{i,2}, ..., e_{i,L}\right) = (p(w_1|i), p(w_2|i), ..., p(w_L|i)) \tag{1}$$

Generally, an image i is composed of N regions, that is, $i = \{R_1, R_2,... R_N\}$, each of
which can be described by the visual features extracted from that region. In this work,
the region is described by color compositions, λ_j. The sense of "color composition"
includes both the characteristics of separate colors that occur in images and the organization for combining these parts into a whole.

Then, we assume that the color composition of a whole image is obtained by combining the respective color compositions of the segmented regions, and that the color
compositions among regions are independent. Accordingly, m_{th} feature of an affective
feature vector $e(i)$, $e_{i,m}$ in Eq.(1) can be rewritten as follows:

$$e_{i,m} = p(w_m|i) = p\left(w_m|(R_1, R_2, ..., R_N)\right)$$

$$= p\left(w_m|(\lambda_1, \lambda_2, ..., \lambda_P)\right) = \prod_{j=1}^{J} p(w_m|\lambda_j) \tag{2}$$

Consequently, $p(w_m|\lambda_j)$ indicates the probability for a color composition λ_j to be
mapped onto an affective word w_m.

The process to compute the PAM consists of following three steps:

- Image segmentation using mean-shift clustering
- Color composition extraction through RAG analysis
- Affect mapping onto human-devised color compositions

2.2.1   Image Segmentation
One of the classical image segmentation approaches is using clustering algorithm.
After performing color quantization to transpose from RGB color space into 130 basic
colors, we segment the mean-shift clustering algorithm [5].

2.2.2   Image Segmentation
In this module, the color compositions between segmented regions are extracted.
Since considering those all combinations is very computationally intensive, we firstly
select more influential regions having the larger importance than other regions, which
is defined as *seed regions*.

We assumed that regions that have larger areas and are closer to the center of the image are more important than others. Thus, the importance, $\phi(R_j)$, is assigned using a criterion based on its area $A(R_j)$ and its Gaussian distance to the image center, $G(R_j)$, such as $\phi(R_j) = A(R_j) \times G(R_j)$, where $A(R_j) = \frac{area\ of\ R_j}{width \times heigh}$ and $G(R_j) = \mathcal{G}_d(\mu, \Sigma) = \frac{1}{2\pi|\Sigma|} e^{-\frac{1}{2}(d-\mu)^T \Sigma^{-1}(d-\mu)}$.

Regions on an image are ranked in decreasing order of their importance values, and then the top M regions among all regions are selected as seed regions, $S(i)$, using threshold T_1.

Thereafter, the RAG is draw from the $S(i)$, and the analysis between seed regions and their adjacent regions is performed. Let denote $\mathcal{C}(\cdot)$ is function to compute color composition between seed region S_j and $\eta(S_j)$, where $\eta(S_j)$ is the set of adjacent regions of S_j. Then, a color composition λ_j is described by a triplet of colors of S_j and $\eta(S_j)$,

$$\Lambda(i) = \{(\lambda_1, \lambda_2, \dots, \lambda_P) \mid \lambda_j = \mathcal{C}(S_j, \eta(S_j))\} \tag{3}$$

Accordingly, in case of a seed region with four neighbors, there are ${}_4C_2$ cases when selecting secondary colors. By summing each color compositions for all seed regions, the $\Lambda(i)$ is finally obtained.

2.2.3 Affect Mapping

We predict the affective features from color compositions obtained from an image. For this, all of color compositions in $\Lambda(i)$ is mapped onto 1170 human-devised color compositions, Ξ, which is developed by Kobayashi [2].

$$\Xi = \{(\xi_1, w_1), (\xi_2, w_2), \dots, (\xi_K, w_K)\} \tag{4}$$

A pair (ξ_k, w_k) is composited of triplet $\xi_k = \{C'_{k,0}, C'_{k,1}, C'_{k,2}\}$, and affective class w_k ($w_k \in W$) which is manually assigned to the corresponding triplet, ξ_k.

Using Mallow distance [6], each color composition $\lambda_j = \{C_{j,0}, C_{j,1}, C_{j,2}\}$, is compared with triplets, ξ_k in Ξ.

$$Sim(\lambda_j, \xi_k) = -min \sum_{x=0}^{2} \sum_{y=0}^{2} \phi(R_{j,x}) \cdot \left\| C_{j,x} - C'_{k,y} \right\|^2 \tag{5}$$

Here, $C_{j,x}$ and $C'_{k,y}(x, y = 0,1,2)$ are colors of λ_j and ξ_k, respectively. The importance value $\phi(R_{j,x})$ of the region with the color $C_{j,x}$ is used as weight for this distance.

A color composition, λ_j, is projected to its nearest neighbor with the highest similarity in all of ξ_{k*}, and then k^*_{th} affective class is assigned to λ_j and the reflection, finally $p(w_{k*}|\lambda_j)$, is calculated based on its importance and its similarity.

$$k^* = arg\max_k Sim(\lambda_j, \xi_k), (1 \leq \xi_k \leq 1170)$$

$$p(w_{k*}|\lambda_j) = \frac{\phi(s_j)}{|Sim(\lambda_j,\xi_{k*})|} , (w_{k*} \in W)$$

(6)

While scanning a whole of $\Lambda(i)$, $p(w_{k*}|\lambda_j)$ are accumulated for affective classes, w_{k*}, thus the $e(i)$ is computed.

3   Image Retrieval

We developed retrieval system of landscape images using human affects. When given a query, q, our system retrieves the results ranked according to similarities between query q and images in database. Then, for the comparison with query, the images in DB were annotated by PAM. Thus, these images were represented as $E = \{e(i) = (e_{i,1}, e_{i,2}, ..., e_{i,L})| i = 1, ..., n\}$, where n and L are the total number of images in DB and that of affective classes, respectively.

Accordingly, when searching the images, the query q, given by a user, is first transformed to an affective feature vector $e(q)$, and then, the $e(q)$ is compared with all of affective vectors in the E.

In this work, the similarities between images and query are calculated using the cosine similarity. Thus, the similarity between query q and image i is

$$s(q,i) = Cos(e(q), e(i)) = \frac{\sum_{m=1}^{L} e_{q,m} \times e_{i,m}}{\sum_{m=1}^{L} e_{q,m}^2 \times e_{i,m}^2} , (e(i) \in E)$$

(7)

The $e_{q,m}$ and the $e_{i,m}$ are the m_{th} affective feature values of query q and image i from DB, respectively. The images are ranked based on these similarities $s(q,i)$, where $i = 1,...,n$. Thus, the higher the similarity is, the more relevant the image is.

Figure 1 shows the scene of our retrieval system on Web. In our system, a user can submit the query using either affective class or example image. In case of text query to denote m_{th} affective class, w_m ($w_m \in W$), only the affective feature corresponding to m_{th} affective class has non-zero value, and others have zero values. For example, when selected a 'romantic' class as query q, the $e(q)$ is represented as $e(q) = (0,0,0,0,0,1,0,0,0,0,0,0,0,0,0)$. On the other hand, for the query by example, the query image q is transformed to an affective vector $e(q)$ by the PAM. Thereafter, for the respective query type, the cosine similarity is performed with images in DB and the relevant images are retrieved. In addition, our system recommends some images that are mostly selected by users within a certain period of time, as shown in below of Figure 1.

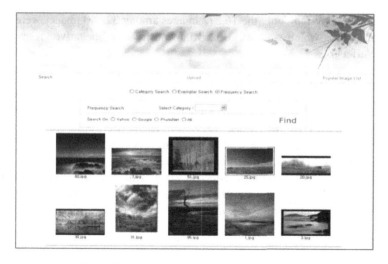

Fig. 1. The scene of our retrieval system on Web

Consequently, through such query schemes, a user can easily search more preferable images.

4   Experiments

To assess the effectiveness of our system, we first assess performance of PAM and then evaluate the retrieved results.

4.1   Image Collection

In this study, two different DBs were selected. The first is image DB from Photo.net, which is one of large online photo sharing community. The second is image DB from Yahoo! search. To collect landscape images from those sites, ten queries to describe the landscape were used such as *coast, desert, field, forest, lake side, mountain, snow scene, sky, sunset and waterfall.*

From Photo.net, 2,000 images were downloaded, 32,280 images were collected using Yahoo! openAPI. Consequently, 34,280 images were used in experiments.

For a complete evaluation of the proposed system, users' affective judgments are needed. For the user study, 52 participants (16 female and 36 male) were asked to participate. It was conducted through a survey system, which was implemented on the Web.

However, as there is wide variability in people's individual feelings, it is very difficult to establish the ground-truth data for affects. As such, we need to consider the varieties between people's sensitivities. Moreover, even the same person can make different judgments according to the mood, context, motivation, etc. Thus, we considered issues of variability within the same person and between persons. To deal with these varieties within a person and between persons, the repeated subject evaluations are adapted during user study and the fuzzy system is used as illustrated in [4].

Through fuzzy system, only the 16,272 images among 32,380 images were used as experimental data, which was categorized into positive and negative groups for the respective affects.

4.2   Experimental Results

The annotation of the PAM was compared with the users' affective judgments, that is, the ground-truth through fuzzy rules. They were categorized into positive and negative ones for the respective affects. Meanwhile, the PAM has a real value ranged 0 to 1, so a threshold value was applied on the output of PAM, for classifying the image into positive and negative one each affective class. This threshold value is set to 0.4 by experiments. Thus, if the output of PAM of certain affective class was larger than 0.4, the image was classified as relevant to the corresponding affective class, otherwise it was classified as irrelevant one.

Table 1. Performance of our method for two DBs

Affective classes	# of images		Photo.net		Yahoo!	
	Photo.net	Yahoo!	Recall	Precision	Recall	Precision
Pretty	81	925	82.72	97.50	93.25	75.00
Colorful	83	995	91.57	95.34	89.31	68.00
Dynamic	81	1,031	88.75	80.00	82.06	67.88
Gorgeous	85	1,029	90.59	89.47	87.30	66.39
Wild	93	1,130	84.62	80.00	89.62	73.33
Romantic	73	1,076	91.78	85.71	89.82	68.72
Natural	88	1,131	81.71	91.89	70.26	80.29
Graceful	81	930	93.83	88.89	71.27	66.67
Quiet	92	909	91.30	83.33	70.16	69.13
Classic	77	879	94.67	71.43	70.46	62.36
Dandy	83	893	86.75	76.67	70.36	65.40
Majestic	89	995	92.13	90.00	80.14	79.11
Pure	91	1,099	90.11	84.62	85.69	80.85
Cool	91	1,122	90.11	88.18	88.00	84.93
Modern	88	856	95.51	76.67	72.68	56.92
Average	1,276	15,000	89.74	85.31	80.69	71.00

Table 1 summarizes the performance of the proposed method in annotating the Photo.net and Yahoo! image DBs, respectively. When comparing the results for DBs, the proposed method showed a better performance of Photo.net DB than that of Yahoo! DB, which was likely caused by a difference in the quality of the images between the two DBs.

When comparing the images in the respective DBs as regards their beauty or artistic merit, the images from the former were rated higher than those from the latter. This difference clearly influenced the results of the human affective judgments, which the variances for the former were much smaller than those for the latter; on average, the values were 1.23 and 3.33, respectively. This difference resulted in greater human ambiguity in the results, thereby decreasing the accuracy of the PAM. Nonetheless, despite the difference in performance for the DBs, the PAM produced on average a recall of 85.22% and precision of 78.16%.

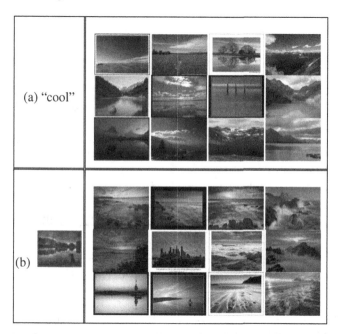

Fig. 2. Top-12 ranked results from our system: (a) for text query "cool" and (b) for query by example image

Top-12 ranked results of our retrieval system for respective query type are shown in Figure 2.

In practice, the important issue in image retrieval is to filter images according to the highest relevant images for a given query. Therefore, two experiments are used to evaluate the effectiveness of the proposed system.

The first experiment was designed to examine a conservative version of relevance for the ranking results. For this experiment, the top 20 images from the proposed system were compared with the top 20 images from the ground-truth. Figure 3 shows the number of intersections between them for 15 affective classes for each dataset.

As shown in Figure 3, the proposed system produced on average a relevance accuracy of 63% and 62% for the two DBs, respectively. In the case of annotation, a significant difference was observed between the results for the two DBs, whereas for retrieval only a slight difference was observed between the two DBs. This was because the goal of retrieval, unlike annotation, is not to order the full set of images, but only to select the best ones to show. That is, the proposed system provided generally satisfactory search results for the user.

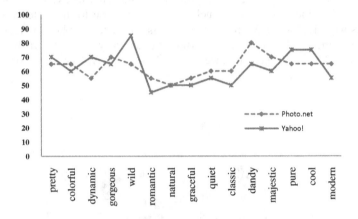

Fig. 3. Relevance graphs at top twenty ranked images

Table 2. Irrelevant images for two DBs

	Photo.net DB	Yahoo! DB
Top-3	0.00	0.13
Top-5	0.20	0.93
Top-10	0.93	2.73
Top-20	3.13	5.93

Meanwhile, minimizing the number of irrelevant images is also important for image retrieval [7]. For this, the second experiment to evaluate the performance of the proposed system is performed by counting the number of irrelevant images in the top 20, top 10, top 5, and top 3 ranked images for each DB. Table 2 shows the results. Among the top 20 images, 3.13 and 5.93 irrelevant results were produced on average for each DB, respectively. However, when looking at the top 3 images, the number of irrelevant images dropped to 0.0 and 0.13, respectively. Although there were some variances in the number of images shown, the proposed retrieval system achieved a good performance.

Consequently, the experimental results demonstrated that the proposed system can be successfully used in affect-based annotation and retrieval of photographic images.

5 Conclusion

In this paper, we suggested retrieving landscape images using human affects. For this, we presented the PAM to automatically predict various human affects from image and used it as annotation process in our retrieval system. Our system was tested on 16,276 online landscape images with the affective judgments of 52 users. Then, it produced the performance of 85.22% in annotation and that of 62.5% in retrieval. To improve the accuracy of the system, we intend to explore about the incorporation of other features such as textures or shapes in future work.

References

1. Datta, R., Joshi, D., Li, J., Wang, J.Z.: Studying Aesthetics in Photographic Images Using a Computational Approach. In: Leonardis, A., Bischof, H., Pinz, A. (eds.) ECCV 2006. LNCS, vol. 3953, pp. 288–301. Springer, Heidelberg (2006)
2. Kobayashi, S.: Color Image Scale. Publishing of Kodansha (1991)
3. Wei, K., He, B., Zhang, T., He, W.: Image emotional classification based on color semantic description. In: Tang, C., Ling, C.X., Zhou, X., Cercone, N.J., Li, X. (eds.) ADMA 2008. LNCS (LNAI), vol. 5139, pp. 485–491. Springer, Heidelberg (2008)
4. Shin, Y., Kim, E.Y., Kim, Y.: Automatic textile image annotation by prediction emotional concepts from visual features. Image and Vision Computing 28, 526–537 (2010)
5. Comaniciu, D., Meer, P.: Mean shift: A robust approach toward feature space analysis. IEEE Trans. Pattern Anal. Mach. Intell. 24(5), 603–619 (2002)
6. Li, J., Wang, J.Z.: Real-time computerized annotation of pictures. IEEE Trans. Pattern Anal. Mach. Intell. 30(6), 985–1002 (2008)
7. Jing, Y., Baluja, S.: Pagerank for product image search. In: ACM WWW 2008, pp. 307–316 (2008)

A Study on Combinative Value Creation in Songs Selection

Hiroko Shoji[1], Jun Okawa[2], Ken Kaji[2], and Ogino Akihiro[3]

[1] Chuo University, Japan
[2] Graduate School, Chuo University, Japan
[3] Kyoto Sangyo University, Japan
hiroko@indsys.chuo-u.ac.jp,
kxexn69@gmail.com,
ogino@cc.kyoto-su.ac.jp

Abstract. Recently, advances in information and communications technology have allowed us to easily download our favorite songs from the Internet. A song in general is more often played in sequence with other various ones than listened separately. The evolution of devices, however, has caused an increased number of portable songs and thus frequent difficulties in nicely combining multiple songs from a flood of songs to make a satisfactory playlist. There are many existing research works on songs search and retrieval, such as a songs each system using affective words and a songs recommendation system in consideration for the user's preference. These existing researches, however, are intended for "selecting a single song suited to the user's image", and never takes into consideration a combination of multiple songs. Therefore, it is difficult that existing systems automatically generate a desired playlist.

Keywords: combination value, playlist, recommendation, onomatopoeia.

1 Introduction

1.1 Combination Value

In the modern information society, we can access information with little to no difficulty. What is difficult, however, is creating beneficial combinations from this vast quantity of available information. Constantly deliberating about the most advantageous combinations would involve strenuous and time-consuming effort. To take a simple example of this kind of combination, every day we combine, out of the clothes we own, a set of clothing to wear. Clothes are not typically worn independently; they are combined with other clothes or accessories such as shoes or bags. The overall value of a set of clothes depends on how well they are coordinated, in addition to the individual value of each item of clothing. Likewise, the overall value of information changes depending on how the individual pieces of information are combined. Combination value is defined as the difference (should there be one) between the sum of

M. Kurosu (Ed.): Human-Computer Interaction, Part V, HCII 2013, LNCS 8008, pp. 372–380, 2013.

the individual values of all pieces of information and the overall value of the information once combined. Combination value may be expressed, using "individual value" and "overall value", as follows:

$$\text{Overall value} = \Sigma \text{ individual value} \pm \text{combination value} \qquad (1)$$

In our study, we focused on combination value, as expressed by Equation 1, and narrowed our scope to the subject of music. People do not typically listen to a single piece of music, but rather to a combination of musical pieces, demonstrating their interest in the combination value of the respective pieces. With respect to music, we intend to enhance overall value by increasing the combination value, and propose, in this paper, a method for doing so.

1.2   Background and Purpose of the Study

A playlist created by a user of music technology tends to reflect that user's mood, as it commonly takes them a significant amount of time to create a combination of music to their liking. In other words, such a playlist typically has a high combination value. However, given that advances in technology have increased the number of songs available to users, carefully selecting playlists that reflect user mood from such a large selection of music may involve significantly more effort on the user's behalf. It might be wiser to use the random play function in order to save time. However, the random play function simply plays music from a designated selection of songs, and does not consider the positional relationship of these songs, or the music's overall flow or concept. One of the tools available on iTunes for creating playlists is an automatic playlist generating function (Smart Playlist). However, this function recommends musical pieces from the user's play record, typically representing a large selection of songs, and can produce a playlist that is not in accord with the user's mood. Studies have already been conducted on music recommendation approaches that consider each piece of music individually: there are music retrieval systems that use sensitive indicators, and music recommendation systems that consider user preferences. However, these existing systems do not consider song combinations, making it difficult for them to automatically generate playlists that meet user expectations. Therefore, our study analyzes how to combine songs based on user mood, and proposes a method for generating playlists that eventually become more similar to those manually produced by the user.

2   How Users Combine Songs

2.1   Combination Analysis Experiment

Questionnaire on Music Impressions. As a preliminary step in the experiment, a questionnaire survey on music impressions was conducted. A set of questions was given to 14 subjects, asking them to answer, using a checklist, how they felt about each piece of music. In our study, onomatopoeias were used as "impression words": words that convey the impression a subject felt or received when they listened to a particular piece of music. Music is made of sound, and thus we believe that onomatopoeias were the most appropriate way to express how a person felt about the music. A total of 52 onomatopoeias were used, including both the onomatopoeias used by Ogino et al. in their recommendation system[1], and onomatopoeias selected from a Japanese onomatopoeia dictionary[2]. The music selected for our study consisted of 135 musical pieces from various genres. One or more impression words were assigned to each of the 135 pieces of music, based on the results of the questionnaire survey.

2.2   Music Combination Experiment

An experiment was conducted to examine what combinations users create. This experiment involved 10 subjects who had each listened to the 135 pieces used in the questionnaire, and was conducted according to the following procedure:

1. Each subject selected 10 or more songs from those evaluated with the word "shimi-jimi (feelingly)" to describe the listeners' impression of the music.
2. Each subject prepared a playlist of 10 songs of their own choice from the songs selected in Step 1.
3. Each subject repeated Steps 1 and 2 for songs evaluated with "wai-wai (excitedly or uproariously)"

Subjects were allowed to listen to the selected songs (hooks only) as many times as they liked during the course of the experiment.

2.3   Experimental Results

Our results indicate that music combinations can largely be classified into three patterns: conceptual, dispersed and intermediate. Each pattern is explained below:

Conceptual Pattern. Some of the songs that subjects included in combinations classified as having a conceptual pattern are shown in Table 1.

This table shows that with respect to songs selected from the group of songs originally evaluated with the word "shimi-jimi", subjects tended to combine songs described by the three onomatopoeias shown in the table. Likewise, a similar conceptual pattern was shown with respect to combinations created from songs that were originally evaluated with the word "wai-wai". Therefore, we conclude that songs combined by users emphasizing conceptual music combinations tend to be uniform in terms of the impressions generated by the respective songs, with user mood typically reflected throughout the entire combination.

Dispersed Pattern. Dispersed pattern combinations are composed of songs that tend to generate a variety of often contradictory impressions, in contrast to the more uniform impressions generated by songs in the conceptual pattern combinations. Table 2 shows examples of songs included in combinations classified as having a dispersed pattern, and that were selected from the group of songs originally evaluated with the word "shimi-jimi".

This table shows that songs frequently evaluated as "wai-wai (excitingly)" and "yuttari (leisurely)" tended to be included in combinations classified as having a dispersed pattern. We infer that a combination of diverse musical pieces allows the user to remain interested, and combinations following the dispersed pattern typically have a relatively balanced mixture of songs associated with a variety of impressions.

Table 1. Conceptual pattern combinations

Song	Shimi-jimi (feelingly)	Jin-wari (gradually)	Shin-miri (quietly)
Itsuka Dokokade (Sometime Somewhere)	4	4	5
Mirai Yosozu II (Future Forecast Diagram II) – 2007 Version	3	3	3
I for You	2	3	3
Shiawasena Ketsumatsu (Happy Ending)	3	3	3
Ienaiyo (Can't Say)	5	6	3
Sobaniiruyo (I'll Be There for You)	3	2	5
Single Bed	4	2	4
Mou Koinante Shinai (Won't Fall in Love Again)	5	4	7
Invitation	1	1	1
Heya to Y shatsu to watashi (Room, Y-shirt and Me)	1	3	2

Table 2. Dispersed pattern combinations

Song	Shimi-jimi (feelingly)	Yuttari (leisurely)	Wai-wai (excitingly)
Love Somebody	1	1	3
KNOCKIN' ON YOUR DOOR	1	0	2
Booty Music	1	2	4
Bye Bye (So So Def Remix)	1	3	0
Lay Up	2	3	0
One Last Try	2	1	1
You Gotta Be	2	2	1
If I Ain't Got You	4	4	0
Take Me Back	2	2	0
Take A Bow	5	3	0

Intermediate Pattern. Combinations classified as having this pattern principally consist of pieces that conform to the combination's core concept impression; however, a smaller number of pieces associated with a wider variety of impressions are also included. Of the song combinations generated during this experiment, the largest number fell into the intermediate pattern category. Table 3 shows examples of songs included in intermediate pattern combinations that were selected from the group of songs originally evaluated with the word "shimi-jimi". The table shows that songs associated with onomatopoeias similar to "shimi-jimi (feelingly)" and "shin-miri (quietly)" were included in the combinations, and that songs associated with quite different onomatopoeias like "nori-nori (willingly)" were selected as well. The choice of songs associated with impressions dissimilar to "shimi-jimi" varied depending on the subject. It was observed that subjects frequently included in their combinations songs described with "wai-wai" or "nori-nori" (which represent impressions nearly opposite to that represented by "shimi-jimi"); however, the balance between uniformity and diversity varied from person to person. Therefore, any method for reproducing intermediate pattern combinations will need to be able to appropriately balance these two factors before it can be included in our music combination system.

Table 3. Intermediate pattern combinations

Song	Shimi-jimi (feelingly)	Shin-miri (quietly)	Nori-nori (willingly)
1_3 No Junjo Na Kanjo (Pure Sentiment of 1_3)	1	1	3
Fragile	4	2	1
M	5	5	0
Mirai Yosozu II (Future Forecast Diagram II) - Version 2007	3	3	0
Zoo	3	4	0
Mou Koinante Shinai (Won't Fall in Love Again)	5	7	0
Watarase Bashi (Watarase Bridge)	4	3	0
Single Bed	4	4	0
Ienaiyo (Can't Say)	5	3	0
Aitai (Wanna See You)	5	5	0

3   Implementation of the Proposed Methods

3.1   Definition of Distance

In order to analyze the combinations created by the subjects, and simulate our proposed methods, each musical piece is assumed to be a point set of the 52nd dimension. The distance between songs is calculated to determine the degree of similarity between them. Euclidean distance is used for distance in this calculation. The distance between a piece i and another piece j, L_{ij} ($i = 1,2,3\ldots135$, $j = 1,2,3\ldots135$), is calculated as follows (using the evaluation x_{ik} of the onomatopoeia k of the song i, and the evaluation x_{jk} of the onomatopoeia k of the song j):

$$L_{ij} = \sqrt{\sum_{k=1}^{52}(x_{ik} - x_{jk})^2} \tag{2}$$

3.2   In Our Study, It Is Understood That Songs Closer to Each Other Have Greater Similarity. Proposal of a Playlist Generation Method

Proposed Method. In Section 2 we reported the details of an experiment on how users combine music. The results revealed that there were users who created combinations with a strong, conceptual pattern, and other users who created combinations of slight diversity with weak conceptual patterns. Section 3 proposes three methods (using the concept of distance defined in 3.1) that allow us to obtain results similar to the combinations created by the subjects. Each method aims at generating combinations that reproduce one of the three patterns identified by the experiment.

- Minimum distance method: This method selects songs closest in distance to those initially selected by users. It is expected to produce combinations similar to the conceptual variety.
- Maximum distance method: This method selects songs farthest in distance from those initially selected by users, and is conceptually opposite to the minimum distance method. It is expected to produce combinations similar to the dispersed variety.
- Hub selection method + minimum distance method (Hub selection method): In this combination of methods, two or more hubs are selected, and then a few songs are selected around the hubs, based on the minimum distance method. In our study, a song of higher centrality is called a "hub". Matrix A (obtained by deleting all songs, from the 135, which have a mutual distance of 8.5 or less) is used to calculate information centrality (C_{inf}). Songs of greater centrality have an increased probability of selection. Information centrality is a centrality indicator that considers the shortest route between vertices contained in the network or the length of each route.

From matrix A = a_{ij}, we then determine matrix B = b_{ij} as diagonal component b_{ii}, which is equal to the sum of the values of sides connected to vertex i plus one:

$$b_{ij} = \begin{cases} 1 & \text{when vertex i and vertex j do not lie side by side} \\ 1 - a_{ij} & \text{when vertex i and vertex j lie side by side} \end{cases}$$

Then, using Matrix B, Cinf is calculated as follows:

$$C_{inf} = {}^{n}/(nT + S_T - 2R) \tag{3}$$

where, the inverse of matrix B having been determined, the vector of the diagonal component of the obtained inverse matrix is T, the trace of the inverse matrix is S_T, and the row sum of any given inverse matrix is $R[3]$. In this study, three hubs (representing three songs) are chosen, and then two more pieces close to the hubs are selected. The hub selection method is expected to produce combinations similar to the intermediate variety.

3.3 Comparative Evaluation

This paragraph details a comparative evaluation of proposed methods, using the results of the combinations created by the subjects. The differences in average distance between the playlists created by the subjects and the corresponding playlists generated by the proposed methods are shown in Table 4. With respect to the minimum distance method, the minimum average distance was recorded for subject G. With the songs initially selected by subject G as respective "centers", the playlist of G, and the corresponding playlist produced by the minimum distance method, are presented graphically in Fig. 1.

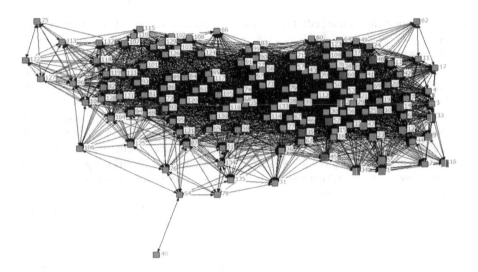

Fig. 1. Comparison between subject G's playlist and the corresponding playlist generated by the minimum distance method

In this figure, red nodes represent the combination created by G, while green nodes represent songs selected by the minimum distance method. It can be seen that the selected songs are typically close to those initially selected by G.

Table 4. Differences in average distance between the subjects' results and those of our proposed methods

Subject	Minimum distance method	Maximum distance method	Hub selection method
A	-2.322	3.992	-1.776
B	-1.97	2.325	0.239
C	-2.189	2.681	-0.702
D	-2.03	3.048	-1.67
E	-5.482	0.623	-5.102
F	-2.844	2.025	-1.069
G	-0.94	3.99	1.255
H	-1.401	1.757	-1.454
I	-1.794	3.266	-0.76
J	-2.203	2.875	-1.852

Table 4 shows that the maximum distance method produced results closest to those of subject E. In Fig. 2, songs included in subject E's playlist are represented in red, while those selected by the maximum distance method are represented in green. Fig. 2 also shows that the combination of songs selected by the subject is similar to the combination produced by the proposed maximum distance method. It is therefore reasonable to conclude that the proposed minimum and maximum distance methods are able to successfully reproduce the conceptual pattern and dispersed pattern combinations, respectively.

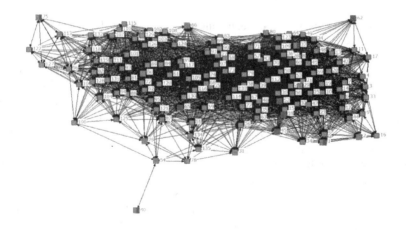

Fig. 2. Comparison between subject E's playlist and the corresponding playlist generated by the maximum distance method

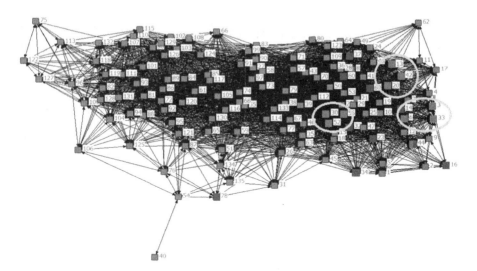

Fig. 3. Comparison between subject B's playlist and the corresponding playlist generated by the hub selection method

The hub selection method produced results similar to those of subjects B, C and I. Fig. 3 shows the songs selected by the hub selection method and those selected by subject B, in different colors. The circled areas show the hubs and the songs selected based on the hubs. It remains unclear which hubs will be selected by the hub selection method, and it can be reasonably concluded that differences in hub selection will lead to varying results.

4   Summary and Future Tasks

This study calculated distances between songs based on an evaluation method using impression words, and proposed a playlist generation method based on song distances. The experimental results could demonstrate that music combinations created by users could be categorized as conceptual pattern combinations (combinations of songs that generate similar impressions) and dispersed pattern combinations (combinations of songs that generate different impressions). Our study attempted to develop a means of creating music combinations with similar patterns using the concept of distance between musical pieces. It is concluded that the proposed minimum distance and maximum distance methods are capable of reproducing the conceptual pattern and diverse pattern combinations. On the other hand, although the hub selection method produced results similar to those of some subjects, this method is designed to stochastically select songs with high centrality, and there is a lack of clarity with respect to the number of hubs, and/or songs close to the hubs, the method selects. In addition, it is necessary to conduct simulations when songs farther from the hubs are selected. Therefore, a future task is to improve the hub selection method, such that it can flexibly reproduce combinations with patterns similar to those generated by users.

References

1. Yoshida, N., Ogino, M.: Proposal of a Music Selection Support System to Conform to User Image. Collection of Papers Prepared for the Japan Society of Kansei Engineering for 13th Convention vol. 13 (2011)
2. Ono, M.: Onomatopoeias and Mimetic Words 4500 Japanese Onomatopoeia Dictionary. Shogakukan Publishing (2007)
3. Suzuki, T.: Network Analysis. Data Science Learned with R, vol. 8. Kyoritsu Shuppan Co., Ltd (2008)
4. Okawa, H., Kaji, M., Nishimura, T., Shoji, Y., Ogino, A.: Experiment and Analysis Concerning the Creation of Music Combination Value. Collection of Papers Prepared for the 14th Convention of the Japan Society of Kansei Engineering (2012)

The Influence of Context Knowledge for Multi-modal Affective Annotation

Ingo Siegert, Ronald Böck, and Andreas Wendemuth

Cognitive Systems Group, Otto von Guericke University Magdeburg, Germany
ingo.siegert@ovgu.de

Abstract. To provide successful human-computer interaction, automatic emotion recognition from speech experienced greater attention, also increasing the demand for valid data material. Additionally, the difficulty to find appropriate labels is increasing.

Therefore, labels, which are manageable by evaluators and cover nearly all occurring emotions, have to be found. An important question is how context influences the annotators' decisions. In this paper, we present our investigations of emotional affective labelling on natural multi-modal data investigating different contextual aspects. We will explore different types of contextual information and their influence on the annotation process.

In this paper we investigate two specific contextual factors, observable channels and knowledge about the interaction course. We discover, that the knowledge about the previous interaction course is needed to assess the affective state, but that the presence of acoustic and video channel can partially replace the lack of discourse knowledge.

Keywords: emotion comparison, affective state, labelling, context influence.

1 Introduction

In future, technical systems should provide more human-like interaction abilities. Therefore, these systems have to be adaptable to the user's individual skills, preferences and current emotional states [20]. To enable such systems to determine a user's affective state, the recognition needs to rely on all signals humans use in the interaction, like speech, facial expressions and gestures. To provide successful human computer interaction, automatic emotion recognition from speech experienced greater attention, also increasing the demand for valid data material.

The recognition of the user's affective state is still a challenging task. Many years the focus was set on acted data e.g. [3], also due the lack of available datasets. While, in acted data, the label (ground-truth) is clearly instructed to the actor, resulting in clear and high expressive emotional recordings, see [21] the application of classifiers trained on acted data material within a realistic or naturalistic human computer interaction shows that these databases do not

M. Kurosu (Ed.): Human-Computer Interaction, Part V, HCII 2013, LNCS 8008, pp. 381–390, 2013.
© Springer-Verlag Berlin Heidelberg 2013

include the variety of natural occurring affective states [9]. So, within research community, the focus changed from acted emotions to more realistic emotional expressions like [7,8,18], because emotion can induce changes in speech that cannot be controlled by the speaker. This recordings mostly have a lower expressiveness of the affective states and raises the problem of a reliable ground-truth generation [6]. So the difficulty, to find appropriate labels is increasing. Whereas in acted emotional data, the *label* is clearly instructed and can assessed via perception tests in realistic recordings the expressions are uncontrolled and not as obvious. The generation of a *label* is persuaded by an annotation process where a large number of annotators is used to assess the observed affective state, by choosing a suitable label.

Therefore, labels, which are manageable by evaluators and cover nearly all occurring emotions, have to be found. But besides the utilized label also the design of the labelling process is important. An important question is how context influences the annotators' decisions [4]. The authors in [5] investigated the influence of the context onto the perception of anger. They argue that traditional associations between tones and attitudes are misleading and that the contextual factor can neutralise the anger perception. The authors perform two studies, where they could show, that neutral uttered wh-words are perceived as anger, when heard without surrounding context. A further study is performed in [13], where the authors investigate the role of channel information onto aggression detection utilizing three different settings: audio only, video only and audio plus video. They stated, that for 46% of their material the annotation of all three sets differs for the same samples.

This study supports our hypothesis, that the context plays an important role within affect recognition. In our study, we want to combine both investigations of surrounding information and channel influence. Furthermore, we investigated both context influences within a realistic human-computer interaction.

The remainder of the paper is structured as follows: In Section 2 the utilized dataset is described in detail. Afterwards,we introduce our research methods in Section 3. The results of our study are presented and discussed in Section 4. Finally, in Section 5, an outlook for further research is given.

2   Dataset

The conducted study utilizes the LAST MINUTE corpus [15]. It contains multi-modal recordings of 130 native German subjects collected in a Wizard-of-Oz experiment. The technical recordings and first classification results are described in detail in [10]. As background information the subjects are told to test a new natural language communication interface. The setup revolves around a journey to an unknown place "Waiuku" that they have won. Using voice commands the subjects have to prepare the journey, assembling the baggage and select clothing. The task is designed to generate affective enriched material from a naturalistic human computer interaction [17]. First results on multi-modal affect recognition can be found in [12,14]. The utilized TTS uses an artificial and mechanical voice, providing a lot of explanation, which leads to long monologues by the system.

During the dialogue critical events provoking negative emotions and could may leading to a break off of the dialogue, are induced. We focus on three key events, where the user should be set in a certain condition: Baseline (BL), Challenge (CH), and Waiuku (WA). All three events are designed in such a way, that the system gives a specific information whereupon the user shall show a specific reaction. At the BL event, occurring after 5-10 minutes, the test person has been adapted to the experimental situation and the first excitement is gone. The person starts packing the luggage. The system only confirms the action requests. The CH event happens when the system creates mental stress by suddenly claiming to reach a previously luggage limit. This event arises after 15-20 minutes of the experiment. In the WA event a second strategy change has to be performed, when claiming a different voyage destination. It is winter instead of summer at the destination. At this point the subject notice, that it has to re-arrange its complete baggage. This event occurs at about 20-25 minutes of the experiment. Neither we can be sure about the real stress factor for the particular subject, nor can we assure the real duration of any higher stress level.

A the expected time effort for labelling is up to four times higher than the material to be labelled, we selected 4 subjects from the whole corpus. This results in a subset of approx 23 minutes. Furthermore, we splitted the events into separate utterances, as it was already mentioned, that the surrounding words are important for a proper assessment [5]. So, we end up with 135 snippets with a length of 3 seconds to 50 seconds. The mean is 11 seconds. An overview about the number of snippets for each event and subject is given in Table 1.

Table 1. Overview of utilized snippets and their distribution for the selected subjects and experimental events

subject	BL	CH	WA
20101006aFM	16	9	9
20101117bMT	15	5	8
20101206bEG	19	7	9
20110126aFW	21	8	8
Total	71	29	34

3   Study Design

As stated in the introduction, we rely on the related work of [4,5,13], all claiming, that surrounding information and observable channels are important to receive a useful annotation. To verify our hypotheses, we design different labelling tasks, where we varied either the different observable channels or the interaction course. Hereby we will support the following hypotheses: labels can only be gathered, when both acoustic and visual information are present, information about interaction development supports the labelling process. Preliminary results onto the influence of system responses where presented in [17].

3.1   Define Dependent Variables

Based on this, we can now define the following two dependent variables and their expressions. The observable channel consists of the values "audio only", "video only" and "audio plus video". The interaction course can be either random or ordered. Therefore, we conducted six different sets, where both variables assume all its defined values. The resulting sets are presented in Table 2.

Table 2. Overview of the labelling sets, generated by the two dependent variables with their expressions

interaction course	observable channels	label set
random	audio only	Set 1
	video only	Set 2
	audio + video	Set 3
ordered	audio only	Set 4
	video only	Set 5
	audio + video	Set 6

3.2   Design the Annotation Process

The design of the annotation process is similar to [19]. Vidrascu proposes several phases to decide the list of labels, annotation scheme, and segment length and afterwards to start with the actual annotation process. To define the segment length, we rely on experience of [1], where an assessment based on a speech chunks is proposed. For our utilized database, these chunks are identical to the subjects utterances, as they are quite short in our material. A shorter length of single words will mislead the labellers, as investigated by [5]. A longer segment length including a complete dialogue turn, consisting of several human utterances and system responses, can be composed of several affective states.

To get the proper affective labels we utilized results of our study presented in [16]. There, we investigated the differences between three labelling methods and the observed emotional labels for a similar human-computer interaction utilizing the NIMITEK Corpus, see [11]. Therefore, we investigated the used list of labels and the annotation method usable for labelling human-computer interaction. The following affects proved to be useful, see Investigation I Table 3. As the NIMITEK corpus was designed to provoke negative emotions [11], whereas the LAST MINUTE corpus tries to investigate possible dialogue break offs [15], we conducted an experiment to gather more suitable emotional labels for the domain of the LAST MINUTE corpus. Therefore, we presented the utilized snippets to six labellers, all of them where psychologist students, utilizing the labels from [16], with the explicit task, to add emotional terms, they need to describe the affective state of the subject. This investigation results in additional labels, see Investigation II in Table 3. The affective labels found in both investigations and are combined into one word list, for the persuaded study, see Table 3 for an overview. Additionally, the labellers could asses (o) *no emotion*, if they assess, that no emotion was observed and we gave them the opportunity to leave a comment.

Table 3. Overview of the utilized affective word list

Investigation I (see [16])	
(a) sadness (b) contempt (c) helplessness (d) interest (e) hope (f) relief	
(g) joy (h) surprise (i) confusion (j) anger	
Investigation II	Additional
(k) shame (l) stress (m) concentration (n) impatientness	(o) no emotion

We utilized 10 labellers, all of them with psychological background. To support the labellers during their annotation process, we used a variant of ikannotate [2]. The labellers could see or hear the actual snippet and could chose one or several words from the presented word list.The order of presented snippets could not be influenced by the labellers. The programme forces them watch the complete snippet and assess it afterwards, a repeated view of the actual one is possible.

4   Results

We evaluated our results on the basis of each set, as described in Table 2. To investigate the influence of the defined variables, we compared the assessed affective states resulting from a majority voting. Only the assesment where five or more labellers agreed on the same affective state, is used as a valid label. In the case, where only five labellers agreed, than the remaining labellers should not agree on the same other affect.

4.1   Influence of Present Channels

Comparing the influence of the available channels in Fig. 1, we notice that the number of majority votes for *concentration* (m) is nearly assesed most for all conditions. We observe an increasing number of votes from the audio-only (1) over the video-only (2) to the audio plus video set (3). The same behavior can be observed for *joy* (g). The opposite effect is noticed for *impatientness* (n). The affective state *surprise* (h) and *relief* (f) gets a rising number of votes comparing

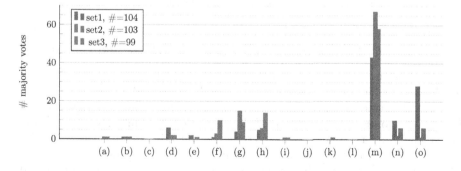

Fig. 1. Number of majority votes for different channel informations

the audio only and video only set, but when both channels are present, the number of votes is decreasing. The affective state *anger* (j) is labelled sufficient only, when both channels are present. Remarkable is the varying number of *no emotion* (o) votes, for Set 1 we have the highest number of 28 votes. This effect was expected, as especially in the WA event we used some snippets, where the subject did not talk. For Set 2, where only the video channel was used, only one item was labelled showing no emotion. Finally, Set 3 having both channel informations, we observe again 5 items voted with *no emotion* (o).

4.2   Influence of Interaction Course

The next aspect, we want to analyse, is the influence of the knowledge about the interaction course. The resulting numbers of the majority votes are given in Fig. 2. It can be noticed, that the distribution of majority votes does not differ much. In Set 6, utilizing the experimental data in an ordered way, whereas Set 3 uses a random order. Additionally, we count the total number of majority votes, reached for each set. The results are given in Fig. 2. Here it can be noticed, that the number increased from 104 to 131 items, where a majority vote could be drawn. We call this a reduction of variety. We attribute this to the influence of the prior knowledge. The labellers know, which affective state they observed before and therefore take that decision into account.

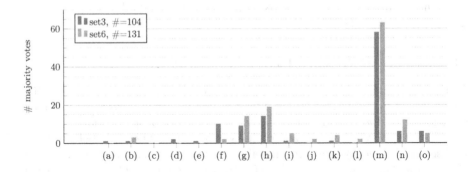

Fig. 2. Number of majority votes for ordered/random interaction courses

4.3   Influence of Both Present Channels and Interaction Course

When comparing the influence on the interaction course, ordered or random, together with a limited channel, we get the results presented in Fig. 3 and Fig. 4. Here we can notice, that the influence, the ordered presentation has, is stronger than for the contrary presentation of channels. Resulting in a reduction of the chosen labels and a shaping of selected ones. The affective states *surprise* (h), *joy* (g), *concentration* (m), *impatientness* (n) and *no emotion* (o) are labelled most. Whereas especially the number of *no emotion*-votes decreases for the audio onyl set, when using the ordered presentation. Whereas, we can notice an increased

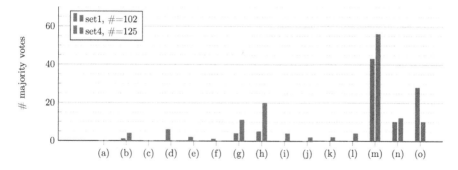

Fig. 3. Number of majority votes for ordered/random interaction courses, while using only the audio channel

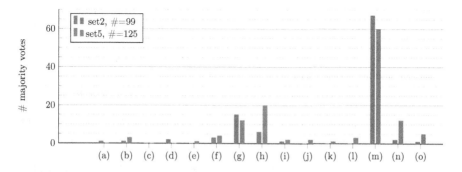

Fig. 4. Number of majority votes for ordered/random interaction courses, while using only the video channel

number of majority votes for the affects *concentration* (m) and *joy* (g) comparing the audio only random/ordered sets, we get the opposite result comparing same the video-only sets. But for the affects *surprise* and *impatientness* (n), we notice the same increasing votes. Also the chosen affects are identical for Set 4 and Set 5 despite *relief* (f), which is labelled only for the video only set (Set 5). Furthermore the effect of an increasing total number of given majority votes can be noticed, the number increased from 102 to 125 for the audio only sets (Set 1, Set 4) and from 99 to 125 for the video only sets (Set 2, Set 5). This is similar to the comparison of Set 3 and Set 6.

4.4   Notes on the Experimental Events

Comparing the labelled affective states regarding the described experimental events, it can be noticed that the affective labels *surprise* (h), *joy* (g), *confusion* (i), and *concentration* (m) are assessed within all investigated experimental events. From the pure experimental design, the presence of *surprise* (h), *confusion* (m), and *joy* (g) were not expected for BL. The distribution of *joy* and

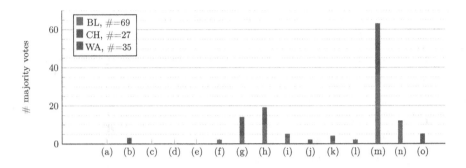

Fig. 5. Number of majority votes for the utilized experimental events

concentration are in conformity with the experimental design, as CH requires *concentration* whereas WA can create *joy*. The less votes for the affect *surprise* (h) does not match the expected assessment.

Affective states only assigned to BL are *relief* (f), *impatientness* (n) and *stress* (i), with *impatientness* assessed most. This confirms the expected reaction, that should be induced due the mechanical voice and the dominance of the system monologues. The occurence of *contempt* (b) and *anger* (j) during CH is also according the the design, but very rare. The assessment of the WA event snippets with *no emotion* (o) and *shame* (k) indicates, that the intended effect was achieved. The monologue about the travel information should not provoke any affect and the information about the wrong assumed destination provokes shame, as the subject could have know the expected destination.

5   Conclusion

We focus on two terms with different conditions, the a) role of available channel information and b) knowledge about the interaction course. These effects where investigated utilizing affective word lists and 10 labellers.

Evaluating the channel information, we can state, that the availability of both channels, audio and video is important. In contrast to [13], we could not get such a confusion between the sets presenting audio only, video only and both channels. This could be due the fact, that our material consist of material where the face is recorded in a frontal view with a very good illumination, so that the labeller could always assess the facial expressions very good.

Evaluating the influence of the interaction course, we can state, that this is important, too. But the differences between the majority votes of the ordered and unordered set, where both channels are present, is quite small. Nevertheless, presenting the interaction in an ordered way can support the labellers in situations, where one channel is partly missing. This is supported, by the comparison of Set 1 and Set 4 or Set 2 and Set 5, respectively. Here the resulting majority votes of the ordered sets is similar to Set 6, presenting both channels in an ordered way.

These results show that the affective states of observed subjects within a naturalistic human computer interaction are assessable, but consist mostly of states with low expressiveness and indicating affects, pointing on a potential problematic dialogue are very rare.

A very interesting additional investigation, we want to follow-up with, is the investigation of the detailed changes of affective state for the different experimental conditions. This could help to get a deeper understanding which affective states are often confused and which influence specific contextual informations have.

References

1. Batliner, A., Seppi, D., Steidl, S., Schuller, B.: Segmenting into adequate units for automatic recognition of emotion-related episodes: A speech-based approach. In: Advances in Human-Computer Interaction 2010 (2010)
2. Böck, R., Siegert, I., Vlasenko, B., Wendemuth, A., Haase, M., Lange, J.: A processing tool for emotionally coloured speech. In: Proc. of the 2011 IEEE International Conference on Multimedia & Expo., Barcelona, Spain (July 11-15, 2011)
3. Burkhardt, F., Paeschke, A., Rolfes, M., Sendlmeier, W., Weiss, B.: A database of german emotional speech. In: Proc. of Interspeech (2005)
4. Callejas, Z., López-Cózar, R.: Influence of contextual information in emotion annotation for spoken dialogue systems. Speech Com. 50, 416–433 (2008)
5. Cauldwell, R.T.: Where did the anger go? the role of context in interpreting emotion in speech. In: Proc. of the ISCA Workshop on Speech and Emotion, Newcastle, Northern Ireland, UK, pp. 127–131 (September 2000)
6. Cowie, R., Cornelius, R.R.: Describing the emotional states that are expressed in speech. Speech Commun. 40(1-2), 5–32 (2003)
7. Douglas-Cowie, E., Campbell, N., Cowie, R., Roach, P.: Emotional speech: towards a new generation of databases. Speech Com. Special Issue Speech and Emotion 40, 33–60 (2003)
8. Douglas-Cowie, E., et al.: The HUMAINE database: Addressing the collection and annotation of naturalistic and induced emotional data. In: Paiva, A.C.R., Prada, R., Picard, R.W. (eds.) ACII 2007. LNCS, vol. 4738, pp. 488–500. Springer, Heidelberg (2007)
9. Douglas-Cowie, E., Devillers, L., Martin, J.C., Cowie, R., Savvidou, S., Abrilian, S., Cox, C.: Multimodal databases of everyday emotion: facing up to complexity. In: European Conference on Speech Com. and Technology, pp. 813–816 (2005)
10. Frommer, J., Michaelis, B., Rösner, D., Wendemuth, A., Friesen, R., Haase, M., Kunze, M., Andrich, R., Lange, J., Panning, A., Siegert, I.: Towards Emotion and Affect Detection in the Multimodal LAST MINUTE Corpus. In: Proc. of the Eight International Conference on Language Resources and Evaluation (LREC 2012), ELRA, Istanbul, Turkey (May 2012)
11. Gnjatović, M., Rösner, D.: On the role of the NIMITEK corpus in developing an emotion adaptive spoken dialogue system. In: Proc. of the Language Resources and Evaluation Conference (LREC 2008), Marrakech, Morocco (2008)
12. Krell, G., Glodek, M., Panning, A., Siegert, I., Michaelis, B., Wendemuth, A., Schwenker, F.: Fusion of Fragmentary Classifier Decisions for Affective State Recognition. In: Schwenker, F., Scherer, S., Morency, L.-P. (eds.) MPRSS 2012. LNCS, vol. 7742, pp. 116–130. Springer, Heidelberg (2013)

13. Lefter, I., Rothkrantz, L.J.M., Burghouts, G.J.: Aggression detection in speech using sensor and semantic information. In: Sojka, P., Horák, A., Kopeček, I., Pala, K. (eds.) TSD 2012. LNCS, vol. 7499, pp. 665–672. Springer, Heidelberg (2012)
14. Panning, A., Siegert, I., Al-Hamadi, A., Wendemuth, A., Rösner, D., Frommer, J., Krell, G., Michaelis, B.: Multimodal affect recognition in spontaneous hci environment. In: IEEE International Conference on Signal Processing, Communications and Computings (ICSPCC), pp. 430–435 (2012)
15. Rösner, D., Friesen, R., Otto, M., Lange, J., Haase, M., Frommer, J.: Intentionality in interacting with companion systems – an empirical approach. In: Jacko, J.A. (ed.) Human-Computer Interaction, Part III, HCII 2011. LNCS, vol. 6763, pp. 593–602. Springer, Heidelberg (2011)
16. Siegert, I., Böck, R., Philippou-Hübner, D., Vlasenko, B., Wendemuth, A.: Appropriate Emotional Labeling of Non-acted Speech Using Basic Emotions, Geneva Emotion Wheel and Self Assessment Manikins. In: Proc. of the IEEE International Conference on Multimedia and Expo., ICME 2011, Barcelona, Spain (2011)
17. Siegert, I., Böck, R., Wendemuth, A.: The influence of context knowledge for multimodal annotation on natural material. In: Proc. of the First Workshop on Multimodal Analyses Enabling Artificial Agents in Human-Machine Interaction (MA3), Santa Cruz, USA (September 2012)
18. Vaughan, B., Kosidis, S., Cullen, C., Wang, Y.: Task-based mood induction procedures for the elicitation of natural emotional responses. In: The 4th International Conference on Cybernetics and Information Technologies, Systems and Applications, Orlando, Florida (2007)
19. Vidrascu, L., Devillers, L.: Real-life emotion representation and detection in call centers data. In: Tao, J., Tan, T., Picard, R.W. (eds.) ACII 2005. LNCS, vol. 3784, pp. 739–746. Springer, Heidelberg (2005)
20. Wendemuth, A., Biundo, S.: A Companion Technology for Cognitive Technical Systems. In: Esposito, A., Esposito, A.M., Vinciarelli, A., Hoffmann, R., Müller, V.C. (eds.) COST 2102. LNCS, vol. 7403, pp. 89–103. Springer, Heidelberg (2012)
21. Zeng, Z., Pantic, M., Roisman, G.I., Huang, T.S.: A Survey of Affect Recognition Methods: Audio, Visual, and Spontaneous Expressions. IEEE Trans. on Pattern Analysis and Machine Intelligence 31, 39–58 (2009)

Generation of Facial Expression Emphasized with Cartoon Techniques Using a Cellular-Phone-Type Teleoperated Robot with a Mobile Projector

Yu Tsuruda, Maiya Hori, Hiroki Yoshimura, and Yoshio Iwai

Graduate School of Engineering, Tottori University
101 Minami 4-chome, Koyama-cho, Tottori, 680-8550 Japan

Abstract. We propose a method for generating facial expressions emphasized with cartoon techniques using a cellular-phone-type teleoperated android with a mobile projector. Elfoid is designed to transmit the speaker's presence to their communication partner using a camera and microphone, and has a soft exterior that provides the look and feel of human skin. To transmit the speaker's presence, Elfoid sends not only the voice of the speaker but also emotional information captured by the camera and microphone. Elfoid cannot, however, display facial expressions because of its compactness and a lack of sufficiently small actuator motors. In this research, facial expressions are generated using Elfoid's head-mounted mobile projector to overcome the problem. Additionally, facial expressions are emphasized using cartoon techniques: movements around the mouth and eyes are emphasized, the silhouette of the face and shapes of the eyes are varied by projection effects, and color stimuli that induce a particular emotion are added. In an experiment, representative face expressions are generated with Elfoid and emotions conveyed to users are investigated by subjective evaluation.

1 Introduction

Video-conferencing and videotelephony are used as tools for communication between people in remote areas. These systems use not only the voice of the speaker but also a video to make communication smoother. A technology called telepresence has been devised to increase the human presence by increasing the resolution of the video or presenting video to an immersive display. In terms of human presence, however, conventional telepresence systems are insufficient because they do not have dimensional information and tactile feedback of the speaker. Therefore, for communication with people in remote locations, robots that have human appearance have been developed. Some studies use a humanoid robot for transmission of human presence. In particular, teleoperated android robots, such as Geminoid F and Geminoid HI-1 [1], have appearances similar to an actual person, and were intended to transfer the presence of actual people. These humanoid robots have high degrees of freedom and can transfer the human presence. However, they are expensive and limited to a specific individual

M. Kurosu (Ed.): Human-Computer Interaction, Part V, HCII 2013, LNCS 8008, pp. 391–400, 2013.

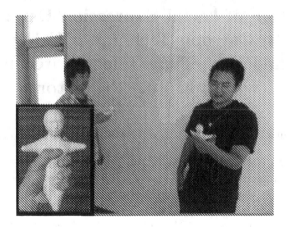

Fig. 1. Communication using Elfoid, which conveys the human presence to remote locations

target. A robot called Telenoid R1 has been developed to reduce the number of actuators and costs involved [2]. Telenoid is not limited to a specific individual target and is designed to appear and to behave as a minimalistic human at the very first glance. A person can easily recognize Telenoid as a human while the Telenoid does not appear to be gender- or age-specific. With this minimal design, Telenoid allows people to feel as if a distant acquaintance is next to them. Moreover, Telenoid's soft skin and doll-like body size make it easy to hold. However, it is difficult to carry by hand on a daily basis.

To allow daily use, a communication medium that downsizes Telenoid and has mobile-phone communication technology is now under development. The resulting Elfoid is easy to hold in the hand like a cellular phone as shown in Fig. 1. When we use such robots for communication, it is important to convey the facial expressions of a speaker and thus increase the modality of communication. In this research, facial expressions are generated using Elfoid's head-mounted mobile projector.

2   Related Work

In communication between people, it is important to convey the emotions of the speaker. There have been a considerable number of studies on basic human emotions [3–6]. Ekman et al. [6] defined basic facial expressions for anger, disgust, fear, happiness, sadness and surprise. This shows that the facial expressions of emotion are not culturally determined but are universal across all human cultures and are thus biological in origin. In communication, these emotions are important in efforts to increase the modality.

Studies [7–9] have used a communication robot to convey emotions. However, it is difficult for communication robots to generate facial expressions like a human face. To overcome this problem, the studies displayed colors with motions of a robot for communication between humans and robots. Elfoid also cannot generate facial expressions like a human face because it has a compact design that cannot be activated intricately. That is, since a priority of Elfoid is portability, the modality of communication is less than that of Telenoid. For this reason, it is necessary to convey emotions some other way.

In this research, facial expressions are generated using Elfoid's head-mounted mobile projector. If the speaker's facial movements estimated employing conventional face-recognition approaches are accurately regenerated with Elfoid, the human presence can be conveyed. However, even if a captured face image is projected directly, details of facial expression cannot be conveyed because the resolution of projection is low. According to the facial action coding system (FACS) [6], which describes relationships between emotion and facial movement, movements around the mouth and eyes play important roles. However, some emotions cannot be conveyed even if the movements around the mouth and eyes are emphasized [10]. In this study, facial expressions are emphasized using cartoon techniques [11]. It is widely recognized that cartoons have a strong advantage in expressing emotions and feelings.

3   Generation of Facial Expression Using Elfoid with a Projector

3.1   Elfoid Characteristics

Elfoid is used as a cellular phone for communication as shown in Fig.1. To convey the human presence, Elfoid has the following functions.

− Elfoid has a body that is easy to hold in the hand.
− Elfoid's design is recognizable at first glance to be nothing more than a human and is capable of being interpreted equally as male or female, old or young.
− Elfoid has a soft exterior that provides a feeling of human skin.
− Elfoid is equipped with a camera and microphone.

Additionally, a mobile projector is mounted in Elfoid's head and a facial expression is generated by projecting images from within the head as shown in Fig.2.

The procedure of the proposed method is as follows. First, individual facial images are captured by a camera mounted within Elfoid. Next, facial expression is recognized using a conventional method such as the use of a point distribution model [12][13]. Effects that induce a particular emotion are added to the image. Finally, the generated image is projected on the face of Elfoid from within.

3.2   Generation of Facial Expressions Using Cartoon Techniques

In this study, facial expressions are emphasized using cartoon techniques [11]. It is widely recognized that cartoons have a strong advantage in expressing emotions and feelings. According to FACS [6], which describes relationships between

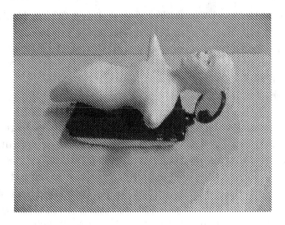

Fig. 2. A mobile projector is mounted in Elfoid's head

emotion and facial movement, features around the mouth and eyes play important roles. In this study, facial expressions associated with the speaker showing basic emotions defined by Ekman et al.[6] are generated according to the following rules.

- Angry expression
  Eyes are moved to the center. Corners of the mouth are lowered. Complexion is changed to red.
- Disgusted expression
  Only one eye is moved to the center. Only one corner of the mouth is raised. Complexion is changed to blue-green.
- Fearful expression
  Eyes are moved to the center and each inside is raised. Mouth is opened. Complexion is changed to blue.
- Happy expression
  Shapes of eyes are varied so that the positions of the cheeks are raised. Corners of the mouth are raised. Cheek color is associated with blushing.
- Sad expression
  Eyes are moved to the center and each inside is raised. Corners of the mouth are lowered. Symbols of tears are added to the eyes.
- Surprised expression
  Eyes are opened wide. Movement of the mouth is increased.

The cartoon techniques that are used in this study are shown in Fig. 3. The movements around the mouth and eyes are emphasized. Moreover, color stimuli that induce a particular emotion are added.

3.3 Generation of Projection Images for Elfoid

In this study, projection images are generated to backproject Elfoid's face with consideration of Elfoid's material. To generate a facial expression with Elfoid,

(a) Angry
expression

(b) Disgusted
expression

(c) Fearful
expression

(d) Happy
expression

(e) Sad
expression

(f) Surprised
expression

Fig. 3. Facial expressions conveyed by cartoon techniques

calibration with respect to both projection position and projection color is needed. As for the projection color, color variation after projection is investigated by projecting various colors in advance. As for the projection position, three-dimensional positions on the surface of Elfoid are investigated by projecting an image that has a number of known two-dimensional positions. The correspondence between the three-dimensional positions on Elfoid's surface and two-dimensional positions in the projection image can be obtained by this calibration. Facial expressions are generated on the surface of Elfoid using the characteristics described in 3.2, and a projection image is generated using the result of the calibration.

4   Experiment

To investigate the emotions that are derived from each facial expression, we conducted experiments with Elfoid.

4.1   Generation of Facial Expressions with Elfoid

In an experiment, facial expressions associated with the speaker showing basic emotions defined by Ekman et al.[6] are generated . The emotions are anger, disgust, fear, happiness, sadness, and surprise. Figure 4 shows the generated projection images. An animation is generated varying from a normal expression

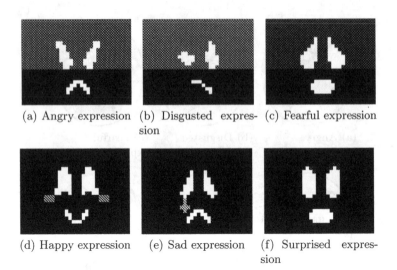

(a) Angry expression (b) Disgusted expression (c) Fearful expression

(d) Happy expression (e) Sad expression (f) Surprised expression

Fig. 4. Projection images

to the generated expression using morphing technology. Figure 5 shows the facial expressions generated with Elfoid.

It was found that the projection is seamless in spite of the complicated surface of Elfoid.

4.2 Subjective Evaluation of Emotional Conveyance

In this experiment, an emotion conveyed to users is investigated by subjective evaluation. Six facial patterns were presented to 28 subjects in random order. Each subject was asked to rate a facial pattern for six emotions: anger, disgust, fear, happiness, sadness, and surprise. Each emotion was rated from 1 (not conveyed at all) to 6 (conveyed extremely strongly). The results of the subjective evaluation process for each facial expression are shown in Fig. 6. The items in these figures are the average score and standard variation of the subjective evaluation. Dunnett's test [14] is used to compare the average scores. Dunnett's test is a multiple-comparison procedure used to compare each of a number of treatments with a single control. The emotions conveyed when generated facial expressions are presented are discussed as follows.

– Angry expression
  Figure 6(a) shows the emotions conveyed when the generated angry expression is displayed. The highest score is observed for the emotion "anger"; the average score of the subjects is 5.25, and Dunnett's test is performed with reference to this score. Dunnett's test indicates a significant difference between the score for "anger" and the scores for all other emotions. Therefore, the angry expression of Elfoid efficiently conveys "anger".

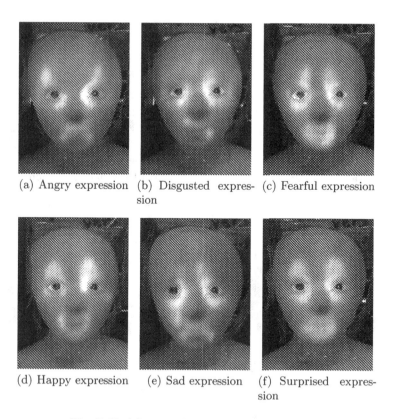

(a) Angry expression (b) Disgusted expression (c) Fearful expression

(d) Happy expression (e) Sad expression (f) Surprised expression

Fig. 5. Facial expressions generated with Elfoid

– Disgusted expression
Figure 6(b) shows the emotions conveyed when the generated disgusted expression is displayed. The highest score is observed for the emotion "disgust"; the average score of the subjects is 4.32, and Dunnett's test is performed with reference to this score. Dunnett's test indicates a significant difference between the score for "disgust" and the scores for all other emotions. Therefore, the disgusted expression of Elfoid efficiently conveys "disgust".

– Fearful expression
Figure 6(c) shows the emotions conveyed when the generated fearful expression is displayed. The highest score is observed for the emotion "fear"; the average score of the subjects is 3.75, and Dunnett's test is performed with reference to this score. Dunnett's test indicates that there was no significant difference between the scores for "disgust" and "sadness". Therefore, the fearful expression of Elfoid does not convey "fear" accurately. This may be because the shape of the eyes in the fearful expression gives a negative impression.

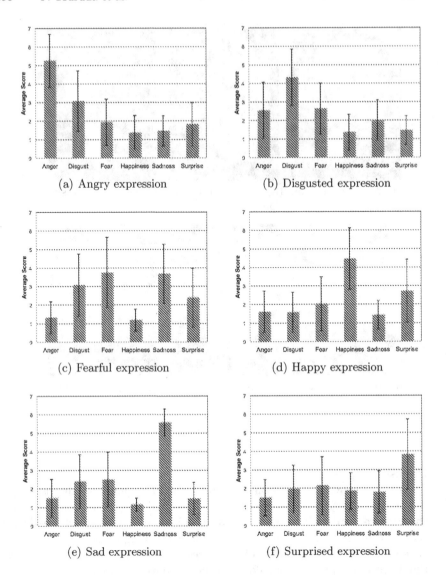

Fig. 6. Results of evaluation

– Happy expression

Figure 6(d) shows the emotions conveyed when the generated happy expression is displayed. The highest score is observed for the emotion "happiness"; the average score of the subjects is 4.46, and Dunnett's test is performed with reference to this score. Dunnett's test indicates a significant difference between the score for "happiness" and the scores for all other emotions. Therefore, the happy expression of Elfoid efficiently conveys "happiness".

− Sad expression

Figure 6(e) shows the emotions conveyed when the generated sad expression is displayed. The highest score is observed for the emotion "sadness"; the average score of the subjects is 5.57, and Dunnett's test is performed with reference to this score. Dunnett's test indicates a significant difference between the score for "sadness" and the scores for all other emotions. Therefore, the sad expression of Elfoid efficiently conveys "sadness".

− Surprised expression

Figure 6(f) shows the emotions conveyed when the generated surprised expression is displayed. The highest score is observed for the emotion "surprise"; the average score of the subjects is 3.82, and Dunnett's test is performed with reference to this score. Dunnett's test indicates a significant difference between the score for "surprise" and the scores for all other emotions. However, since the average score for "surprise" is low, the surprised expression needs to be improved.

5   Conclusion

We proposed a method for generating facial expressions emphasized with cartoon techniques using Elfoid with a mobile projector. In experiments, facial expressions were generated by backprojecting facial patterns to Elfoid's face. Five facial expressions, but not a fearful expression, can be conveyed as the intended emotion. In the case of the fearful expression, negative emotions, such as "sadness" and "disgust", are conveyed co-instantaneously with "fear". In future work, we will implement an entire system to convey emotions with Elfoid according to the results of facial recognition studies.

Acknowledgment. This research was supported by the JST CREST (Core Research for Evolutional Science and Technology) research promotion program "Studies on cellphone-type tele-operated androids transmitting human presence".

References

1. Asano, C.B., Ogawa, K., Nishio, S., Ishiguro, H.: Exploring the uncanny valley with geminoid HI-1 in a real-world application. In: Proc. Int'l Conf. of Interfaces and Human Computer Interaction, pp. 121–128 (2010)
2. Ogawa, K., Nishio, S., Koda, K., Balistreri, G., Watanabe, T., Ishiguro, H.: Exploring the natural reaction of young and aged person with Telenoid in a real world. Jour. of Advanced Computational Intelligence and Intelligent Informatics 15(5), 592–597 (2011)
3. Arnold, M.B.: Emotion and Personality: Psychological aspects. Columbia University Press (1960)

4. Plutchik, R.: Emotion: A Psychoevolutionary Synthesis. Harper & Row (1980)
5. Izard, C.E.: The Psychology of Emotions. Springer (1991)
6. Ekman, P., Frisen, W.: Facial action coding system: A technique for the measurement of facial movement. Consulting Psychologists Press (1978)
7. Sugano, S., Ogata, T.: Emergence of mind in robots for human interface - research methodology and robot model. In: IEEE Int'l Conf. Robotics and Automation, pp. 1191–1198 (1996)
8. Hiraiwa, A., Hayashi, K., Manabe, H., Sugimura, T.: Alter-ego interface technology. NTT Technical Review 1(8), 72–76 (2003)
9. Ariyoshi, T., Nakadai, K., Tsujino, H.: Effect of facial colors on humanoids in emotion recognition using speech. In: Int'l Workshop on Robot and Human Interactive Communication, pp. 59–64 (2004)
10. Hori, M., Takakura, H., Yoshimura, H., Iwai, Y.: Generation of facial expression for communication using elfoid with projector. In: Schwenker, F., Scherer, S., Morency, L.-P. (eds.) MPRSS 2012. LNCS, vol. 7742, pp. 27–34. Springer, Heidelberg (2013)
11. Thomas, F., Johnston, O.: The Illusion of life: Disney animation. Disney Press (1995)
12. Cootes, T.F., Edwards, G.J., Taylor, C.J.: Active appearance models. IEEE Trans. on Pattern Analysis and Machine Intelligence 23(6), 681–685 (2001)
13. Saragih, J.M., Lucey, S., Cohn, J.F.: Deformable model fitting by regularized landmark mean-shift. Int'l Jour. of Computer Vision 91(2), 200–215 (2011)
14. Dunnett, C.: New tables for multiple comparisons with a control. Biometrics 20(3), 482–491 (1964)

Part IV

Biophysiological Aspects of Interaction

A Biofeedback Game for Training Arousal Regulation during a Stressful Task: The Space Investor

Olle Hilborn[1], Henrik Cederholm[1], Jeanette Eriksson[2], and Craig Lindley[1]

[1] Blekinge Institute of Technology, Karlskrona, Sweden
{firstname.lastname}@bth.se
[2] Malmö University, Malmö, Sweden
jeanette.eriksson@mah.se

Abstract. Emotion regulation is a topic that has considerable impact in our everyday lives, among others emotional biases that affect our decision making. A serious game that was built in order to be able to train emotion regulation is presented and evaluated here. The evaluation consisted of a usability testing and then an experiment that targeted the difficulty of the game. The results suggested adequate usability and a difficulty that requires the player to engage in managing their emotion in order to have a winning strategy.

1 Introduction

This work was a part of the Europeans project xDELIA (Xcellence in Decision-making through Enhanced Learning in Immersive Applications, www.xdelia.org), which tried to improve decision making by training emotion regulation, thus lowering impact from emotional biases. The game presented here, Space Investor, is a redesign of a simple shooting game that was affected by the players' arousal level, called the Aiming Game [1]. The Space Investor game was created in order to support emotion regulation training, using the concepts presented here. The other theme of this work was to evaluate if Space Investor was difficult in a way that actually could support training of emotion regulation capabilities.

Biofeedback is the concept of displaying specific bodily signals to the subject in order to make them explicit, so that the subject may change his or her behaviour. This has been a successful approach for teaching a variety of concepts such as e.g. pain regulation [2], lowering of blood pressure [3], and anger management [4]. Recently, studies involving games with biofeedback has been carried out where the main goal was to increase the quality of interaction by modifying the game in real time [5, 6]. These two concepts came together in the work described here by facilitating learning with a game that will encourage, by changing difficulty, a specific bio-physiological state.

Emotions can generally be classified by the independent components arousal and valence [7], where arousal represents excitement level and valence defines whether the arousal is positive or negative. This means that emotions can be visualized where arousal and valence defines each axis, and both need to be controlled in order to control emotions. There are methods for extracting and interpreting valence from e.g.

M. Kurosu (Ed.): Human-Computer Interaction, Part V, HCII 2013, LNCS 8008, pp. 403–410, 2013.
© Springer-Verlag Berlin Heidelberg 2013

electromyography (EMG) measuring devices [8], as well as arousal from e.g. heart rate (HR) [9] or galvanic skin response (GSR) [10].

In order to make Space Investor a convenient, usable, and affordable game, heart rate was chosen for Space Investor to measure arousal and alter the gameplay. It is a signal that is easily measured, due to the large electric activity from the heart. Also, the sensors are very unlikely to fall off or be affected by movement since various heart rate monitors are designed on the same principles (waistband and wireless connection) that are used in sports (e.g. [11]). The reason to include only arousal was to avoid an unnecessarily complex experience for the players, i.e. a multi-factored affective game system. The sensor system used is the Movisens and xAffect system [12], which calculates arousal and feed it into the game.

2   Game Description

The aim of the game is to navigate the spaceship from one planet to another in space, collecting resources and destroying obstacles on the way. The ship is carrying goods, which are to be delivered on the destination planet, together with the resources the player managed to pick up on the way there. The game is structured with different levels, where each level consists of a travel route between two planets. The game is presented in 3D and the camera is set to the front of a spaceship, which means that the player will not actually see the ship (First person view, see Fig. 1). Because Space Investor is a serious game that is developed through EU-funding, a very neutral setting was desirable. The choice of space and asteroids was deemed sufficient for this, since no hostile action can be taken against another living thing.

Fig. 1. View of game screen

2.1   Gameplay

Space Investor is a single-player game in which the player tries to avoid getting hit by asteroids by shooting them down. The player cannot die in the game, only lose resources. The spaceship is constantly moving forward through space and is frequently approached by asteroids that must be shot down in order for the ship not to get hit. Occasionally the player will encounter resources which are collected automatically when they hit the ship, but are immediately destroyed if the player shoots at them. There is a third, hybrid type of asteroids (resource asteroids) that need to be shot at before they turn into resources.

2.2   Input

When playing the game, the ship is automatically moving forward in the space environment without the need for player input. The main goal is to aim and shoot at different object. This is done with the mouse buttons, left for the primary weapon and right for the secondary. The indirect (meaning: not under direct conscious control) input from the player is the heart rate that is registered and then calculated into an arousal value with the Movisens system [12]. This arousal value will distort the game and make it harder when the arousal is too high.

2.3   Difficulty Level

When developing a serious game, such as the Space Investor game, it must be taken into consideration the fact that the target group may not be experienced game players or have the incentive to go through a steep learning curve. Therefore, the game should be designed to be playable by all types of people, ranging from hard core gamers to completely inexperienced players. In the case of the Space Investor game there is a delicate line where players must feel pressured and perceive the game as hard no matter how experienced they may be, at the same time as not perceiving the game to be too difficult for them to even try. When progressing between levels two factors will be affected.

— *Asteroid spawn rate* will be lower, meaning there will be shorter distance between asteroids spawning, resulting in more asteroids in less time.
— *Various types of asteroids* and resources will appear, making it harder to distinguish the ones that will give resources and the ones that need to be shot down.

2.4   Arousal Effects

In order to train emotion regulation during the game, it is important that the game is sufficiently challenging in the aspect of emotional control, in order to elicit an emotional response. In the game these effects are explained with the bio-physiological interface of the ship and when the physical state of the player is undesired, the ship's functions are working at a suboptimal level. The higher arousal, the bigger each of the effects will be. The elements affected are

— *Speed:* The ship's speed is increasing if high arousal levels are detected. This makes the game harder since asteroids will hit the ship faster and also spawn with higher frequency (in time). The normal speed of the ship is 100 space units per second, with up to another 100 space units depending on arousal level. This means that the speed will double if the player has an arousal level on a maximum distance from the wanted value. The time it takes to reach the next planet is thus shorter when the player is aroused.

— *Aiming Offset:* A randomized offset from the aiming point will be present when unwanted levels of arousal occur. This is meant to make the game harder since the targets will be harder to hit in general. With the help of a double cosine function, a fractal movement pattern is created. The arousal value determines the amount the fractal function is allowed to affect the aiming. The functions are $X_n = X_{n-1}+arousal*50(3*cos(0.6t)+cos(\pi t))$ for movement along the x-axis and $Y_n=Y_{n-1}+arousal*50(3*cos(0.75t)+cos(\pi t))$ for movement along the y-axis, where t is the time in milliseconds since the last frame.

— *Blur.* The blur is primarily a means to make it hard for the player to distinguish between resources and asteroids, rendering the player to mistakenly shoot down resources.

— *Shields:* The spaceship has shields in order to protect it from asteroids; these are weakened as a result of arousal levels and damage levels will increase. Damage will make the player to lose resources and the camera will shake (more damage equals longer shake time).

An overview of the difficulty and the specific settings used can be found in Table 1, where the arousal effects for each level can be found, as well as the spawn rate for asteroids. The spawn rate worked as when 600 space units had been travelled (for level 0) a new asteroid was spawned.

Table 1. The difficulty settings for each level. Spawn rate is measured in space units, resources have a constant spawn rate of 500.

Level	Arousal effects	Asteroid spawn rate	Resource asteroid spawn rate
0	None	600	n/a
1	Speed	560	n/a
2	Speed, aiming	540	n/a
3	Speed, aiming, blur	520	n/a
4	Speed, aiming, blur, shields	500	520

2.5   Weapons

To enhance the support for an immersive experience and the drive for self enhancement in the game, various weapons existed. This was also so that players' could choose their own strategy when playing. During gameplay, the player was equipped with two weapons (one for left mouse button, and one for the right one) at all times. This made it a strategic choice about which guns the player should choose and use. Each weapon has the two attributes damage and cooldown time, which is the amount of time before the weapon can be fired again.

— *Main Cannon:* The main cannon was the default weapon that all players received from the very beginning of the Space Investor game. It is designed to have a rather standardized functionality, which should be recognizable by most semi-experienced players, a low damage with a low cooldown time.

— *Bomber:* The Bomber is a slow but powerful weapon that acts as an auxiliary weapon. It can destroy any obstacle in one hit but has a long cool down period before it can be used again making it rather powerless against multiple targets.

3   Testing of the Game

The game was tested in two ways, a small usability play testing with the focus on what players thought about the game and a larger that tested different hypotheses in the xDELIA project as well as data directly relevant to the game. The same game setup was used in both cases (see Table 1.), consisting of five levels with a length of each 18.000 space units (a maximum of 180 seconds play time, depending on arousal). It is worth noting that level four had resource asteroids that will damage the ship if not shop down, but will turn into resources if they are shot down.

3.1   Usability Test

The play testing evaluation of the Space Investor game was done with five participants, all students at Blekinge Institute of Technology. All players filled in the game experience questionnaire (GEQ) [13] and the system usability scale (SUS) [14]. They have both been used in earlier work to assess serious games [1]. The answers in both questionnaires range from "1 – not at all" to "5 – completely". See Fig. 2 for GEQ (scores can range from 1 to 5). The average result across all players regarding the SUS was 81.5, out of a total of 100.

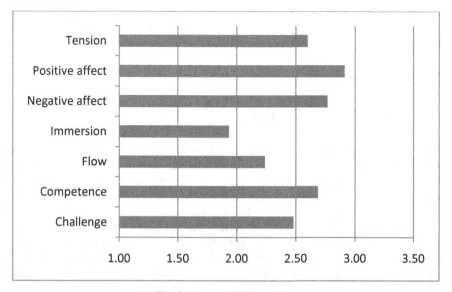

Fig. 2. Average GEQ results

3.2   Experiment Testing

The usability and game experience was deemed sufficient to use the game for a full scale study. Sixty participants played the game once a week for three weeks. The participants' arousal levels were recorded and classified according to the following: Arousal 0 (≤0.00), Arousal 25 (0.01-0.25), Arousal 50 (0.26-0.50), Arousal 75 (0.51-0.75), and Arousal 99 (0.76-0.99). The proportion of the asteroids that hit a player during each of the arousal states were calculated (see Fig. 3).

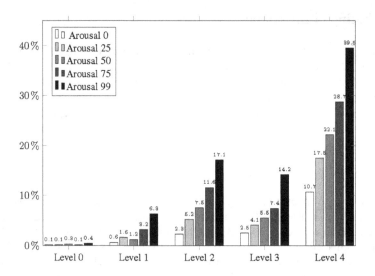

Fig. 3. The proportion of the subjects being hit by an asteroid, given level and arousal

A Kruskal-Wallis test over the three different play sessions showed that people performed significantly better during level 1 to 4 the consecutive times they played. Moreover, it could be seen for at least two arousal states during each of these levels. Spearman's correlation suggested significant negative correlations between number of play sessions and the probability of being hit by asteroids.

4   Discussion and Conclusion

Based on the SUS and GEQ, the game had adequate usability, but improvements could be made in regards to game experience. In this study it was deemed sufficient since the main purpose was not to create a game with the best game experience, but a game where emotion regulation could be practiced. Further work may go into either why some people like it or how it can be more inclusive.

From the proportion (see Fig. 3) of players that got hit by asteroids, it can be seen that the game got successively harder at higher levels, but most apparent that higher arousal values made the game harder. For each level that was modified by arousal (i.e. 1 to 4), it is evident that the optimal strategy was to lower arousal levels in order to perform better. This is in line with the purpose of the game; to encourage emotion regulation training.

Regarding the Kruskal-Wallis test together with Spearman's correlation it was interpreted as the game became easier the more times it was played, meaning that the game is not too easy and there is room for improvement. It cannot be distinguished between if players get less stressed by the game or if they are better at regulating their emotions, since these variables are interdependent.

When Space Investor was designed, the purpose was that it should punish high arousal levels, which it very much does according to the gameplay data. During level one to four, people are twice as likely to be hit by an asteroid when experiencing high arousal compared with low. It is also progressively harder the longer the game is played. A next step in research would be to find out at what difficulty levels people start feeling too frustrated and give up, as well as when the optimal training is achieved and tune the game variables accordingly.

The conclusion drawn from this work was that while Space Investor might not be the game with impactful game experience, it is very usable. Also, it fulfils its purpose by having the optimal strategy being that players need to regulate their arousal levels in order to succeed. Learning effects were seen, but they cannot be attributed to either game adaption or a better emotion regulation capability.

References

1. Cederholm, H., Hilborn, O., Lindley, C., Sennersten, C.: The Aiming Game: Using a Game with Biofeedback for Training in Emotion Regulation. In: Proceeding of DiGRA 2011 Conference: Think Design Play (2011)
2. deCharms, R.C., Maeda, F., Glover, G.H., Ludlow, D., Pauly, J.M., Soneji, D., Gabrieli, J.D.E., Mackey, S.C.: Control over brain activation and pain learned by using real-time functional MRI. Proceedings of the National Academy of Sciences of the United States of America 102, 18626–18631 (2005)
3. Goldstein, D.S., Ross, R.S., Brady, J.V.: Biofeedback Heart Rate Training during Exercise. Biofeedback and Self-Regulation 2, 107–125 (1977)
4. Achmon, J., Granek, M., Golomb, M., Hart, J.: Behavioral treatment of essential hypertension: a comparison between cognitive therapy and biofeedback of heart rate. Psychosomatic Medicine 51, 152–164 (1989)
5. Dekker, A., Champion, E., Arts, M., Box, P.O.: Please Biofeed the Zombies: Enhancing the Gameplay and Display of a Horror Game Using Biofeedback, 550–558 (2007)
6. Rani, P., Sarkar, N., Liu, C.: Maintaining Optimal Challenge in Computer Games through Real-Time Physiological Feedback Mechanical Engineering. In: Proceedings of the 11th International Conference on Human Computer Interaction (2005)
7. Russel, J.A.: A circumplex model of affect. Journal of Personality and Social Psychology 39, 1161–1178 (1980)
8. Cacioppo, J.T., Petty, R.E., Losch, M.E., Kim, H.S.: Electromyographic activity over facial muscle regions can differentiate the valence and intensity of affective reactions. Journal of Personality and Social Psychology 50, 260–268 (1986)
9. Anttonen, J., Surakka, V.: Emotions and heart rate while sitting on a chair. In: Proceedings of the SIGCHI Conference on Human Factors in Computing Systems, pp. 491–499 (2005)

10. Winton, W.M., Putnam, L.E., Krauss, R.M.: Facial and autonomic manifestations of the dimensional structure of emotion. Journal of Experimental Social Psychology 20, 195–216 (1984)
11. Gamelin, F.X., Berthoin, S., Bosquet, L.: Validity of the polar S810 heart rate monitor to measure R-R intervals at rest. Medicine and Science in Sports and Exercise 38, 887–893 (2006)
12. Schaaff, K., Müller, L., Kirst, M., Heuer, S.: xAffect – A Modular Framework for Online Affect Recognition and Biofeedback Applications. In: 7th European Conference on Technology Enhanced Learning (2012)
13. IJsselsteijn, W.A., Poels, K., De Kort, Y.A.W.: The game experience questionnaire: Development of a self-report measure to assess player experiences of digital games. FUGA technical report, deliverable 3.3, TU Eindhoven, Eindhoven, Netherlands (2008)
14. Brooke, J.: SUS: A "quick and dirty" usability scale. In: Jordan, P.W., Thomas, B., Weerdmeester, B.A., McClelland, I.L. (eds.) Usability Evaluation in Industry, pp. 189–194. Taylor & Francis, London (1996)

Responses Analysis of Visual and Linguistic Information on Digital Signage Using fNIRS

Satoru Iteya[1], Atsushi Maki[2], and Toshikazu Kato[3]

[1] Graduate School of Science and Engineering Chuo University 1-13-27 Kasuga,
Bunkyo-ku, Tokyo, 112-8551 Japan
robinboy.lj@gmail.com
[2] Hitachi, Ltd. 1-18-13, Soto-Kanda, Chiyoda-ku, Tokyo, 101-8608 Japan
atsushi.maki.nn@hitachi.com
[3] Faculty of Science and Engineering Chuo University 1-13-27 Kasuga,
Bunkyo-ku, Tokyo, 112-8551 Japan
kato@indsys.chuo-u.ac.jp

Abstract. When customers receive recommended information through digital signage, it is important not only to choose suitable commodities matching each customer's preferences, but also to choose suitable information media to express their features. This paper proposes a method to estimate their preferences on information media by measuring brain activity. First step in order to achieve our final goal, we disclose that there are significant differences in brain activity in case subjects receive recommended information. The result of analysis shows there are significant differences in brain activity, especially visual cortex and language area.

Keywords: fNIRS, Preference on Commodities and Information Parts, Information Recommendation.

1   Introduction

Digital signage has become popular in real shops and in web sites as advertising media for sales promotion [1]. Customers receive much information on commodities through digital signage in shopping. It is important not only to choose suitable commodities matching each customer's preferences, but also to choose suitable information media to express their features. If we can estimate his personal preference on information media, which describe some specific feature of commodities, he can provide more effective message to the customer regarding the specific commodities.

In society, only page view logs are adopted to estimate each customer's personal preference of commodities, which is insufficient to analyze the reasons of his behavior. We expect that some indices of brain activity are a key to estimate the degree of his attention to the information media as well as to the commodity. If some messages describe a specific feature of the preferred commodity, the customer pays attention to messages, otherwise does not.

Therefore, we measure the customer's response on visual and linguistic information on display by fNIRS to detect the degree of his attention.

M. Kurosu (Ed.): Human-Computer Interaction, Part V, HCII 2013, LNCS 8008, pp. 411–420, 2013.
© Springer-Verlag Berlin Heidelberg 2013

2   Methods of Our Research

These days, technologies of measuring brain activities have been developing, and there are some experiments that these technologies have been applying to marketing research to understand consumer's psychological state in shopping [2]. Previous studies showed that if image information was displayed to a subject, there were significant activities in his visual cortex [3]. There are some measuring systems, and in our study, we choose functional near-infrared spectroscopy (fNIRS), because fNIRS has high spatial resolution where there are significant activities in brain to same degree and lower physical restraint to a subject than other measuring systems [4].

We assume two behaviors in shopping; first behavior is that when a customer finds a commodity correspond with his preference, he pays attention to a commodity, and second behavior is that when he is received information about his preferred commodity, he pays attention to part of preferred information. Therefore in our research, we are developing a control mechanism to alter information message styles according to each customer's preference on a commodity (Fig. 1).

Our current purpose is that in case customers view some information about a commodity according to their preference, there will be significant differences in brain activities between displayed information. In this paper, we especially focus on clothes, and classify information into two categories; first category holds mainly commodity's visual messages, such as coordinates with other commodities. The other category holds mainly commodity's linguistic explanation, such as statements about its features.

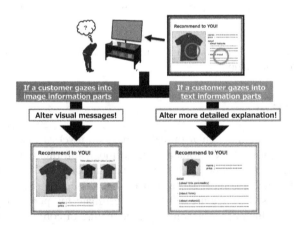

Fig. 1. Recommendation message styles of our suggestion

We assume two behaviors. (1) When a customer finds a commodity correspond with his preference, he pays attention to a commodity. (2) When he is received information about his preferred commodity, he pays attention to part of preferred information.

3   Expriment: Measuring Responses of Information on Display by Fnirs

We used the multichannel fNIRS optical topography system ETG-4000 (HITACHI Medical Cooperation) and adopted the event-related design in our experiment. In our experiments, subjects viewed recommended information about a commodity, and responded to all commodities by rating scale five degrees about preference. Our procedure we experimented is summarized in Fig. 2.

We set the fNIRS probes to cover the visual cortex and language area according to the international 10-20 systems in electroencephalography [5], and showed in Fig. 3. Especially in this study, the placement on language area was left side of the brain. In visual area, Oz was consistent with the optical electrode middle second of the lower row of 3×3, the bottom line was along a well-balanced to horizontal reference curve. In language area, T3 was consistent with the optical electrode middle second of the lower row of 3×3, and the bottom line was along a well balanced to horizontal reference curve. For spatial profiling of fNIRS data, we employed virtual registration to register the data to MNI standard brain space [7][8][9].

Profiles of ten subjects are male, 22-24 years old and right-handed. The reason of right-handed is that the significant brain is different whether they are right-handed or left-handed [6]. In experiment, we divided a task into two stages; former was subjects understood a commodity, later was they evaluated their preference. The reason was to make sure if there are significant activities in brain, that significant activities are resulted from the process whether they understood commodities or they evaluated their preference.

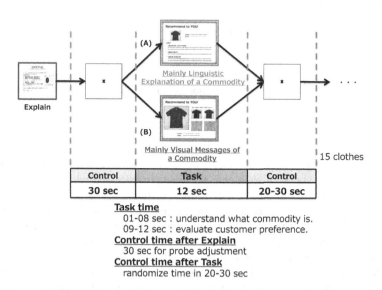

Fig. 2. The event-related design in our experiment

(A) Type of commodity's linguistic explanation, (B) Type of commodity's visual messages

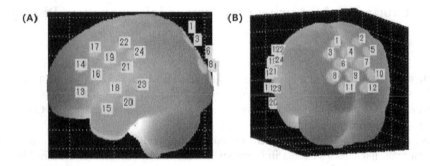

Fig. 3. The placement of our study based on international 10-20 system [7][8][9]

(1) The placement of fNIRS probes on language area, especially Broca's area, (2) The placement of fNIRS probes on visual cortex

4 Hypothesises in Our Methods and Analysis

Our hypotheses are described below. First hypothesis is that when subjects receive commodity's visual messages, oxy-Hb values in visual cortex were increasing more than those of language area. Second hypothesis is that when they receive commodity's linguistic explanation, oxy-Hb values in language area were increasing more than those of visual cortex. In this paper, we adopted oxy-Hb, because oxy-Hb is more reliable parameter than other parameters measured by fNIRS.

4.1 Method of Integral Analysis

Measurement data are revised by integral analysis, and we calculated arithmetic mean of all subject's revised data on each type of linguistic explanation and visual messages. We tested statistical analysis by t-test and described in Fig. 4.

4.2 Method of Comparing Period's Mean

Measurement data are preprocessed by the following way. First, each subject's measurement data were preprocessed with a trend revision. Second, individual timeline data for the oxy-Hb signal of each channel were preprocessed with a band pass filter using cut-off frequencies of 0.012 Hz to remove drift and 0.8 Hz to filter out heartbeat pulsations. Third, we set 3 periods in timeline: pre period was 5 sec before stimulus periods, stimulus period was 12 sec during task, and post period was 15 sec. after stimulus period. According to this, total timeline for each commodity was 32 sec. Finally, pre period was set to baseline and we calculated amount of change from baseline. After preprocessed, we obtained each period's mean from preprocessed timeline: pre period's mean was calculated from 0-5 sec, stimulus period's mean was calculated from 11-23 sec, post period's mean was calculated in 27-32 sec (Fig. 5).

Through this method, we tested statistical analysis in two cases. First case was to compare stimulus period' mean with pre/post period's mean in same recommended type, and we tested statistical analysis by non-repeated measures four-way ANOVA and Bonferroni method. The other case was to compare stimulus period's mean between visual messages' experiment and linguistic explanation's experiment, and we tested statistical analysis by non-repeated measures five-way ANOVA.

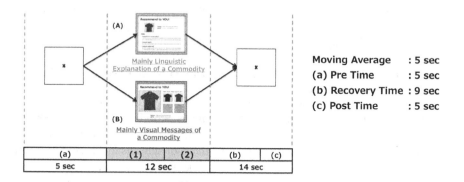

Fig. 4. Outline of integral analysis

(1) Subjects understand what commodity is. (2) Subjects evaluate their preference about a commodity. (A) Type of commodity's linguistic explanation, (B) Type of commodity's visual messages

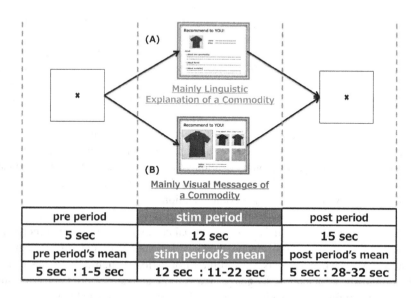

Fig. 5. Outline of calculating each period's mean for analysis

(A) Type of commodity's linguistic explanation, (B) Type of commodity's visual messages

5   Current Results

5.1   Results of Integral Analysis

We found significant differences on additional averages on total data. In case of received commodity's visual messages, when subjects evaluated their preferences degree about commodities, Δoxy-Hb values in visual cortex increased more than those of language area. Also, in case of received commodity's linguistic explanation, when subjects evaluated their preferences, Δoxy-Hb values in language area increased more than those of visual cortex (Fig. 6, 7, 8).

5.2   Results of Comparing Each Period's Mean on Same Recommended Type

According to non-repeated measures four-way ANOVA and Bonferroni method, we found some significant differences between stimulus period's mean and pre/post period's mean in each recommended types. In case of received commodity's linguistic explanation, there were some significant main effects of subjects and period-type, and some significant interactions, such as interaction between subjects and period-types. In case of received commodity's visual messages, there were some significant main effects of subjects and period-type, and some significant interactions, such as interaction between subjects and period-types.

5.3   Results of Comparing Stimulus Period's Mean between Visual Messages and Linguistic Explanations

According to non-repeated measures five-way ANOVA, we found some significant differences between visual messages' experiment and linguistic explanation's experiment and showed in Fig. 9. There were some significant main effects of subjects and recommended types in CH 4, 6, 7, 9, 14, 16, 17, 19. Also, there were some significant interactions, such as interaction between subjects and recommended types. However, there weren't significant indications about preference. We showed Δoxy-Hb mean values each recommended types in Table 1, which there were significant differences. In channels that there were significant difference, when subjects view visual message's recommendation, stimulus mean values in visual cortex were bigger than subjects view visual message's recommendation. Also, when subjects view linguistic explanation's recommendation, stimulus mean values in language area were bigger than subjects view visual message's recommendation.

Therefore, these results showed that when subjects view some information which holds mainly commodity's visual messages, visual cortex is more activated more than when subjects view some information which holds mainly commodity's linguistic explanation. Also, when subjects view some information, which holds mainly commodity's linguistic explanation, language area is more activated more than when subjects view some information, which holds mainly commodity's visual messages.

We showed anatomical labeling and spatial probability were estimated for the significantly activated channels (Table 2). For example, CH9 in visual cortex was estimated Visual Association Cortex by Brodmann Area, and CH14 in language area was estimated pars triangularis Broca's area.

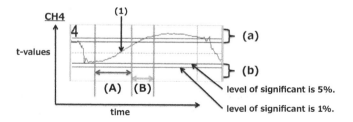

Fig. 6. How to look results of analysis

(1)T-values in t-test detecting differences of activation level by (oxy-Hb values in visual cortex) – (oxy-Hb values in language area) (A) Subjects understand what commodity is. (B) Subjects evaluate their preference about a commodity. (a) This value showed visual cortex was more activated than language area. (b) This value showed language area was more activated than visual cortex.

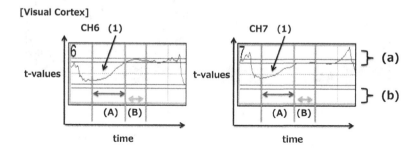

Fig. 7. Part of results in visual cortex

When subjects evaluated their preferences degrees about commodities, there were significant differences in visual cortex. (1) T-values in t-test detecting differences of activation level by (oxy-Hb values in visual cortex) – (oxy-Hb values in language area) (A) Subjects understand what commodity is. (B) Subjects evaluate their preference about a commodity. (a) This value showed visual cortex was activated more than language area. (b) This value showed language area was activated more than visual cortex.

[Language Area]

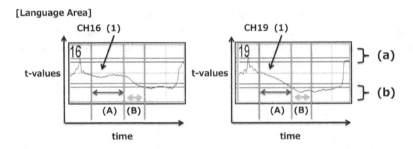

Fig. 8. Part of results in language area

When subjects evaluated their preferences, oxy-Hb values in language area were increasing more than those of language area. (1) T-values in t-test detecting differences of activation level by (oxy-Hb values in visual cortex) – (oxy-Hb values in language area) (A) Subjects understand what commodity is. (B) Subjects evaluate their preference about a commodity. (a) This value showed visual cortex was activated more than language area. (b) This value showed language area was activated more than visual cortex.

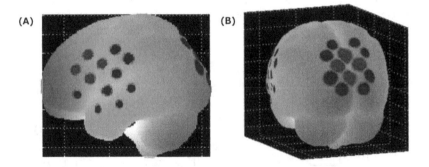

Fig. 9. The result comparing each period's mean [7][8][9]

(A) The result in language area, (B) The result in visual cortex Red circles showed there were significant differences in channels based on analysis. Blue circles showed there weren't significant differences.

Table 1. Stimulus mean values on each recommended types

In visual cortex	CH4**	CH6**	CH7**	CH9**	p-value	
visual messages	0.0393	0.0361	0.0517	0.0242	***	0.001
linguistic explanation	-0.0093	0.0041	0.0093	-0.0139	**	0.01
In language area	CH14*	CH16*	CH17**	CH19***	*	0.05
visual messages	-0.0164	-0.0091	-0.0068	-0.0011		
linguistic explanation	0.0161	0.0308	0.0413	0.0507		

Table 2. Spatial and functional profiles of the channels [7][8][9]

CH	Position			SD (mm)	BrodmanArea(Chris rorden' MRIcro)	Percentage
	x	y	z			
4	-1.804	-81.46	47.522	10	7 - Somatosensory Association Cortex	0.601
					19 - V3	0.399
6	-15.122	-91.367	37.197	10	19 - V3	0.609
					18 - Visual Association Cortex (V2)	0.391
7	14.41	-91.307	36.77	10	19 - V3	0.614
					18 - Visual Association Cortex (V2)	0.386
9	-2.657	-97.748	22.707	10	18 - Visual Association Cortex (V2)	0.608
					17 - Primary Visual Cortex (V1)	0.392
14	-51.131	36.234	21.921	10	45 - pars triangularis Broca's area	1
16	-59.022	20.75	13.201	10	45 - pars triangularis Broca's area	0.415
					44 - pars opercularis, part of Broca's area	0.355
					48 - Retrosubicular area	0.119
					6 - Pre-Motor and Supplementary Motor Cortex	0.11
17	-50.422	22.797	37.424	10	44 - pars opercularis, part of Broca's area	0.759
					45 - pars triangularis Broca's area	0.175
					9 - Dorsolateral prefrontal cortex	0.066
19	-60.98	7.164	28.252	10	6 - Pre-Motor and Supplementary Motor Cortex	0.596
					44 - pars opercularis, part of Broca's area	0.195
					43 - Subcentral area	0.178
					4 - Primary Motor Cortex	0.031

6   Conclusion

Our current purpose was that in case customers view some information about a commodity according to their preference, there were significant differences in brain activities between displayed information. Especially, we focused on clothes, and classify information into two categories, which were visual messages and linguistic explanation. In experiment, subjects were experimented two types: first experiment was subjects viewed recommended information, which holds mainly commodity's visual messages. Second experiment was subjects viewed recommended information, which holds mainly commodity's linguistic explanation. According to compare each experiment's result, we aimed to achieve our current goal. Through above methods, results were when subjects view some information, which hold mainly commodity's visual messages, visual cortex is more activated than when subjects view some information, which hold mainly commodity's linguistic explanation. Also, when subjects view some information, which hold mainly commodity's linguistic explanation, language area is more activated than when subjects view some information, which hold mainly commodity's visual messages.

Through analysis on comparing same recommended types, there were not significant differences in visual cortex and language area whether subjects have preferences on commodities or not. Thus, we found when subjects view some information about commodities, their brain activities in visual cortex and language area is same activity degrees whether they have preferences on commodities or not. We observed idea in order to achieve our final goal, it is important for us to focus on preferences and demonstrate there are significant differences in brain activities.

Acknowledgement. This work was partially supported by JSPS KAKENHI grants, "Effective Modeling of Multimodal KANSEI Perception Processes and its Application to Environment Management" (No.24650110). "Robotics modeling of diversity of multiple KANSEI and situation understanding in real space" (No. 19100004) and TISE Research Grant of Chuo University, "KANSEI Robotics Environment".

References

1. Fujiwara, K.: The Cutting Edge In-Store Promotion using Digital Signage. Journal of the Japan Society of Mechanical Engineers 114(1110), 353–355 (2011)
2. Fujisawa, T., Matsui, T., Kazai, K., Furuya, S., Katayose, H.: Music in Our Brain. Information Processing Society of Japan Magazine 50(8), 764–770 (2009)
3. Taya, S., Maehara, G., Kojima, H.: Hemodynamic responses corresponding to the stimulated visual field. Technical report of The Institute of Electronics, Information and Communication Engineers. HIP 106(328), 49–52 (2006)
4. Maki, A., Sato, D., Obata, A.: Optical Topography and Brain Science. Journal of the Japan Society of Precision Engineering 74(11), 1147–1151 (2008)
5. Monden, Y., Dan, H., Nagashima, M., Dan, I., Kyutoku, Y., Okamoto, M., Yamagata, T., Momoi, M.Y., Watanabe, E.: Clinically-oriented monitoring of acute effects of methylphenidate on cerebral hemodynamics in ADHD children using fNIRS. Clinical Neurophysiology 123(6), 1147–1157 (2012), doi:10.1016/j.clinph.2011.10.006.Epub (November 15, 2011)
6. Tanaka, S.: Cerebral Lateralization. Journal of Japanese Society for Artificial Intelligence 20(4), 486–491 (2005)
7. Tsuzuki, D., Jurcak, V., Singh, A., Okamoto, M., Watanabe, E., Dan, I.: Virtual spatial registration of stand-alone fNIRS data to MNI space. Neuroimage 34(4), 1506–1518 (2007) PMID: 17207638
8. Singh, A.K., Okamoto, M., Dan, H., Jurcak, V., Dan, I.: Spatial registration of multichannel multi-subject fNIRS data to MNI space without MRI. Neuroimage 27(4), 842–851 (2005)
9. Rorden, C., Brett, M.: Stereotaxic display of brain lesions. Behavioural Neurology 12(4), 191–200 (2000)

A Method for Promoting Interaction Awareness by Biological Rhythm in Elementary School Children

Kyoko Ito[1,2], Kosuke Ohmori[2], and Shogo Nishida[2]

[1] Center for the Study of Communication-design, Osaka University
ito@sys.es.osaka-u.ac.jp
[2] Graduate School of Enginering Science, Osaka University, Osaka, 5608531, Japan

Abstract. Recently, in Japan, education about the ability to make decisions as part of a group composed of children with different ways of thinking has become more important. Therefore, discussion activities have been adopted in elementary school education. This study considers a method that supports discussion activities by making children aware of the "state" (i.e., atmosphere, progress) of their group during discussion, and of the ways they are influencing this state themselves. We developed a system which allows us to visualize the entrainment of the biological rhythm to present the group's state. An experiment using this system was conducted to clarify whether the children were aware of the group state during discussion, and how they were affected by this awareness. We found that this system has the potential to support children when considering ways of participating in the discussion. Also, it was found that the system can act as an interface, encouraging children to think about the importance of their listening to others in the group.

Keywords: Education support, Elementary school education, discussion activity, interaction, biological rhythm.

1 Introduction

Recently in Japan, education about the ability to make decisions in groups composed of children with different ways of thinking has become more important[1]. This is difficult to achieve with the traditional teaching style, where the teacher teaches while the students listen[2], therefore, discussion activities have been adopted[3]. In such an activity, a group composed of 5~6 children discusses a complex theme, such as an energy/environmental issue. Taking into consideration the critical period, it has been found appropriate to adopt this activity as a part of elementary school education. The critical period is a theory about the specific period during the development of a child when he/she can construct a neural circuit efficiently and utilize it effectively in the long term[4].

In this type of activity, children learn how to make a decision in a group through communicating with each other. To heighten the learning effect of this activity, children need to realize the meaning of this activity by themselves[5].

M. Kurosu (Ed.): Human-Computer Interaction, Part V, HCII 2013, LNCS 8008, pp. 421–430, 2013.

Therefore, it is important for them to think about ways to participate in the discussion. The purpose of this kind of activity is to educate children about how to make group decisions, therefore they need to be aware of the "state" (i.e., atmosphere, progress) of the group during discussion when they decide how to participate in the discussion. In other words, each child should be aware of how he/she relates to the group's state. However, the state of a group changes continually during discussion, and since these changes are invisible, it is difficult for the children to be aware of them. Therefore, it is necessary to support discussion activities by making children "see" the group's state, and make them aware of the role they play in it.

The purpose of this study is to investigate a method that supports discussion activities by making children aware of the group's state and of the ways they themselves influence it. We will also clarify how the method affects the children's attitude during the discussion.

2   Related Study

In this section, we shall introduce several related studies about children's discussion activities, and several studies about various methods for supporting children's discussion using information technology.

Isomura investigates the meaning of the one-to-many communication between children and teacher in the lower grades of elementary school[5], showing how teachers educate children's abilities to speak up, and how this affects children. Matsuo looks into how experienced teachers initiate discussion activities[6]. Also, this study shows methods used by teachers to make and implement discussion rules. Sakai shows that giving a role to each child in the discussion is an effective strategy to make the activity proceed smoothly[7].

Other studies use electronic whiteboards or tabletop interfaces for supporting the discussion activity. Otsuki suggests encouraging children's participation in the discussion by using an interactive electronic whiteboard[8]. The purpose of using the whiteboard is to elicit group competition, which stimulates the desire of the children to take part in the activity. Kitahara suggests supporting discussion activities by using a tabletop interface[9]. This study uses the interface to make children's exchange of information proceed smoothly.

As mentioned in section I., children should think about ways to participate in the discussion activity by themselves. Creating rules and routines, as well as using tabletop interfaces can make discussion smooth, while using an electronic whiteboard may improve children's participation. Nevertheless, these methods do not make children aware of the group's state during discussion. This is where our study will bring something new and different from the above-mentioned approaches.

3   Awareness of Group Condition

In order to support children's discussions, this study proposes a method to make children understand clearly the way the atmosphere in the group changes, and the role they themselves play in their group's general state. To achieve this, we take into consideration visualizing the information concerning the group's state. This section first describes what makes up the "state of the group", and then suggests a method to visualize it.

3.1   Group's State during Discussion

The ways children participate in a discussion can vary. For example, some children are good at speaking, some are good at listening, and some are good at thinking. Therefore, a child's discussion ability is one factor which decides his/her participation in the discussion. Children should learn how to use their abilities; They should also know that the best way is not the same for all, or in all situations. They should be taught to adopt one communication strategy or another based on how their ability affects the group's state. When teaching children how to make a group decision, each child should be taught to think about the best way to participate in the discussion by paying attention to the relation between his participation and the group.

So, what makes up the group's state? One component is the way children interact with each other in the discussion. Children should be aware that their interaction is an important factor in shaping the general atmosphere of the group discussion.

Usually, one participates in a discussion by speaking, listening, and thinking. Therefore, the interaction between children in discussion has the following components:

- Interaction through speaking - listening;
- Emotional interaction.

Information changes continually in discussion, and these changes are invisible. Therefore, this study tries to visualize the information. A method for extracting the interaction between children will be considered in 3.2.

3.2   A Method for Extracting Information about Children's Interaction

The interaction between children in discussion has the following components:

- Interaction through speaking - listening;
- Emotional interaction.

The emotional interaction between children includes not only external information, but also internal information. Therefore, it is not enough to visualize the statements, the volume of the voice, facial expressions or nutation to show interaction. This study takes into consideration biomarkers, which show one's internal status[10]. Some biomarkers, such as breath or heart rate, can be used to measure the biological rhythm. It is said that biological rhythm reflects one's internal status, and each person has their own rhythm. Watanabe shows that entrainment of biological rhythms occurs between two persons when they are communicating or having the same feelings. This is why in this study we will use information pertaining to the biological rhythm entrainment.

3.3   Preliminary Research toward Using Biological Rhythm Entrainment

We conducted a preliminary research to find a method of processing the biological rhythm entrainment to show the interaction between children.

The participants were 15 graduate students, divided into 5 groups of 3. A 45 minute discussion session was held in each group, and the heart rate of each participant was measured during the discussion. After the discussion, each participant was asked to give details about how the discussion was going when their heart rate entrainment shows special features. Heart rate entrainment was calculated based on equation (1), as a correlation coefficient. Based on the 9 sets of data obtained in an interval of 3 minutes from every participant in the experiment, we calculated the following correlation coefficient.

$$C_{xy} = \frac{\sum_{i=1}^{n}(x_i - \bar{x})(y_i - \bar{y})}{\sqrt{\sum_{i=1}^{n}(x_i - \bar{x})^2}\sqrt{\sum_{i=1}^{n}(y_i - \bar{y})^2}} \tag{1}$$

x_i, y_i : heart rate, $\bar{x}, \bar{y}$: average of x_i, y_i, n : number of data

The participants mentioned the following instances of an increase in the number of heart rate entrainment sets:

- Listening to the speaker;
- Forming group opinion.

These results show that visualizing the number of sets of biological rhythm entrainments can make participants aware of the group's state in the discussion, as mentioned in 3.1. Based on these results, we decided to use the correlation coefficient (formula (1)) to calculate the heart rate entrainment. From the analysis of the correlation coefficient data of each pair of participants, the threshold to determine whether the heart rate is entraining or not was set at 0.3.

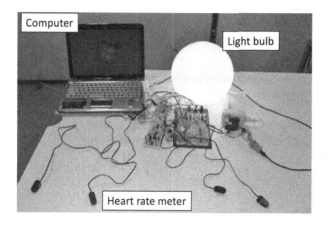

Fig. 1. Overview of the system*
*The greater the entrainment of children's heart rates,
the brighter the light bulb

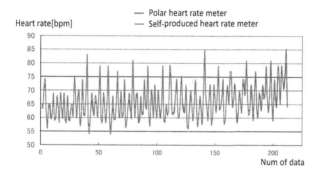

Fig. 2. Comparison between RS800CX and our heart rate meter

3.4 Visualizing the Entrainment of the Biological Rhythm

To visualize the entrainment of the biological rhythm, we developed the system shown in fig.1. Each participant's heart rate can be measured by placing a heart rate meter on their ear, which is noninvasive. The number of entrainment sets is calculated by a computer and made visible by changing the intensity of a light bulb.

We built the heart rate meter ourselves, and also conducted accuracy measurements comparing it with the heart rate meter made by Polar Electro (RS800CX). The result of the comparison for 3 minutes is shown in fig.2. The error was ± 10%.

The purpose of this study is to propose a method that supports discussion activities by making children aware of the group's state and of the role they play in shaping it, and to clarify how the method affects the children's attitude during the discussion. An experiment using this system was conducted.

4   Experiment

4.1   Purpose

The purpose of this experiment is to clarify the following issues, while using our system in the discussion:

(A) How the children were aware of the group's state in discussion;
(B) How the children were affected by their awareness.

4.2   Method

- **Participants**
  Thirty-three children (15 from 5th grade, 18 from 6th grade) and 2 teachers participate in the experiment. The children are divided into 3 groups for each grade.
- **Experiment Flow**
  The teachers decide the topics for discussion.
  (i) Explanation (10 minutes): Children hear an explanation on the experiment, the topic of discussion, and the system.
  (ii) Discussion (15 minutes): Every child wears a heart rate meter on the ear. Each group uses the system in the discussion. After 15 minutes, each group has to make a decision.
  (iii) Interview with children (10 minutes): After the discussion, to investigate how children were aware of the group's state, each group is interviewed. The interviewer asks the following questions:
     — How was the discussion going when the bulb was bright?
     — How was the discussion going when the bulb was dark?
  (iv) Presentation of the group decision (10 minutes): Each group presents their decision, then the teacher draws the conclusions.
  (v) Interview with the teacher (10 minutes): After the discussion, each teacher is interviewed about whether the children were different from usual. The interviewer asks the following questions:
     — What did the children think about the presentation of the system?
     — How were the children reacting to the presentation of the system?
     — How were the children different from everyday discussion?
- **Method of Analysis**
  The interview data from the teacher is analyzed by a qualitative method called "modified grounded theory approach(M-GTA). This is a modified version of the "grounded theory approach"(GTA)[15]. M-GTA is a method for reconstructing the process of real scientific hypothesis testing from the results. First, the data that can explain a concept is chosen from interview data, and some concepts are extracted from it. The concepts that are extracted are divided into categories based on the relations between them[11][12][13]. This method has come to be used in recent research about interface fields[14].

Fig. 3. Group in 5th grade using the system

Table 1. Results of interview with the children

Questions	Results	Groups
How was the discussion going when the bulb was bright?	Everyone was laughing	3
	Everyone was sighing at the same time	2
	It was bright even when I was not speaking	1
	Everyone was thinking something	3
	Everyone was thinking the same thing	3
	Everyone was paying attention to one person	1
How was the discussion going when the bulb was dark?	No one was speaking	3
	No one was listening to what was being said	4
	Talking about unrelated things	2
	Thinking different things from what was being said	3

4.3 Results

The themes of discussion in each class were the following:

– 5th grade: How to reduce waste;
– 6th grade: How to save electricity.

Fig.3 shows a group in 5th grade using the system in discussion.

Children's Awareness of the Group's State. Table 1 shows the results of the interview with the children. In this table, the question, result and number of the group that had the same answer are shown in this order.

From these results, it seems that the children were aware of how they were interacting with the others by listening or speaking. Also, it seems that they were aware of how their thoughts or feelings were relating with each other's. In addition, the appearance of keywords such as 'everyone' or 'at the same time' in the results of the interview shows that children were aware of how they were relating to the group's state.

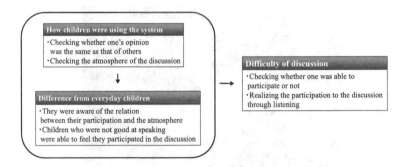

Fig. 4. Conceptual diagram obtained from the M-GTA analysis

Analysis of M-GTA. The length of the interview with the teacher was 10 minutes, and the data from the interview contained about 500 words. The results of the interview are below.

— Children were aware of the existence of a general atmosphere of the discussion, which is difficult to observe usually without using the system.
— Children were aware of the relation between their participation and the atmosphere of the discussion.
— Usually, children feel that they are not participating to the discussion if they do not speak and only listen, but using the system, they felt that they were participating to the discussion by listening, too.
— Children were checking whether their opinion was the same as that of the others.

Fig.4 shows the conceptual diagram which is the result of the analysis by M-GTA.

From the analysis and the result of the conceptual diagram, we obtained the following explanation of the discussion using the system. In a usual discussion, it is difficult to check whether one is able to participate in the discussion or not. Also, it is difficult to be aware of one's the participation to discussion through listening. On the other hand, using the proposed system to check the difference or concurrence of opinion, or to check the atmosphere of discussion, children were able to overcome these difficulties. They noticed the relation between their participation and the atmosphere of the discussion. In addition, being aware of the relation between the general discussion atmosphere and their listening, children who were not good at speaking were able to feel that they were participating in the discussion, too.

These results are pointing to two main effects of making children aware of the group's state.

1. Children were aware of the relation between the atmosphere of discussion and their participation.
2. Children realized that listening is a way to participate in discussion.

4.4   Consideration

The purpose of the system was to make children aware of the group's state in discussion and of how they relate to it. Through the experiment, we shed light on several ways that the children were affected by being aware of the group's state. Two main effects were brought to our attention:

The first effect was that the children were aware of the relation between the atmosphere of discussion and their participation. In this study, we explained that in order to consider what is the best way to participate in a discussion one needs to be aware of the relation between one's way of participating and the state of the group during discussion. Therefore, making children aware of the group's state in discussion can help them when they are trying to find and choose a way to participate in discussion.

Another effect was that children realized that listening is a way to participate in discussion. Speaking is an easy way to participate in a discussion because one can affect others or the group with their utterances. On the other hand, one does not affect others or the group directly by listening, so it is difficult to become aware of one's participation in discussion. Nevertheless, a number of studies about listening[16][17] show that it is just as important as speaking. According to previous research, one's listening attitude affects the speaking attitude of the others[16]. In addition, it has also been proven that a good listening attitude affects one's learning[17]. Being aware of the group's state in discussion, and being aware of how one's way of participating affects the group's state will give children an opportunity to think about the effects listening has on the discussion.

5   Conclusion

Our work has considered a method that supports discussion activities by making children aware of the group's state (i.e., atmosphere or progress) during discussion, and also of the way they are influencing this state themselves. In order to present the group's state, we developed a system which allows us to visualize the entrainment of the biological rhythms. An experiment using this system was conducted to clarify whether the children were aware of the group during discussion, and how they were affected by this awareness.

Our findings suggest that it is indeed possible for this system to support children in considering ways of participating to the discussion by themselves. Also, it was found that the system can act as an interface which encourages children to think about the importance of their listening to others in the group.

Our future work includes conducting additional investigations in order to theorize system-assisted discussion methods. We also intend to clarify the effects of the system as an interface which helps children to think about the importance of their listening to others.

430   K. Ito, K. Ohmori, and S. Nishida

References

1. Ministry of Education, Culture, Sports, Science and Technology: New Education Ministry guidelines (2011) (in Japanese)
2. Cazden, C.B.: Classroom discourse, 2nd edn. Heinemann, Portsmouth (2001)
3. Ichiyanagi, T.: How Do Students Listen to Others in Classroom Discussions?: Differences Across Classrooms and Subjects. Japanese Association of Educational Psychology 57, 361–372 (2009) (in Japanese)
4. Nagae, S.: Psychological study on early education and critical period of brain: school education and brain (1). In: Proceedings of Fukuoka University of Education, No.4, pp.95–101 (2009) (in Japanese)
5. Isomura, R., Machida, T., Muto, T.: A Turning Point of Classroom Communication in Lower Elementary School: Introducing "Everyone" to the Class. Japan Society of Development Psychology 16(1), 1–14 (2005) (in Japanese)
6. Matsuo, G., Maruno, S.: How Does an Expert Teacher Create Lessons So That Children Think Subjectively and Learn from Each Other? Students' Sharing of the Ground Rules for Classroom Discussion. Japanese Association of Educational Psychology 55, 93–105 (2007) (in Japanese)
7. Sakai, T.: Developing the method of educational guidance for discussion activity. The National Association of College Teachers for Japanese Language and Literature Education 119, 234–237 (2010) (in Japanese)
8. Otsuki, Y., Bandoh, H., Kato, N., Nakagawa, M.: An Educational Software Supporting Learning through Group Competition Using an Interactive Electronic Whiteboard and Its Effect. Information Processing Society of Japan 44(6), 1635–1644 (2003) (in Japanese)
9. Kitahara, K., Inoue, T., Shigeno, H., Okada, K.: A Tabletop Interface for Supporting Collaborative Learning. Information Processing Society of Japan 47(11), 3054–3062 (2006) (in Japanese)
10. Watanabe, T., Okubo, M., Kuroda, T.: Biological Signal Analysis of Entrainment in Face-to-Face Communication. The 52th National Convention of Information Processing Society of Japan, 419–420 (1996) (in Japanese)
11. Kinoshita, Y.: M-GTA Practice of Grounded Theory Approach, Kobundo (2005) (in Japanese)
12. Kinoshita, Y.: M-GTA Practice of Grounded Theory Approach in Field Study, Kobundo (2005) (in Japanese)
13. Kinoshita, Y.: All of M-GTAFMethod to practice Field Study, Kobundo (2007) (in Japanese)
14. Ando, M.: A Qualitative Approach for Psychological Factors in Use of Interactive Products. Journal of Human Interface Society 12(4), 345–356 (2010) (in Japanese)
15. Saiki, S.C.: Practice of Grounded Theory Approach Shinyosya (2008) (in Japanese)
16. Nagata, S., Kawakami, A.: The Influence of Cognitive Empathy on Small Group Learning. Journal of Japan Society for Educational Technology 32, 141–144 (2008) (in Japanese)
17. Igarashi, R., Maruno, S.: Development and Application of a Method for the Analysis of Classroom Discussion: Visualizing Mutual Links between Utterances in Transcripts. Journal of Japan Society for Educational Technology 32, 89–98 (2008)

Internet Anxiety: Myth or Reality?

Santosh Kumar Kalwar, Kari Heikkinen, and Jari Porras

Department of Software Engineering and Information Management,
Lappeenranta University of Technology, Lappeenranta, Finland
{santosh.kalwar,kari.heikkinen,jari.porras}@lut.fi

Abstract. The purpose of this paper is to determine if Internet anxiety is a myth or reality using literature, questionnaires, and analysis of the collected data. Results showed that the Internet anxiety phenomenon is mostly reality. By placing strong emphasis on the existent Internet anxiety phenomenon, the HCI community could constructively build effective tools and techniques to mitigate users' anxiety.

Keywords: Internet, anxiety, concept, qualitative, myth, reality.

1 Introduction and Background

The rush to remain up-to-date with popular social networking services (e.g. Twitter, Facebook, Google+) is not only crafting unrealistic expectations of these tools but also driving social changes in our daily lives [2]. Ten years ago, applications and services on the Internet were starting to emerge as an integral part of modern life and today internet services are evolving rapidly and extending to ever more areas. Such profound change can clearly be expected to have psychological impact. The research question that we seek to answer in this paper: Is the Internet anxiety phenomenon: myth or reality? To conceptualize and measure Internet anxiety phenomenon it is advantageous to delve deeper into the Internet anxiety literature [1, 14-15, 21]. The existence of many different types of Internet anxiety has been claimed [4, 5]; for example, a user might suffer from an inability to understand certain terms on the Internet, Internet terminology anxiety (ITA), or a user might suffer from general Internet failure anxiety (GIFA). These anxieties taken together can lead to generalized Internet anxiety, unease regarding all aspects of Internet usage. This paper presents a thorough review of the area of Internet anxiety and the current literature in this field. Although many informal surveys [6, 9] in social and behavioral sciences have reported that Internet anxiety (IA) has a significant adverse impact on users' willingness to use the Internet, the problem seems severe as some claims might suggests [2, 9-10]. In reality, the Internet anxiety phenomenon is difficult to understand and quantify. People often report various symptoms of experiencing IA phenomena, e.g., impatience, frustration, irritability, anger and concentration difficulties.

The major contribution in this paper is analysis of the Internet anxiety phenomenon. The method described is based on ongoing research in this area and utilization of qualitative research methodologies. The paper is organized as follows: Section 2 describes the materials and methodologies used. Section 3 describes the literature

M. Kurosu (Ed.): Human-Computer Interaction, Part V, HCII 2013, LNCS 8008, pp. 431–440, 2013.
© Springer-Verlag Berlin Heidelberg 2013

study and Section 4 the experiment and observations; i.e., qualitative user feedback and responses from participants. Section 5 discusses and presents the findings. Section 6 concludes the paper and discusses future work.

2   Methodology

To develop a comprehensive description of Internet anxiety, a qualitative research method was used, supported by semi-structured interviews, participants' feedback and criticism. Use of qualitative research methodologies entails in-depth analysis of the phenomenon under study [16].

2.1   Participants

The participants in this study were technical university students. The study, started with only thirty-seven participants. The research concluded with an extensive survey with five hundred participants with "open-ended" questions. However, the aim of this paper limits on large survey results since the objective is to understand only the Internet anxiety phenomenon (myth or reality) and report on qualitative user responses from the observed participants.

2.2   Data Gathered

Several methods of data collection were used e.g., semi-structured interviews, and survey. These methods were used to identify and acquire a general perspective on the Internet anxiety phenomenon. Various nonverbal and verbal cues of behavioral symptoms were also identified (e.g. user frustration, lack of concentration, impatience).

3   Literature Study

The Internet anxiety literature study presents various claims, anxiety types, and possible symptoms (see Table 1). For example, "If a teenager is trying to have a conversation on an e-mail chat line while doing algebra, she'll suffer a decrease in efficiency, compared to if she just thought about algebra until she was done [1, p. 1238]." This claim could be classified into one or many IA types e.g., Internet terminology anxiety (ITA) (ITDA) (EA), and (EEA) etc. [4, 5]. Possible symptoms could be e.g., lack of focus, lack of concentration, and attention deficiency. This might be assumed as a myth because the task in question is very complex, and the brain would have limited capability to process such tasks [1].

Table 1. Internet anxiety "myth or reality" observed

CLAIMS	INTERNET ANXIETY (TYPES)	POSSIBLE SYMPTOMS	IINITIAL ASSUMPTIONS
"If a teenager is trying to have a conversation on an e-mail chat line while doing algebra, she'll suffer a decrease in efficiency, compared to if she just thought about algebra until she was done [1, p. 1238]."	Could be classified into one or many types; e.g., Internet terminology anxiety (ITA), Internet time-delay anxiety (ITDA), Experience anxiety (EA), and Environment and attraction	E.g. lack of focus, lack of concentration, attention deficiency	This is a myth. The task in question is very complex, and the brain would have limited capability to process such tasks [1].
There is no gender difference in "overall usage of the Internet" [8, p.374].	Could be classified into one or many types; e.g., General Internet failure (GIFA), Usage anxiety (UA) etc. [4, 5]	E.g., "impatience", "frustration", "anger"	This is a reality. Based on present study, and the current data, we postulate that there is gender difference of overall usage of the Internet, in-line with recent findings of Joiner, R., et al. [3].
People on the Internet could be addicted [11, 12], and anxious. For example, Internet user will not suffer from general Internet failure anxiety.	We have found various types of Internet anxiety, e.g., General Internet failure anxiety, Net search anxiety etc. [5]	E.g., Irritation, edgy, Narcissism [7]	This is a reality. Internet user might suffer from general Internet failure anxiety and many other types of Internet anxieties [4, 5, 10].
Internet user very often finds what they are looking for on the Internet. Similarly, people often understand every single word/terms used on the site(s) [5, 13].	Net Search anxiety (NSA), Internet terminology anxiety (ITA) [4, 5]	E.g., Tension, worriness, frustration Inability to understand terms, defeating oneself, upset	This is both myth and reality. It is reality because users report on difficulty for not finding what they are looking for, especially while searching e.g. varies and depends on many factors (e.g. search term/time)[5]. It is a myth because the user does not necessarily understand every single word/terms on the site(s)
Is social networking really causing anxiety? For example, "Is the Web Driving Us Mad?" [2, 9, 10]	Could be linked to any of IA types, e.g. Internet terminology anxiety, Usage anxiety etc. [4-5, 10]	E.g., Over usage, annoyance, sleep deprivation, distracted, defeating oneself, uneasy feeling	This is a reality. A recent study conducted by anxiety UK states that, "about half of the survey's 298 participants, all of whom identified themselves as social media users, say that their use of social networks like Facebook and Twitter makes their lives worse. In particular, participants noted that their own accomplishments to those of their online friends [6]."

4   User Feedback and Observations

The following user feedback or responses were received from the respondents when asked questions about the "Internet anxiety" phenomenon. The limited quoted data (i.e. words) from the participants are shown in bold and highlighted to aid interpretation, and to determine commonalities in the observations.

Some of the questions were rephrased and re-worded in the light of initial responses and after careful consideration. For example, a user response was:

> 1st.   *"Some of the **questions were quite unclear** and would have benefited from **better wording**."*

As the response from the participants was *"Questions were unclear"* and *"better wording"* reframing and rewording of the questions were considered. The question, *"what is the Internet anxiety?"* was replaced by asking *how do you feel about using the Internet and does this have impact on your real life?* The participant (2nd) went beyond the "anxiety" phenomenon and started to relate with "Internet addicts." Furthermore, the same participant also wished for and reported on possible Internet anxiety symptom to manage time (e.g. *"I'd like to spend less time"*) on the Internet. Another participant described an Internet anxiety symptom; e.g. *"Heart rate goes high"* (3rd) while experiencing a sports program on the Internet. Example responses from these participants are:

> 2nd.   *"Although I don't consider myself as an Internet addict (especially compared to others) **I'd like to spend less time** on the Internet (or on computer)."*
>
> 3rd.   *"Many of the effects are not really Internet related but more on content. For example, I watch football from Internet that causes my **heart rate goes high**. That would also happen if I were to watch the game live on stadium or from TV...Internet is a tool for me, not a life."*

Unstructured and semi-structured interview pools of the participants felt the questions asked were not actually related with the Internet but focused more on Internet contents. Furthermore, participants expressed concern about possible effects of social networking sites (SNS). The participant (4th) reported that she is *"not addicted to social networking"* but somehow feels *"concerned,"* and is *"often not happy"* about her close friends and families spending *"enormous amount"* of time on the SNS. The same participant revealed possible Internet anxiety symptom; e.g. unpleasant/ unhappiness (i.e. *"not happy"*) feeling. For example, the participant responded:

> 4th.   *"I personally am **not addicted to social networking** however **I have found myself raising concerns about close associates and relatives who spend enormous amount** of precious **facebooking**. I think it affects their time to work and make money and I am **often not happy** about it. Also sometimes such **individuals post contents** which concern me e.g. my photos, what I am up to, etc on facebook and I am **not happy** about this."*

Related to "information overload" and SNS anxiety, a participant (5th) responded that he has "*skipped Social networks*" and believed that these SNS "*mainly annoy*" him. Interestingly, the same participant narrated that he is "*not interested in paying any-thing*" on the Internet. Moreover, he blocks "*ads from browser*" and any eye-catching application or services (e.g. "*blink applications*"). Another participant (6th) considers that SNS are "*bullshit*" and declared that SNS are "*commercial selling tool.*" Similar-ly, another participant (7th) expressed concern over "*commercials*" on the Internet. The example responses from these participants were:

> 5th.   "*I'm mostly concerned about information glut. **I've skipped Social networks** (FB, G+ etc), and those mainly **annoy** me if they pose **hindrance**, like required access to FB to get to content, but I can live without such content or find respective in-formation from somewhere else. Also, I'm **not interested in paying** about anything in Internet. I'v blocked mots on **ads from browser** and generally **blocks anything distracting colorful "blink blink" applications**.*"

> 6th.   "*Social media are **bullshit**; they are just another **commercial selling tool**.*"

> 7th.   "*Most annoying part of the internet is **commercials** which make sound and/or cover the main page content.*"

The participant described possible Internet anxiety symptoms; e.g. "*Losing internet connection*" (8th), encountering "*delays in accessing internet content*" (9th), "*fru-strated because of time delays*" (10th), and major concerns over "*low internet band-width*" (11th). The example responses from these participants were:

> 8th.   "*I need Internet for my work and, consequently, **losing Internet connection** is of-ten **problematic**.*"

> 9th.   "*It is unclear if I should answer **how often I face** e.g. **delay in accessing Internet content** (and feel anxiety about it) or if I am anxious if a delay occurs.*"

> 10th.   "*Nowadays, I actually get **frustrated** because of **time delays** when using **mobile devices**.*"

> 11th.   "*I don't like **low Internet bandwidth**.*"

But, one participant (12th) reported that "*using the Internet doesn't bother*" but, while performing time critical task(s), if there is no Internet connection than they might get anxious ("*sometime there's no connection...I do get anxious*"). Another participant (13th) self-corrects on general Internet usage-to-usage frustration and reports symp-toms such as, "*frustration*". The example responses from these participants were:

> 12th.   "*Using the Internet doesn't bother me. It doesn't make me anxious. It's just so **normal** to use it everywhere that it has become a norm for me. Then when **sometimes there's no connection and I need to check something** (e.g. timetables) **I do get anxious**, but it's not the Internet's fault.*"

> 13th.   "***If 'usage anxiety' were 'usage frustration', my answer would be often.** I usually become **frustrated** but not anxious. (I hope I understood the term 'anxiety' correctly.)*"

Feelings associated with Internet use resulted in some comments from the partici-pants. A participant (14th) acknowledges possibilities of using the Internet (e.g. *"serves me well...work and during free time"*) and expresses that he is *"not emotionally attached"* with the Internet. The example response from the participant is:

14th.   *"The Internet **serves me well both at work and during free time** but I'm **not emotionally involved** with it!"*

Furthermore, some participants were concerned about Internet search and keeping-in-touch with friends and families (15th) whereas other participants (16th) reported that they *"do not necessarily get anxious"* but believed in improving and *"optimizing search engines."* The example responses from these participants are:

15th.   *"I try to use the Internet **only for searching** modern information in the area of my research work. Also it **helps me to be in touch with my relatives and friends..."***

16th.   *"**I do not necessarily get anxious** but **I feel that there is still a lot more to be done to optimize search engines**, etc, especially when searched content cannot be found."*

The example response from a participant is: *"**Good survey** for the **young people**, around ages 18 to 25."* Table below (see Table 2) shows the collected data (in the first column) and initial results.

Table 2. Commonly highlighted participant responses (in the second column) and possible Internet anxiety types (in the third column)

Participants no.	Common highlighted participant responses	Possible Internet anxiety types	Ref.
16	*"I feel that there is still a lot more to be done to optimize search engines"*	Net search anxiety (NSA)	[4, 5]
6, 15	*"Helps me to be in touch with my relatives and friends," "Bullshit,"*	Experience anxiety (EA)	[5, 17]
13	*"If 'usage anxiety' were 'usage frustration', my answer would be often"*	Internet terminology anxiety (ITA), Usage anxiety (UA)	[4,5, 17]
12	*"Sometimes there's no connection and I need to check something."*	General Internet failure anxiety (GIFA)	[5]
11, 10	*"Low internet bandwidth, time delays..."*	Internet time-delay anxiety (ITDA)	[4,5]
5	*"Distracting colorful "blink blink" applications, not interested in paying"*	Environment and attraction anxiety (EEA)	[5, 17]

5   Results and Discussion

The results based on the data gathered from various qualitative coding, experiment and observations are summarized in Table 3.

Table 3. Initial participant assumptions, possible responses, final outcome, and justifications

Partic-ipant no.	Initial assump-tion	User Feedback or Responses	Final Outcome	Justifications
2	Myth	".. I'd like to spend less time ..."	Reality	Spending more time on the Internet can affect social, personal, and academic performance [1, 2].
3	Reality	"...heart rate goes high..."	Myth	It's normal to feel that heart rate goes high, a user has a sense of psy-cho-physiological change in her body in a particular context or situation
4	Myth	"…raising concerns about close associates and relatives...", "not addicted to Facebooking"	Reality	People have lower at-tention span and those who spend more time on SNSs, usually spend less time on social ac-tivities [7, 20-21].
5-7	Reality	"..annoyance", "distraction", "ads from browser", "commercials"	Reality	Popular social network-ing tools might distract users, and might affect on academic perfor-mance [1], various other social implications [5-7].
8-11	Myth	"losing internet connection, "delay in accessing Inter-net contents"	Reality	Users show possible symptoms when the Internet is not working [5, 17].
12-13	Myth	"it's not the Internet fault", "frustrated but not anxious"	Reality	Self-blame is sign of Internet anxiety symp-toms.

Table 3. (*continued*)

14	Myth	"not emotionally involved"	Myth and/or Reality	It depends. Some user might report of being "emotionally involved" and other "not emotionally involved".
15-16	Reality	"only for searching", "lot more to be done to optimize search engine"	Reality	Search engines are far from being perfect and 'search anxiety' is reality not because something is easy to find but because it is hard to analyze and trust.

Suggested by the data collection and analysis, the results of the Internet anxiety phenomenon seems to modulate more to reality than myth. Although researchers have looked deeper into various demographic variables; e.g. gender, personality, beliefs etc. [3, 15], this study paints a distinct picture in postulating a general claim of the Internet anxiety phenomenon as a real phenomenon. Although the issue of Internet anxiety is not something new in the scholarly community [17, 18], the new knowledge in this paper is a possible mapping of various Internet anxiety types with reported symptoms through data collection and the results. The implication of this study shows that one has to consider possible hazardous implications of using the Internet (or not using the Internet) [20] and the applications therein, as there is an ongoing debate in the HCI community on possible impacts of the Internet on the lives of children and younger people [19], e.g., "Facebook and academic performance" [1, 7], "problematic Internet use" [4, 6], and the "obsession with Technology" [2]. As our modern life is spent more online, we seem to have less time for personal facial interaction and there is more virtual interaction on the Internet [2]. This type of behavior seems to negatively influence our personality, relationships, and well-being [15]. The pervasiveness of Internet use can generate high levels of Internet anxiety [17, 18], and we need to take a short break to reset our mental and physical faculties [2, 10]. Let us now conclude and present few recommendations based on what we have learned from this study.

6   Conclusions and Recommendations

In this paper, we considered the issue of the present day Internet anxiety phenomenon: myth or reality. The comparison data suggest that the Internet anxiety phenomenon is mostly reality. Based on an experiment and observations, we found that the Internet anxiety phenomenon experienced by the participants can be an existent phenomenon. The reason we came to this decision is twofold: firstly, the sampling of our participants and the coded data revealed possible deleterious symptoms of using the Internet. Secondly, similar possible symptoms were also discovered in present Internet anxiety literature. The finding from the collected data, experiment and literature sources validate the findings. Therefore, one could easily argue that the Internet anxiety phenomenon is a real thing. However, there are few limitations in this work. First, the data collected might not represent the entire Internet population generally and second, the results are the subjective evaluation of the participant's responses and do not necessarily give an objective measure for the phenomenon under study. In our future work, we plan to create a module with the collection of algorithms (i.e. Feelcalc) to mitigate real Internet anxiety phenomenon. Possible strategies for mitigating the Internet anxiety phenomenon include:

- As recommended by researchers [2, 20, 21], take short breaks and avoid using the Internet in the middle of night or at the dinner table. This is to avoid distraction by technology and mental disturbance or an adverse effect on sleeping patterns.
- Be responsible, careful, and smart when using social media applications or sites (e.g. Facebook and Twitter). Researchers have reported the possible adverse impact of using these sites with narcissistic [7], loneliness, sexting, and pornographic, antisocial, cyber-bullying, and addictive behavioral symptoms.
- The results of Internet anxiety research can be useful, if we can build and design effective sites and contents on the Internet considering reported user's anxiety symptoms; e.g. impatience, frustration, depression, tension, and anxiety.
- Multi-tasking is a myth [1]. Users are not actually multitasking on the Internet but constantly switching between various tasks. Instead of doing many things concurrently, one has to consider doing one thing at the right time.
- The Internet anxiety phenomenon is real and how we can cope up with this new reality is all in our hands.

Acknowledgements. This work has been partially supported by ECSE (East Finland Graduate School in Computer Science and Engineering), the Foundation of Nokia Corporation, LUT Foundation grant (Lauri ja Lahja Hotisen rahasto, Väitöskirjan viimeistelyapurahat), and the Finnish Foundation for Technology Promotion (Tekniikan edistämissäätiö).

References

1. Kirschner, P.A., Karpinski, A.C.: Facebook® and academic performance. Computers in Human Behavior 26(6), 1237–1245 (2010) ISSN 0747-5632
2. Rose, L.D.: iDisorder: Understanding Our Obsession with Technology and Overcoming Its Hold on Us. Palgrave Macmillan (2012)
3. Joiner, R., Gavin, J., Brosnan, M., Cromby, J., Gegory, H., Guiller, J., Maras, P., Moon, A.: Gender, Internet experience, Internet identification and Internet anxiety: a ten year follow-up. Cyberpsychology, Behavior, and Social Networking 15(7), 370–372 (2012)
4. Presno, C.: Taking the byte out of internet anxiety: Instructional techniques that reduce computer/internet anxiety in the classroom. Journal of Educational Computing Research 18(2), 147–161 (1998)
5. Kalwar, S.K.: Comparison of Human Anxiety Based on Different Cultural Backgrounds. CyberPsychology, Behaviour and Social Networking 13(4), 443–446 (2010)

6. Torevell, T.: (July 9, 2012), http://www.anxietyuk.org.uk/2012/07/for-some-with-anxiety-technology-can-increase-anxiety/
7. Ryan, T., Xenos, S.: Who uses Facebook? An investigation into the relationship between the Big Five, shyness, narcissism, loneliness, and Facebook usage. Computers in Human Behavior 27(5), 1658–1664 (2011) ISSN 0747-5632
8. Jackson, L.A., Ervin, K.S., Gardner, P.D., Schmitt, N.: Gender and the Internet: Women Communicating and Men Searching. Sex Roles 44(5), 363–379 (2001)
9. Dokupil, T.: Is the Web Driving Us Mad? (July 9, 2012), http://www.thedailybeast.com/newsweek/2012/07/08/is-the-internet-making-us-crazy-what-the-new-research-says.html
10. Carr, N.: The Shallows: What the Internet Is Doing to Our Brains. W. W. Norton & Company (2010)
11. Grohol, J.M.: Too much time online: Internet addiction or healthy social interactions. Cyberpsychol. Behav. 2(5), 395–401 (1999)
12. Griffiths, M.D.: Internet addiction: Does it really exist? In: Gackenbach, J. (ed.) Psychology and the Internet: Intrapersonal, Interpersonal, and Transpersonal Implications. Academic Press, New York (1998)
13. Widyanto, L., Griffiths, M.D., Brunsden, V., Mcmurran, M.: The psychometric properties of the Internet related problem scale: a pilot study. International Journal of Mental Health and Addiction 6(2), 205–213 (2007)
14. Kalwar, S.K., Heikkinen, K., Porras, J.: Conceptual Framework for Assessing Human Anxiety on the Internet. Procedia-Social and Behavioral Sciences 46, 4907–4917 (2012)
15. Thatcher, J.B., Loughry, M.L., Lim, J., McKnight, D.H.: Internet anxiety: An empirical study of the effects of personality, beliefs, and social support. In: Proceedings of Mathematical Lecture Note Series, pp. 353–363 (2007)
16. Peshkin, A.: Understanding complexity: A gift of qualitative inquiry. Anthropology & Education Quarterly 19(4), 416–424 (2009)
17. Kalwar, S.K., Heikkinen, K.: Study of human anxiety on the internet. In: Jacko, J.A. (ed.) HCI International 2009, Part I. LNCS, vol. 5610, pp. 69–76. Springer, Heidelberg (2009)
18. Brosnan, M., Joiner, R., Gavin, J., Crook, C., Maras, P., Guiller, J., Scott, A.J.: The Impact of Pathological Levels of Internet-Related Anxiety on Internet Usage. Journal of Educational Computing Research 46(4), 341–356 (2012)
19. Amichai-Hamburger, Y., Barak, A.: Internet and well-being. In: Amichai-Hamburger, Y. (ed.) Technology and Well-being, pp. 34–76. Cambridge University Press (2009)
20. Laghi, F., Schneider, B.H., Vitoroulis, I., Coplan, R.J., Baiocco, R., Amichai-Hamburger, Y., Hudek, N., Koszycki, D., Miller, S., Flament, M.: Knowing when not to use the Internet: Shyness and adolescents' on-line and off-line interactions with friends. Computers in Human Behavior 29(1), 51–57 (2013) ISSN 0747-5632
21. Stepanikova, I., Nie, N.H., Xiaobin, H.: Time on the Internet at home, loneliness, and life satisfaction: Evidence from panel time-diary data. Computers in Human Behavior 26(3) (May 2010)

Brain Function Connectivity Analysis for Recognizing Different Relation of Social Emotion in Virtual Reality

Jonghwa Kim[1], Dongkeun Kim[2], Sangmin Ann[1],
Sangin Park[1], and Mincheol Whang[2,*]

[1] Department ofEmotion Engineering, SangmyungGraduate School, Seoul, Korea
{rmx2003,eusm36,ini0630}@naver.com
[2] Department of Digital Media, SangmyungUniversity, Seoul, Korea
{dkim,whang}@smu.ac.kr

Abstract. Social emotions are emotion that can be induced from human social relationships when people are interacting with others. In this study, we are aim to analyze a brain function connectivityin terms of different relations of social emotions. The brain function connectivity can be used to observe the neural responses with features of EEG coherences during a cognitive process.In this study, the EEG coherence is measured according to different social emotion evocations. The auditory and visual stimulus for inducing social emotions was presented to participants during 20.5 sec (±3.1 sec). The participants were asked to imagine and explain about similar emotion experience after watching each video clips. The measured EEG coherencewas grouped into two different social emotion categories;the information sharing relation and emotion sharing relation, and compared with the results of subjective evaluation and independent T-test.The information sharing relation was related with the brain connectivity oftherighttemporo-occipitalposition associated with a language memory. The emotion sharing relation was related with the brain connectivity of the left fronto-right parietal position associated with a visual information processing area.

Keywords: Emotion, Social emotion, Emotion relation, EEG coherence, Brain function connectivity.

1 Introduction

In this study, social emotions are emotion that can be induced from human social relationships when people are interacting with others. Social emotion makes notice the intention of others and thus helps a more fluent information interaction to SNS (social networking service) user.

Social emotion makes notice the intention of others and thus helps a more fluent information interactionto SNS (social networking service) user.

The objective of this study is to analyze the brain functional connectivity by the different relation of a SNS. The cerebral activity that responds to cognitive function is

---

* Corresponding author.

M. Kurosu (Ed.): Human-Computer Interaction, Part V, HCII 2013, LNCS 8008, pp. 441–447, 2013.
© Springer-Verlag Berlin Heidelberg 2013

able to analyze by brain functional connectivity, so it is important to the new HCI method.

EEG coherence has been used in the analysis of cognitive function. Rietschel(2012) used brain functional connectivity to analyze and to assess the cognitive process. Differences of task difficulty were compared in brain functional connectivity using EEG coherence. According to the result, the brain functional connectivity in the sensory motor increased when the given task was more difficult [1].

The EEG coherence has been used in order to observe the cognitive function in response to the motion of an object using brain functional connectivity. Experiment was to show visual simulating the collision of moving objects. Connectivity of the left sensory motor circuitthat is between the left occipital lobe and the parietal lobe increased when visual focus was needed [2]. So, according these studies results, EEG coherence was useful to analyzing cognitive function.

Reiser(2012) used the video and voice which expressed an actor's emotion for evoke social emotion. Variation of brain functional connectivity was analyzed using EEG coherence. Connectivity between the prefrontal cortex and the posterior cortical regions was recorded. The social emotion stimuli such as sadness, anxiety and neutrality represented to the subjects. According to the result, decrease of the brain connectivity in some subjects meant a lower degree of attention on emotional information. For other subjects, increase of the brain connectivity in the right hemisphere showed the focus on the social emotional information [3]. Thus brain functional connectivity changed between the prefrontal cortex and the posterior lobe area was related with emotional recognition.

Ye(2011) analyzed brain functional connectivity change by social relation.The images of persons who were different social relation with participants were presented and the subjects were instructed to recognize them. According to the result, frontotemporal connectivity was activated when social emotion was evoked [4].

Through these studies, the right cerebral hemisphere associated with emotion by the social relation was able to know how the brain reacts. Especially, connectivity between the frontal and temporal lobes was correlated with emotion by the social relation. However, the previous studies were difficult to see as a reaction to the process such as acquisition of information, emotion evoking, and emotion expression that occur within the SNS. Therefore, the objective of this paper is to analyze connectivity of brain function using EEG coherence in different relation with a similar situation within a SNS.

2   Method

2.1   Independent Variable and Dependent Variable

The independent variables were consisted as conformation as information sharing relation and emotion sharing relation. To evoke social emotion, the scenarios defined as visual stimuli (text and representative images) and auditory stimuli (prerecorded audio of a scenario) were presented simultaneously. Research of social emotion has been undertaken with a variety of independent variables such as image, text and recorded sounds noted. [4-10].

For the dependent variable, EEG was measured at eight points (F3, F4, T3, T4, P3, P4, O1, O2) shown as figure 1 and the sampling rate was 400Hz. An EEG100C (Biopac, USA) was used for measurement and impedance was less than 5kΩ.

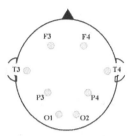

Fig. 1. Eight points of EEG measurement

2.2   Participants and Procedure

Twenty seven right-handed normal subjects (male: 13, female: 14) performed this experiment. The average age was 24.8(±1.3). The Subjects were paid about $50 for participation. They were instructed to have a deep sleep before the experiment, and were also instructed not to drink any caffeinated beverages, in order to limit natural arousal stimuli. When the subjects arrived in the experiment room, they were given 10-minutes rest in order to accustom them to the environment. Afterwards, they were informed about the experiment objective and process of test.

The EEG data was measured during the experiments of social emotion evocation. Stimulus within the experiment is a scenario which is the explanation of relation. This auditory and visual stimulus was presented to the participants during 22.5 sec. (±2.1 sec.). The participants were asked to imagine and describe a similar experience. The average time of imagination was 18.8 sec. (±2.5 sec.) and the vocal expression was 31.3 sec. (±3.1 sec.).

This experimental protocol is an improved method on the existing experimental protocol which has been used in various research purposes for the reaction according to the social emotion [9-14].

3   Analysis and Results

3.1   The Coherence of EEG

By using 8 measuring points, a total of 28 coherences (between the points) were analyzed (e.g. F3-F4, F3-T3, F3-T4...). EEG coherence measured the synchronization of two signals as a value between 0 and 1. If the coherence value was one, the two signals were identical. EEG coherence between two points can be measured with formula (1). In formula 1, R_x (w) and R_y (w) represents the spectrum power of each signal, and R_{xy} (w) represents the cross spectrum power. The FFT method was used for spectrum analysis, by applying a sliding window. The window size of the sliding window consists of 1600 (4 sec) samples, while the interval in which the slide progressed forward is at a rate of 800 samples (2 sec).

The EEG coherence result, divided as spectrum: delta(0.5~4Hz), theta(4~7Hz), alpha low(8~10Hz), alpha high(10~12Hz), beta low(12~15Hz), beta mid(15~18Hz), beta high(18~30Hz), and Gamma(26~100Hz).

$$C_{xy} = \frac{|R_{xy}(w)|^2}{R_x(w) \times R_y(w)} \tag{1}$$

3.2   Normalized EEG Coherence

Normalized EEG coherence was used in a number of studies as a way to clarify the comparison between the stimuli [15- 17]. Entire individual coherence measurements without the distinction of social emotional stimuli were normalized using formula 2. In Figure 2, the normalization of the results before and after the F3-F4 alpha high coherence values can be compared.

As shown in Figure 2, the maximum value of coherence and the top 5% of the value of individual differences were not significant when normalization was not applied. Also shown in Figure 2, it can be seen that a decrease in the top 5% of the value of individual differences, but maintains the maximum change in the value of the data when normalization was applied. In addition, the mean value is fixed at 0. So, Normalized EEG coherence gave the advantages of easier analysis of the pattern of response to stimulation by the individual subject.

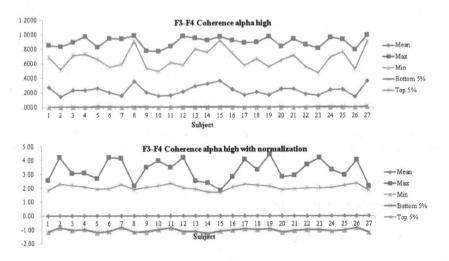

Fig. 2. Normalized EEG coherence of F3-F4 alpha high by individual (Top: Non normalized EEG coherence, Bottom: Normalized EEG coherence)

$$Z - \text{normalization}(x_i) = \frac{x_i - Mean}{Standard\ Deviation} \tag{2}$$

3.3   Statistical Verification and Results

According to the experiment process, Normalized EEG coherence data was divided as stimulus, imagination, and expression and each individual data was separated as the emotion sharing relation or information sharing relation. The separated results were compared using an independent T-test by SPSS 19. Some normalized EEG coherence showed significant differences and is displayed in figure 3(p<0.01). Only the common responses of the three processes from the experiment (stimulus, imagination, and expression) were arranged. Because of the result, one process had a chance of response, except an emotion response such as a difference of visual stimulus or an auditory volume.

As shown in figure 3, right temporo-occipitalipsilateral connectivity changed at alpha low(8~10Hz) and translateral connectivity between left frontal and right parietal changed at alpha high(10~12Hz). When the social emotion of emotion sharing relation was evoked, alpha low was decreased and alpha high was increased during stimulation, imagination and expression process.Otherwise, alpha low was increased and alpha high was decreased when the relation was information sharing.

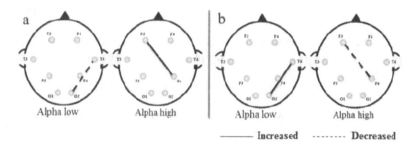

Fig. 3. Significant EEG coherence of relation (a: Emotion sharing relation, b: Information sharing relation)

4   Conclusion and Discussion

The research of social emotion within a SNS is invaluable because social emotion helps fluent information transaction and also gives a clue to the invention of a new HCI method. But less is known about the body's reaction of social emotion and basic emotion. So the purpose of this study was to find a central nervous response by relation of social emotion.

The method of relation stimuli such as the information sharing and the emotion sharing was used visuals and audio.And three steps such as stimulus, imagination, and expressions were performed by the participants. During stimulus, 8ch EEG was recorded and converted as EEG coherence. EEG coherence was transformed as normalized value to decrease individual difference. The central nervous response by relation was compared using an independent T-test.

According to the result, when the social emotion of emotion sharing relation was evoked,the translateral connectivity between left frontal and right parietal was increased during stimulation, imagination and expression process.Otherwise, right temporo-occipitalipsilateral connectivity was increased when the relation was information sharing.

This connectivity was related to language memory process and the visual information processing. The information sharing response which is right temporo-occipital connectivity was associated with the language memory [18, 19]. And reaction of emotion sharing which is the connectivity between left frontal and right parietal was associated with the visual information processing area [20].Therefore, language memory was enhanced when the relation was information sharing. And visual information process was activated when emotion sharing relation.

Using this coherence result of this paper, the relation between persons in SNS could be recognizable. Furthermore, we could figure out the dominant cognitive process when social emotion was evoked.

Acknowledgements. This work was supported by the Global Frontier R&D Program on <Human-centered Interaction for Coexistence> funded by the National Research Foundation of Korea grant funded by the Korean Government (MEST) (NRF-2012M3A6A3054312).

References

1. Rietschel, J.C., Miller, M.W., Gentili, R.J., Goodman, R.N., McDonald, C.G., Hatfield, B.D.: Cerebral-cortical networking and activation increase as a function of cognitive-motor task difficulty. Biol. Psychol. 90(2), 127–133 (2012)
2. Spape, M.M., Serrien, D.J.: Prediction of collision events, and an EEG coherence analysis. Clin. Neurophysiol. 122(5), 891–896 (2011)
3. Reiser, E.M., Schulter, G., Weiss, E.M., Fink, A., Rominger, C., Papousek, I.: Decrease of prefrontal-posterior EEG coherence: lose control during social-emotional stimulation. Brain Cogn. 80(1), 144–154 (2012)
4. Ye, Z., Kopyciok, R., Mohammadi, B., Krämer, U.M., Brunnlieb, C., Heldmann, M., Samii, A., Münte, T.F.: Androgens Modulate Brain Networks of Empathy in Female-to-Male Transsexuals: An fMRI Study. Zeitschrift für Neuropsychologie 22(4), 263–277 (2011)
5. Alba-Ferrara, L., Hausmann, M., Mitchell, R.L., Weis, S.: The neural correlates of emotional prosody comprehension: disentangling simple from complex emotion. PloS One 6(12), 28701 (2011)
6. Baron-Cohen, S., Wheelwright, S., Jolliffe, A.T.: Is there a language of the eyes? Evidence from normal adults, and adults with autism or Asperger syndrome. Visual Cognition 4(3), 311–331 (1997)
7. Critchley, H., Daly, E., Phillips, M., Brammer, M., Bullmore, E., Williams, S., Van Amelsvoort, T., Robertson, D., David, A., Murphy, D.: Explicit and implicit neural mechanisms for processing of social information from facial expressions: a functional magnetic resonance imaging study. Human Brain Mapping 9(2), 93–105 (2000)
8. Critchley, H.D., Daly, E.M., Bullmore, E.T., Williams, S.C.R., Van Amelsvoort, T., Robertson, D.M., Rowe, A., Phillips, M., McAlonan, G., Howlin, P.: The functional neuroanatomy of social behaviour changes in cerebral blood flow when people with autistic disorder process facial expressions. Brain 123(11), 2203–2212 (2000)
9. Adolphs, R., Baron-Cohen, S., Tranel, D.: Impaired recognition of social emotions following amygdala damage. Journal of Cognitive Neuroscience 14(8), 1264–1274 (2002)
10. Adolphs, R., Sears, L., Piven, J.: Abnormal processing of social information from faces in autism. Journal of Cognitive Neuroscience 13(2), 232–240 (2001)

11. Abe, N., Suzuki, M., Mori, E., Itoh, M., Fujii, T.: Deceiving others: distinct neural responses of the prefrontal cortex and amygdala in simple fabrication and deception with social interactions. Journal of Cognitive Neuroscience 19(2), 287–295 (2007)

12. Burnett, S., Blakemore, S.J.: Functional connectivity during a social emotion task in adolescents and in adults. European Journal of Neuroscience 29(6), 1294–1301 (2009)

13. Bauminger, N.: The facilitation of social-emotional understanding and social interaction in high-functioning children with autism: Intervention outcomes. Journal of Autism and Developmental Disorders 32(4), 283–298 (2002)

14. Cooke, T.P., Apolloni, T.: Developing positive social-emotional behaviors: A study of training and generalization effects. Journal of Applied Behavior Analysis 9(1), 65 (1976)

15. Locatelli, T., Cursi, M., Liberati, D., Franceschi, M., Comi, G.: EEG coherence in Alzheimer's disease. Electroencephalography and Clinical Neurophysiology 106(3), 229–237 (1998)

16. Knott, V., Mohr, E., Mahoney, C., Ilivitsky, V.: Electroencephalographic Coherence in Alzheimer's Disease: Comparisons with a Control Group and Population Norms. Journal of Geriatric Psychiatry and Neurology 13(1), 1–8 (2000)

17. Miskovic, V., Schmidt, L.A., Georgiades, K., Boyle, M., Macmillan, H.L.: Adolescent females exposed to child maltreatment exhibit atypical EEG coherence and psychiatric impairment: linking early adversity. The Brain, and Psychopathology, Development and Psychopathology 22(2), 419–432 (2010)

18. Meister, I.G., Weidemann, J., Foltys, H., Brand, H., Willmes, K., Krings, T., Thron, A., Töpper, R., Boroojerdi, B.: The neural correlate of very long term picture priming. European Journal of Neuroscience 21(4), 1101–1106 (2005)

19. Mickley Steinmetz, K.R., Kensinger, E.A.: The effects of valence and arousal on the neural activity leading to subsequent memory. Psychophysiology 46(6), 1190–1199 (2009)

20. Finke, K., Bublak, P., Zihl, J.: Visual spatial and visual pattern working memory: Neuropsychological evidence for a differential role of left and right dorsal visual brain. Neuropsychologia 44(4), 649–661 (2006)

A Mobile Brain-Computer Interface
for Freely Moving Humans

Yuan-Pin Lin, Yijun Wang, Chun-Shu Wei, and Tzyy-Ping Jung

Swartz Center for Computational Neuroscience,
Institute for Neural Computation, University of California, San Diego, USA
{yplin,yijun,cswei,jung}@sccn.ucsd.edu

Abstract. Recent advances in mobile electroencephalogram (EEG) systems featuring dry electrodes and wireless telemetry have promoted the applications of brain-computer interfaces (BCIs) in our daily life. In the field of neuroscience, understanding the underlying neural mechanisms of unconstrained human behaviors, *i.e.* freely moving humans, is accordingly in high demand. The empirical results of this study demonstrated the feasibility of using a mobile BCI system to detect steady-state visual-evoked potential (SSVEP) of the participants during natural human walking. This study considerably facilitates the process of bridging laboratory-oriented BCI demonstrations into mobile EEG-based systems for real-life environments.

Keywords: EEG, BCI, SSVEP, moving humans.

1 Introduction

A steady-state visual-evoked potential (SSVEP) is a frequency-modulated brain signal in response to a periodic visual flickering. Signals acquired from the parieto-occipital region over the visual cortex commonly provide the highest signal-to-noise ratio (SNR) than from other scalp locations. An SSVEP-based brain-computer Interface (BCI) has recently gained much attention since it requires minimal user training and provides high information transfer rate (ITR) [3]. The SSVEP BCI has thus become a promising modality for patients with severe motor disabilities to directly communicate with the environment through recognizing the frequencies of the acquired signals.

SSVEP has also been widely used in clinical diagnostic medicine. In 1959, Golla and Winter [4] first reported that migraineurs have distinct brain activities in response to photic stimulation, *i.e.* SSVEP, compared to healthy controls. Furthermore, Chen *et al.* [5] showed significantly different SSVEPs between interictal and peri-ictal periods in migraineurs. That is, SSVEP might be a useful tool for predicting the headache attacks. A mobile and wearable SSVEP-based BCI system is critical for continuously and robustly monitoring migraineurs' EEG activities in natural head/body positions and movements. Recently, rapid advances in mobile electroencephalogram (EEG) systems featuring dry and zero-prep electrodes, miniature electronics, wireless telemetry [1], and/or cell-phone-based platforms [2] have promoted the translation of a BCI system from a laboratory setting to real-world practices. Particularly, previous studies [1-2] have shown that a

M. Kurosu (Ed.): Human-Computer Interaction, Part V, HCII 2013, LNCS 8008, pp. 448–453, 2013.

mobile and non-tethered EEG system can reliably detect SSVEP signals. However, the SSVEP studies were conducted in well-controlled laboratories, where participants were strictly instructed to restrict any body movements while gazing at the flickering visual targets – largely out of fear of introducing non-brain artifacts into the EEG data records. This restriction hinders the long-term, continuous and routine EEG monitoring in the workplace or at home. Until recently, Debener *et al.* [6] explored the feasibility of assessing BCI-related EEG tasks, *e.g.* P300 event-related potential (ERP), in walking humans. However, the feasibility of acquiring SSVEP signals in hostile recording conditions has not been fully explored.

This study systematically tests the feasibility of using a mobile and wireless BCI system to detect SSVEP of the participants during natural walking. A treadmill was adopted to create a speed-adjustable walking platform for eliciting different degrees of head/body movements, *i.e.* increasing the walking speed would accompany larger head and body sways. Canonical correlation analysis (CCA), which has been proved robust in detecting SSVEP frequencies [7-8], was used to explore the SSVEP detectability under different walking speeds.

2   Material and Method

2.1   Experiment Setup

To acquire EEG data during walking, this study employed an adjustable-speed treadmill to mimic natural human walking in daily life (Fig. 1(a)). Participants were instructed to gaze at a flickering stimulus of 11 Hz/12 Hz for 60 seconds while walking on the treadmill with three speeds of 1, 2 and 3 mile (s) per hour (MPH). The flickering stimuli (7.5cm x 6.0cm) were presented at the center of an LCD monitor with a 60Hz refresh rate. The stimulus program was developed under Microsoft Visual C++ using the Microsoft DirectX 7.0 framework [9]. For comparison, this study also acquired the EEG signals while participants were standing on the treadmill (0 MPH). A variable time interval of 10 to 20 seconds was interleaved with visual flickering stimuli to avoid visual and/or motion fatigue.

2.2   EEG Data Acquisition

Ten healthy participants (8 males and 2 females; 23-31 years of age; mean age: 27.5 years) with normal or corrected-to-normal vision participated in this experiment. Each participant signed an informed consent approved by the UCSD Human Research Protections Program before the experiment.

This study used a 32-channel EEG system (Cognionics, Inc.) featuring dry electrodes and wireless telemetry to record SSVEP signals with a sampling rate of 250 Hz. The headset is made from soft fabric and completely encloses the system's electronics (Fig. 1(b)). This study only used two four-electrode straps (eight electrodes: P3, P1, P2, P4, PO3, PO1, PO2 and PO4) over the parietal and occipital areas to record SSVEP signals (Fig. 1(c)). Dry electrodes at O1 and O2 were substituted by wets ones for performing a wet-dry comparison of the SSVEP performance (against adjacent dry electrodes: PO1 and PO2). For each participant, the dataset consisted of 8 60-s EEG segments (four walking speeds x two flickering frequencies).

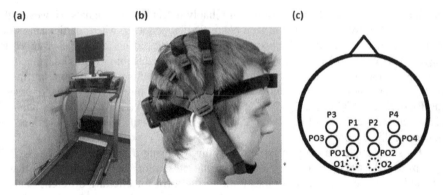

Fig. 1. (a) Experiment setup, (b) a 32-channel wireless EEG system, and (c) electrode locations used for extracting SSVEP signals

2.3   SSVEP Analysis

This study adopted CCA [7-8], a widely used algorithm in SSVEP-based BCIs, to detect frequencies of the SSVEP signals. CCA aims to maximize the correlation between the recorded EEG signals and the sinusoidal templates corresponding to the flickering frequencies. Applying the coefficients of CCA as spatial filters to multi-channel EEG time series returned SNR-enhanced SSVEP signals. Note that CCA calculation in this study only relied on the fundamental frequency of template signals.

In addition, this study systematically assessed two important parameters: (1) the length of an EEG epoch, and (2) the number of channels, from the recorded EEG signals to explore a better way for detecting SSVEP in freely moving humans. The 8-channel EEG data were first filtered by a 5-50 Hz band-pass Chebyshev Type I filter to remove low-frequency signal drifts and high-frequency motion artifacts. Each 60-s EEG time series was then segmented into N-s epochs (N=1-4). Epochs contaminated by severely transient motion artifacts were removed from further analysis. EEG data from one participant were excluded from the analysis because the remaining number of epochs was very limited after trial rejection.

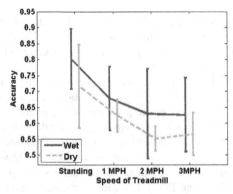

Fig. 2. Averaged SSVEP detection accuracy using two wet (O1 and O2) and dry (PO1 and PO2) electrodes across different walking speeds

This study performed three tests: (1) a comparison between the SSVEPs measured by dry versus wet electrodes; (2) the effect of the number of electrodes (four electrodes on the parieto-occipital strap versus eight electrodes on both the parieto-occipital and parietal straps in Fig. 1(b) and (c)) on the detectability of SSVEPs; and (3) the effect of the length of an epoch on the detectability of SSVEPs. This study evaluated the detectability of SSVEP by calculating the binary classification accuracy of single-epoch SSVEPs, *i.e.* the percentage of correctly recognized trials, at 11Hz and 12Hz under different walking conditions.

3   Results

To compare the quality of SSVEP measured by dry versus wet electrodes, two wet electrodes were placed at O1 and O2, next to the dry electrodes at PO1 and PO2, to simultaneously measure the EEG signals during experiments. Figure 2 shows the detection accuracy of 1-s SSVEPs measured by wet and dry electrodes under different walking speeds. Using dry electrodes provided an averaged accuracy of 71.57±13.09% while participants gazed at the flickering in the standing condition, as compared to the wet electrodes (80.19±9.45%). The detection accuracy using either the dry or wet electrodes tended to degrade as the speed of the treadmill increased from 1 to 3 MPH (1 MPH: 62.41±5.14% vs. 67.78±9.97%, 2 MPH: 55.19±3.84% vs. 62.96±14.13%, 3 MPH: 56.57±6.62% vs. 62.59±11.61%), which were however all above the chance level (50%).

Figure 3 shows the effects of the epoch length and the number of channel on the detection accuracy of SSVEPs under different walking speeds. The results clearly showed that in general CCA returned better SSVEP detectability using longer epochs and more channels, *i.e.* the accuracy increased diagonally from the upper-left to the lower-right corner. Using 8-channel 4-s EEG data to detect SSVEP resulted in a maximal accuracy at any given walking speed, except for the condition of standing. Interestingly, EEG signals acquired from four channels placed on the parieto-occipital strap were able to detect SSVEP with a comparable accuracy of 91.85±8.35% against that of using 8 channels (91.11±8.35%). The accuracy declined as walking speed increased (1 MPH: 84.07±11.03%, 2 MPH: 75.56±18.09%, and 3 MPH: 74.81±16.25%).

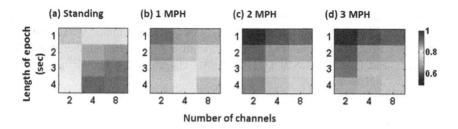

Fig. 3. Averaged SSVEP detection accuracy by CCA with different epoch lengths (1-4 sec) and numbers of channels across different walking speeds (a) standing, (b) 1 MPH, (c) 2 MPH, and (d) 3 MPH

452 Y.-P. Lin et al.

4 Discussions and Conclusion

This study tested the feasibility of using a mobile and wireless BCI system featuring dry electrodes and wireless telemetry to detect SSVEP during natural human walking with emphasis on 1) evaluating the SSVEP quality acquired by the dry electrodes, 2) exploring an optimal set of parameters to detect SSVEP.

To evaluate the feasibility of using dry electrodes to acquire SSVEP, this study performed a wet-dry comparison (wet: O1 and O2, dry: PO1 and PO2). Using wet electrodes to acquire SSVEPs outperformed that of using dry ones by 9% in standing and ~6% in walking conditions (c.f. Fig. 2). This study further optimized SSVEP detectability by using progressively longer multi-channel EEG data. CCA tended to return better classification accuracy with longer data epochs from more channels. The accuracy was found improved by at least 13% across different walking speeds when CCA used 4-s EEG data from eight electrodes over the parietal and the occipital regions, compared to that of using only two electrodes at PO1 and PO2. This finding was consistent with previous works [7-8] that using longer epochs and more channels obtained better CCA-based SSVEP results in movement-restricted participants. The high detection accuracy could be attributed to the fact that CCA could improve SNR by using multivariate covariance information, the imported data with more channels and data points would be beneficial to the detection of the SSVEPs.

The SSVEP detectability was found monotonically decreased as walking speed increased (c.f. Fig. 3). It is very likely due to the fact that the fast walking involved large head/body sway movements, resulting in widespread motion artifacts over multiple sensors. A natural next step of this study is to apply the artifact removal approaches to enhance the SNR of SSVEP and thereby increase the reliability of the mobile and wireless BCI system for freely moving and unconstrained human subjects.

Conclusively, although the SSVEP detectability was found degraded as walking speed increased, this study demonstrated the feasibility of using a mobile EEG system (with dry, non-prep sensor and wireless telemetry) to monitor SSVEP under hostile recording conditions. This demonstration could greatly improve the practicability of SSVEP or BCI applications for continuous, long-term health care in the hospital and at home.

Acknowledgements. This work was supported by Office of Naval Research (N00014-08-1215), Army Research Office (under contract number W911NF-09-1-0510), Army Research Laboratory (under Cooperative Agreement Number W911NF-10-2-0022), and DARPA (USDI D11PC20183). The authors acknowledge Melody Jung for editorial assistance.

References

1. Chi, Y.M., Wang, Y.T., Wang, Y., Maier, C., Jung, T.P., Cauwenberghs, G.: Dry and Non-contact EEG Sensors for Mobile Brain-Computer Interfaces. IEEE Transactions on Neural Systems and Rehabilitation Engineering 20, 228–235 (2012)
2. Wang, Y.T., Wang, Y., Jung, T.P.: A Cell-Phone-Based Brain-Computer Interface for Communication in Daily Life. Journal of Neural Engineering 8, 025018 (2011)

3. Wang, Y., Gao, X., Hong, B., Jia, C., Gao, S.: Brain-Computer Interfaces Based on Visual Evoked Potentials: Feasibility of Practical System Designs. IEEE Engineering in Medicine and Biology Magazine 27, 64–71 (2008)
4. Golla, F.L., Winter, A.L.: Analysis of Cerebral Responses to Flicker in Patients Complaining of Episodic Headache. Electroencephalography and Clinical Neurophysiology 11, 539–549 (1959)
5. Chen, W.T., Wang, S.J., Fuh, J.L., Lin, C.P., Ko, Y.C., Lin, Y.: Peri-ictal Normalization of Visual Cortical Excitability in Migraine: An MEG Study. Cephalalgia 29, 1202–1211 (2009)
6. Debener, S., Minow, F., Emkes, R., Gandras, K., de Vos, M.: How about Taking a Low-cost, Small, and Wireless EEG for a Walk? Psychophysiology 49, 1617–1621 (2012)
7. Bin, G.Y., Gao, X.R., Yan, Z., Hong, B.: An Online Multi-Channel SSVEP-Based Brain-Computer Interface Using a Canonical Correlation Analysis Method. Journal of Neural Engineering 6, 046002 (2009)
8. Lin, Z., Zhang, C., Wu, W., Gao, X.: Frequency Recognition Based on Canonical Correlation Analysis for SSVEP-Based BCIs. IEEE Transactions on Biomedical Engineering 53, 2610–2614 (2006)
9. Wang, Y., Wang, Y.T., Jung, T.P.: Visual Stimulus Design for High-Rate SSVEP BCI. Electronics Letters 46, 1057–1058 (2010)

The Solid Angle of Light Sources
and Its Impact on the Suppression
of Melatonin in Humans

Philipp Novotny[1,2], Peyton Paulick[1,3], Markus J. Schwarz[4], and Herbert Plischke[1,2]

[1] Generation Research Program – Human Science Centre –
Ludwig Maximilians University, Munich, Germany
{novotny,plischke}@grp.hwz.uni-muenchen.de
[2] Network Aging Research – Graduate Program Dementia – Ruprecht Karls University,
Heidelberg, Germany
[3] Department of Biomedical Engineering – University of California Irvine, USA
ppaulick@uci.edu
[4] Clinic of Ludwig Maximilians University, Psychatric Clinic, Section on
Psychoneuroimmunology and Therapeutic Drug Monitoring, Munich, Germany
Markus.Schwarz@med.uni-muenchen.de

Abstract. Our group conducted a preliminary study to examine the influence of different sizes of light sources, and therefore different illuminance levels, at the retina. Six participants were exposed to two lighting scenarios and saliva samples were collected to determine melatonin levels throughout the experiment. Melatonin levels were analyzed to compare the efficacy of each lighting scenario and its ability to suppress melatonin period. Our data is showing a trend that both lighting scenarios are capable of suppressing melatonin. Moreover, the preliminary data show that the lighting scenario with the large solid angle is more effective at suppressing melatonin compared to the lighting scenario with the small solid angle lighting scenario period. Further testing with a larger patient population will need to be done to prove statistical significance of our findings. Our further studies will repeat this experiment with a larger test group and modifying the time frame between different lighting scenarios period.

Keywords: light, health, melatonin, suppression, optimal healing environment, chronodisruption, circadian rhythm, shift work, dementia, light therapy.

1 Introduction

Light has been a treatment for a variety of disorders, in combination with medication or without. Light eases seasonal affective disorders (SAD) [1], helps to calm down demented elderly in the evening from the sundowning syndrome [2] or can simply be used to ease recovery from jet lag period [3-5].

In the year 2001, two independent research groups, discovered that the melatonin suppression in humans could be induced by the exposure to light of the blue spectra [6, 7]. Through this complex circadian regulation system, our day and night cycle is

M. Kurosu (Ed.): Human-Computer Interaction, Part V, HCII 2013, LNCS 8008, pp. 454–463, 2013.
© Springer-Verlag Berlin Heidelberg 2013

synchronized to a 24h cycle period. Our research interests are focused on exploring alternative methods to suppress melatonin during the day and facilitate a synchronized chrono rhythm. Artificial light sources, those which have a sufficient amount of blue spectral parts and intensity can also support the chrono rhythm of the human body. This is especially important in environments where an insufficient amount or even the absence of natural light is present.

The topic we are interested in exploring is, how individuals should be exposed to artificial light sources. Researchers have begun to study the sensitivity of the human retina and its relation to the suppression of melatonin [8-14]. Our research group proposed an optimized lighting environment to alleviate the symptoms of chronodisruption. We are especially interested in chronodisruption among shift workers, where research has linked this working style to a variety of cancers [15, 16], as well as the regulation of the circadian rhythm in demented elderly. Regarding the population of demented elderly a variety of studies have been conducted to analyze the effect of specific lighting conditions on behavior, cognition and circadian rhythmicity of demented elderly [17-20]. These studies have also taken a look at the circadian rhythmicity of their caretakers[21].

In previous studies the amount of light exposed to the eye was measured at eye level. There is no information that can help to determine the amount of light exposed to the retina that is necessary to suppress melatonin. The characteristic of the light source and the method in which light is delivered to the eye modifies the illumination of the retina, which is ultimately the deciding factor in melatonin suppression. Therefore, it is not clear if the threshold level that is published is valid for every lighting condition, since these values report illumination at eye level and not illumination at the retina [9, 12, 22]. Our group believes more information about the lighting sources needed to produce accurate threshold levels for melatonin suppression.

Our feasibility study compared the solid angle of two different light sources. The solid angle measures how large the light source appears to the observer and how it is mapped on the retina. The type and location of a light source has a direct effect on the illuminance level on the retina and thus suppression of melatonin.

2   Methods

2.1 General Study Setup

Our preliminary study had a total of six participants (1 female / 5 male). The testing was performed during the night from 09:00 pm to 01:00 am. Our testing setup asked patients to sit in a darkened room (lined with black fabric to absorb any scattered light) and watch a television program during the testing. This testing room was outfitted with a variety of lighting sources that the patients would be exposed to while watching television. The television was used to keep the participants entertained and maintain a stable gazing direction. From 09:00 pm to 10:00 pm, participants had to wear glasses that absorbed the blue spectral parts of the artificial light sources in the laboratory. Blocking out exposure to blue spectral parts of artificial light and thus should eliminate melatonin suppression. From 10:00 pm to 01:00 am, participants were placed in a separate room in front of a television with two different light sources. To prevent an influence of blue light emitted by the television, a yellow foil was added in front of the television. From 09:00 pm to 11:00 pm, there was a Washout Phase (WO), were no melatonin suppressing light was exposed to the participant, in order to allow the natural release of melatonin. From 11:00 pm to midnight, the participant was exposed to one of the two lighting scenarios. From midnight to 01:00 am, the other lighting scenario was started. To avoid sequential errors we switched lighting scenarios with each participant. This allows for the elimination of any error due to the order of the lighting scenario. For example, the first participant was exposed first to the small light source (sLs) and afterwards to the large light source (lLs). The second participant was exposed first to the lLs and then to sLs and so on.

The testing room layout is depicted in figure 1. Distance from the participants eye to the back wall was approximately 90 cm. Distance from the participants eye to the ground was approximately 120 cm.

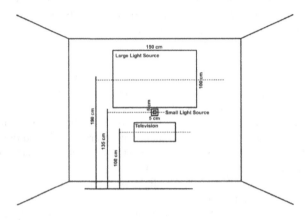

Fig. 1. Sketch of study setup. Arrangement of the objects.

2.2 Light Sources

Both light sources use light emitting diodes (LED) (OSRAM GmbH; Munich, Germany). The LEDs used are white (*Golden Dragon Plus ultra white*) and blue (*Golden Dragon Plus blue*). The specification of bot LED-Types can be seen in the following table:

Table 1. Specification of the used LEDs

Golden Dragon Plus ultra white	Luminous Flux: 116 lm at 350 mA up to 273 lm at 1000 mA Color: Cx = 0.31, Cy = 0.32 acc. to CIE 1931 (white); Optical efficiency (max.): 146 lm/W at 100 mA
Golden Dragon Plus blue	Luminous Flux: 28 lm at 350 mA up to 55 lm at 1000 mA Color: blue (467 nm) Optical efficiency (max.): 35 lm/W at 100 mA

Efficacy of the combined LEDs in suppressing melatonin is presented by the $a_{mel,v}$ value. This value indicates the melanopic efficacy rating. How effective a light source is to suppress melatonin compared to sunlight ($a_{mel,v,sun} = 1$)[23].

Small Light Source. The small light source has a dimension of 5 x 5 cm. Luminance level for the small light source was 33000 cd/m^2 measured in the center and 15000 cd/m^2 measured in the corners. The LEDs were placed in a circle behind a scattering glass create a diffuse light surface. The small light source has an $a_{mel,v}$ of 1.14.

Large Light Source. The large light source has a dimension of 100 x 150 cm. Luminance level for the large light source was about 200 cd/m^2 measured in the center and 40 cd/m^2 measured in the corners. The LEDs were placed in two bars, which were shining against a diffuse white painted surface in the dimensions described in figure 1. The large light source has an $a_{mel,v}$ of 1.33. The slightly higher $a_{mel,v}$ value is due to a slightly different spectrum of the large light source, compared to the small light source.

2.3 Melatonin Level

To determine how effective the small or the large light sources are, in suppressing melatonin, a total of 14 saliva samples were taken throughout the four hour experiment. From 07:00 pm, participants were not allowed to eat, drink, or smoke except non-carbonated water until the end of the testing to prevent contamination of the saliva samples. During the testing participants had to washout their mouth, during testing participants were provided with non-carbonated water. At 09:30 pm, participants had to give the first saliva sample. From 10:00 pm, every 15 minutes a saliva sample was taken (Salivette Cortisol, code blue; SARTEDT GmbH; Nümbrecht, Germany). Participants had to chew on a synthetic swab for at least 30 seconds. The saliva samples then were frozen (-25 °C) until they were sent to the laboratory for analysis.

Melatonin level was determined out of saliva samples with an enzyme immunoassay (ELISA).

2.4 Evaluation of Melatonin Suppression

To compare the efficacy of each lighting condition for suppressing melatonin, we compare the gradient of melatonin levels of each lighting condition in respect to the gradients of melatonin levels of the Washout Phase and the gradient of melatonin levels of the other lighting condition. (See Fig. 2)

2.5 Illuminance Level on the Retina

The information about the illuminance level mostly refers to the illuminance level measured at eye level, but the real amount of light, that is exposed to the retina, is regulated by the pupil diameter. With the specification about the dimension of the light source, the distance to the eye, the illuminance level on the cornea, and the pupil diameter, the amount of light, exposed to the retina can be calculated with the following formula [24]:

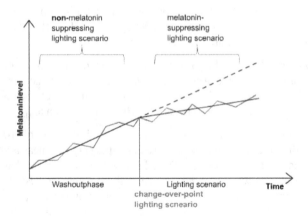

Fig. 2. Sketch of theoretical progress of melatonin level and light intervention. Dashed black line indicates the progress of melatonin level without light intervention. Drawn black line indicates the theoretical change in melatonin level.

$$E_{Retina} = E_{Cornea} * \frac{D_p^2}{k} * \frac{h^2}{A}$$

E_{Retina}: Illuminance at the retina in lux
E_{Cornea}: Illuminance at the eye in lux
D_P: Pupil diameter in mm
k: Factor for different solid angle
h: Distance from the light source to the eye in m
A: Area of the light source in m²

Taking this formula into account, we measured the pupil diameter in each lighting conditions two times, at the beginning and in the middle of each lighting scenario with an infrared camera eye-tracking-system (SensoMotoric Instrument GmbH; Teltow, Germany).

We also measured the illuminance level at eye level with a lux-meter (Digitales Luxmeter MS-1500; Voltcraft) of each participant for each lighting condition.

3   Results[1]

3.1 Melatonin Levels

The following results represent absolute melatonin level in pg/ml. Since the order of lighting scenarios switched with each participant, time is marked with *.

---

[1] In this paper we use Arabic numerals and decimal comma.

Table 2. Melatonin level during Washout Phase

	21:30	22:00	22:15	22:30	22:45	23:00
P 1	0,67	3,63	6,63	9,12	12,69	14,21
P 2	5,52	8,40	11,84	12,03	21,48	16,60
P 3	9,78	10,10	11,59	14,27	12,77	13,19
P 4	1,73	4,86	4,42	5,48	6,36	8,58
P 5	2,21	6,78	8,68	12,13	28,03	35,32
P 6	14,27	56,23	51,66	54,32	44,36	81,78

Table 3. Melatonin level during the lighting scenario with the small light source

	23:15	23:30	23:45	00:00
Or *	00:15	00:30	00:45	01:00
P 1	26,93	19,91	23,51	20,76
P 2 *	21,62	20,03	25,21	31,09
P 3	14,51	12,10	12,28	12,23
P 4 *	13,92	14,32	13,82	15,98
P 5	39,82	50,95	38,52	34,65
P 6 *	60,91	55,17	71,77	80,26

Table 4. Melatonin level during the lighting scenario with the large light source

	23:00	23:15	23:30	24:00
Or *	00:00	00:15	00:30	01:00
P 1 *	26,93	19,91	23,51	20,76
P 2	19,03	17,39	20,63	26,54
P 3 *	13,64	15,25	13,42	13,95
P 4	8,35	9,57	12,48	13,19
P 5 *	40,85	56,82	34,65	64,07
P 6	46,94	58,83	53,48	61,44

3.2 Gradient of Melatonin Level

Table 5. Gradients of melatonin level for each participant for the Washout Phase (WO) and the two lighting scenarios (small light source sLs, large light source lLs)

	WO	sLs	lLs
P 1	2,78	2,89	-1,49
P 2	2,71	3,36	2,58
P 3	0,79	-0,66	-0,09
P 4	1,14	0,57	1,74
P 5	6,65	-2,80	4,75
P 6	8,70	7,46	3,82

Table 6. T-Test of the the gradients of melatonin level, Significance level is < ,05

	Diff Mean	Sign.
WO < > sLs	-1,99	,25
WO < > lLs	-1,91	,09
sLs < > lLs	-0,08	,97

3.3 Illuminance Level

The mean illuminance level for each lighting condition including the WO is shown in the following table.

Table 7. Mean value of illuminance level in each lighting scenario including WO measured at eye level in lux

Lighting Scenario	Mean value in lx
Washout Phase	0,36
Small light source	91,9
Large light source	87,1

3.4 Pupil Diameter and Retinal Illuminance Level

While the pupil diameter varied very strong we calculated a range for possible illuminance levels.

Table 8. Recalculated retinal illuminance level on the retina for lLs

Pupil Diameter in mm	Retinal illuminance in lx
2,5	0,9
6,5	6,1

Table 9. Recalculated retinal illuminance level on the retina for sLs

Pupil Diameter in mm	Retinal illuminance in lx
2,5	414
6,5	2801

3.5 Effect Size

Calculations of effect size have been done for Cohen's d and are listed in the following table.

Table 10. Cohen's d fort he different lighting scenarios in their time window. D: 0,2 = small effect; 0,5 = medium effect; 0,8 = large effect

	WO → sLs	WO → lLs	sLs → sLs
23:00 – 00:00	0,67	0,27	0,50
00:00 – 01:00	0,43	0,62	0,31

4   Discussion

Our data is showing a trend that both lighting scenarios are capable of suppressing melatonin. Moreover, the preliminary data show that the lighting scenario with the large solid angle is more effective at suppressing melatonin compared to the lighting scenario with the small solid angle lighting scenario period. Although the trend in our data is apparent due to our small population size this difference is not significant. We believe with a larger testing size statistical significance can be proven. Further testing with a larger patient population will need to be done to prove statistical significance of our findings. Nevertheless, our findings, even intended as a feasibility study, seem to support the general assumption that larger illumination areas are more effective in suppressing melatonin. Our further studies will repeat this experiment with a larger test group and modifying the time frame between different lighting scenarios period. Based on our preliminary results that larger lighting sources are more effective in suppressing melatonin onset, our group feels this lighting scheme can be easily integrated into current working and home environmental infrastructure. Walls and ceilings can be effective light diffusers to maintain a large solid angle of a light source. These findings compel our research group to continue the exploring the most optimal way to deliver light in order to regulate chronobiology and promote healthy living.

5   Recommendation

Our first recommendation based on this pilot study is to split lighting scenarios and test participants with each lighting scenario individually. We suggest that that way a greater effect will be visible. Our second recommendation is to increase the amount of participants. Furthermore, we recommend testing participants with a similar chronotype (morning vs. evening type). Lastly, our group would like to integrate a schedule entrainment were patients sleep wake cycles are regulated for a two week period. Our group hopes to give participants guidelines wake/sleep times. This would reduce variation in melatonin onset due to varying schedules, different daily activity levels and rest periods.

Acknowledgement. We want to thank DIN FNL 27 for their support and knowledge and OSRAM AG for the hardware support. We also want to thank the Robert Bosch Foundation for supporting the first author. The authors would also like to thank the Whitaker International Fellows Program who supports our second author Peyton Paulick.

References

1. Pjrek, E., Winkler, D., Stastny, J., Konstantinidis, A., Heiden, A., Kasper, S.: Bright light therapy in seasonal affective disorder–does it suffice? European Neuropsychopharmacology: the Journal of the European College of Neuropsychopharmacology 14(4), 347–351 (2004)

2. Dowling, G.A., Mastick, J., Hubbard, E.M., Luxenberg, J.S., Burr, R.L.: Effect of timed bright light treatment for rest-activity disruption in institutionalized patients with Alzheimer's disease. Int. J. Geriatr. Psychiatry 20(8), 738–743 (2005)

3. Arendt, J.: Managing jet lag: Some of the problems and possible new solutions. Sleep Medicine Reviews 13(4), 249–256 (2009)

4. Coste, O., Lagarde, D.: Clinical management of jet lag: what can be proposed when performance is critical? Travel Medicine and Infectious Disease 7(2), 82–87 (2009)

5. Eastman, C.I., Burgess, H.J.: How To Travel the World Without Jet lag. Sleep Medicine Clinics 4(2), 241–255 (2009)

6. Brainard, G.C., Hanifin, J.P., Greeson, J.M., Byrne, B., Glickman, G., Gerner, E., Rollag, M.D.: Action spectrum for melatonin regulation in humans: evidence for a novel circadian photoreceptor. J. Neurosci. 21(16), 6405–6412 (2001)

7. Thapan, K., Arendt, J., Skene, D.J.: An action spectrum for melatonin suppression: evidence for a novel non-rod, non-cone photoreceptor system in humans. J. Physiol. 535(Pt. 1), 261–267 (2001)

8. Adler, J.S., Kripke, D.F., Loving, R.T., Berga, S.L.: Peripheral vision suppression of melatonin. J. Pineal Res. 12(2), 49–52 (1992)

9. Aoki, H., Yamada, N., Ozeki, Y., Yamane, H., Kato, N.: Minimum light intensity required to suppress nocturnal melatonin concentration in human saliva. Neurosci. Lett. 252(2), 91–94 (1998)

10. Glickman, G., Hanifin, J.P., Rollag, M.D., Wang, J., Cooper, H., Brainard, G.C.: Inferior retinal light exposure is more effective than superior retinal exposure in suppressing melatonin in humans. J. Biol. Rhythms 18(1), 71–79 (2003)

11. Lasko, T.A., Kripke, D.F., Elliot, J.A.: Melatonin suppression by illumination of upper and lower visual fields. J. Biol. Rhythms 14(2), 122–125 (1999)

12. McIntyre, I.M., Norman, T.R., Burrows, G.D., Armstrong, S.M.: Human melatonin suppression by light is intensity dependent. J. Pineal Res. 6(2), 149–156 (1989)

13. Ruger, M., Gordijn, M.C., Beersma, D.G., de Vries, B., Daan, S.: Nasal versus temporal illumination of the human retina: effects on core body temperature, melatonin, and circadian phase. J. Biol. Rhythms 20(1), 60–70 (2005)

14. Visser, E.K., Beersma, D.G., Daan, S.: Melatonin suppression by light in humans is maximal when the nasal part of the retina is illuminated. J. Biol. Rhythms 14(2), 116–121 (1999)

15. Erren, T.C., Pape, H.G., Reiter, R.J., Piekarski, C.: Chronodisruption and cancer. Die Naturwissenschaften 95(5), 367–382 (2008)

16. Franzese, E., Nigri, G.: Night work as a possible risk factor for breast cancer in nurses. Correlation between the onset of tumors and alterations in blood melatonin levels. Professioni Infermieristiche 60(2), 89–93 (2007)

17. van Hoof, J., Aarts, M.P.J., Rense, C.G., Schoutens, A.M.C.: Ambient bright light in dementia: Effects on behaviour and circadian rhythmicity. Building and Environment 44(1), 146–155 (2009)

18. Riemersma-van der Lek, R.F., Swaab, D.F., Twisk, J., Hol, E.M., Hoogendijk, W.J., Van Someren, E.J.: Effect of bright light and melatonin on cognitive and noncognitive function in elderly residents of group care facilities: a randomized controlled trial. Jama 299(22), 2642–2655 (2008)

19. Haffmans, P.M., Sival, R.C., Lucius, S.A., Cats, Q., van Gelder, L.: Bright light therapy and melatonin in motor restless behaviour in dementia: a placebo-controlled study. Int. J. Geriatr. Psychiatry 16(1), 106–110 (2001)

20. Skjerve, A., Bjorvatn, B., Holsten, F.: Light therapy for behavioural and psychological symptoms of dementia. Int. J. Geriatr. Psychiatry 19(6), 516–522 (2004)
21. Friedman, L., Spira, A.P., Hernandez, B., Mather, C., Sheikh, J., Ancoli-Israel, S., Yesavage, J.A., Zeitzer, J.M.: Brief morning light treatment for sleep/wake disturbances in older memory-impaired individuals and their caregivers. Sleep Med. 13(5), 546–549 (2012)
22. Rea, M.S., Bullough, J.D., Figueiro, M.G.: Phototransduction for human melatonin suppression. J. Pineal Res. 32(4), 209–213 (2002)
23. DIN: DIN SPEC 5031-100:2012 Optical radiation physics and illuminating engineering - Part 100: Non-visual effects of ocular light on human beings - Quantities, symbols and action spectra. Beuth Verlag (2012)
24. Schierz: Ist die Beleuchtungsstärke am Auge die richtige Größe für biologische Lichtwirkungen? 4. DIN-Expertenforum, pp. 7–17. Beuth Verlag, Beuth (2010)

Facial Electromyogram Activation
as Silent Speech Method

Lisa Rebenitsch and Charles B. Owen

Media and Entertainment Technologies Laboratory (METLAB)
Michigan State University
East Lansing, MI, 48824
rebenits@msu.edu, cbowen@cse.msu

Abstract. A wide variety of alternative speech-free input methods have been developed, including speech recognition, gestural commands, and eye typing. These methods are beneficial not only for the disabled, but for situations where the hands are preoccupied. However, many of these methods are sensitive to noise, tolerate little movement, and require it to be the primary focus of the environment. Morse code offers an alternative when background noise cannot be managed. A Morse code-inspired application was developed employing electromyograms. Several muscles were explored to determine potential electrode sites that possessed good sensitivity and were robust to normal movement. The masseter jaw muscle was selected for later testing. The prototype application demonstrated that the jaw muscle can be used as a Morse "key" while being robust to normal speech.

Keywords: Silent Speech, Human computer interaction, User interfaces.

1   Introduction

Most devices used today for nonverbal communication, such as texting and internet browsing, typically require the use of one's hands. However, there are many instances where the hands mobility is limited. These instances can be either when the hands are otherwise occupied or when the hands lack the degree of control normally assumed. A few examples are operating machinery, driving, and working with disabilities. Attempting to temporarily shift the hands from the primary task can also be dangerous. The US Department of Transportation reports nearly half a million people were injured due to a distracted driving in 2009 [1]. This includes not only cell phone usage, but stereo control and air conditioning. New interface designs and technology attempt to decrease focus time on the interface, but interfaces still routinely require use of one's hands.

Many alternatives have been developed to permit hands-free input, such as speech recognition, gesture recognition, and eye focus/typing. However, speech recognition has limited accuracy, and its performance is degraded wherever there is background noise. Gestures are severely limited when the hands must be excluded, and eye focus/typing has the "Midas touch" issue wherein normal gaze movement are perceived

M. Kurosu (Ed.): Human-Computer Interaction, Part V, HCII 2013, LNCS 8008, pp. 464–473, 2013.

as actual commands [2, 3]. Morse code-inspired systems are an option where the hands are unavailable and have the added advantage of being robust to noise in the environment. Morse decoders can employ wireless systems to allow greater range of movement, and have been developed previously for environmental control systems [4]. However, this environmental control system required the use of a traditional Morse key, and therefore, the use of one's hands. As an alternative, the tightening and relaxing of muscles not on the hands or arms and their related bioelectrical (electromyogram) signal activations can be interpreted as the Morse signal.

We present a prototype system that uses a single electromyogram (EMG) electrode with a rate adaptive Morse decoder to create a hands-free typing interface in a partially mobile environment. To produce Morse code, the selected muscle must have a fair degree of control, limited interference from normal movement, easy detection, and a location where the sensing electrode is unobtrusive. The jaw fulfills these requirements.

One inherent feature of the jaw muscle is that the signal can bleed to other nearby electrodes due to its proximity to the skin surface and relative size to other nearby muscles. This implies that if multiple muscles are to be employed, nearby EMG signals from other muscles would require decoupling. A single muscle site was chosen to avoid this issue, disallowing the "2-button" Morse switch. The resulting "tighten and relax" sequence of the EMG is then analyzed by modeling the signal as a random zero-mean Gaussian signal with amplitude proportional to clench strength [5].

The Application Environment and Morse Decoding Algorithm are discussed in Section 3, while the Signal Processing and Signal Speed are discussed in Section 4. Hardware Specifications and Discussion are in Sections 5 and 6, respectively.

2   Background

The intent of the prototype was to determine the feasibility of an EMG based command system in a larger, partially mobile environment where retuning to the computer to adjust the system would be impractical. Other requirements were that the commands were to be a secondary focus so the commands would not interfere with the primary application, would be resistant to noise from normal motion, and would avoid false positives more strongly than false negatives. An application with similar requirements would be textual input in a web browser.

EMGs are inherently noisy signals. Morse code has been used as a communications medium for nearly two centuries due to its robustness in the presence of noise. Automatic decoding with reasonable accuracy has been available since as early as 1959 [6, 7]. More recently, machine decoding has demonstrated 96-98% accuracy rates, which is sufficiently accurate for command tasks [7]. Morse code by experienced operators has been reported at 35 words per minute (wmp) [7]. This is faster than reports on eye typing which often report speeds of 5-10 wmp with higher rates requiring word prediction methods [2, 3]. This is too slow for a command system and a much higher wmp with a "one-finger typing" restriction of eye typing should not be expected. Soukoreff and MacKenzie calculated the theoretical wmp bounds of using a stylus on a QWERTY keyboard, or "one-finger typing," to be 8.9 to 30.1 wmp [8]. Eye typing will slower to compensate for the Midas effect, although there are attempts to mitigate it. Isokoski explains that a primary reason for the Midas effect is that the textual input and the application overlap, and therefore, both have primary focus at all times [9]. Isokoski moved the gaze switches to the sides of a monitor to remove this overlap which allowed for a dwell time under 100ms. While this method permits textual input to be a secondary task, it does not permit partial mobility.

The bounds calculate by Soukoreff and MacKenzie can also can be applied to Morse code since the bounds were calculated using the bits (choices/clicks) per second [8]. Those "clicks" can be considered to be Morse key presses, and given the shorter distance between "clicks," the slightly higher than theoretical upper bounds of 35 wmp for Morse code is unsurprising. Therefore, these theoretical bounds were used as an objective feasibility measure of our Morse code-inspired system. However, to keep the hands free, a different input source is required. We used the jaw's EMG in our system. The selection criteria of muscle are discussed more in 3.1.

Other methods have been used for hand-free interfaces. MacKenzie provides an overview of one key (button) keyboard interfaces and their issues [10]. These interfaces are often intended for the disabled who cannot use a normal keyboard or for small devices such as some cell phones that cannot support a full keyboard. There are two primary categories: scanning the keys until the wanted character is reached, and a hierarchal method where a subset of options is repeatedly selected until only one character/command remains. Morse code, while consisting of one or two switches, does not require a keyboard counterpart and thus allows direct input into the device. As a consequence the character codes must be memorized rather than searched for visually. The advantage to this method is that it can be much faster overall.

Systems similar to our own use EMGs as input. Felzer and Nordmann created a Morse code-inspired system that uses the eyebrow's EMG [11]. Their encoding uses the direction of the cursor, which is set by the number of EMG pulses. While the system is fairly intuitive in that the state is easily seen by cursor motion, it requires more muscle clenches than Morse code. For example, the letter A would require 4 clenches while Morse code would require 2. The worst case would require 8 clenches. Nilas, Rani, and Sarka present another option that uses the eyebrow EMG as the Morse code signal [12]. They restricted the set of characters to decrease learning time and mapped the characters to high level robot commands such as "move left." However, their method would require it to be the primary focus to avoid false positives. Park et. al. developed a system similar to our own designed for disabled users [13]. An EMG on the masseter was used as input into a Morse decoder. However, there are sparse details on the implementation, and whether it could be used as a secondary input in a partially mobile environment is ambiguous.

3   Application Environment

Our application is a GUI coded in C# and C using a single EMG signal. The Morse decoding application is comprised of several state variables, such as the letters decoded thus far, the approximate words per minute, and a feedback character of the current state as an aid to new users. An example of the decoder window is available in Fig. 1. The Morse element categories of the decoder are dot(dit), dash, the different spaces, and undefined. A chart of the feedback character for each element is available in Table 1. A detailed explanation of the Hardware Specifications is available in section 5.

Fig. 1. Morse decoder application screen shot after approximately an hour of practice

Table 1. Feedback Character for each Morse Element

Element	Feedback Character
Dot	.
Dash	–
Inter-element_space	I
Character_space	C
Word_space	W
Undefined	no feedback

3.1 Electrode Placement

When selecting potential muscles for the sensing electrode placement, a few require-ments are enforced: the muscle may not be on the arms or hands, the muscle is large enough to allow for minor electrode displacement, the electrode is not inhibitive to task, the electrode causes minimal discomfort, the electrode is robust to minor move-ment, and the muscle must be capable of a fair degree of control and strength.

Clenching the jaw muscle (masseter) fit these requirements. It has a fair degree of control, yields a strong signal, and the electrode placed on the rear of the jaw is less obtrusive. Other candidates were the neck, eyebrow, and eye movement. However, the neck and eye movement resulted in numerous false positives from normal motion, and thus would have required textual input to be the primary focus at all times. The activations from normal speech were sufficiently lower than that of a clenched jaw to avoid most of these false positives. The eyebrow had fewer, but still many, false posi-tives, and was the most obtrusive of all candidate locations. The positive electrode was placed on the rear of the left jaw on the masseter muscle with the negative elec-trode placed immediately below, as shown in Fig. 2. The ground electrode was placed on the back of the right hand or forearm for convenience.

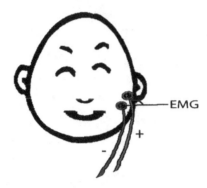

Fig. 2. Placement of negative and positive electrodes

3.2   Morse Decoding Algorithm

The Morse decoder is a multistate system derived from Kyriazis "xdemorse" program, and is distributed under the terms of the GNU General Public License [14]. The decoder code was extracted and adapted for use in our system. The input signal of original project is replaced by muscle activity to represent the "on/off" switch. The training and dot-dash detection sections were then modified to better accommodate the jaw EMG signal. The dot-dash detection and training algorithms were isolated to later allow potential substitution of other Morse decoding algorithms and contextual rules.

The decoder breaks the incoming signal into a string of 250 bits per second, called "fragments." Each fragment is encoded as a true (jaw clenched) or false (jaw relaxed). Each fragment is then fed to the decoder to update the decoder state. A broad overview of the state machine is provided in Fig. 2. Each state has the ability to either wait or progress to a new state depending on the incoming fragment, and return a decoded letter or feedback, if required.

Each fragment is parsed individually. If the fragment does not cause a change in state, feedback is provided as an aid for those new to Morse code. In the MARK_SIGNAL, CHAR_SPACE, and WAIT_WORD_SPACE states, the length of the sequence is checked to determine the current element (e.g. dot, character space) and is then displayed to the user as feedback. If the fragment does cause a change in state, the sequence is decoded. In the feedback states mentioned before and ELEM_SPACE and WORD_SPACE, the length of the current input string is compared against the stored threshold lengths for each Morse element.

Once an element is decided, it is encoded into a binary string and then read as a hexadecimal value. The initial string is 0x01, and then if the element is a dot, a 1 is shifted in. If it is a dash, a 0 is shifted in. For example, an "E" (single dot), is encoded as a 0x03 and a "T" (single dash), is encoded as a 0x02. This encoding is then indexed into an array of characters to decode the element set. A space element is stored separately as the length of time.

Upon decoding the sequence, the current lengths of the elements are stored for later use by the adaptation algorithm, and the lengths are reset to 0. Upon reaching the end of a word, the adaptation algorithm is called. The algorithm is described in more detail in section 4.3.

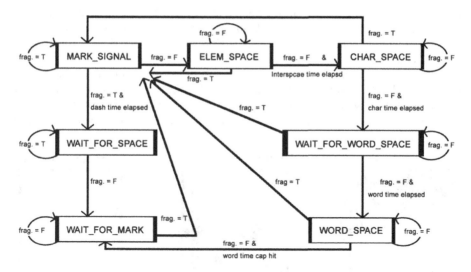

Fig. 3. State change diagram Morse decoder algorithm

3.3   Further Concerns

When assessing the training algorithm, the jaw demonstrated a slower response time and greater variance in signal lengths then those expected from hand generated code. In an effort to counter the short Morse spacing, the jaw tended to remain slightly tensed which quickly resulted in fatigue. Thus, the algorithm was modified to permit greater spacing between the dots and dashes, and the threshold lengths were modified to be the midpoint between two element lengths. In addition, previously trained threshold lengths were loaded rather than beginning with default settings in which the operator was trying to "force" a certain response.

4   Signal Processing

The incoming signal is an EMG signal at 250 samples per second, with a gain of 1000. The EMG signal is modeled as a zero-mean Gaussian distribution noise signal with amplitude proportional to the clench strength [5]. However, cyclical noise from neighboring wires, poor electrode contact, signal shifts, speech, and other noise sources remain in the incoming signal resulting in the raw amplitude being unreliable. Therefore, the signal must be preprocessed to remove the outside noise, and then smoothed for detection of activation time and length.

4.1   Outside Noise Removal

Temporal whitening has been shown to improve the amplitude estimation for EMG signals [5, 15] and has the added benefit of removing cyclical noise. The outside cyclical 60 Hz information from nearby wires and electrode drift is first removed from the EMG signal. This is performed through an adaptive Wiener whitening filter derived from the algorithm described by Baldwin [16]. For a 250 Hz signal, the parameters were a window size of 35 and a filter feedback constant of .05. This filter adapts from a zeroed initial state to remove stable cyclical noise from signal within a few seconds. Afterwards, mild to moderate noise shifts in the incoming signal are removed rapidly. This whitening filter has the added advantage of making the signal have a mean of 0, which effectively removes electrode drift in one sweep of the signal. An image of the initial signal and the whitening filter output is shown in Fig. 4.

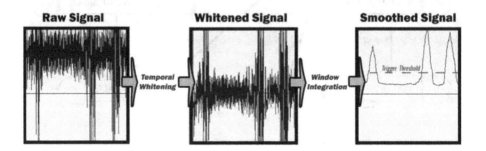

Fig. 4. Signal processing progression (The threshold line is added for visual clarity)

4.2   Threshold Smoothing

After the signal is temporally whitened, it is smoothed so that it can be compared against a threshold limit. To permit greater time for improved decoding algorithms, the EMG signal is smoothed via a fast window integration function with a width of approximately 150 milliseconds, or 35 samples at the 250 Hz sample rate. The window size was chosen empirically applying the tradeoff of signal-to-noise ratio as described by St-Amant, Rancourt, and Clancy [17]. The output signal is then compared directly to an activation threshold. Should the signal amplitude surpass the threshold, the muscle state is changed to "on" until the amplitude drops below the threshold. The threshold is static and is manually chosen by the user. The threshold normally only changes if the skill level of the user changes. For example, when training there is a tendency to strongly clench the jaw. After the Morse decoder has been trained, there is not the perceived need to be as forceful. While a changing threshold, such as the method provided by Nilas, Rani, and Sarka, which reacts much more strongly to the initial increases could be used, it carries a risk of many false positives during normal speech [12].

4.3   Signal Speed Adaptation

A user's Morse rate changes with practice, fatigue, and focus on the Morse code versus the environment. Although the system provides an option for static element lengths for practice, a rate adapting algorithm is essential. The original algorithm adjusted the basic element threshold length (inter-element_space) at the end of characters and words. If both dots and dashes were present, the basic element threshold is updated by analyzing the fragments since the last threshold update. The higher the estimated character/word count, the shorter the next threshold. The other element thresholds were then set to multiples of the `inter-element_space`.

However, this algorithm was found to have poor accuracy for the jaw EMG signal. The algorithm is modified so that it is performed at the end of the word, but only when both dots and dashes were present to assure a comparison of the elements. The adaptation algorithm then determines the average length of the all Morse elements in the current word. It then uses the weighted average of the original element thresholds and current averaged element thresholds to update. However, to prevent the elements from overlapping, each element threshold is given a minimum and maximum cap based on the next smallest related element. These lengths are then used in the decoding process as described in section 3.

5   Hardware Specifications

The Morse decoder application was run on a Toshiba Satellite 505 with 4GB of memory running Windows 7. The electromyogram amplifier (EMG100C) of a BIOPAC MP150 system, a bioelectric suite from Biopac Systems, Inc., was used. The gain on the BIOPAC module was set to 1000, the low pass band filer was set to 100 Hz, the high pass filter was set to 1Hz, and the 100Hz high pass filter was turned off. Foam electrodes with clip leads from Biopac Systems were used. The operator had the option to sit or stand in front of the monitor at a distance of their choosing.

6   Discussion

A short informal experiment was performed to examine the feasibility of the system. The electrodes were attached as explained in section 3.1, and the applicant was trained for a few minutes by inputting SOS, and then the element lengths were saved. After approximately one hour of practice with a reference chart, the novice attempted printing the alphabet while stopping approximately 3/4 through to talk normally. An example result is available in screen shot in Fig. 1. A few mistakes were made (some not shown due to an added erase command). Surprisingly, normal speech had little effect on the output beyond a rare "E". A novice's input is slow at only about 5 wmp which is below the lower bound of 8.9 calculated by Soukoreff and MacKenzie [8]. This lower value may be due to the slower response time of the jaw, and the user's unfamiliarity with Morse code. There were many delays due to referencing the Morse code chart. Had the codes been memorized, the wmp would likely have doubled.

Learning Morse code takes time and practice, but was chosen since it is a well known standard code that handles noisy environments well. The time can be dramatically decreased when using a restricted set of input codes as suggested by Nilas, Rani, and Sarka [12]. A set of five to ten input commands would be easier to learn and could consist of commands such as stop, start new file, previous, next, select, and other application specific commands. These commands are available in most standard computer software, but are often not available in non-traditional interfaces.

Areas where the hands are otherwise engaged would receive the greatest benefit, such as in cockpit simulators, machine operation where the hands must stay on the controls and noise limits speech recognition, and disabled users whose hands do not have sufficient control over traditional interfaces.

The Morse decoder for the jaw muscle currently employs an exact pattern match for character retrieval and a fast weighting scheme for updating. Better decoding algorithms, such as fuzzy logic, word completion, and spell checking for code sequences with errors would increase accuracy. Changing the update algorithm to store previous examples of elements and/or characters should also improve accuracy and usability of the system.

References

1. US Department of Transportation, National Highway Traffic Safety Administration, Distracted Driving 2009 (September 2010),
   `http://www.distraction.gov/research/PDF-Files/`
   `Distracted-Driving-2009.pdf` (accessed January 2012)
2. Majaranta, P., Räihä, K.-J.: Twenty Years of Eye Typing: Systems and Design Issues. In: Proceedings of Eye Tracking Research and Applications, pp. 15–22 (2002)
3. Hansen, D.W., Hansen, J.P., Niels, M., Johansen, A.S.: Eye Typing using Markov and Active Appearance Models. In: Proceedings of the Sixth IEEE Workshop on Applications of Computer Vision, pp. 132–136 (2002)
4. Yang, C.-H., Chuang, L.-Y., Yang, C.-H., Luo, C.-H.: Morse Code Application for Wireless Environmental Control Systems for Severely Disabled Individuals. IEEE Transactions on Rehabilitation Engineering 11(4), 463–469 (2003)
5. Clancy, E.A., Hogan, N.: Single Site Electromyograph Amplitude Estimation. IEEE Transactions on Biomedical Engineering 41(2), 159–167 (1994)
6. Gold, B.: Machine Recognition of Hand-Sent Morse Code. IRE Transactions on Information Theory 5(1), 17–24 (1959)
7. Blair, C.R.: On Computer Transcription of Manual Morse. Journal of the ACM 6(3), 429–442 (1959)
8. Soukoreff, R.W., MacKenzie, I.S.: Theoretical Upper and Lower Bounds on Typing Speed Using a Stylus and Soft Keyboard. Behaviour & Information Technology 14, 370–379 (1995)
9. Isokoski, P.: Text Input Methods for Eye Trackers Using. In: Proceedings of the 2000 Symposium on Eye Tracking Research & Applications, pp. 15–21 (November 2000)
10. MacKenzie, I.S.: The One-Key Challenge: Searching for a Fast One-Key Text Entry Method. In: Proceedings of the 11th International ACM SIGACCESS Conference on Computers and Accessibility, New York (2009)
11. Felzer, T., Nordmann, R.: Alternative Text Entry Using Different Input Methods. In: Proceedings of the 8th International ACM SIGACCESS Conference on Computers and Accessibility, pp. 10–17 (October 2006)
12. Nilas, P., Rani, P., Sarkar, N.: An innovative high-level human-robot interaction for disabled persons. In: Proceedings of Internation. Conference on Robotics and Automation, New Orleans (2004)
13. Park, H.-J., Kwon, S.-H., Kim, H.-C., Park, K.-S.: Adaptive EMG-driven communication for the disabled. In: Proceedings of BMES/EMBS Conference Serving Humanity, Advancing Technology, Atlanta (1999)

14. Kyriazis, N.: Linux Morse Code Decoding Software (2002),
    http://www.qsl.net/5b4az/pages/morse.html (accessed 2011)
15. Clancy, E.A., Farry, K.A.: Adaptive Whitening of the Electromyogram to Improve Amplitude Estimation. IEEE Transactions on Biomedical Engineering 47(6), 709–719 (2000)
16. Baldwin, R.G.: An Adaptive Whitening Filter in Java (November 1, 2005),
    http://www.developer.com/java/other/article.php/3560501
    (accessed 2010)
17. St-Amant, Y., Rancourt, D., Clancy, E.A.: Influence of Smoothing Window Length on Electromyogram Amplitude Estimates. IEEE Transactions on Biomedical Engineering 45(6), 795–799 (1998)

The Impact of Gender and Sexual Hormones on Automated Psychobiological Emotion Classification

Stefanie Rukavina, Sascha Gruss, Jun-Wen Tan, David Hrabal, Steffen Walter, Harald C. Traue, and Lucia Jerg-Bretzke

University Clinic for Psychosomatic Medicine and Psychotherapy,
Medical Psychology, Ulm
Stefanie.Rukavina@uni-ulm.de

Abstract. It is a challenge to make cognitive technical systems more empathetic for user emotions and dispositions. Among channels like facial behavior and nonverbal cues, psychobiological patterns of emotional or dispositional behavior contain rich information, which is continuously available and hardly willingly controlled. However, within this area of research, gender differences or even hormonal cycle effects as potential factors in influencing the classification of psychophysiological patterns of emotions have rarely been analyzed so far.

In our study, emotions were induced with a blocked presentation of pictures from the International Affective Picture System (IAPS) and Ulm pictures. For the automated emotion classification in a first step 5 features from the heart rate signal were calculated and in a second step combined with two features of the facial EMG. The study focused mainly on gender differences in automated emotion classification and to a lesser degree on classification accuracy with Support Vector Machine (SVM) per se. We got diminished classification results for a gender mixed population and also we got diminished results for mixing young females with their hormonal cycle phases. Thus, we could show an improvement of the accuracy rates when subdividing the population according to their gender, which is discussed as a possibility of incrementing automated classification results.

Keywords: emotion classification, gender, hormonal cycle, heart rate, facial EMG.

1   Introduction

Living in the age of information technology (IT), everyone is nowadays surrounded by technical systems, spreading in all corners of daily living. Users tend to show patterns of behavior when they communicate with technical surfaces including emotional components, generally observed within human-human-interactions [1]. User Interfaces are developed to improve the interaction between humans and the system, however an improvement presumes the identification of the user`s emotion and disposition particular in stressful situations or with people from vulnerable populations. Within

M. Kurosu (Ed.): Human-Computer Interaction, Part V, HCII 2013, LNCS 8008, pp. 474–482, 2013.
© Springer-Verlag Berlin Heidelberg 2013

the research field of affective computing the aim is to make technical systems more empathetic for user emotions and dispositions presuming the possibility of reliable and valid emotion classification.

Emotions can be described by a three dimensional concept with the dimensions of valence, arousal and dominance (VAD) [2]. Valence reflects the pleasantness of a stimulus, whereas arousal refers to the emotional activation. Because dominance is often found to correlate with valence we used in this study only the two-dimensional theory and did not consider dominance further [3]. Psychophysiological changes occur during emotional behavior and are part of emotions per definition. Therefore, one possibility to recognize emotional behavior is using these psychophysiological parameters. Other 'emotional channels' could be speech (semantics, pitch) [4] or mimic expressions [5]. As a disadvantage speech and even mimic expressions can be willingly influenced, whereas psychophysiological behaviors are generally autonomic-regulated and as a second advance they are constantly available. Interestingly, in the field of interactive systems and affective computing the impact of gender has been disregarded, although from a psychophysiological perspective gender and accordingly the menstrual cycle must be a crucial co-factor. To our knowledge, only a few classification studies considered gender as important and used gender specific classification vectors [6-7]. Extending this topic, the hormonal cycle has been shown to result in different emotional responding between young females [8-11]. Therefore, the consideration of differences between males and females can improve the identification of emotions using psychophysiological parameters and should be investigated more precisely.

In the current study we used a Support Vector Machine (SVM) classification based on 5 heart rate parameters (RMSSD, mean NN, SDNN, mean HR, SD HR) and 2 facial EMG parameters (mean amplitude of Corrugator supercilii and Zygomaticus major). The combination of these two physiological parameters has been chosen because they are often found to describe the valence dimension e.g. [12] [13]. The analysis about the influence of gender and especially hormonal cycle effects concerning classification results is the first study done to our knowledge in the field of affective computing.

2   Material and Methods

For emotion elicitation we used pictures of the International Affective Picture System and Ulm pictures [14] [15]. All pictures were rated according to the emotional dimensions valence, arousal and dominance, however we focus only the VA-dimensional concept. The valence dimension can be separated roughly in positive (High Valence (HV)), negative (Low Valence (LV)) and neutral quadrants (see Fig. 1).

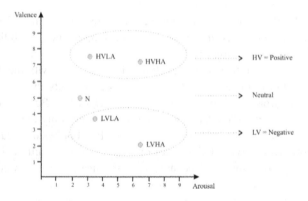

Fig. 1. Valence - Arousal concept with the ratings of the used stimuli

According to their arousal they can be distinguished between high and low arousal. To intensify the emotion induction we used a prolonged presentation according to [16-17]. One stimuli sequence (picture block) consisted of 10 pictures with similar valence and arousal ratings, with each picture being displayed for 2 sec without a pause between the following pictures. In total, 100 pictures were displayed in terms of 10 picture blocks, including 5 different affective states (see Fig. 2).

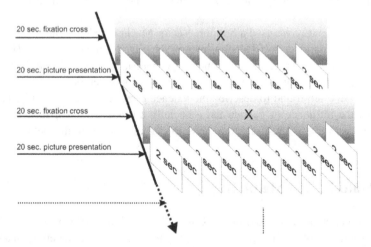

Fig. 2. Stimuli presented in a blocked version, with each block consisting of 10 pictures, shown without a pause within between

3   Subjects

All right-handed subjects we recruited over bulletins scattered around the University of Ulm and the University of Magdeburg for participation in our experiment.

They were healthy and had normal vision or corrected normal vision. The experiment was designed and implemented according to the ethical guidelines of the University of Ulm (Approval by the Ethical Committee #245/08-UBB/se).

Our sample in total n= 122 subjects was subdivided in different subgroups according to different objectives (to measure gender and menstrual cycle differences as well as age differences). We recruited 73 young participants (n=50 females; n=25 follicular (Ø age=23.72 years, SD=3.52); n=25 luteal women (Ø age = 25.88 years, SD= 4.94)) and 49 older subjects (n=28 females (Ø age = 61.5 years, SD=6.14) and n=21 males (Ø age = 68.29 years, SD=6.40)). For measuring the influence of the menstrual cycle, all young females took part at a specific day according to their menstrual cycle phase. For further information about determining the menstrual cycle phase and recruiting young females please see [18].

4   Procedure

All physiological signals were recorded with a NeXus-32 (*NeXus-32, Mind Media, The Netherlands*) and the trigger data was recorded with the software Biobserve Spectator (*BIOBSERVE GmbH, Germany*).

The heart rate information was measured with a BVP (blood volume pulse)-sensor, which was attached to the left middle-finger of every right handed subject and measured the blood volume running through the blood vessels via infrared light within each heart period (plethysmography).

The facial EMG signals were captured by using bipolar miniature silver/silver chloride (Ag/AgCl) skin electrodes with a 4 mm diameter. Both electrodes were placed on participants' left *corrugator supercilii* and *zygomaticus major* muscle regions, according to the guidelines for EMG placement recommended by Fridlund and Cacioppo [19].

Previous to the start of each experiment, each biosignal was visualized with the BioTrace software (appertained to the NeXus-32) and corrected to avoid bad signals or other influences.

5   Signal Processing and Feature Extraction

In total, 19 subjects had to be excluded, either because of technical problems (n=2) or due to a large amount of artifacts within the heart rate recordings (n=13). If the HR during one stimuli block contained artifacts, the whole subject was removed. Additionally, the final EMG subject group was conformed to the final HR population, thus 4 subjects had to be excluded, which lead to a total population of n=53 young subjects (n=37 females) and n=36 older subjects (n=21 females). However, for this gender study only the younger participants have been used for the analysis.

a) Heart Rate
All parameters were calculated after the offline identification of the Inter-Beat-Interval of the following NN-intervals. A Matlab script extracting these intervals almost automatically was developed, however every signal segment was displayed on the screen for visual correction and for having the chance to correct the identification

points of the NN-intervals or to delete the signal segment for excluding artifacts. We extracted the following parameters of the time-domain of the heart rate signal:
Mean NN [ms], SDNN [ms], mean HR [bpm], SD HR [bpm], RMSSD [ms].

$$\text{Mean NN [ms]: } mean\ NN = \frac{1}{n}\sum_{k=1}^{N}IBI_k \quad (1)$$

$$\text{SDNN [ms]:} SDNN = \sqrt{\frac{1}{n}\sum_{k=1}^{N}(NN_k - \overline{NN})^2} \quad (2)$$

$$\text{RMSSD [ms]: } RMSSD = \sqrt{\frac{1}{n}\sum_{j=1}^{N}(NN_{j+1} - \overline{NN_j})^2} \quad (3)$$

Mean HR [bpm] and SD HR [bpm] were calculated from the mean NN. These parameters served as the initial basis for different classification topics. We did not use frequency domain features, due to problems with the validity of short HRV recordings [20].

b) Facial EMG
The raw facial EMG of the *corrugator supercilii* and *zygomaticus major* were offline filtered by a 20–250 Hz band-pass Butterworth filter (order = 4) to exclude motion related components and an adaptive filter was applied to deal with 50 Hz power line interference [21]. The signals were then rectified and smoothed by the root mean square (RMS) technique with a 125 ms sliding window. Facial EMG changes were derived from subtracting baseline activity (i.e., the mean of the RMS of two seconds before each picture block onset) from the respective picture block viewing periods (i.e., the mean of the RMS). Subsequently, we standardized (i.e., Z score) EMG changes within each participant and within each site (according to [22]) to remove variability that might exist in different stimulus conditions. This was done to directly compare the signals from the distinct groups.

All the processing and analyses were conducted using the MATLAB software package (*version R2009a, Mathworks Inc., USA*). Every classification was performed with the data mining software RapidMiner (http://www.rapid-i.com).

6   Results

At first, the five extracted raw heart rate features (mean HR, SD HR, mean NN, SD NN and RMSSD) were used to classify between the valence dimensions positive vs. negative. For each classification we used a 10fold validation (see Table 1).

Table 1. Classification results with the use of 5 heart rate features

Group	classes	HR	Acc. [%]
follicular	pos/neg	5	60.88
luteal	pos/neg	5	59.38
young females	pos/neg	5	54.19
young males	pos/neg	5	58.81
gender mixed	pos/neg	5	53.90

As you can see in Table 1 the classification rates differ in respect to gender. For example, the valence dimension could be identified within males for 58%, whereas the accuracy for females was about 54%. Interestingly, the classifications separately for the specific menstrual cycle phases were higher than the classification for all young females. The same effect can be seen in mixing gender, where only 53.9% accuracy rate could be achieved.

However, these classification results were still not satisfying, considering the chance level of about 50% for two classes. Therefore, we calculated for another classification each heart rate parameter again but baseline corrected, which reflected the cardiac reactivity compared to the baseline situation, see Table 2 (left third). Before we tried the combination of heart rate and facial EMG features, we classified the data with solely EMG. The classification accuracy was even worse, except for the male group. Afterwards, the combination of both feature sets were tested for each subgroup but the results differed for each group. Whereas the accuracy for the follicular group was improved, the accuracy for the luteal and male group got worse.

Table 2. Comparison of different classifications between positive and negative with different feature combinations (left column: HR_change features; middle: EMG features; right: combination of both features)

Group	HR	Acc. [%]	EMG	Acc.[%]	HR+EMG	Acc. [%]
follicular	5	60.38	2	54.40	7	64.51
luteal	5	59.05	2	63.14	7	59.76
Young females	5	54.10	2	59.46	7	52.74
Young males	5	60.83	2	69.49	7	63.21
Gender mixed	5	53.07	2	60.83	7	55.38

In the last step, we classified the valence dimension again, but we used the raw heart rate features and combined them with the EMG data (see Table 3).

These two feature sets resulted in the best classification results, with accuracy rates above 60% for each subgroup. The classification achieved still within the follicular group the highest rates and mixing gender (gender mixed) or both menstrual cycle phase (females) the accuracy drops. Classifying the menstrual cycle phase within all negative and in another classification within all positive resulted in the highest accuracy rates of about 72-72%.

Table 3. Final classification with 5 HR and 2 EMG features

Group	classes	HR and EMG	Acc. [%]
follicular	pos/neg	7	68.52
luteal	pos/neg	7	67.38
young females	pos/neg	7	61.78
young males	pos/neg	7	66.35
gender mixed	pos/neg	7	61.56
all negative	follicular/luteal	7	74.29
all positive	follicular/luteal	7	72.86

7   Discussion

Interactions between humans and technological systems happen nowadays more frequently, because humans are surrounded by technical systems like computers or automats. Although a lot have been done to improve these interactions (e.g. specific user interfaces) the identification of the users´ emotions and dispositions is still not clearly solved. One possibility to improve such classifications or the emotion identification is to consider gender and within young females the menstrual cycle phase as important co-factors. We analyzed the impact of these two factors on psychobiological emotion classification with a SVM and the use of heart rate and facial EMG features.

At first, we could show that for the use of heart rate features it is not necessary to take baseline-corrected features, in contrary this lead to a diminished identification of the valence information between positive and negative. We could also show that the use of solely heart rate features or solely EMG features to discriminate the valence is not enough, although both parameters have been found to be valence sensitive. Interestingly we found a higher accuracy rate for the male group with the EMG features, which indicates a specific EMG pattern for males but it somehow is not consistent for the follicular group. This demonstrates the need for a gender specific use of feature extraction.

General differences concerning gender were that the classification of men and women seems to differ. The reason for the diminished accuracy rate within young females seems to be their menstrual cycle phase. Because classifying positive vs. negative conditions within the follicular and luteal group lead to higher classification results of about 67% or 68%. Combining the female group the accuracy rate drops to about 62%. The same happens when gender is mixed in the gender-mixed- classification.

Another interesting result was that we achieved the highest accuracy rates for distinguishing the menstrual cycle phase within all negative/positive conditions. It means that the differentiation between the two cycle phases is easier than the differentiation between positive and negative for the SVM, which confirms our hypotheses of being influencing on psychobiological classification.

To improve the presented classification rates, another psychobiological parameter like the skin conductance should be included in the future. This parameter is innervated only by the sympathetic nervous system and therefore should serve well for predicting the arousal dimension. During the further procedure, a feature selection is planned to analyze gender specific features. Maybe the so far used features can be reduced specifically for gender and menstrual cycle phases. In respect to the aim of real-time classification we tried to use as less features as possible achieving still satisfying identification results. However, one has to keep in mind that under such circumstances the classification results might be not as high as after processes taking too much time (e.g. feature selection processes, different calculations of parameters).

Acknowledgements. This research was supported by the Transregional Collaborative Research Center SFB/TRR 62 "Companion-Technology for a Cognitive Technical System" funded by the German Research Foundation (DFG), a PhD scholarship by the University of Ulm for Stefanie Rukavina and a doctoral scholarship by the China Scholarship Council (CSC) for Jun-Wen Tan.

References

1. Reeves, B., Nass, C.: The Media Equation: How People Treat Computers, Television, and New Media Like Real People and Places. Center for the Study of Language and Information. Cambridge University Press, New York (1996)
2. Lang, P.J., et al.: Looking at pictures: affective, facial, visceral, and behavioral reactions. Psychophysiology 30(3), 261–273 (1993)
3. Burriss, L., Powell, D.A., White, J.: Psychophysiological and subjective indices of emotion as a function of age and gender. Cognition & Emotion 21(1), 182–210 (2007)
4. Cowie, J.M., et al.: A review of Clinical Terms Version 3 (Read Codes) for speech and language record keeping. Int. J. Lang. Commun. Disord. 36(1), 117–126 (2001)
5. Limbrecht, K., et al.: Advantages of mimic based pre-analysis for neurophysiologic emotion recognition in human-computer interaction. In: F.-M.A. (ed.) Emotional Expression: The Brain and The Face. UFP Press (2012)
6. Frantzidis, C.A., et al.: Toward emotion aware computing: an integrated approach using multichannel neurophysiological recordings and affective visual stimuli. IEEE Transactions on Information Technology in Biomedicine 14(3), 589–597 (2010)
7. Bailenson, J.N., et al.: Real-time classification of evoked emotions using facial feature tracking and physiological responses. International Journal of Human-Computer Studies 66, 303–317 (2007)
8. Ossewaarde, L., et al.: Neural mechanisms underlying changes in stress-sensitivity across the menstrual cycle. Psychoneuroendocrinology 35(1), 47–55 (2010)
9. Goldstein, J.M., et al.: Sex differences in stress response circuitry activation dependent on female hormonal cycle. Journal of Neuroscience 30(2), 431–438 (2010)
10. Andreano, J.M., Cahill, L.: Menstrual cycle modulation of medial temporal activity evoked by negative emotion. Neuroimage 53(4), 1286–1293 (2010)
11. Farage, M.A., Osborn, T.W., MacLean, A.B.: Cognitive, sensory, and emotional changes associated with the menstrual cycle: a review. Archives of Gynecology and Obstetrics 278(4), 299–307 (2008)
12. Tan, J., et al.: Repeatability of facial electromyography (EMG) activity over corrugator supercilii and zygomaticus major on differentiating various emotions. J. Ambient Intell. Human Comput. 3(3), 3–10 (2012)
13. Bradley, M.M., et al.: Emotion and motivation II: sex differences in picture processing. Emotion 1(3), 300–319 (2001)
14. Lang, P.J., Bradley, M.M., Cuthbert, B.N.: International affective picture system (IAPS): Technical manual and affective ratings. University of Florida, Center for Research in Psychophysiology, Gainesville (1999)
15. Walter, S., et al.: The influence of neuroticism and psychological symptoms on the assessment of images in three-dimensional emotion space. Psychosoc. Med. 8, Doc04 (2011)
16. Smith, J.C., Bradley, M.M., Lang, P.J.: State anxiety and affective physiology: effects of sustained exposure to affective pictures. Biological Psychology 69(3), 247–260 (2005)
17. Valenza, G., Lanata, A., Scilingo, E.P.: The Role of Nonlinear Dynamics in Affective Valence and Arousal Recognition. IEEE Transactions on Affective Computing 99 (2011) (PrePrints)
18. Rukavina, S., et al.: The influence of the menstrual cycle on gender differences in personality items. In: Annual International Conference of Cognitive and Behavioral Psychology (CBP). Global Science & Technology Forum (GSTF), Singapore (2012)

19. Fridlund, A.J., Cacioppo, J.T.: Guidelines for human electromyographic research. Psychophysiology 23(5), 567–589 (1986)
20. Nussinovitch, U., et al.: Reliability of Ultra-Short ECG Indices for Heart Rate Variability. Ann. Noninvasive Electrocardiol. 16(2), 117–122 (2011)
21. Andrade, A.O., et al.: EMG Decomposition and Artefact Removal, in Computational Intelligence in Electromyography Analysis - A Perspective on Current Applications and Future Challenges, I.-.-.-.-. Ganesh R. Naik, Editor. 2012, InTech.
22. Neta, M., Norris, C.J., Whalen, P.J.: Corrugator muscle responses are associated with individual differences in positivity-negativity bias. Emotion 9(5), 640–648 (2009)

Evaluation of Mono/Binocular Depth Perception Using Virtual Image Display

Shys-Fan Yang-Mao, Yu-Ting Lin, Ming-Hui Lin,
Wen-Jun Zeng, and Yao-lien Wang

Industrial Technology Research Institute, Taiwan
{Yangmao,stevenytli,lmw,moir,Yao_lien_Wang}@itri.org.tw

Abstract. Augmented reality (AR) is a very popular technology in various applications. It allows the user to see the real world, with virtual objects composited with or superimposed upon the real world. The usability of interactive user interface based on AR relies heavily on visibility and depth perception of content, virtual image display particularly. In this paper, we performed several basic evaluations for a commercial see-through head mounted display based on those factors that can change depth perception: binocular or monocular, viewing distance, eye dominance, content changed in shape or size, indicated by hand or reference object. The experiment results reveal many interesting and fascinating features. The features will be user interface design guidelines for every similar see-through near-eye display systems.

Keywords: augmented reality, virtual image display, see-through near-eye display, user interface, depth perception.

1   Introduction

Augmented reality (AR) is one part of the general area of mixed reality according to the reality-virtuality continuum, as show in Fig.1, and it provides local virtuality [4]. Not like virtual environment (VE) technologies that completely immerse a user inside a synthetic environment and the user cannot see the real world around him, in contrast, AR allows the user to see the real world, with virtual objects composited with or superimposed upon the real world [5]. As a result, the AR technology functions by enhancing and enriching one's current perception of reality [3]. The natural and intuitional characteristics of AR increase its popularity in various applications, such as personal assistance, medication, education, industry, navigation and entertainment [1].

AR is also the technology to create a "next generation, reality-based interface" [6] and is moving from laboratories around the world into consumer markets. Current dominance head-worn AR technologies can be divided into three main types, projective AR, video AR and optical head-worn AR [1]. The optical head-worn AR (also known for see-through near-eye display) is the best choice of head-worn personal assistance for the advantages of delay-free (user can see the world around him without any transmission delay) and not required for special material for projection.

M. Kurosu (Ed.): Human-Computer Interaction, Part V, HCII 2013, LNCS 8008, pp. 483–490, 2013.

However, see-through near-eye display has many disadvantages: low contrast, low brightness, low resolution, inaccurate depth perception, high power consuming and eye strain caused by dynamic refocus (changes from virtual objects and the real world). One difficult registration problem is accurate depth perception of virtual content. Naturally, stereoscopic has higher depth perception accuracy than that binocular has. However, additional problems including accommodation-convergence conflicts or low resolution and dim displays cause object to appear further away than they should be [1]. This is an essential issue for AR applications, since AR allows the user to see the real world and virtual objects simultaneously, and inaccurate depth perception will case the misalignment of the virtual objects.

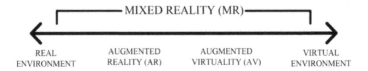

Fig. 1. Reality-virtuality continuum

In order to build a high usability user interface for optical head-worn AR system, factors that can change depth perception should be confirmed. We performed several basic evaluations based on those factors: binocular or monocular, viewing distance, eye dominance, content changed in shape or size, indicated by hand or reference object. The experiment results reveal many interesting and fascinating features. The features will be user interface design guidelines for every similar see-through near-eye display systems.

2   Depth Perception

2.1   Equipment

In this paper, we use the VUZIX (type STAR 1200) for depth perception evaluation. VUZIX is a mono/binocular see-through near-eye virtual image display and provides two separated and identical virtual screens for observer. It also provides a virtual screen (43 inches in diagonal) with high-resolution (852x480 WVGA) that appears in front of the user (3m). Since virtual screen is refracted from a very small LCD display through the well-designed optical apparatus, as show in Fig.2, it has very low power consumption.

2.2   Test Environment Set Up

In test environment of depth perception evaluation, as shown in Fig.3, moderator will display images in one or two screens in VUZIX and ask evaluator try to indicate the exactly location of the virtual floating object, using their finger (of their normal used hand) or a reference object (very small real object that connected to a sliding ruler). The location will be measured and recorded by a precision measuring device, which the scale can be read directly to 0.01mm. Evaluator will be fixed at a chin holder to the center of the background, which is a 22 inches screen in those evaluations. The field of view (FOV) through VUZIX will be limited inside the background. Ambient light will be fixed to the luminous recommendation of office lighting (320-500 Lux, under European law UNI EN 12464) to simulate indoor usage. For binocular test, we set the convergence angle of both eye to the viewing angle of a near eye object at distances of 30, 40 and 50 cm. The viewing angle is measured and confirmed by another precision eye-tracking device before evaluations.

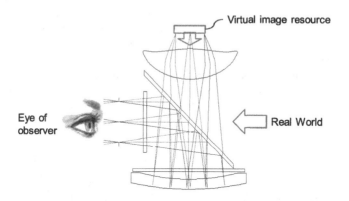

Fig. 2. Sample of optical configuration of virtual image display

2.3   Binocular and Monocular Setup on See-Through Near-Eye Display

With properly content design, we can simulate the viewing distance changing of virtual objects for depth perception evaluations. Since the convergence angle of eye will be automatic adjusted when viewing distance changed, when we focus on a near object, the convergence angle of eye is bigger than that we focus on a far object. In VUZIX, the virtual image is seen by two eyes separately, and the convergence angle of eyes will be fixed into a predefined angle, as a result, a floating virtual screen appeared in front of user at predefined viewing distance (43-inch virtual screen at 3m). In order to create a virtual object at particular position in front of user, we change the location of virtual images on both virtual screens. For example, shift the image in left screen toward to right and shift the image in right screen toward to left, the virtual object will be visually closer to user, since the convergence angle is going bigger. The image in tests is a simple white circle or white square on black background without any textures or other monocular depth perception cues. For monocular tests, we display only one image in left screens or right screens.

2.4   Participations

Total 30 evaluators (10 females) with age 18-35 participated in this study. We verified that no one was color blind and all evaluators have normal or corrected-to-normal vision. We also confirmed that none of them were stereo blindness.

2.5   Evaluations

Factors that could affect the depth perception will be evaluated in tests: binocular or monocular, viewing distance of virtual objects, eye dominance, content changed in shape or size, indicated by hand or reference object. The evaluation includes 9 stages, which are 5 stages in monocular and 4 stages in binocular, and performed in randomly order. Fig.4. shows one example of task list for evaluation. At first and second stage, we display one white circle on left (1st) and right (2nd) virtual screen in VUZIX and ask evaluator to indicate the location of the circle by finger. At third stage, we display one white circle on left screen and then display one bigger white circle at the same location and ask evaluator to indicate the location of the bigger circle by finger. At fourth stage, we display one white circle on left screen and then display one white square with the same size at the same location and ask evaluator to indicate the location of the square by finger. At fifth stage, we display one white circle on left screen and ask evaluator to indicate the location of the circle by moving a reference object. At sixth to ninth stages, we display one white circle on both screens to simulate seeing a near eye object at viewing distances of 30, 40 and 50 cm. Then we ask evaluator to indicate the location by their finger (at 30, 40 and 50 cm) or by a reference object (at 40 cm).

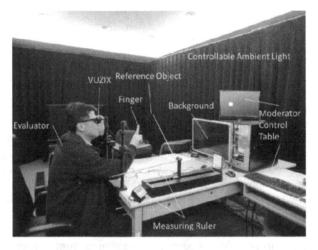

Fig. 3. Test environment set up for depth perception evaluation

Task list

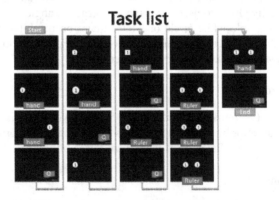

Fig. 4. Task list example for depth perception evaluation

3   Experiment Results

3.1   Monocular and Binocular

The experiment results of depth perception evaluations reveal many interesting and fascinating features, as shown in Fig.5. In monocular tests, it displays only one small circle or square in one eye, the observer can't sense the viewing distance directly from the virtual object in those monocular stages. The depth perception might come from the user experience of observer by compared their finger size and the size of virtual circle. It is obviously and naturally that the binocular depth perception is more accurate than monocular, as show in Fig.6. The standard deviation is smaller in binocular than that in monocular, as show in Table.1.

Table 1. Results of paired samples t-test for different property

Property		Paired Differences				t	df	Sig. (2-tailed)
				95% Confidence Interval of the Difference				
		Mean	Std. Deviation	Lower	Upper			
Size change	small - big	44.15167	89.46459	10.74504	77.55829	2.703	29	.011
Shape change	circle - square	-9.34233	94.70232	-44.70476	26.02009	-.540	29	.593
Binocular	hand - ruler	-34.73067	63.62760	-58.48960	-10.97173	-2.990	29	.006
Monocular	hand - ruler	-83.25967	135.22556	-133.75372	-32.76561	-3.372	29	.002
Eye domination	left - right	-5.97700	109.00321	-46.67947	34.72547	-.300	29	.766

3.2   Viewing Distances

The binocular depth perception accuracy is various at different viewing distances, as show in Fig.5. Depth perception is accurate at viewing distance 400mm when it was

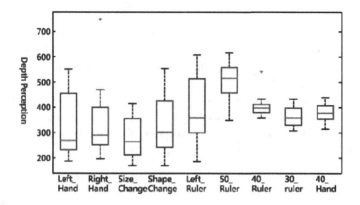

Fig. 5. Experiment results of depth perception evaluations

indicated by finger or reference object. Only one of the evaluators says that he cannot make well convergence of the binocular images at viewing distances 500 mm and 300 mm. Depth perception is very inaccurate at viewing distance 500 mm and 300 mm, 500 mm particularly. This result may be contributed by hand lengths of evaluators. The usual and comfortable location caused by hand length will change the depth perception at different distances. It requires further tests to determinate the best operation zone for virtual objects interactive user interface.

3.3   Eye Domination

Since approximately two-thirds of the population is right-eye dominant and one-third left-eye dominant [2], we need to confirm the influence of eye domination on depth perception. Paired sample t-test was used for eye domination (image on left screen only and image on right screen only), as show in table.1. The experiment result indicated that there was no different in eye domination (Eye domination, $p > 0.05$) for depth perception in our tests.

3.4   Content Change in Size or Shape

Content change may affect the depth perception. We use paired sample t-tests for content change in size (small to big) and shape (circle to square), as show in table.1. Significant differences between small circle and big circle were noted (Size change, $p < 0.05$). Depth perception will naturally increase when virtual image going bigger, as show in Fig.7. Experiment result also indicated that there was no different in change shape (Shape change, $p > 0.05$).

3.5   Indicated by Hand or Reference Object

The tools for distance measurement of virtual object will affect the depth perception. In this study, we let evaluator use their finger to indicate the location of virtual object or operate a sliding ruler for indication in both monocular stage and binocular stage. The reference object attached on sliding ruler is a small needle that smaller than virtual image and that will not cover or overlapping with virtual object. Otherwise, finger can cover on or overlap with virtual object in those tests. Paired sample t-tests were used for indicator change in monocular (hand to ruler) and binocular (hand to ruler), as show in table.1. Significant differences between hand and ruler in binocular and monocular ($p < 0.05$) were noted. The results revealed that depth perception on ruler was further than that on finger. Depth perception will increase when it was indicated by a reference object, as show in Fig.8. Therefore, we can confirm the collision of virtual objects will affect the depth perception. Even the virtual image should be closer than real world or finger for user's eye, as show in Fig. 2.

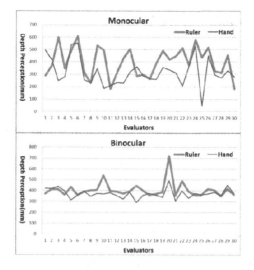

Fig. 6. Experiment results from monocular stage with image on left eye and binocular stage with virtual object at 400 mm

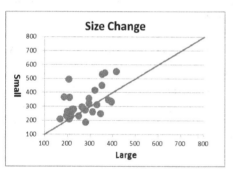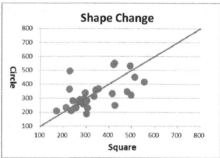

Fig. 7. Correlation of size and shape changing

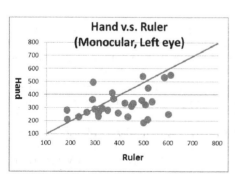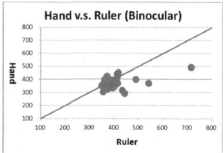

Fig. 8. Correlation of measurement tool changing for monocular and binocular test

4   Conclusions and Future Works

Augmented reality is a very popular technology in various applications. The usability of user interface based on AR relies heavily on visibility of content and depth perception of interactive surface. In order to build a high usability user interface, factors that can change depth perception should be confirmed. We performed several basic evaluations based on those factors. The experiment results reveal many interesting and fascinating results. Significant differences between binocular or monocular, content changed in shape or size, indicated by hand or reference object were noticed. Otherwise, depth perception is accuracy at viewing distance 400 mm that indicated by finger or reference objects.

Here we list several suggestions for user interface design:

1. Displaying content in binocular optical head-wear AR with properly design based on convergence angles can simulate viewing distance changing.
2. The viewing distance of interactive virtual floating control panel should be not too close or too far from user, 400 mm is recommended.
3. Applications of optical head-wear AR that require precision depth perception, binocular is recommended.
4. Fault tolerance design for depth perception of binocular applications at viewing distance 40cm, offset ± 5cm is recommended.
5. The size of virtual object is a better cue for depth perception rather than shape.
6. Multi-dimensional control panel is required on monocular virtual image display since the low precision depth perception.

Those suggestions can not only be the user interface design guidelines for VUZIX but also for every similar see-through near-eye display system. In the future, there still remain various properties, such as content color, texture gradient, contrast, resolution, stereoscopic rendering, object motion and real time feedback that might affect the usability of UI and all should be confirmed through a lot of work.

References

1. van Krevelen, D.W.F., Poelman, R.: A Survey of Augmented Reality Technologies, Applications and Limitations. The International Journal of Virtual Reality 9(2), 1–20 (2010)
2. Ehrenstein, W.H., Arnold-Schulz-Gahmen, B.E., Jaschinski, W.: Eye preference within the context of binocular functions. Graefes Arch. Clin. Exp. Ophthalmol. 243(9), 926–932 (2005)
3. Graham, M., Zook, M., Boulton, A: Augmented reality in urban places: contested content and the duplicity of code. Transactions of the Institute of British Geographers (2012), doi:10.1111/j.1475-5661.2012.00539.x 2012
4. Milgram, P., Kishino, F.: A taxonomy of mixed reality visual displays. IEICE Trans. Information and Systems E77-D(12), 1321–1329 (1994)
5. Azuma, R.T.: A Survey of Augmented Reality. Presence: Teleoperators and Virtual Environments, 355–385 (1997)
6. Jebara, T., Eyster, C., Weaver, J., Starner, T., Pentland, A.: Stochasticks: Augmenting the billiards experience with probabilistic vision and wearable computers. In: First International Symposium on Wearable Computers (1997)

Visual Image Reconstruction from fMRI Activation Using Multi-scale Support Vector Machine Decoders

Yu Zhan[1], Jiacai Zhang[1], Sutao Song[2], and Li Yao[1,3]

[1] School of Information Science and Technology, Beijing Normal University, Beijing, China
[2] School of Education and Psychology, University of Jinan, Shandong, China
[3] State Key Laboratory of Cognitive Neuroscience and Learning, Beijing Normal University
zhanyu@mail.bnu.edu.cn, {jiacai.zhang,yaoli}@bnu.edu.cn,
Sep_songst@ujn.edu.cn

Abstract. The correspondence between the detailed contents of a person's mental state and human neuroimaging has yet to be fully explored. Previous research reconstructed contrast-defined images using combination of multi-scale local image decoders, where contrast for local image bases was predicted from fMRI activity by sparse logistic regression (SLR). The present study extends this research to probe into accurate and effective reconstruction of images from fMRI. First, support vector machine (SVM) was employed to model the relationship between contrast of local image and fMRI; second, additional 3-pixel image bases were considered. Reconstruction results demonstrated that the time consumption in modeling the local image decoder was reduced to 1% by SVM compared to SLR. Our method also improved the spatial correlation between the stimulus and reconstructed image. This finding indicated that our method could read out what a subject was viewing and reconstruct simple images from brain activity at a high speed.

Keywords: Image Reconstrucion, fMRI, Multi-scale, SVM.

1 Introduction

Functional magnetic resonance imaging (fMRI) provides a convenient tool for scientists to determine what a person perceives from his/her brain activity [1-6]. Researches about visual information reading decoded brain activity at three levels: classification, identification and reconstruction. Classification is to predict which category the present image belongs to from the brain pattern of activity [8-12]. Beyond classification, image identification established the computation model to identify the image that the subject was viewing out of a set of potential images from brain activity measurements [1, 2]. Kay and colleagues utilized a Gabor wavelet function to capture the visual stimuli characteristics related to fMRI activity, and characterized the relationship between visual stimuli and fMRI activity in early visual areas with quantitative receptive-field models. Their study showed that these receptive-field models make it possible to perform image identification [1]. More recently, researchers have moved a more forward step to reconstruct the visual image composed of flickering

M. Kurosu (Ed.): Human-Computer Interaction, Part V, HCII 2013, LNCS 8008, pp. 491–497, 2013.

checkerboard patterns or even more complicated actual natural images that were seen or movie experience, rather than simply choosing the image from a known set [4, 5].

Thirion et al. build the inverse model of the retinotopy of the visual cortex to infer the visual content of real or imaginary scenes from the brain activation patterns [6]. Another representative work of image reconstruction was the study of Miyawaki et al., they established the method of multi-scale local image decoder to directly model the relationship between the image stimulus and fMRI activity at the specific time when image was presented to subjects [3]. In their study, reconstructed images were modeled by a linear combination of local image bases. Miyawaki fixed the shape and size of image bases in his study, and he utilized local image of four scales: 1×1, 1×2, 2×1, and 2×2 patch areas [3]. These are totally 361 image bases for a 10×10 flickering checkerboard patterns. The local decoders based on sparse logistic regression approach were defined to predict the mean contrast of each local image bases. Each of the 361 local decoders was individually trained to classified fMRI data samples into a class corresponding to contrast level. After the decoders' output, a linear combination of the 361 image bases was applied to reconstruct the predicted images.

Miyawaki's research stands for the highest level of visual decoding studies that have emerged over the years. The spatial correlation between the presented stimulus and reconstruction image even came to 0.68 ± 0.16 (mean ± s.d.) for individuals. However, the time and space complexity of their method is extremely high because of the extremely laborious computation in sparse logistic regression model.

To further investigate the effective image reconstruction method in visual information decoding from brain activities, we are trying to find a method raising the training speed with little precision loss. First, the heavy work in training the contrast decoder model with SLR for each element image in Miyawaki's work was reduces by classifiers designed with SVM. As each local decoder consisted of a multi-class classifier, Instead of the sparse logistic regression method, we could use an SVM model to classify fMRI data samples into discrete contrast levels. Second, we will also use the 1×3 and 3×1 image bases, as well as the fixed image bases used in Miyawaki's work. Using Bayesian based canonical correlation analysis (CCA), Fujiwara proposed a method to automatically find a set of image bases from the fMRI data, and reconstruction results illustrated that this set of 3-pixel image bases improved the reconstruction performance [7].

2   Method

2.1   Dataset

We used the same dataset from Miyawaki et al., where fMRI signals were measured in two independent sessions. In each session, the subject viewed visual images consisting of contrast-defined 10×10 patches. In the random image session, a total of 440 flickering checkerboard spatially random pictures were presented, each stimulus block was 6 seconds long followed by 6 s rest period. In the figure image session, a total of 120 pictures were presented. Each stimulus block was 12 seconds long followed by a 12 s rest period. Stimulus pictures were geometric shapes (i.e. "square", "plus", "X") or alphabet letters ("n", "e", "u", "r", "o"). Each picture had been shown to the subject for 4 or 8 times. The Supplemental Data can be found online at http://www.neuron.org/supplemental/S0896-6273(08)00958-6.

2.2   Multi-scale Image Bases

As with Miyawaki's research, we assume that an image is represented by a linear combination of local image elements of multiple scales. The local image bases used in Miyawaki's work were 1×1, 1×2, 2×1 and 2×2 patch areas. They were placed at every location in the image with overlaps. For example, the 1×2 scale bases will cover the 9 rectangles (1 1)-(1 2), (1 2)-(1 3), ... , (1 9)-(1 10) in the first row of the 10×10 image. So there are a total of 10×9 image bases of scale 1×2. Similarly, there are respectively 10×10, 9×10, 9×9 image bases of scale 1×1, 2×1 and 2×2. For each stimuli images, we counted the number of flickering grids in an image base, 1×1 scale yields value 0 for all patches staying gray and 1 for whole element image flicking. For the 2×1 scale image bases, there are three values, 0 means no flickering grid, 1 means one flickering grid, and 2 means all the 2 grid in image bases are flickering. Similarly, 1×2 yield contrast values 0, 1, 2 and 2×2 yields 0, 1, 2, 3, 4. The present study used extra image elements of 1×3 and 3×1 patch areas, thus yielded another 160 image bases (10×8 and 8×10 for 1×3 and 3×1 patch areas respectively). The flickering grids number for 3-patch image ranges from 0 to 3. The mean contrast value of each local image base was defined as the total number of patches in that local image divided by the number of flickering patches (represented as white girds).

2.3   SVM Model

Instead of applying sparse logistic regression method, we used support vector machine to predict the mean contrast value for each local element. In this study, we only used the fMRI activity of V1 area to build the SVM models. The sample features are the activation of voxels in V1 area; the sample label is the mean contrast value in each image base. All the local image decoders except those for 1×1 patches, the mean contrast level belonged to more than 2 classes, so we used multi-class SVM models. The SVM code used here was implemented by Lin Chih-Jen in Taiwan University. The source code is available online at http://www.csie.ntu.edu.tw/~cjlin/. Here, linear SVM models were trained with samples in the random image session. And we evaluated the model with test dataset from the figure image session.

2.4   Linear Combination

This procedure makes up the reconstructed image by adding all local image bases altogether. In Miyawaki's work, least square error method was used to obtain the coefficient of each image base. In our work, the prediction accuracy on the training set are 100% for all 1×1 image bases, all the 1×1 image bases will have coefficients 1. To simplify this problem, we assume that all the pixels in the 10×10 image are predicted by summing up the class labels of correlated image bases, and the coefficients of all the image bases are 1. The mainframe of this approach is shown in Fig. 1.

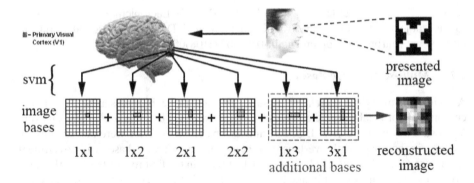

Fig. 1. The fMRI signals were measured while subject viewed the stimulus images. The SVM models were trained using V1 area data of the random image session. After the SVM assigned the class labels to each image base for the figure image session data, the predicted value of a specific pixel was calculated by summing up the class labels of all the image bases that covered the pixel.

3 Results

3.1 Reconstruction Performance

The stimuli images were reconstructed using the element images, whose contrast values predicted by SVM models trained with all block-average data (average of 6s or 3-volumns fMRI data, TR=2s) in the random image session. Reconstruction was performed on block-averaged data (average of 12s or 6-volumns fMRI data) in the figure image session. Although model training used only random images, the reconstructed images of test dataset showed obvious similarity between presented images and reconstructed images for stimulus shapes or letters. Using the combination of 1×1, 1×2, 2×1 and 2×2 image bases, the spatial correlation was 0.6643 ± 0.1207 (mean ± s.d. data not shown). By adding the 1×3 and 3×1 image bases, the spatial correlation increased to 0.6934 ± 0.1165 (mean ± s.d.). Reconstruction images from all trials of the figure image session are illustrated in Fig.2.

3.2 Computational Expenses

The SVM model training and testing time cost for all 6 scales are listed in Table.1. The comparison suggested that the computing complexity using SVM was far lower than that of sparse logistic regression.

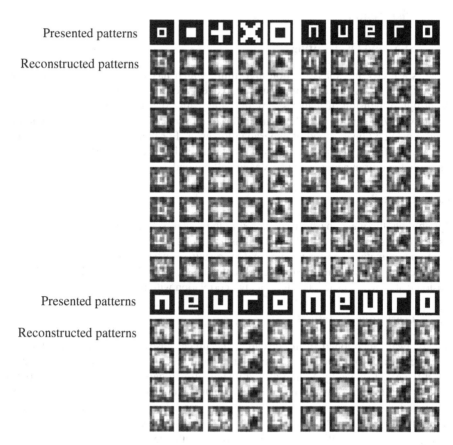

Fig. 2. Reconstructed visual images of chareacters. The reconstruction results of all the trials for the subject are shown with presented images from the figure image session (a total of 120 images). The 10 stimuli images in the first row repeated 8 times and the 10 stimuli in the 10th row repeated 4 times. The reconstructed images are sorted in descending order of the spatial correlation. No post processing was applied.

Table 1. The total time (in seconds) including training and testing for support vector machine (SVM) and sparse logistic regression (SLR) under different scales were shown below. System Information: Windows 7 SP1(64bit), CPU: Intel(R) Core(TM)2 CPU 6600@ 2.40GHz 2.39GHz, Ram: 4.00GB.

Scale	1×1	1×2	2×1	2×2	1×3	3×1
method						
SVM	85	103	97	106	92	98
SLR	11831	21661	18739	58229	--	--

4   Discussion

4.1   SVM and Over-Fitting

SVM is a pretty fast classification algorithm, which has its application in many research fields. In our study, more than 1700 voxels are used to train the local decoders. Such high dimension of features may probably cause the problem of over-fitting. As we know, SVM implements the structural risk minimization principle and it searches to minimize an upper bound of generalization error. For this reason, SVM reduces the risk of over-fitting and provides a relatively stable classification performance. In this research, the output accuracies of SVM are all beyond guess probability.

4.2   Image Bases

By intuition, the image bases more closed to the visual recognition of human-beings the less the reconstruction error will be. There is no verdict in what shapes of image bases are more approximate to our visual system. Our results showed that using extra 3-pixel image bases produces slightly better reconstructed images, consistent with Fujiwara's conclusion that 3-pixel image bases have a high correlation with the visual cortex activities.

4.3   Least Square Method

As Miyawaki mentioned in his research, the image bases are not orthogonal which means even if all the image base decoders output the right results, we cannot reconstruct the exact image by adding all the image bases together. To deal with this problem, he used the least square method to calculate the coefficients of each image bases. We found this process is technically not necessary. First, if all the SVM decoders output right, the reconstructed images by adding all image bases altogether have an average spatial correlation of 0.9551 ± 0.0217 with the original ones. Second, the training set, which has only 440 random images, is relatively small for predicting a total of 521 least square coefficients. While it is true that using different coefficients (not LSM) may yield better performance, little improvement could be achieved. We assign unit value to each coefficient and this simple strategy seemed work well.

5   Conclusions

The results reported here provide an efficient and accurate method to reconstruct visual stimulus from fMRI signals. Furthermore, the improved spatial correlation using 3-pixel image bases suggests that these image bases may provide more supplementary visual information. The main features of the stimulus were emerged in the reconstructed images, which indicated that SVM could exactly map the activation of visual cortex (V1 area) to the contrast stimulus patterns. Here we also drew the same conclusion with Miyawaki that the outputs of local decoders in the center of an image are more accurate than that in the edges or corners, demonstrating that the visual attention is likely to be concentrated in the center of sight. Further research can be applied to investigate how to accurately predict these surrounding areas.

Acknowledgments. We thank Yoichi Miyawaki in National Institute of Information and Communications Technology, Kyoto, Japan for sharing their fMRI data used in this study. This work is supported by the National High-tech R&D Program (863 Program) under Grant Number 2012AA011600 and the Fundamental Research Funds for the Central Universities. This work is also supported by the Major Research Plan of the National Natural Science Foundation of China (NSFC) Key Program 60931003.

References

1. Kay, K., Naselaris, T., Prenger, R., Gallant, J.: Identifying natural images from human brain activity. Nature 452(7185), 352–355 (2008)
2. Mitchell, T., Shinkareva, S., Carlson, A., Chang, K., Malave, V., Mason, R., Just, M.: Predicting human brain activity associated with the meanings of nouns. Science 320(5880), 1191 (2008)
3. Miyawaki, Y., Uchida, H., Yamashita, O., Sato, M., Morito, Y., Tanabe, H., Sadato, N., Kamitani, Y.: Visual image reconstruction from human brain activity using a combination of multiscale local image decoders. Neuron 60(5), 915–929 (2008)
4. Naselaris, T., Prenger, R.J., Kay, K.N., Oliver, M., Gallant, J.L.: Bayesian reconstruction of natural images from human brain activity. Neuron 63(6), 902–915 (2009)
5. Nishimoto, S., Vu, A.T., Naselaris, T., Benjamini, Y., Yu, B., Gallant, J.L.: Reconstructing Visual Experiences from Brain Activity Evoked by Natural Movies. Current Biology 21, 1641–1646 (2011)
6. Thirion, B., Duchesnay, E., Hubbard, E., Dubois, J., Poline, J., Lebihan, D., Dehaene, S.: Inverse retinotopy: inferring the visual content of images from brain activation patterns. NeuroImage 33(4), 1104–1116 (2006)
7. Fujiwara, Y., Miyawaki, Y., Kamitani, Y.: Estimating image bases for visual image reconstruction from human brain activity. Advances in Neural Information Processing Systems 22, 576–584 (2009)
8. Haxby, J., Gobbini, M., Furey, M., Ishai, A., Schouten, J., Pietrini, P.: Distributed and overlapping representations of faces and objects in ventral temporal cortex. Science 293(5539), 2425–2430 (2001)
9. Cox, D., Savoy, R.: Functional magnetic resonance imaging (fMRI)"brain reading": detecting and classifying distributed patterns of fMRI activity in human visual cortex. NeuroImage 19(2), 261–270 (2003)
10. Spiridon, M., Kanwisher, N.: How distributed is visual category information in human occipito-temporal cortex? An fMRI study. Neuron 35(6), 1157–1165 (2002)
11. Carlson, T., Schrater, P., He, S.: Patterns of activity in the categorical representations of objects. Journal of Cognitive Neuroscience 15(5), 704–717 (2003)
12. O'toole, A., Jiang, F., Abdi, H., Haxby, J.: Partially distributed representations of objects and faces in ventral temporal cortex. Journal of Cognitive Neuroscience 17(4), 580–590 (2005)

Alterations in Resting-State after Motor Imagery Training: A Pilot Investigation with Eigenvector Centrality Mapping

Rushao Zhang[1], Hang Zhang[2], Lele Xu[1], Mingqi Hui[3], Zhiying Long[3],
Yijun Liu[2], and Li Yao[1,3,*]

[1] School of Information Science and Technology, Beijing Normal University,
Xin Jie Kou Wai Da Jie 19#, Beijing, China, 100875
[2] Department of Biomedical Engineering, Peking University,
No.5 Yiheyuan Road Haidian Dis-trict, Beijing, China, 100871
[3] State Key Laboratory of Cognitive Neuroscience and Learning,
Beijing Normal University, Xin Jie Kou Wai Da Jie 19#, Beijing, China, 100875
{zhangrushao,xulelebnu}@gmail.com, yaoli@bnu.edu.cn,
kevinhangpku@foxmail.com, yjliufl@gmail.com,
{iry230,friskying}@163.com

Abstract. Motor training, including motor execution and motor imagery training, has been indicated to be effective in mental disorders rehabilitation and motor skill learning. In related neuroimaging studies, resting-state has been employed as a new perspective besides task-state to examine the neural mechanism of motor execution training. However, motor imagery training, as another part of motor training, has been few investigated. To address this issue, eigenvector centrality mapping (ECM) was applied to explore resting-state before and after motor imagery training. ECM could assess the computational measurement of eigenvector centrality for capturing intrinsic neural architecture on a voxel-wise level without any prior assumptions. Our results revealed that the significant increases of eigenvector centrality were in the precuneus and medial frontal gyrus (MFG) for the experimental group but not for the control group. These alterations may be associated with the sensorimotor information integration and inner state modulation of motor imagery training.

Keywords: Motor imagery, functional magnetic resonance imaging (fMRI), ECM, precuneus, medial frontal gyrus (MFG).

1   Introduction

Motor training, including motor execution and motor imagery training, has been indicated to be effective in mental disorders rehabilitation and motor skill learning [1, 2]. Additional to the improvement of movement in spatial and temporal accuracy, motor training is also accompanied with alterations in human brain's activity. In the last

---

* Corresponding author.

M. Kurosu (Ed.): Human-Computer Interaction, Part V, HCII 2013, LNCS 8008, pp. 498–504, 2013.
© Springer-Verlag Berlin Heidelberg 2013

decade, the neural mechanism underlying motor training has attracted increasing attention in the neuroimaging explorations. Researchers have conducted extensive investigations on both motor execution and imagery in task state, and have indicated that motor execution and imagery may induce different alterations within brain in spite of their similar neural substrates [3-5]. Recently, many researchers have suggested that resting state may contain rich information on the neural mechanism of motor training and may offer the possibility of more complete detection of specific neural system [6, 7].

Identifying the neural alterations induced by motor training from the resting state aspect has been the focus of many neuroimaging studies in recent years. Xiong et al. (2009) revealed that 4 weeks finger sequential training could induce a significant increase of rCBF (regional cerebral blood flow) in M1 [6]. Albert et al. (2009) found that the fronto-paritial resting state network and the cerebellar resting state network were altered by a visuomotor tracking task [8]. Ma et al. (2011) found dynamic changes (increase-first-then-decreased) in resting state functional connectivity in the rPCG and rSMG during a 4 weeks finger sequential training [7]. Taubert et al. (2011) suggested that learning a challenging motor task could lead to long-lasting changes in resting state network, and the changes occurred in SMA and prefrontal were directly correlated with structural grey matter changes [9]. All these studies on the resting human brain have been focused on motor execution training. However, motor imagery training, as another part of motor training, has been few investigated.

In this study, using functional magnetic resonance imaging (fMRI), we applied a novel graph-based network analysis technique named eigenvector centrality mapping (ECM) to explore the influence of motor imagery training on resting state across the whole brain. ECM has been indicated to be a data-driven method without any prior assumptions, and it could assess the computational measurement of eigenvector centrality for capturing intrinsic neural architecture on a voxel-wise level [10]. We hypothesized that motor imagery training could alter the eigenvector centrality of resting state.

2   Materials and Methods

2.1   Subjects

Fourteen right hand-dominant subjects (seven males, mean age: 22±2 years) participated in the training, and another twelve right hand-dominant subjects (five males, mean age: 24±2 years) were recruited as a control group. Participants with histories of neurological disorders, psychiatric disorders, experience with typewriters, or any experience learning to play musical instruments were excluded. All participants provided written consent according to the guidelines set by the MRI Center of Beijing Normal University.

2.2   Experiment Procedure

The experiment procedure included a pre- resting-state session, two pre-task sessions, a motor imagery training period (experimental group)/a no-training period (control group), a post- resting-state session and two post-task sessions. Here, only the resting-state data was examined. In each 10-min resting-state session, subjects were instructed to keep their eyes closed, relax their mind, and remain motionless as much as possible. In the training period, all participants were instructed that from their index to little finger, each of the four fingers of their right hand represented a single digit number: one, two, three, and four. Fourteen motor imagery practice sessions were employed over 14 consecutive days to make sure the sufficient training. Each training session consisted of two 15-min sections, metronome-pacing and self-pacing respectively. In each section, participants were instructed to imagine tapping sequence 4–2–3–1–3–4–2 with their right hand fingers repeatedly as fast as the pace of the metronome or the pace controlled by themselves for 30 seconds with an interval of 30-s rest. The training period were only performed in the experimental group while participants did not attend any training during the 14 days in the control group.

2.3   fMRI Data Acquisition

Brain scans were performed at the MRI Center of Beijing Normal University using a 3.0-T Siemens whole-body MRI scanner. A single-shot T2*-weighted gradient-echo, EPI sequence was used for the functional imaging acquisition, with the parameters: TR/TE/flip angle = 3000ms/ 40ms/ 90°, the acquisition matrix was 64×64, the field of view (FOV) was 240 mm and slice thickness =5mm with no inter-slice gap. 32 axial slices parallel to the AC-PC line were obtained in an interleaved order to cover the whole cerebrum and cerebellum.

2.4   Image Preprocessing and Analyses

The functional images of both groups were first realigned, spatially normalized into standard stereotaxic space (EPI template provided by the Montreal Neurologic Institute, MNI), re-sliced to 3×3×4mm voxels and smoothed with a 8×8×8 full-width at half maximum (FWHM) Gaussian kernel by SPM8 software (Statistical Parametric Mapping; http://www.fil.ion.ucl.ac.uk/spm). The first five images of each series were removed from further analysis. Then the data was analyzed with a novel voxel-based method named eigenvector centrality mapping (ECM) according to a recent study (see Lohmann 2010) [10]. ECM attributes an eigenvector centrality value to each voxel in the brain such that a voxel receives a larger value if it is more strongly correlated with many other voxels which are central within the network themselves. The ECM analyses included the following four steps. First, a whole brain mask was defined according to a prior anatomical automatic labeling (AAL) atlas containing 90 areas of n = 40,743 voxels. Second, the time series were extracted from each voxel in the defined mask for each subject. Linear correlation which proposed as a metric of functional connectivity was calculated between any two voxels as follows:

$$r_{ij} = \frac{\sum_{t=1}^{T}\left[x_i(t)-\overline{x_i}\right]\left[x_j(t)-\overline{x_j}\right]}{\sqrt{\sum_{t=1}^{T}\left[x_i(t)-\overline{x_i}\right]^2}\sqrt{\sum_{t=1}^{T}\left[x_j(t)-\overline{x_j}\right]^2}} \tag{1}$$

where $x_i(t), x_j(t)$ $(t=1,...,T=200)$ are the time series of voxel i and j, $\overline{x_i}, \overline{x_j}$ are the mean of the two series. In this way, a similarity matrix A was obtained for each subject in the two resting scans and each r in A was substituted by $\tilde{r}=r+1$ as the similarity matrix should be positive according to the previous study. Then, the eigenvector centrality value x_i of a voxel i is defined as the i-th entry in the normalized eigenvector x belonging to the largest eigenvalue λ of the similarity matrix A, and the formula is as follows:

$$Ax=\lambda x, \text{ equivalent to } x=\frac{1}{\lambda}Ax, \text{ and } x_i=\mu\sum_{j=1}^{n}a_{ij}x_j \qquad (2)$$

where $\mu=\frac{1}{\lambda}$, and a_{ij} represents the element in the i-th row and j-th column of A.

For each subject in the two resting scans, an ECM containing the eigenvector centrality value of each voxel in the mask was obtained here. At last, a paired t-test was performed between the ECMs of pre- and post- resting scan. At the statistical analysis level, a voxel-cluster threshold correction was used to control the Type I error rate in the whole-brain statistics, yielding an overall corrected alpha rate of $P < 0.05$. The correction threshold was determined from a Monte Carlo simulation in AFNI and required a voxel-wise threshold of $P < 0.005$ within a minimum 3D cluster of 41 contiguous significant voxels (minimum cluster volume = 1476μl; FWHM autocorrelation estimate = 8.0 mm). The cluster-level inference was performed within SPM8.

3   Results

Fig. 1 shows the whole brain mask defined in this study, containing 90 anatomical automatic labeling (AAL) areas of 40,743 voxels. The group averages of eigenvector centrality maps are illustrated in Fig. 2. For the experimental group, Fig. 2a and 2b show the results of the pre- and post- resting scans respectively. For the control group, the group averages of ECMs in the pre- resting scan are shown in Fig. 2c, and Fig. 2d displays the results of the post- resting scan. After motor imagery training, the significantly increased eigenvector centrality was detected in the precuneus and medial frontal gyrus (MFG) for the experimental group while no significant alterations were found for the control group (see details in Fig.3 and Table 1).

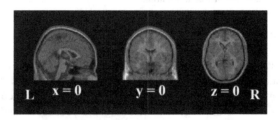

Fig. 1. The mask used in this study, containing 40,743 voxels

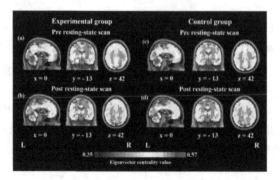

Fig. 2. Group averages of eigenvector centrality maps. (a) pre- resting scan, experimental group; (b) post- resting scan, experimental group; (c) pre- resting scan, control group; (d) post-resting scan, control group.

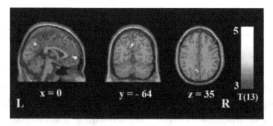

Fig. 3. Statistical parametric map of regions showing eigenvector centrality increased in experimental group induced by motor imagery learning. The statistical threshold was set at p<0.05 corrected for multiple comparisons at the cluster level.

Table 1. Brain regions where eigenvector centrality significantly increased in the resting-state after motor imagery training (p < 0.05 corrected for multiple comparisons at the cluster level)

Region	L/R	BA	x	y	z	t_{max}
Experimental group						
precuneus	L	7	-3	-64	38	4.27
medial frontal gyrus	L	10	-6	59	2	4.45
Control group						
None						

Note. MNI coordinates; BA—Brodmann's area.

4 Discussion

To address whether motor imagery training affects resting state, we used ECM to investigate the neural changes underlying the resting state associated with 14 consecutive days of motor imagery training with fMRI. Our results revealed that the eigenvector centrality of two brain regions, including the precuneus and medial frontal gyrus (MFG), were significantly increased by the motor imagery training.

Numerous of neuroimaging studies have suggested the important role of precune-
nus and MFG in motor training. The role of precunues in motor imagery has been
suggested by Sakai et al. (1998) that it could be activated when subjects learned sequences of finger movements, indicating that precuneus may be related to spatial
motor sequence information integration and retrieval [12]. A Magnetoencephalogra-
phy (MEG) study by Ogiso et al. (2000) confirmed that the precuneus may involve in
retrieval of spatial information and/or setting up spatial attributes for motor imagery
[13]. Our previous study also found the activation in precuneus increased after motor
imagery training, indicating that the precuneus may be engaged in mental representation and episodic memory retrieval [11]. Therefore, the alteration of precuneus in the
current study may due to spatial information processing and retrieval.

Previous studies have suggested that the role of MFG may be involved in inner
state modulation and motor planning, as well as complex nonmotor tasks such as
decision making, discrimination, computation, and reasoning [14, 15]. The MFG was
implied to be associated with the ability to reflect on one's own mental states and self-
referential processing such as mediate less-deliberate, emotion-driven influences on
action selection [16, 17]. It was also suggested to be important in allowing subject to
guide actions by internal or overarching plans so as to achieve an optimal behavior
performance [14]. Thus, we proposed that the changes in MFG may be the result of
the modulation of subjects' inner state to get an optimal behavior and decisions about
motor plans.

The results observed in this study confirmed our hypothesis that there are alterations in the resting state induced by motor imagery training. As a pilot investigation,
our result indicated that motor imagery training, as an important part of motor training, is worthy to be further investigated in more details.

5   Conclusion

We used a method named eigenvector centrality mapping to explore the resting-state
before and after a 2-week motor imagery training. We found that through motor imagery training, the significant increases of eigenvector centrality were detected in the
precuneus and medial frontal gyrus (MFG). These alterations may be related to the
spatial information integration or retrieval and inner state modulation during motor
imagery training, which further provided new insights into the understanding of the
neural mechanism underlying motor imagery training.

Acknowledgements. This work is supported by the Key Programs of National Natural Science Foundation of China under Grant Nos. 61131003 and 60931003.

References

1. Sharma, N., Pomeroy, V.M., Baron, J.: Motor Imagery A Backdoor to the Motor System
   After Stroke? Stroke 37, 1941–1952 (2006)
2. Olsson, C.J., Jonsson, B., Nyberg, L.: Learning by doing and learning by thinking: an fMRI
   study of combining motor and mental training. Frontiers in Human Neuroscience 2 (2008)

3. Stippich, C., Ochmann, H., Sartor, K.: Somatotopic mapping of the human primary senso-rimotor cortex during motor imagery and motor execution by functional magnetic reson-ance imaging. Neuroscience Letters 331, 50–54 (2002)

4. Beck, S., Taube, W., Gruber, M., Amtage, F., Gollhofer, A., Schubert, M.: Task-specific changes in motor evoked potentials of lower limb muscles after different training interven-tions. Brain Research 1179, 51–60 (2007)

5. Rodriguez, M., Llanos, C., Gonzalez, S., Sabate, M.: How similar are motor imagery and movement? Behavioral Neuroscience 122, 910 (2008)

6. Xiong, J., Ma, L., Wang, B., Narayana, S., Duff, E.P., Egan, G.F., Fox, P.T.: Long-term motor training induced changes in regional cerebral blood flow in both task and resting states. Neuroimage 45, 75–82 (2009)

7. Ma, L., Narayana, S., Robin, D.A., Fox, P.T., Xiong, J.: Changes occur in resting state network of motor system during 4weeks of motor skill learning. NeuroImage 58, 226–233 (2011)

8. Albert, N.B., Robertson, E.M., Miall, R.C.: The resting human brain and motor learning. Current Biology 19, 1023 (2009)

9. Taubert, M., Lohmann, G., Margulies, D.S., Villringer, A., Ragert, P.: Long-term effects of motor training on resting-state networks and underlying brain structure. Neuroimage 57, 1492–1498 (2011)

10. Lohmann, G., Margulies, D.S., Horstmann, A., Pleger, B., Lepsien, J., Goldhahn, D., Schloegl, H., Stumvoll, M., Villringer, A., Turner, R.: Eigenvector centrality mapping for analyzing connectivity patterns in fMRI data of the human brain. PloS One 5, e10232 (2010)

11. Zhang, H., Xu, L., Wang, S., Xie, B., Guo, J., Long, Z., Yao, L.: Behavioral improvements and brain functional alterations by motor imagery training. Brain Research 1407, 38–46 (2011)

12. Sakai, K., Hikosaka, O., Miyauchi, S., Takino, R., Sasaki, Y., Pütz, B.: Transition of brain activation from frontal to parietal areas in visuomotor sequence learning. The Journal of Neuroscience 18, 1827–1840 (1998)

13. Ogiso, T., Kobayashi, K., Sugishita, M.: The precuneus in motor imagery: a magnetoence-phalographic study. Neuroreport. 11, 1345–1349 (2000)

14. Roca, M.A., Torralva, T., Gleichgerrcht, E., Woolgar, A., Thompson, R., Duncan, J., Manes, F.: The role of area 10 (BA10) in human multitasking and in social cognition: a le-sion study. Neuropsychologia 49, 3525–3531 (2011)

15. Talati, A., Hirsch, J.: Functional specialization within the medial frontal gyrus for percep-tual go/no-go decisions based on? "what", "when", and "where" related information: an fMRI study. Journal of Cognitive Neuroscience 17, 981–993 (2005)

16. Ridderinkhof, K.R., van den Wildenberg, W.P., Segalowitz, S.J., Carter, C.S.: Neurocogni-tive mechanisms of cognitive control: the role of prefrontal cortex in action selection, re-sponse inhibition, performance monitoring, and reward-based learning. Brain and Cogni-tion 56, 129–140 (2004)

17. Johnson, M.K., Raye, C.L., Mitchell, K.J., Touryan, S.R., Greene, E.J., Nolen-Hoeksema, S.: Dissociating medial frontal and posterior cingulate activity during self-reflection. Social Cog-nitive and Affective Neuroscience 1, 56–64 (2006)

Author Index